W9-CWT-625

VOLUME FOUR HUNDRED AND EIGHTY-TWO

# METHODS IN ENZYMOLOGY

## Cryo-EM, Part B

### 3-D Reconstruction

# METHODS IN ENZYMOLOGY

VOLUME FOUR HUNDRED AND EIGHTY-TWO

# METHODS IN ENZYMOLOGY

# Cryo-EM, Part B

## 3-D Reconstruction

EDITED BY

GRANT J. JENSEN
*Division of Biology and Howard Hughes Medical Institute*
*California Institute of Technology*
*Pasadena, California, USA*

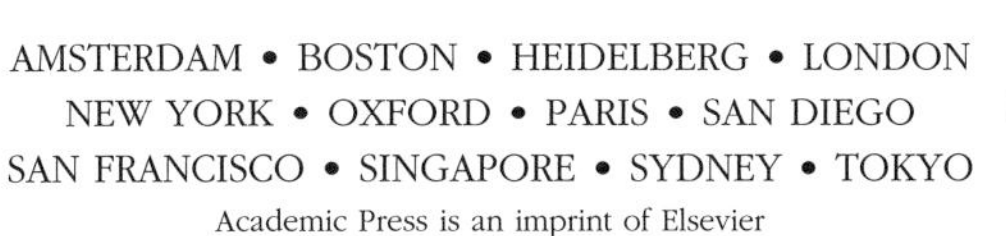

AMSTERDAM • BOSTON • HEIDELBERG • LONDON
NEW YORK • OXFORD • PARIS • SAN DIEGO
SAN FRANCISCO • SINGAPORE • SYDNEY • TOKYO
Academic Press is an imprint of Elsevier

Academic Press is an imprint of Elsevier
525 B Street, Suite 1900, San Diego, CA 92101-4495, USA
30 Corporate Drive, Suite 400, Burlington, MA 01803, USA
32 Jamestown Road, London NW1 7BY, UK

First edition 2010



For information on all Academic Press publications
visit our website at elsevierdirect.com

ISBN: 978-0-12-384991-5
ISSN: 0076-6879

Printed and bound in United States of America
10 11 12 10 9 8 7 6 5 4 3 2 1

# Contents

# Contributors

**Fernando Amat**[1]
Department of Electrical Engineering, Stanford University, Stanford, California, USA

**Marcel Arheit**
C-CINA, Biozentrum, University of Basel, Basel, Switzerland

**Timothy S. Baker**
Department of Chemistry & Biochemistry, and Division of Biological Sciences, University of California, San Diego, La Jolla, California, USA

**Jose-Maria Carazo**
Biocomputing Unit, Centro Nacional de Biotecnología – CSIC, Cantoblanco, Madrid, Spain

**Bridget Carragher**
The National Resource for Automated Molecular Microscopy, Department of Cell Biology, The Scripps Research Institute, La Jolla, California, USA

**Daniel Castaño-Diez**
C-CINA, Biozentrum, and Center for Cellular Imaging and Nanoanalytics, Department of Structural Biology and Biophysics, University of Basel, Basel, Switzerland

**Yao Cong**
National Center for Macromolecular Imaging, The Verna and Marrs McLean Department of Biochemistry and Molecular Biology, Baylor College of Medicine, Houston, Texas, USA

**Ruben Diaz**
Cryo-electron Microscopy Facility, New York Structural Biology Center, New York, USA

**Peter C. Doerschuk**
Department of Biomedical Engineering, Cornell University, Weill Hall, Ithaca, New York, USA

[1]Current address: Janelia Farm Research Campus, Howard Hughes Medical Institute, Ashburn, Virginia, USA

**Edward H. Egelman**
Department of Biochemistry and Molecular Genetics, University of Virginia, Charlottesville, Virginia, USA

**Bryant Gipson**
C-CINA, Biozentrum, University of Basel, Basel, Switzerland

**Mark Horowitz**
Department of Electrical Engineering, Stanford University, Stanford, California, USA

**Grant J. Jensen**
Division of Biology, Howard Hughes Medical Institute, California Institute of Technology, Pasadena, California, USA

**Albert Lawrence**
National Center for Microscopy and Imaging Research, Center for Research in Biological Structure, University of California at San Diego, La Jolla, California, USA

**Peter A. Leong**
Department of Applied Physics, California Institute of Technology, Pasadena, California, USA

**Andres Leschziner**
Department of Molecular and Cellular Biology, Harvard University, Cambridge, Massachusetts, USA

**Steven J. Ludtke**
National Center for Macromolecular Imaging, The Verna and Marrs McLean Department of Biochemistry and Molecular Biology, Baylor College of Medicine, Houston, Texas, USA

**Farshid Moussavi**
Department of Electrical Engineering, Stanford University, Stanford, California, USA

**Elena V. Orlova**
Crystallography and Institute of Structural Molecular Biology, Birkbeck College, London, United Kingdom

**Pawel A. Penczek**
Department of Biochemistry and Molecular Biology, The University of Texas, Houston Medical School, Houston, Texas, USA

**Clinton S. Potter**
The National Resource for Automated Molecular Microscopy, Department of Cell Biology, The Scripps Research Institute, La Jolla, California, USA

**William, J. Rice**
Cryo-electron Microscopy Facility, New York Structural Biology Center, New York, USA

**Helen R. Saibil**
[illegible]
London, United Kingdom

**Andreas D. Schenk**
Department of Cell Biology, Harvard Medical School, Boston, Massachusetts, USA

**Sjors H. W. Scheres**[2]
Biocomputing Unit, Centro Nacional de Biotecnología – CSIC, Cantoblanco, Madrid, Spain

**Fred J. Sigworth**
Department of Cellular and Molecular Physiology, Yale University, New Haven, Connecticut, USA

**Robert S. Sinkovits**
Department of Chemistry & Biochemistry, University of California, San Diego, La Jolla, California, USA

**Ross Smith**
Center for Health Informatics and Bioinformatics and, Department of Cell Biology, New York University School of Medicine, New York, New York, USA

**Henning Stahlberg**
C-CINA, Biozentrum, University of Basel, Basel, Switzerland

**David L. Stokes**
Cryo-electron Microscopy Facility, New York Structural Biology Center, and Skirball Institute, Department of Cell Biology, New York University School of Medicine, New York, USA

**Jinghua Tang**
Department of Chemistry & Biochemistry, University of California, San Diego, La Jolla, California, USA

**Neil R. Voss**
The National Resource for Automated Molecular Microscopy, Department of Cell Biology, The Scripps Research Institute, La Jolla, California, USA

**Hanspeter Winkler**
Institute of Molecular Biophysics, Florida State University, Florida, USA

[2]Current address: MRC Laboratory of Molecular Biology, Hills Road, Cambridge, UK.

**Xuekui Yu**
Department of Microbiology, Immunology and Molecular Genetics, The California NanoSystems Institute, University of California Los Angeles, Los Angeles, California, USA

**Xiangyan Zeng**
Department of Mathematics and Computer Science, Fort Valley State University, Fort Valley, Georgia, and EON Corporation, California, USA

**Z. Hong Zhou**
Department of Microbiology, Immunology and Molecular Genetics, The California NanoSystems Institute, University of California Los Angeles, Los Angeles, California, USA

# Preface

In this, the fifty-fourth year of *Methods in Enzymology*, we celebrate the discovery and initial characterization of thousands of individual enzymes, the sequencing of hundreds of whole genomes, and the structure determination of tens of thousands of proteins. In this context, the architectures of multienyzme/multiprotein complexes and their arrangement within cells have now come to the fore. A uniquely powerful method in this field is electron cryomicroscopy (cryo-EM), which in its broadest sense, is all those techniques that image cold samples in the electron microscope. Cryo-EM allows individual enzymes and proteins, macromolecular complexes, assemblies, cells, and even tissues to be observed in a "frozen-hydrated," near-native state free from the artifacts of fixation, dehydration, plastic-embedding, or staining typically used in traditional forms of EM (Chapter 3, Vol. 481). This series of volumes is therefore dedicated to a description of the instruments, samples, protocols, and analyses that belong to the growing field of cryo-EM.

The material could have been organized well by two schemes. The first is by the symmetry of the sample. Because the fundamental limitation in cryo-EM is radiation damage (Chapter 15, Vol. 481), a defining characteristic of each method is whether and how low-dose images of identical copies of the specimen can be averaged. In the most favorable case, large numbers of identical copies of the specimen of interest, like a single protein, can be purified and crystallized within thin "two-dimensional" crystals (Chapter 1, Vol. 481). In this case, truly *atomic* resolution reconstructions have been obtained through averaging very low dose images of millions of copies of the specimen (Chapter 11, Vol. 481; Chapter 4, Vol. 482; and Chapters 5 and 6, Vol. 483). The next most favorable case is helical crystals, which present a range of views of the specimen within a single image (Chapter 2, Vol. 481 and Chapter 7, Vol. 483), and can also deliver atomically interpretable reconstructions, although through quite different data collection protocols and reconstruction mathematics (Chapters 5 and 6, Vol. 482). At an intermediate level of (60-fold) symmetry, icosahedral viruses have their own set of optimal imaging and reconstruction protocols, and are just now also reaching atomic interpretability (Chapters 7 and 14, Vol. 482). Less symmetric particles, such as many multienyzme/multiprotein complexes, invite yet another set of challenges and methods (Chapters 3, 5, and 6, Vol. 481; Chapters 8–10, Vol. 482). Many are conformationally heterogeneous, requiring that images of different particles be first classified and then averaged (Chapters 10 and 12, Vol. 482; Chapters 8 and 9, Vol. 483).

Heterogeneity and the precision to which these images can be aligned have limited most such reconstructions to "sub-nanometer," where the folds of proteins are clear but not much more (Chapter 1, Vol. 483). Finally, the most challenging samples are those which are simply unique (Chapter 8, Vol. 481), eliminating any chance of improving the clarity of reconstructions through averaging. For these, tomographic methods are required (Chapter 12, Vol. 481; Chapter 13, Vol. 482), and only nanometer resolutions can be obtained (Chapters 10–13, Vol. 483).

But instead of organizing topics according to symmetry, following a wonderful historical perspective by David DeRosier (Historical Perspective, Vol. 481), I chose to order the topics in experimental sequence: Sample preparation and data collection/microscopy (Vol. 481); 3-D reconstruction (Vol. 482); and analyses and interpretation, including case studies (Vol. 483). This organization emphasizes how the relatedness of the mathematics (Chapter 1, Vol. 482), instrumentation (Chapters 10 and 14, Vol. 482), and methods (Chapter 15, Vol. 482; Chapter 9, Vol. 481) underlying all cryo-EM approaches allow practictioners to easily move between them. It further highlights how in a growing number of recent cases, the methods are being mixed (Chapter 13, Vol. 481), for instance, through the application of "single particle-like" approaches to "unbend" and average 2-D and helical crystals (Chapter 6, Vol. 482), but also average subvolumes within tomograms. Moreover, different samples are always more-or-less well-behaved, so the actual resolution achieved may be less than theoretically possible for a particular symmetry, or to the opposite effect, extensively known constraints may allow a more specific interpretation than usual for a given resolution (Chapters 2–4 and 6, Vol. 483). Nevertheless, within each section, the articles are ordered as much as possible according to the symmetry of the sample as described above (i.e., methods for preparing samples proceed from 2-D and helical crystals to sectioning of high-pressure-frozen tissues; Chapter 8, Vol. 481). The cryo-EM beginner with a new sample must then first recognize its symmetry and then identify the relevant chapters within each volume.

As a final note, our field has not yet reached consensus on the placement of the prefix "cryo" and other details of the names of cryo-EM techniques. Thus, "cryo-electron microscopy" (CEM), "electron cryo-microscopy" (ECM), and "cryo-EM" should all be considered synonyms here. Likewise, "single particle reconstruction" (SPR) and "single particle analysis" (SPA) refer to a single technique, as do "cryo-electron tomography" (CET), "electron cryo-tomography" (ECT), and cryo-electron microscope tomography (cEMT).

Grant J. Jensen

# Methods in Enzymology

Volume I. Preparation and Assay of Enzymes
*Edited by* Sidney P. Colowick and Nathan O. Kaplan

Volume II. Preparation and Assay of Enzymes
*Edited by* Sidney P. Colowick and Nathan O. Kaplan

Volume III. Preparation and Assay of Substrates
*Edited by* Sidney P. Colowick and Nathan O. Kaplan

Volume IV. Special Techniques for the Enzymologist
*Edited by* Sidney P. Colowick and Nathan O. Kaplan

Volume V. Preparation and Assay of Enzymes
*Edited by* Sidney P. Colowick and Nathan O. Kaplan

Volume VI. Preparation and Assay of Enzymes *(Continued)*
Preparation and Assay of Substrates
Special Techniques
*Edited by* Sidney P. Colowick and Nathan O. Kaplan

Volume VII. Cumulative Subject Index
*Edited by* Sidney P. Colowick and Nathan O. Kaplan

Volume VIII. Complex Carbohydrates
*Edited by* Elizabeth F. Neufeld and Victor Ginsburg

Volume IX. Carbohydrate Metabolism
*Edited by* Willis A. Wood

Volume X. Oxidation and Phosphorylation
*Edited by* Ronald W. Estabrook and Maynard E. Pullman

Volume XI. Enzyme Structure
*Edited by* C. H. W. Hirs

Volume XII. Nucleic Acids (Parts A and B)
*Edited by* Lawrence Grossman and Kivie Moldave

Volume XIII. Citric Acid Cycle
*Edited by* J. M. Lowenstein

Volume XIV. Lipids
*Edited by* J. M. Lowenstein

Volume XV. Steroids and Terpenoids
*Edited by* Raymond B. Clayton

Volume XVI. Fast Reactions
*Edited by* Kenneth Kustin

Volume XVII. Metabolism of Amino Acids and Amines (Parts A and B)
*Edited by* Herbert Tabor and Celia White Tabor

Volume XVIII. Vitamins and Coenzymes (Parts A, B, and C)
*Edited by* Donald B. McCormick and Lemuel D. Wright

Volume XIX. Proteolytic Enzymes
*Edited by* Gertrude E. Perlmann and Laszlo Lorand

Volume XX. Nucleic Acids and Protein Synthesis (Part C)
*Edited by* Kivie Moldave and Lawrence Grossman

Volume XXI. Nucleic Acids (Part D)
*Edited by* Lawrence Grossman and Kivie Moldave

Volume XXII. Enzyme Purification and Related Techniques
*Edited by* William B. Jakoby

Volume XXIII. Photosynthesis (Part A)
*Edited by* Anthony San Pietro

Volume XXIV. Photosynthesis and Nitrogen Fixation (Part B)
*Edited by* Anthony San Pietro

Volume XXV. Enzyme Structure (Part B)
*Edited by* C. H. W. Hirs and Serge N. Timasheff

Volume XXVI. Enzyme Structure (Part C)
*Edited by* C. H. W. Hirs and Serge N. Timasheff

Volume XXVII. Enzyme Structure (Part D)
*Edited by* C. H. W. Hirs and Serge N. Timasheff

Volume XXVIII. Complex Carbohydrates (Part B)
*Edited by* Victor Ginsburg

Volume XXIX. Nucleic Acids and Protein Synthesis (Part E)
*Edited by* Lawrence Grossman and Kivie Moldave

Volume XXX. Nucleic Acids and Protein Synthesis (Part F)
*Edited by* Kivie Moldave and Lawrence Grossman

Volume XXXI. Biomembranes (Part A)
*Edited by* Sidney Fleischer and Lester Packer

Volume XXXII. Biomembranes (Part B)
*Edited by* Sidney Fleischer and Lester Packer

Volume XXXIII. Cumulative Subject Index Volumes I-XXX
*Edited by* Martha G. Dennis and Edward A. Dennis

Volume XXXIV. Affinity Techniques (Enzyme Purification: Part B)
*Edited by* William B. Jakoby and Meir Wilchek

Volume XXXV. Lipids (Part B)
*Edited by* John M. Lowenstein

Volume XXXVI. Hormone Action (Part A: Steroid Hormones)
*Edited by* Bert W. O'Malley and Joel G. Hardman

Volume XXXVII. Hormone [illegible]
*Edited by* Bert W. O'Malley and Joel G. Hardman

Volume XXXVIII. Hormone Action (Part C: Cyclic Nucleotides)
*Edited by* Joel G. Hardman and Bert W. O'Malley

Volume XXXIX. Hormone Action (Part D: Isolated Cells, Tissues, and Organ Systems)
*Edited by* Joel G. Hardman and Bert W. O'Malley

Volume XL. Hormone Action (Part E: Nuclear Structure and Function)
*Edited by* Bert W. O'Malley and Joel G. Hardman

Volume XLI. Carbohydrate Metabolism (Part B)
*Edited by* W. A. Wood

Volume XLII. Carbohydrate Metabolism (Part C)
*Edited by* W. A. Wood

Volume XLIII. Antibiotics
*Edited by* John H. Hash

Volume XLIV. Immobilized Enzymes
*Edited by* Klaus Mosbach

Volume XLV. Proteolytic Enzymes (Part B)
*Edited by* Laszlo Lorand

Volume XLVI. Affinity Labeling
*Edited by* William B. Jakoby and Meir Wilchek

Volume XLVII. Enzyme Structure (Part E)
*Edited by* C. H. W. Hirs and Serge N. Timasheff

Volume XLVIII. Enzyme Structure (Part F)
*Edited by* C. H. W. Hirs and Serge N. Timasheff

Volume XLIX. Enzyme Structure (Part G)
*Edited by* C. H. W. Hirs and Serge N. Timasheff

Volume L. Complex Carbohydrates (Part C)
*Edited by* Victor Ginsburg

Volume LI. Purine and Pyrimidine Nucleotide Metabolism
*Edited by* Patricia A. Hoffee and Mary Ellen Jones

Volume LII. Biomembranes (Part C: Biological Oxidations)
*Edited by* Sidney Fleischer and Lester Packer

Volume LIII. Biomembranes (Part D: Biological Oxidations)
*Edited by* Sidney Fleischer and Lester Packer

Volume LIV. Biomembranes (Part E: Biological Oxidations)
*Edited by* Sidney Fleischer and Lester Packer

Volume LV. Biomembranes (Part F: Bioenergetics)
*Edited by* Sidney Fleischer and Lester Packer

Volume LVI. Biomembranes (Part G: Bioenergetics)
*Edited by* Sidney Fleischer and Lester Packer

Volume LVII. Bioluminescence and Chemiluminescence
*Edited by* Marlene A. DeLuca

Volume LVIII. Cell Culture
*Edited by* William B. Jakoby and Ira Pastan

Volume LIX. Nucleic Acids and Protein Synthesis (Part G)
*Edited by* Kivie Moldave and Lawrence Grossman

Volume LX. Nucleic Acids and Protein Synthesis (Part H)
*Edited by* Kivie Moldave and Lawrence Grossman

Volume 61. Enzyme Structure (Part H)
*Edited by* C. H. W. Hirs and Serge N. Timasheff

Volume 62. Vitamins and Coenzymes (Part D)
*Edited by* Donald B. McCormick and Lemuel D. Wright

Volume 63. Enzyme Kinetics and Mechanism (Part A: Initial Rate and Inhibitor Methods)
*Edited by* Daniel L. Purich

Volume 64. Enzyme Kinetics and Mechanism (Part B: Isotopic Probes and Complex Enzyme Systems)
*Edited by* Daniel L. Purich

Volume 65. Nucleic Acids (Part I)
*Edited by* Lawrence Grossman and Kivie Moldave

Volume 66. Vitamins and Coenzymes (Part E)
*Edited by* Donald B. McCormick and Lemuel D. Wright

Volume 67. Vitamins and Coenzymes (Part F)
*Edited by* Donald B. McCormick and Lemuel D. Wright

Volume 68. Recombinant DNA
*Edited by* Ray Wu

Volume 69. Photosynthesis and Nitrogen Fixation (Part C)
*Edited by* Anthony San Pietro

Volume 70. Immunochemical Techniques (Part A)
*Edited by* Helen Van Vunakis and John J. Langone

Volume 71. Lipids (Part C)
*Edited by* John M. Lowenstein

Volume 72. Lipids (Part D)
*Edited by* John M. Lowenstein

Volume 73. Immunochemical Techniques (Part B)
*Edited by* John J. Langone and Helen Van Vunakis

Volume 74. Immunochemical Techniques (Part C)
*Edited by* John J. Langone and Helen Van Vunakis

Volume 75. Cumulative Subject Index Volumes XXXI, XXXII, XXXIV–LX
*Edited by* Edward A. Dennis and Martha G. Dennis

Volume 76. Hemoglobins
*Edited by* Eraldo Antonini, Luigi Rossi-Bernardi, and Emilia Chiancone

Volume 77. Detoxication and Drug Metabolism
*Edited by* William B. Jakoby

Volume 78. Interferons (Part A)
*Edited by* Sidney Pestka

Volume 79. Interferons (Part B)
*Edited by* Sidney Pestka

Volume 80. Proteolytic Enzymes (Part C)
*Edited by* Laszlo Lorand

Volume 81. Biomembranes (Part H: Visual Pigments and Purple Membranes, I)
*Edited by* Lester Packer

Volume 82. Structural and Contractile Proteins (Part A: Extracellular Matrix)
*Edited by* Leon W. Cunningham and Dixie W. Frederiksen

Volume 83. Complex Carbohydrates (Part D)
*Edited by* Victor Ginsburg

Volume 84. Immunochemical Techniques (Part D: Selected Immunoassays)
*Edited by* John J. Langone and Helen Van Vunakis

Volume 85. Structural and Contractile Proteins (Part B: The Contractile Apparatus and the Cytoskeleton)
*Edited by* Dixie W. Frederiksen and Leon W. Cunningham

Volume 86. Prostaglandins and Arachidonate Metabolites
*Edited by* William E. M. Lands and William L. Smith

Volume 87. Enzyme Kinetics and Mechanism (Part C: Intermediates, Stereo-chemistry, and Rate Studies)
*Edited by* Daniel L. Purich

Volume 88. Biomembranes (Part I: Visual Pigments and Purple Membranes, II)
*Edited by* Lester Packer

Volume 105. Oxygen Radicals in Biological Systems
*Edited by* Lester Packer

Volume 106. Posttranslational Modifications (Part A)
*Edited by* Finn Wold and Kivie Moldave

Volume 107. Posttranslational Modifications (Part B)
*Edited by* Finn Wold and Kivie Moldave

Volume 108. Immunochemical Techniques (Part G: Separation and Characterization of Lymphoid Cells)
*Edited by* Giovanni Di Sabato, John J. Langone, and Helen Van Vunakis

Volume 109. Hormone Action (Part I: Peptide Hormones)
*Edited by* Lutz Birnbaumer and Bert W. O'Malley

Volume 110. Steroids and Isoprenoids (Part A)
*Edited by* John H. Law and Hans C. Rilling

Volume 111. Steroids and Isoprenoids (Part B)
*Edited by* John H. Law and Hans C. Rilling

Volume 112. Drug and Enzyme Targeting (Part A)
*Edited by* Kenneth J. Widder and Ralph Green

Volume 113. Glutamate, Glutamine, Glutathione, and Related Compounds
*Edited by* Alton Meister

Volume 114. Diffraction Methods for Biological Macromolecules (Part A)
*Edited by* Harold W. Wyckoff, C. H. W. Hirs, and Serge N. Timasheff

Volume 115. Diffraction Methods for Biological Macromolecules (Part B)
*Edited by* Harold W. Wyckoff, C. H. W. Hirs, and Serge N. Timasheff

Volume 116. Immunochemical Techniques
(Part H: Effectors and Mediators of Lymphoid Cell Functions)
*Edited by* Giovanni Di Sabato, John J. Langone, and Helen Van Vunakis

Volume 117. Enzyme Structure (Part J)
*Edited by* C. H. W. Hirs and Serge N. Timasheff

Volume 118. Plant Molecular Biology
*Edited by* Arthur Weissbach and Herbert Weissbach

Volume 119. Interferons (Part C)
*Edited by* Sidney Pestka

Volume 120. Cumulative Subject Index Volumes 81–94, 96–101

Volume 121. Immunochemical Techniques (Part I: Hybridoma Technology and Monoclonal Antibodies)
*Edited by* John J. Langone and Helen Van Vunakis

Volume 122. Vitamins and Coenzymes (Part G)
*Edited by* Frank Chytil and Donald B. McCormick

Volume 123. Vitamins and Coenzymes (Part H)
*Edited by* Frank Chytil and Donald B. McCormick

Volume 124. Hormone Action (Part J: Neuroendocrine Peptides)
*Edited by* P. Michael Conn

Volume 125. Biomembranes (Part M: Transport in Bacteria, Mitochondria, and Chloroplasts: General Approaches and Transport Systems)
*Edited by* Sidney Fleischer and Becca Fleischer

Volume 126. Biomembranes (Part N: Transport in Bacteria, Mitochondria, and Chloroplasts: Protonmotive Force)
*Edited by* Sidney Fleischer and Becca Fleischer

Volume 127. Biomembranes (Part O: Protons and Water: Structure and Translocation)
*Edited by* Lester Packer

Volume 128. Plasma Lipoproteins (Part A: Preparation, Structure, and Molecular Biology)
*Edited by* Jere P. Segrest and John J. Albers

Volume 129. Plasma Lipoproteins (Part B: Characterization, Cell Biology, and Metabolism)
*Edited by* John J. Albers and Jere P. Segrest

Volume 130. Enzyme Structure (Part K)
*Edited by* C. H. W. Hirs and Serge N. Timasheff

Volume 131. Enzyme Structure (Part L)
*Edited by* C. H. W. Hirs and Serge N. Timasheff

Volume 132. Immunochemical Techniques (Part J: Phagocytosis and Cell-Mediated Cytotoxicity)
*Edited by* Giovanni Di Sabato and Johannes Everse

Volume 133. Bioluminescence and Chemiluminescence (Part B)
*Edited by* Marlene DeLuca and William D. McElroy

Volume 134. Structural and Contractile Proteins (Part C: The Contractile Apparatus and the Cytoskeleton)
*Edited by* Richard B. Vallee

Volume 135. Immobilized Enzymes and Cells (Part B)
*Edited by* Klaus Mosbach

Volume 136. Immobilized Enzymes and Cells (Part C)
*Edited by* Klaus Mosbach

Volume 137. Immobilized Enzymes and Cells (Part D)
*Edited by* Klaus Mosbach

Volume 138. Complex Carbohydrates (Part E)
*Edited by* Victor Ginsburg

Volume 157. Biomembranes (Part Q: ATP-Driven Pumps and Related Transport: Calcium, Proton, and Potassium Pumps)
*Edited by* Sidney Fleischer and Becca Fleischer

Volume 158. Metalloproteins (Part A)
*Edited by* James F. Riordan and Bert L. Vallee

Volume 159. Initiation and Termination of Cyclic Nucleotide Action
*Edited by* Jackie D. Corbin and Roger A. Johnson

Volume 160. Biomass (Part A: Cellulose and Hemicellulose)
*Edited by* Willis A. Wood and Scott T. Kellogg

Volume 161. Biomass (Part B: Lignin, Pectin, and Chitin)
*Edited by* Willis A. Wood and Scott T. Kellogg

Volume 162. Immunochemical Techniques (Part L: Chemotaxis and Inflammation)
*Edited by* Giovanni Di Sabato

Volume 163. Immunochemical Techniques (Part M: Chemotaxis and Inflammation)
*Edited by* Giovanni Di Sabato

Volume 164. Ribosomes
*Edited by* Harry F. Noller, Jr., and Kivie Moldave

Volume 165. Microbial Toxins: Tools for Enzymology
*Edited by* Sidney Harshman

Volume 166. Branched-Chain Amino Acids
*Edited by* Robert Harris and John R. Sokatch

Volume 167. Cyanobacteria
*Edited by* Lester Packer and Alexander N. Glazer

Volume 168. Hormone Action (Part K: Neuroendocrine Peptides)
*Edited by* P. Michael Conn

Volume 169. Platelets: Receptors, Adhesion, Secretion (Part A)
*Edited by* Jacek Hawiger

Volume 170. Nucleosomes
*Edited by* Paul M. Wassarman and Roger D. Kornberg

Volume 171. Biomembranes (Part R: Transport Theory: Cells and Model Membranes)
*Edited by* Sidney Fleischer and Becca Fleischer

Volume 172. Biomembranes (Part S: Transport: Membrane Isolation and Characterization)
*Edited by* Sidney Fleischer and Becca Fleischer

Volume 190. Retinoids (Part B: Cell Differentiation and Clinical Applications)
*Edited by* Lester Packer

Volume 191. Biomembranes (Part V: Cellular and Subcellular Transport: Epithelial Cells)
*Edited by* Sidney Fleischer and Becca Fleischer

Volume 192. Biomembranes (Part W: Cellular and Subcellular Transport: Epithelial Cells)
*Edited by* Sidney Fleischer and Becca Fleischer

Volume 193. Mass Spectrometry
*Edited by* James A. McCloskey

Volume 194. Guide to Yeast Genetics and Molecular Biology
*Edited by* Christine Guthrie and Gerald R. Fink

Volume 195. Adenylyl Cyclase, G Proteins, and Guanylyl Cyclase
*Edited by* Roger A. Johnson and Jackie D. Corbin

Volume 196. Molecular Motors and the Cytoskeleton
*Edited by* Richard B. Vallee

Volume 197. Phospholipases
*Edited by* Edward A. Dennis

Volume 198. Peptide Growth Factors (Part C)
*Edited by* David Barnes, J. P. Mather, and Gordon H. Sato

Volume 199. Cumulative Subject Index Volumes 168–174, 176–194

Volume 200. Protein Phosphorylation (Part A: Protein Kinases: Assays, Purification, Antibodies, Functional Analysis, Cloning, and Expression)
*Edited by* Tony Hunter and Bartholomew M. Sefton

Volume 201. Protein Phosphorylation (Part B: Analysis of Protein Phosphorylation, Protein Kinase Inhibitors, and Protein Phosphatases)
*Edited by* Tony Hunter and Bartholomew M. Sefton

Volume 202. Molecular Design and Modeling: Concepts and Applications (Part A: Proteins, Peptides, and Enzymes)
*Edited by* John J. Langone

Volume 203. Molecular Design and Modeling: Concepts and Applications (Part B: Antibodies and Antigens, Nucleic Acids, Polysaccharides, and Drugs)
*Edited by* John J. Langone

Volume 204. Bacterial Genetic Systems
*Edited by* Jeffrey H. Miller

Volume 205. Metallobiochemistry (Part B: Metallothionein and Related Molecules)
*Edited by* James F. Riordan and Bert L. Vallee

Volume 206. Cytochrome P450
*Edited by* Michael R. Waterman and Eric F. Johnson

Volume 207. Ion Channels
*Edited by* Bernardo Rudy and Linda E. Iverson

Volume 208. Protein–DNA Interactions
*Edited by* Robert T. Sauer

Volume 209. Phospholipid Biosynthesis
*Edited by* Edward A. Dennis and Dennis E. Vance

Volume 210. Numerical Computer Methods
*Edited by* Ludwig Brand and Michael L. Johnson

Volume 211. DNA Structures (Part A: Synthesis and Physical Analysis of DNA)
*Edited by* David M. J. Lilley and James E. Dahlberg

Volume 212. DNA Structures (Part B: Chemical and Electrophoretic Analysis of DNA)
*Edited by* David M. J. Lilley and James E. Dahlberg

Volume 213. Carotenoids (Part A: Chemistry, Separation, Quantitation, and Antioxidation)
*Edited by* Lester Packer

Volume 214. Carotenoids (Part B: Metabolism, Genetics, and Biosynthesis)
*Edited by* Lester Packer

Volume 215. Platelets: Receptors, Adhesion, Secretion (Part B)
*Edited by* Jacek J. Hawiger

Volume 216. Recombinant DNA (Part G)
*Edited by* Ray Wu

Volume 217. Recombinant DNA (Part H)
*Edited by* Ray Wu

Volume 218. Recombinant DNA (Part I)
*Edited by* Ray Wu

Volume 219. Reconstitution of Intracellular Transport
*Edited by* James E. Rothman

Volume 220. Membrane Fusion Techniques (Part A)
*Edited by* Nejat Düzgüneş

Volume 221. Membrane Fusion Techniques (Part B)
*Edited by* Nejat Düzgüneş

Volume 222. Proteolytic Enzymes in Coagulation, Fibrinolysis, and Complement Activation (Part A: Mammalian Blood Coagulation Factors and Inhibitors)
*Edited by* Laszlo Lorand and Kenneth G. Mann

Volume 223. Proteolytic Enzymes in Coagulation, Fibrinolysis, and Complement Activation (Part B: Complement Activation, Fibrinolysis, and Nonmammalian Blood Coagulation Factors)
*Edited by* Laszlo Lorand and Kenneth G. Mann

Volume 224. Molecular Evolution: Producing the Biochemical Data
*Edited by* Elizabeth Anne Zimmer, Thomas J. White, Rebecca L. Cann, and Allan C. Wilson

Volume 225. Guide to Techniques in Mouse Development
*Edited by* Paul M. Wassarman and Melvin L. DePamphilis

Volume 226. Metallobiochemistry (Part C: Spectroscopic and Physical Methods for Probing Metal Ion Environments in Metalloenzymes and Metalloproteins)
*Edited by* James F. Riordan and Bert L. Vallee

Volume 227. Metallobiochemistry (Part D: Physical and Spectroscopic Methods for Probing Metal Ion Environments in Metalloproteins)
*Edited by* James F. Riordan and Bert L. Vallee

Volume 228. Aqueous Two-Phase Systems
*Edited by* Harry Walter and Göte Johansson

Volume 229. Cumulative Subject Index Volumes 195–198, 200–227

Volume 230. Guide to Techniques in Glycobiology
*Edited by* William J. Lennarz and Gerald W. Hart

Volume 231. Hemoglobins (Part B: Biochemical and Analytical Methods)
*Edited by* Johannes Everse, Kim D. Vandegriff, and Robert M. Winslow

Volume 232. Hemoglobins (Part C: Biophysical Methods)
*Edited by* Johannes Everse, Kim D. Vandegriff, and Robert M. Winslow

Volume 233. Oxygen Radicals in Biological Systems (Part C)
*Edited by* Lester Packer

Volume 234. Oxygen Radicals in Biological Systems (Part D)
*Edited by* Lester Packer

Volume 235. Bacterial Pathogenesis (Part A: Identification and Regulation of Virulence Factors)
*Edited by* Virginia L. Clark and Patrik M. Bavoil

Volume 236. Bacterial Pathogenesis (Part B: Integration of Pathogenic Bacteria with Host Cells)
*Edited by* Virginia L. Clark and Patrik M. Bavoil

Volume 237. Heterotrimeric G Proteins
*Edited by* Ravi Iyengar

Volume 238. Heterotrimeric G-Protein Effectors
*Edited by* Ravi Iyengar

Volume 239. Nuclear Magnetic Resonance (Part C)
*Edited by* Thomas L. James and Norman J. Oppenheimer

Volume 240. Numerical Computer Methods (Part B)
*Edited by* Michael L. Johnson and Ludwig Brand

Volume 241. Retroviral Proteases
*Edited by* Lawrence C. Kuo and Jules A. Shafer

Volume 242. Neoglycoconjugates (Part A)
*Edited by* Y. C. Lee and Reiko T. Lee

Volume 243. Inorganic Microbial Sulfur Metabolism
*Edited by* Harry D. Peck, Jr., and Jean LeGall

Volume 244. Proteolytic Enzymes: Serine and Cysteine Peptidases
*Edited by* Alan J. Barrett

Volume 245. Extracellular Matrix Components
*Edited by* E. Ruoslahti and E. Engvall

Volume 246. Biochemical Spectroscopy
*Edited by* Kenneth Sauer

Volume 247. Neoglycoconjugates (Part B: Biomedical Applications)
*Edited by* Y. C. Lee and Reiko T. Lee

Volume 248. Proteolytic Enzymes: Aspartic and Metallo Peptidases
*Edited by* Alan J. Barrett

Volume 249. Enzyme Kinetics and Mechanism (Part D: Developments in Enzyme Dynamics)
*Edited by* Daniel L. Purich

Volume 250. Lipid Modifications of Proteins
*Edited by* Patrick J. Casey and Janice E. Buss

Volume 251. Biothiols (Part A: Monothiols and Dithiols, Protein Thiols, and Thiyl Radicals)
*Edited by* Lester Packer

Volume 252. Biothiols (Part B: Glutathione and Thioredoxin; Thiols in Signal Transduction and Gene Regulation)
*Edited by* Lester Packer

Volume 253. Adhesion of Microbial Pathogens
*Edited by* Ron J. Doyle and Itzhak Ofek

Volume 254. Oncogene Techniques
*Edited by* Peter K. Vogt and Inder M. Verma

Volume 255. Small GTPases and Their Regulators (Part A: Ras Family)
*Edited by* W. E. Balch, Channing J. Der, and Alan Hall

Volume 256. Small GTPases and Their Regulators (Part B: Rho Family)
*Edited by* W. E. Balch, Channing J. Der, and Alan Hall

Volume 257. Small GTPases and Their Regulators (Part C: Proteins Involved in Transport)
*Edited by* W. E. Balch, Channing J. Der, and Alan Hall

Volume 258. Redox-Active Amino Acids in Biology
*Edited by* Judith P. Klinman

Volume 259. Energetics of Biological Macromolecules
*Edited by* Michael L. Johnson and Gary K. Ackers

Volume 260. Mitochondrial Biogenesis and Genetics (Part A)
*Edited by* Giuseppe M. Attardi and Anne Chomyn

Volume 261. Nuclear Magnetic Resonance and Nucleic Acids
*Edited by* Thomas L. James

Volume 262. DNA Replication
*Edited by* Judith L. Campbell

Volume 263. Plasma Lipoproteins (Part C: Quantitation)
*Edited by* William A. Bradley, Sandra H. Gianturco, and Jere P. Segrest

Volume 264. Mitochondrial Biogenesis and Genetics (Part B)
*Edited by* Giuseppe M. Attardi and Anne Chomyn

Volume 265. Cumulative Subject Index Volumes 228, 230–262

Volume 266. Computer Methods for Macromolecular Sequence Analysis
*Edited by* Russell F. Doolittle

Volume 267. Combinatorial Chemistry
*Edited by* John N. Abelson

Volume 268. Nitric Oxide (Part A: Sources and Detection of NO; NO Synthase)
*Edited by* Lester Packer

Volume 269. Nitric Oxide (Part B: Physiological and Pathological Processes)
*Edited by* Lester Packer

Volume 270. High Resolution Separation and Analysis of Biological Macromolecules (Part A: Fundamentals)
*Edited by* Barry L. Karger and William S. Hancock

Volume 271. High Resolution Separation and Analysis of Biological Macromolecules (Part B: Applications)
*Edited by* Barry L. Karger and William S. Hancock

Volume 272. Cytochrome P450 (Part B)
*Edited by* Eric F. Johnson and Michael R. Waterman

Volume 273. RNA Polymerase and Associated Factors (Part A)
*Edited by* Sankar Adhya

Volume 274. RNA Polymerase and Associated Factors (Part B)
*Edited by* Sankar Adhya

Volume 275. Viral Polymerases and Related Proteins
*Edited by* Lawrence C. Kuo, David B. Olsen, and Steven S. Carroll

Volume 276. Macromolecular Crystallography (Part A)
*Edited by* Charles W. Carter, Jr., and Robert M. Sweet

Volume 277. Macromolecular Crystallography (Part B)
*Edited by* Charles W. Carter, Jr., and Robert M. Sweet

Volume 278. Fluorescence Spectroscopy
*Edited by* Ludwig Brand and Michael L. Johnson

Volume 279. Vitamins and Coenzymes (Part I)
*Edited by* Donald B. McCormick, John W. Suttie, and Conrad Wagner

Volume 280. Vitamins and Coenzymes (Part J)
*Edited by* Donald B. McCormick, John W. Suttie, and Conrad Wagner

Volume 281. Vitamins and Coenzymes (Part K)
*Edited by* Donald B. McCormick, John W. Suttie, and Conrad Wagner

Volume 282. Vitamins and Coenzymes (Part L)
*Edited by* Donald B. McCormick, John W. Suttie, and Conrad Wagner

Volume 283. Cell Cycle Control
*Edited by* William G. Dunphy

Volume 284. Lipases (Part A: Biotechnology)
*Edited by* Byron Rubin and Edward A. Dennis

Volume 285. Cumulative Subject Index Volumes 263, 264, 266–284, 286–289

Volume 286. Lipases (Part B: Enzyme Characterization and Utilization)
*Edited by* Byron Rubin and Edward A. Dennis

Volume 287. Chemokines
*Edited by* Richard Horuk

Volume 288. Chemokine Receptors
*Edited by* Richard Horuk

Volume 289. Solid Phase Peptide Synthesis
*Edited by* Gregg B. Fields

Volume 290. Molecular Chaperones
*Edited by* George H. Lorimer and Thomas Baldwin

Volume 291. Caged Compounds
*Edited by* Gerard Marriott

Volume 292. ABC Transporters: Biochemical, Cellular, and Molecular Aspects
*Edited by* Suresh V. Ambudkar and Michael M. Gottesman

Volume 293. Ion Channels (Part B)
*Edited by* P. Michael Conn

Volume 294. Ion Channels (Part C)
*Edited by* P. Michael Conn

Volume 295. Energetics of Biological Macromolecules (Part B)
*Edited by* Gary K. Ackers and Michael L. Johnson

Volume 296. Neurotransmitter Transporters
*Edited by* Susan G. Amara

Volume 297. Photosynthesis: Molecular Biology of Energy Capture
*Edited by* Lee McIntosh

Volume 298. Molecular Motors and the Cytoskeleton (Part B)
*Edited by* Richard B. Vallee

Volume 299. Oxidants and Antioxidants (Part A)
*Edited by* Lester Packer

Volume 300. Oxidants and Antioxidants (Part B)
*Edited by* Lester Packer

Volume 301. Nitric Oxide: Biological and Antioxidant Activities (Part C)
*Edited by* Lester Packer

Volume 302. Green Fluorescent Protein
*Edited by* P. Michael Conn

Volume 303. cDNA Preparation and Display
*Edited by* Sherman M. Weissman

Volume 304. Chromatin
*Edited by* Paul M. Wassarman and Alan P. Wolffe

Volume 305. Bioluminescence and Chemiluminescence (Part C)
*Edited by* Thomas O. Baldwin and Miriam M. Ziegler

Volume 306. Expression of Recombinant Genes in Eukaryotic Systems
*Edited by* Joseph C. Glorioso and Martin C. Schmidt

Volume 307. Confocal Microscopy
*Edited by* P. Michael Conn

Volume 308. Enzyme Kinetics and Mechanism (Part E: Energetics of Enzyme Catalysis)
*Edited by* Daniel L. Purich and Vern L. Schramm

Volume 309. Amyloid, Prions, and Other Protein Aggregates
*Edited by* Ronald Wetzel

Volume 310. Biofilms
*Edited by* Ron J. Doyle

Volume 311. Sphingolipid Metabolism and Cell Signaling (Part A)
*Edited by* Alfred H. Merrill, Jr., and Yusuf A. Hannun

Volume 312. Sphingolipid Metabolism and Cell Signaling (Part B)
*Edited by* Alfred H. Merrill, Jr., and Yusuf A. Hannun

Volume 313. Antisense Technology
(Part A: General Methods, Methods of Delivery, and RNA Studies)
*Edited by* M. Ian Phillips

Volume 314. Antisense Technology (Part B: Applications)
*Edited by* M. Ian Phillips

Volume 315. Vertebrate Phototransduction and the Visual Cycle (Part A)
*Edited by* Krzysztof Palczewski

Volume 316. Vertebrate Phototransduction and the Visual Cycle (Part B)
*Edited by* Krzysztof Palczewski

Volume 317. RNA–Ligand Interactions (Part A: Structural Biology Methods)
*Edited by* Daniel W. Celander and John N. Abelson

Volume 318. RNA–Ligand Interactions (Part B: Molecular Biology Methods)
*Edited by* Daniel W. Celander and John N. Abelson

Volume 319. Singlet Oxygen, UV-A, and Ozone
*Edited by* Lester Packer and Helmut Sies

Volume 320. Cumulative Subject Index Volumes 290–319

Volume 321. Numerical Computer Methods (Part C)
*Edited by* Michael L. Johnson and Ludwig Brand

Volume 322. Apoptosis
*Edited by* John C. Reed

Volume 323. Energetics of Biological Macromolecules (Part C)
*Edited by* Michael L. Johnson and Gary K. Ackers

Volume 324. Branched-Chain Amino Acids (Part B)
*Edited by* Robert A. Harris and John R. Sokatch

Volume 325. Regulators and Effectors of Small GTPases
(Part D: Rho Family)
*Edited by* W. E. Balch, Channing J. Der, and Alan Hall

Volume 326. Applications of Chimeric Genes and Hybrid Proteins
(Part A: Gene Expression and Protein Purification)
*Edited by* Jeremy Thorner, Scott D. Emr, and John N. Abelson

Volume 327. Applications of Chimeric Genes and Hybrid Proteins
(Part B: Cell Biology and Physiology)
*Edited by* Jeremy Thorner, Scott D. Emr, and John N. Abelson

Volume 328. Applications of Chimeric Genes and Hybrid Proteins (Part C: Protein–Protein Interactions and Genomics)
*Edited by* Jeremy Thorner, Scott D. Emr, and John N. Abelson

Volume 329. Regulators and Effectors of Small GTPases (Part E: GTPases Involved in Vesicular Traffic)
*Edited by* W. E. Balch, Channing J. Der, and Alan Hall

Volume 330. Hyperthermophilic Enzymes (Part A)
*Edited by* Michael W. W. Adams and Robert M. Kelly

Volume 331. Hyperthermophilic Enzymes (Part B)
*Edited by* Michael W. W. Adams and Robert M. Kelly

Volume 332. Regulators and Effectors of Small GTPases (Part F: Ras Family I)
*Edited by* W. E. Balch, Channing J. Der, and Alan Hall

Volume 333. Regulators and Effectors of Small GTPases (Part G: Ras Family II)
*Edited by* W. E. Balch, Channing J. Der, and Alan Hall

Volume 334. Hyperthermophilic Enzymes (Part C)
*Edited by* Michael W. W. Adams and Robert M. Kelly

Volume 335. Flavonoids and Other Polyphenols
*Edited by* Lester Packer

Volume 336. Microbial Growth in Biofilms (Part A: Developmental and Molecular Biological Aspects)
*Edited by* Ron J. Doyle

Volume 337. Microbial Growth in Biofilms (Part B: Special Environments and Physicochemical Aspects)
*Edited by* Ron J. Doyle

Volume 338. Nuclear Magnetic Resonance of Biological Macromolecules (Part A)
*Edited by* Thomas L. James, Volker Dötsch, and Uli Schmitz

Volume 339. Nuclear Magnetic Resonance of Biological Macromolecules (Part B)
*Edited by* Thomas L. James, Volker Dötsch, and Uli Schmitz

Volume 340. Drug–Nucleic Acid Interactions
*Edited by* Jonathan B. Chaires and Michael J. Waring

Volume 341. Ribonucleases (Part A)
*Edited by* Allen W. Nicholson

Volume 342. Ribonucleases (Part B)
*Edited by* Allen W. Nicholson

Volume 343. G Protein Pathways (Part A: Receptors)
*Edited by* Ravi Iyengar and John D. Hildebrandt

Volume 344. G Protein Pathways (Part B: G Proteins and Their Regulators)
*Edited by* Ravi Iyengar and John D. Hildebrandt

VOLUME 345. G Protein Pathways (Part C: Effector Mechanisms)
*Edited by* RAVI IYENGAR AND JOHN D. HILDEBRANDT

VOLUME 346. Gene Therapy Methods
*Edited by* M. IAN PHILLIPS

VOLUME 347. Protein Sensors and Reactive Oxygen Species (Part A: Selenoproteins and Thioredoxin)
*Edited by* HELMUT SIES AND LESTER PACKER

VOLUME 348. Protein Sensors and Reactive Oxygen Species (Part B: Thiol Enzymes and Proteins)
*Edited by* HELMUT SIES AND LESTER PACKER

VOLUME 349. Superoxide Dismutase
*Edited by* LESTER PACKER

VOLUME 350. Guide to Yeast Genetics and Molecular and Cell Biology (Part B)
*Edited by* CHRISTINE GUTHRIE AND GERALD R. FINK

VOLUME 351. Guide to Yeast Genetics and Molecular and Cell Biology (Part C)
*Edited by* CHRISTINE GUTHRIE AND GERALD R. FINK

VOLUME 352. Redox Cell Biology and Genetics (Part A)
*Edited by* CHANDAN K. SEN AND LESTER PACKER

VOLUME 353. Redox Cell Biology and Genetics (Part B)
*Edited by* CHANDAN K. SEN AND LESTER PACKER

VOLUME 354. Enzyme Kinetics and Mechanisms (Part F: Detection and Characterization of Enzyme Reaction Intermediates)
*Edited by* DANIEL L. PURICH

VOLUME 355. Cumulative Subject Index Volumes 321–354

VOLUME 356. Laser Capture Microscopy and Microdissection
*Edited by* P. MICHAEL CONN

VOLUME 357. Cytochrome P450, Part C
*Edited by* ERIC F. JOHNSON AND MICHAEL R. WATERMAN

VOLUME 358. Bacterial Pathogenesis (Part C: Identification, Regulation, and Function of Virulence Factors)
*Edited by* VIRGINIA L. CLARK AND PATRIK M. BAVOIL

VOLUME 359. Nitric Oxide (Part D)
*Edited by* ENRIQUE CADENAS AND LESTER PACKER

VOLUME 360. Biophotonics (Part A)
*Edited by* GERARD MARRIOTT AND IAN PARKER

VOLUME 361. Biophotonics (Part B)
*Edited by* GERARD MARRIOTT AND IAN PARKER

Volume 362. Recognition of Carbohydrates in Biological Systems (Part A)
*Edited by* Yuan C. Lee and Reiko T. Lee

Volume 363. Recognition of Carbohydrates in Biological Systems (Part B)
*Edited by* Yuan C. Lee and Reiko T. Lee

Volume 364. Nuclear Receptors
*Edited by* David W. Russell and David J. Mangelsdorf

Volume 365. Differentiation of Embryonic Stem Cells
*Edited by* Paul M. Wassauman and Gordon M. Keller

Volume 366. Protein Phosphatases
*Edited by* Susanne Klumpp and Josef Krieglstein

Volume 367. Liposomes (Part A)
*Edited by* Nejat Düzgüneş

Volume 368. Macromolecular Crystallography (Part C)
*Edited by* Charles W. Carter, Jr., and Robert M. Sweet

Volume 369. Combinational Chemistry (Part B)
*Edited by* Guillermo A. Morales and Barry A. Bunin

Volume 370. RNA Polymerases and Associated Factors (Part C)
*Edited by* Sankar L. Adhya and Susan Garges

Volume 371. RNA Polymerases and Associated Factors (Part D)
*Edited by* Sankar L. Adhya and Susan Garges

Volume 372. Liposomes (Part B)
*Edited by* Nejat Düzgüneş

Volume 373. Liposomes (Part C)
*Edited by* Nejat Düzgüneş

Volume 374. Macromolecular Crystallography (Part D)
*Edited by* Charles W. Carter, Jr., and Robert W. Sweet

Volume 375. Chromatin and Chromatin Remodeling Enzymes (Part A)
*Edited by* C. David Allis and Carl Wu

Volume 376. Chromatin and Chromatin Remodeling Enzymes (Part B)
*Edited by* C. David Allis and Carl Wu

Volume 377. Chromatin and Chromatin Remodeling Enzymes (Part C)
*Edited by* C. David Allis and Carl Wu

Volume 378. Quinones and Quinone Enzymes (Part A)
*Edited by* Helmut Sies and Lester Packer

Volume 379. Energetics of Biological Macromolecules (Part D)
*Edited by* Jo M. Holt, Michael L. Johnson, and Gary K. Ackers

Volume 380. Energetics of Biological Macromolecules (Part E)
*Edited by* Jo M. Holt, Michael L. Johnson, and Gary K. Ackers

Volume 399. Ubiquitin and Protein Degradation (Part B)
*Edited by* Raymond J. Deshaies

Volume 400. Phase II Conjugation Enzymes and Transport Systems
*Edited by* Helmut Sies and Lester Packer

Volume 401. Glutathione Transferases and Gamma Glutamyl Transpeptidases
*Edited by* Helmut Sies and Lester Packer

Volume 402. Biological Mass Spectrometry
*Edited by* A. L. Burlingame

Volume 403. GTPases Regulating Membrane Targeting and Fusion
*Edited by* William E. Balch, Channing J. Der, and Alan Hall

Volume 404. GTPases Regulating Membrane Dynamics
*Edited by* William E. Balch, Channing J. Der, and Alan Hall

Volume 405. Mass Spectrometry: Modified Proteins and Glycoconjugates
*Edited by* A. L. Burlingame

Volume 406. Regulators and Effectors of Small GTPases: Rho Family
*Edited by* William E. Balch, Channing J. Der, and Alan Hall

Volume 407. Regulators and Effectors of Small GTPases: Ras Family
*Edited by* William E. Balch, Channing J. Der, and Alan Hall

Volume 408. DNA Repair (Part A)
*Edited by* Judith L. Campbell and Paul Modrich

Volume 409. DNA Repair (Part B)
*Edited by* Judith L. Campbell and Paul Modrich

Volume 410. DNA Microarrays (Part A: Array Platforms and Web-Bench Protocols)
*Edited by* Alan Kimmel and Brian Oliver

Volume 411. DNA Microarrays (Part B: Databases and Statistics)
*Edited by* Alan Kimmel and Brian Oliver

Volume 412. Amyloid, Prions, and Other Protein Aggregates (Part B)
*Edited by* Indu Kheterpal and Ronald Wetzel

Volume 413. Amyloid, Prions, and Other Protein Aggregates (Part C)
*Edited by* Indu Kheterpal and Ronald Wetzel

Volume 414. Measuring Biological Responses with Automated Microscopy
*Edited by* James Inglese

Volume 415. Glycobiology
*Edited by* Minoru Fukuda

Volume 416. Glycomics
*Edited by* Minoru Fukuda

VOLUME 433. Lipidomics and Bioactive Lipids: Specialized Analytical Methods and Lipids in Disease
*Edited by* H. ALEX BROWN

VOLUME 434. Lipidomics and Bioactive Lipids: Lipids and Cell Signaling
*Edited by* H. ALEX BROWN

VOLUME 435. Oxygen Biology and Hypoxia
*Edited by* HELMUT SIES AND BERNHARD BRÜNE

VOLUME 436. Globins and Other Nitric Oxide-Reactive Protiens (Part A)
*Edited by* ROBERT K. POOLE

VOLUME 437. Globins and Other Nitric Oxide-Reactive Protiens (Part B)
*Edited by* ROBERT K. POOLE

VOLUME 438. Small GTPases in Disease (Part A)
*Edited by* WILLIAM E. BALCH, CHANNING J. DER, AND ALAN HALL

VOLUME 439. Small GTPases in Disease (Part B)
*Edited by* WILLIAM E. BALCH, CHANNING J. DER, AND ALAN HALL

VOLUME 440. Nitric Oxide, Part F Oxidative and Nitrosative Stress in Redox Regulation of Cell Signaling
*Edited by* ENRIQUE CADENAS AND LESTER PACKER

VOLUME 441. Nitric Oxide, Part G Oxidative and Nitrosative Stress in Redox Regulation of Cell Signaling
*Edited by* ENRIQUE CADENAS AND LESTER PACKER

VOLUME 442. Programmed Cell Death, General Principles for Studying Cell Death (Part A)
*Edited by* ROYA KHOSRAVI-FAR, ZAHRA ZAKERI, RICHARD A. LOCKSHIN, AND MAURO PIACENTINI

VOLUME 443. Angiogenesis: *In Vitro* Systems
*Edited by* DAVID A. CHERESH

VOLUME 444. Angiogenesis: *In Vivo* Systems (Part A)
*Edited by* DAVID A. CHERESH

VOLUME 445. Angiogenesis: *In Vivo* Systems (Part B)
*Edited by* DAVID A. CHERESH

VOLUME 446. Programmed Cell Death, The Biology and Therapeutic Implications of Cell Death (Part B)
*Edited by* ROYA KHOSRAVI-FAR, ZAHRA ZAKERI, RICHARD A. LOCKSHIN, AND MAURO PIACENTINI

VOLUME 447. RNA Turnover in Bacteria, Archaea and Organelles
*Edited by* LYNNE E. MAQUAT AND CECILIA M. ARRAIANO

CHAPTER ONE

# Fundamentals of Three-Dimensional Reconstruction from Projections

Pawel A. Penczek

## Contents



## Abstract

Three-dimensional (3D) reconstruction of an object mass density from the set of its 2D line projections lies at a core of both single-particle reconstruction technique and electron tomography. Both techniques utilize electron microscope to collect a set of projections of either multiple objects representing in principle the same macromolecular complex in an isolated form, or a subcellular structure isolated *in situ*. Therefore, the goal of macromolecular electron microscopy is to invert the projection transformation to recover the distribution of the mass density of the original object. The problem is interesting in that in its discrete form it is ill-posed and not invertible. Various algorithms have been proposed to cope with the practical difficulties of this inversion problem and their differ widely in terms of their robustness with respect to noise in the data, completeness of the collected projection dataset, errors in projections orientation parameters, abilities to efficiently handle large datasets, and other obstacles typically encountered in molecular electron microscopy. Here, we

Department of Biochemistry and Molecular Biology, The University of Texas, Houston Medical School, Houston, Texas, USA

*Methods in Enzymology*, Volume 482
ISSN 0076-6879, DOI: 10.1016/S0076-6879(10)82001-4

review the theoretical foundations of 3D reconstruction from line projections followed by an overview of reconstruction algorithms routinely used in practice of electron microscopy.

## 1. Introduction

The two electron microscope (EM)-based techniques that provide insight into three-dimensional (3D) organization of biological specimens, namely single particle reconstruction (SPR) and electron tomography (ET), are often stated as a problem of 3D reconstruction from 2D projections. While one can argue that the main challenge of these techniques is the establishment of the orientation parameters of 2D projections, which is necessary condition for 3D reconstruction, there is no doubt that the 3D reconstruction itself plays a central role in understanding how SPR and ET lead to 3D visualization of the specimen. Indeed, many steps of the respective data processing protocols for ET and SPR are best understood in terms of the 3D reconstruction problem. For example, the *ab initio* determination of a 3D model in SPR, both using random conical tilt approach (Radermacher *et al.*, 1987) and common lines methodology (Crowther *et al.*, 1970b; Goncharov *et al.*, 1987; Penczek *et al.*, 1996; van Heel, 1987), is deeply rooted in the nature of the relationship between an object and its projections. Refinement of a 3D structure, as done using either real-space 3D projection alignment (Penczek *et al.*, 1994) or Fourier space representation (Grigorieff, 1998), depends on correct accounting for 3D reconstruction idiosyncrasies. The principle and limitations of resolution estimation of the reconstructed 3D object using Fourier shell correlation (FSC) or spectral signal-to-noise ratio (SSNR) methodology can only be explained within the context of 3D reconstruction algorithms (Penczek, 2002). Therefore, even a brief introduction into SPR and ET principles has to include an explanation of the foundations of the methods of 3D reconstruction from projections.

Imaging the density distribution of a 3D object using a set of its 2D projections can be considered an extension of computerized tomography (from Greek τέμνειν, to section), which was originally developed for reconstructing 2D cross-sections (slices) of a 3D object from its 1D projections. The first practical application of the methods proposed for the reconstruction of an interior mass distribution of an object from a set of its projections dates back to 1956 when Bracewell reconstructed sun spots from multiple projection views of the Sun from the Earth (Bracewell, 1956). Bracewell is also credited with the introduction of the first semi-heuristic reconstruction algorithm known as filtered backprojection (FBP), which is still commonly used in CT scanners. In 1967, DeRosier and Klug performed 3D reconstruction of a virus capsid by taking advantage of its icosahedral symmetry (Crowther *et al.*, 1970b; DeRosier and Klug, 1968). In 1969, Hoppe proposed 3D high-resolution electron microscopy of nonperiodic biological specimens (Hoppe, 1969), and his ideas later matured into what we today know as ET (reconstruction of individual objects *in situ*) and SPR (reconstruction of macromolecular structures by averaging projection images of multiple, but structurally identical isolated single particles). Tomography became a household name with the development of computed axial tomography (CAT) by Hounsfield and Cormack (independently) in 1972. CAT was the first noninvasive technique that allowed visualization of the interior of living organisms.

Reconstruction algorithms are an integral part of the tomographic technique; without computer analysis and merging of the data it is all but impossible to deduce structure of an object from the set of its projections. Initially, the field was dominated by algorithms based on various semiheuristic principles, and the lack of solid theoretical foundations caused some strong controversies. A question of both theoretical and practical interest concerns the minimum number of projections required for a faithful reconstruction, as algebraic and Fourier techniques seemed to lead to different conclusions (Bellman *et al.*, 1971; Crowther and Klug, 1971; Gordon *et al.*, 1970). Considering the quality of an early ET reconstruction of a ribosome, some claims might have been exaggerated (Bender *et al.*, 1970). However, it was eventually recognized that mathematical foundations for modern tomography predate the field by decades and were laid out in purely theoretical terms: Radon (1986) considered the invertibility of an integral transform involving projections, and Kaczmarz (1993) sought an iterative solution of a large system of linear equations. Currently, the mathematics of reconstruction from projections problems is very well developed and understood (Herman, 2009; Natterer, 1986; Natterer and Wübbeling, 2001).

Even though the mathematical framework of tomography is well laid out, the field remains vibrant both in terms of theoretical progress and development of new algorithms. The main challenge is that the reconstruction of an object from a set of its projections belongs to a class of inverse problems, so one has to find practical approximations of the solution. While a continuous Radon transform with an infinite number of projections is invertible, in practice only a finite number of discretized projections can be measured. It is well known that in this case the original object cannot be uniquely recovered (Rieder and Faridani, 2003). To illustrate this statement, we represent the problem as an algebraic one: we denote all projection data by a vector $\mathbf{g}$, the original object by a vector $\mathbf{d}$, and a projection matrix by $\mathbf{P}$, which is almost always rectangular. The inverse reconstruction is stated as: given $\mathbf{g}$ and $\mathbf{P}$, find $\hat{\mathbf{d}}$ such that $\mathbf{P}\hat{\mathbf{d}} = \mathbf{g}$. The inverse problem is called well-posed if it is uniquely solvable for each $\mathbf{g}$ and if the solution depends continuously on $\mathbf{g}$. Otherwise, it is called ill-posed. Since $\mathbf{P}$ is rectangular in tomography, the inversion of projection transformation is necessarily ill-posed and is either over- or underdetermined depending on the number of projections. Moreover, even if an approximate inverse (reconstruction algorithm) can be found, it is not necessarily continuous, which means that the solution of $\mathbf{P}\hat{\mathbf{d}} = \mathbf{g}$ need not be close to the solution of $\mathbf{P}\hat{\mathbf{d}} = \mathbf{g}^{\varepsilon}$ even if $\mathbf{g}^{\varepsilon}$ is close to $\mathbf{g}$. A further difficulty is caused by the existence of ghosts in tomography. For discretized tomography, the set of projections is finite and data is sampled, and it can be shown that for any set of projection directions there exists a nontrivial object $\hat{\mathbf{d}}_0 \neq 0$ whose projections calculated in the directions of measured data are exactly zero, that is, $\mathbf{P}\hat{\mathbf{d}}_0 = 0$ (Louis, 1984). This implies that the reconstructed object can differ significantly from the original object but still agree perfectly with the given data; in other words, the discrete integral projection transformation is not invertible, even in the absence of noise. In practice, ghosts happen to be very high frequency objects (Louis, 1984; Maass, 1987) and their impact can be all but eliminated by a proper regularization of the solution.

It follows from the previous discussion that there are a number of practical difficulties associated with the problem of 3D reconstruction from 2D projections. The results of various reconstruction algorithms can differ noticeably, so the choice of reconstruction method in SPR or ET can affect the results significantly. In addition, within the context of a specific algorithm, the regularization methods used to minimize artifacts are somewhat arbitrary.

We tend to consider the problem of reconstruction from projections in SPR and ET to be a part of a broader field of tomographic reconstruction. However, it has to be stressed that what we typically encounter in EM differs considerably from the reconstruction problems in other fields and while the general theory holds, many issues arising within the context of SPT and ET do not have sound theoretical or even practical solutions. First, SPR and double-tilt ET are rare examples of "true" 3D reconstruction problems. In most applications, including single-axis ET, the 3D reconstruction from 2D projections can be decomposed into a series of semi-independent 2D reconstructions (slices of the final 3D object) from 1D projections. In this case, the problem is reduced to that of a 2D reconstruction and one can take advantage of a large number of well-established and understood algorithms. Second, in SPR the data collection geometry cannot be controlled and the distribution of projections is random. This requires the reconstruction algorithm to be flexible with parameters that can be adjusted based on the encountered distribution of data. In SPR and double-tilt ET, the distribution of projection directions tend to be extremely uneven, so an effective algorithm must be found to account for the varying SSNR conditions within the same target object. Furthermore, we also note that the problem of varying SSNR is compounded by the extremely low SNR of the data. Third, the uneven distribution of projection directions and experimental limitations of the maximum tilt in ET result in gaps in Fourier space, which not only makes the problem not invertible, but also calls for additional regularization to minimize the artifacts. Fourth, unlike in other fields, orientation parameters in SPR and ET are known only approximately and the errors are both random and systematic. It follows that the 3D reconstruction algorithm should be considered within the context of the structure refinement procedure. Finally, to ensure a desirable SNR of the reconstructed object, the number of projections is typically much larger than their size (expressed in pixels). However, this dramatic oversampling places extreme computational demands on the reconstruction algorithms and therefore eliminates certain methods from consideration.

In this text, I will provide a general overview of reconstruction techniques while focusing on issues unique to EM applications. I will first briefly describe the theory of reconstruction from projections and then analyze reconstruction algorithms practically used in EM with a focus on those that are implemented in major software packages and are routinely used. This is followed by a discussion of issues unique to 3D reconstruction of EM data, and in particular, those of correcting for the effects of the contrast transfer function (CTF) of the microscope.

## 2. The Object and its Projection

In structural studies of biological specimens, the settings of EM are selected such that the collected 2D images are parallel beam 2D projections $g(\mathbf{x})$ of a 3D specimen $d(\mathbf{r})$. For a continuous distribution of projection directions $\boldsymbol{\tau}$, $g$ forms an integral transformation of $d$ called a *ray transform* (Natterer and Wübbeling, 2001):

$$g(\mathbf{x}_\tau) = \int d(\mathbf{r})d\boldsymbol{\tau}, \quad \boldsymbol{\tau}\perp\mathbf{x}. \tag{1.1}$$

Ray transform is an integral over straight line and transforms an $n$D function into a $(n-1)$D function. Ray transform is not to be confused with a Radon transform, which is realized as 1D projections of a $n$D function over $(n-1)$ D hyperplanes. For 2D functions, ray and Radon transforms differ only in the notation. Here, we will concern ourselves with an inverse problem of recovering a 3D function from its 2D ray transform.

The projection direction $\boldsymbol{\tau}$ is defined as a unit vector and in EM it is parametrized by two Eulerian angles $\boldsymbol{\tau}(\varphi,\theta)$ and the projection is formed on the plane $\mathbf{x}$ perpendicular to $\boldsymbol{\tau}$. Most cryo-EM software packages (SPIDER, IMAGIC, MRC, FREALIGN, EMAN2, and SPARX) use the *ZYZ* convention of Eulerian angles which expresses the rotation of a 3D rigid body by the application of three matrices:

$$\begin{bmatrix} \cos\psi & \sin\psi & 0 \\ -\sin\psi & \cos\psi & 0 \\ 0 & 0 & 1 \end{bmatrix} \begin{bmatrix} \cos\theta & 0 & -\sin\theta \\ 0 & 1 & 0 \\ \sin\theta & 0 & \cos\theta \end{bmatrix} \begin{bmatrix} \cos\varphi & \sin\varphi & 0 \\ -\sin\varphi & \cos\varphi & 0 \\ 0 & & 1 \end{bmatrix} \tag{1.2}$$

Angle $\psi$ is responsible for the in-plane rotation of the projection (Fig. 1.1) and is considered trivial as it does not affect information content of the ray transform. The two Eulerian angles $\varphi,\theta$ that define projection direction with the in-plane rotation $\psi$ and in-plane translations $t_x,t_y$ are in EM jointly referred to as projection *orientation parameters*. The distribution of projection directions may be conveniently visualized as a set of points on a unit half-sphere (Fig. 1.2).

In Fourier space, the relationship between an object and its projection is referred to as the central section theorem: the Fourier transformation $G$ of projection $g$ of a 3D object $d$ is the central (i.e., passing though the origin of reciprocal space) 2D plane cross-section of the 3D transform $D$ and is perpendicular to the projection vector (Bracewell, 1956; Crowther *et al.*, 1970b; DeRosier and Klug, 1968):

$$\mathcal{F}[g(\mathbf{x}_{\boldsymbol{\tau}})] = D(\mathbf{s}_{\tau}), \quad \mathbf{s}_{\tau} \perp \boldsymbol{\tau}. \tag{1.3}$$

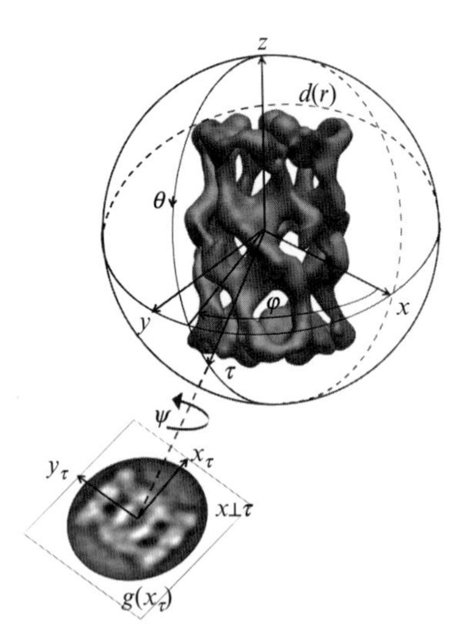


**Figure 1.1** The projection sphere and projection $g(\mathbf{x}_{\tau})$ of $d(\mathbf{r})$ along $\boldsymbol{\tau}$ onto the plane $\boldsymbol{\tau} \perp \mathbf{x}$. The convention of Eulerian angles as in Eq. (1.2). (For color version of this figure, the reader is referred to the Web version of this chapter.)

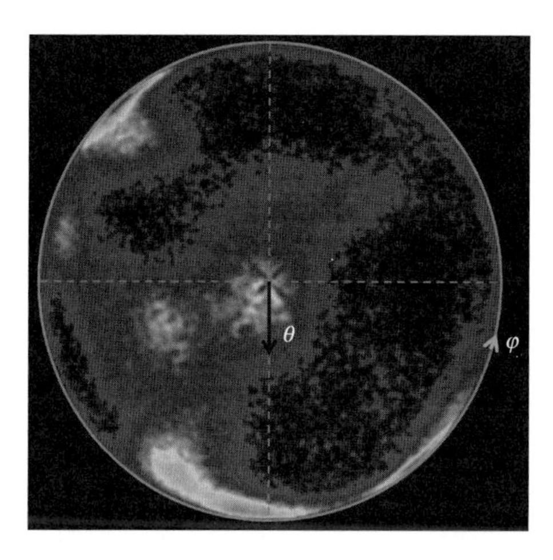


**Figure 1.2** Distribution of projection directions $\boldsymbol{\tau}(\varphi,\theta)$ mapped on a half-sphere. The value of each point on the surface of the unitary sphere is equal to the number of projections whose directions given as vectors $\boldsymbol{\tau}(\varphi,\theta)$ (Eq. (1.2)) fall in its vicinity. All directions are mapped on a half-sphere $0 \leq \theta \leq 90$, $0 \leq \varphi < 360$, with directions $90 < \theta \leq 180$ mapped to mirror-equivalent positions $\theta' = 180 - \theta$, $\varphi' = 180 + \varphi$. The 322,688 angles are taken from the 3D reconstruction of a *Thermus thermophilus* ribosome complexed with EF-Tu at 6.4 Å resolution (Schuette *et al.*, 2009) and the occupancies are color-coded with red corresponding to maximum (130) and dark blue to minimum (0). There are no gaps in angular coverage at the angular resolution of the plot. (See Color Insert.)

where by capital letter we denoted Fourier transform of a function and **s** is a vector of spatial frequencies. From Eq. (1.3) it follows that the inversion of the 3D ray transform is possible if there is a continuous distribution of projections such that their 2D Fourier transforms fill without gaps 3D Fourier transform of the object. This requirement can be expressed as Orlov's condition, according to which the minimum distribution of projections necessary for inversion of Eq. (1.1) is given by the motion of the unit vector over any continuous line connecting the opposite points on the unit sphere (Orlov, 1976a). In 3D, single-axis tilt geometry used in ET with the full range of projection angles is an example of a distribution of projection that ensures the invertibility of the ray transform.

Unlike the case in ET, imaged objects in SPR are randomly and nonuniformly oriented and the distribution of their orientations is beyond our control, so the practical impact of the Orlov's condition in limited. Moreover, in EM the problem is compounded by the high noise level of the data. It is easy to see that single-axis tilt geometry would result in a highly nonuniform distribution of the SSNR in the reconstructed 3D object (Penczek, 2002; Penczek and Frank, 2006). Hence, generally speaking, it is more desirable in SPR to have a possibly uniform distribution of projection directions as this will yield uniform distribution of SSNR. It clearly follows that uniform distribution of projection directions being redundant fulfills Orlov's condition.

## 3. Taxonomy of Reconstruction Methods

The inversion formulae for a continuous ray transform are known: the solution for the 2D Radon transform was originally given by Radon (1986) and his work was subsequently extended to higher dimensions and various data collection geometries (Natterer and Wübbeling, 2001). However, the applicability of these solutions toward the design of algorithms is limited. Simple discretization of analytical solutions is either not possible or does not lead to stable numerical algorithms. Moreover, as we have already pointed out, the discrete ray transform is not invertible, so practical ways of obtaining stable solutions have to be found. It is also necessary to account for particular data collection geometries and experimental limitations, such as limited or truncated view of the object.

The two main groups of reconstruction methods are the algebraic and the transform methods. The algebraic methods begin from discretization of the integral transformation equation (1.1). This leads to a system of linear equations that are typically overdetermined and whose solution yields the desired reconstructed object. The advantages of this approach are numerous: (1) the problem is immediately cast as an algebraic one, so one can take advantage of well-established linear algebra numerical methods to find a solution; (2) discretization, sampling, and interpolation appear naturally and are simply embedded into the projection matrix **P**; and (3) there is no need to design methods to account for particular distribution of projection directions. Since the inverse of **P** normally does not exist, algebraic methods are with few exceptions iterative. While they are easy to implement, the significant disadvantage is that they are also extremely computationally intensive, especially in their applications to EM where the number of 2D projections is very large. It is also difficult to provide precise rules concerning the determination of the number of iterations required to obtain a desired quality of reconstruction.

The transform methods are based on the central section theorem Eq. (1.3) and tend to rely on Fourier space analysis of the problem. The analysis of the problem is done using continuous functions and discretization is considered only during interpolation between polar and Cartesian systems of coordinates. Within the class of transform methods, there are further subdivisions. In the first subgroup of methods one constructs a Fourier space linear filter to account for particular distribution of projections, applies it in Fourier space to projection data, and reconstruction is typically performed in real space using filtered data and a simple backprojection algorithm. These algorithms are easy to implement, reasonably fast and straightforward to use, but for uneven distribution of projections, particularly in 3D, the filter functions are rather crude approximations of optimum solutions, so the results tend to be inferior. The second subgroup contains algorithms that directly and exclusively operate in Fourier space by casting the problem as one of Fourier space interpolation between polar system of coordinates of the projection data and Cartesian system of coordinates of the reconstructed object; collectively, these algorithms are known under the name of *direct Fourier inversion*. Their implementation tends to be challenging, but direct Fourier inversion is computationally very efficient and the 'gridding' algorithm from this group is currently considered the most accurate reconstruction method.

Reconstruction algorithms have to meet a number of requirements, some of which are in contradiction to each other, in order to be successfully applicable to EM data. The algorithm should be capable of handling very large datasets in excess of $10^5$ projection images that can be as large as $256^2$ pixels or more. It should account for an uneven distribution of projection directions that is irregular and cannot be approximated by an analytical function and still be sufficiently fast to permit structure refinement by adjustments of projection directions. The algorithm should also minimize artifacts in the reconstructed object that are due to gaps in coverage of Fourier space; such artifacts tend to adversely affect refinement procedures. It should be capable of correcting for the CTF effects and of accounting for the uneven distribution of spectral SNR in the data and in 3D Fourier space. For larger structures, the algorithm should be capable of correcting for the defocus spread within the object. The reported resolution of a 3D reconstruction depends on the linear relations between parts of the structure and between reconstructions done from different subsets of the data. To ensure a meaningful resolution measure, the algorithm must be linear and shift invariant for otherwise the relative densities within the structure would be distorted, thus leading to incorrect interpretations of and in extreme cases meaningless reported resolutions. Lastly, when applied to ET problems, the algorithm should be able to properly account for slab geometry of the object and for the problem of partial and truncated views.

It is extremely unlikely that a single algorithm would yield optimum results with respect to a set of such diverse requirements. In most cases, the tradeoff is the computational efficiency, that is, the running time. As we will see, iterative algebraic algorithms are the closest to the ideal, but they are very slow and require adjustments of numerous parameters whose values are not known in advance. In addition, some of the problems listed do not have convincing theoretical solutions. Taken together, it all assures that the field of 3D reconstruction from projections will remain fertile ground for the development of new methods and new algorithms.

## 4. Discretization and Interpolation

In digital image processing, space is represented by a multi-dimensional discrete lattice. The 2D projections $g_n(\mathbf{x})$ are sampled on a Cartesian grid $\{\mathbf{k}a : \mathbf{k} \in \mathbf{Z}^n, -(\mathbf{K}/2) \leq \mathbf{k} < \mathbf{K}/2\}$, where $n$ is the dimensionality of the grid of the reconstructed object, $n - 1$ is the dimensionality of projections, $\mathbf{K} \in \mathbf{Z}^n_+$ is the size of the grid, and $a$ is the grid spacing. In EM, the units of $a$ are either Ångstroms or nanometers and we assume that the data is appropriately sampled, that is, the pixel size is less than or equal to the inverse of twice the maximum spatial frequency present in the data $a \leq (1/2s_{\max})$. Since the latter is not known in advance, a practical rule of thumb is to select the pixel size to be about one-third of the expected resolution of the final structure, so that the adverse effects of interpolation are reduced.

The EM data (projections of the macromolecule) are discretized on a 2D Cartesian grid; however, each projection has a particular orientation in polar coordinates. Hence, an interpolation is required to relate the measured samples to the voxel values on the 3D Cartesian grid of the reconstructed structure. The backprojection step can be visualized as a set of rays with base $a^{n-1}$ and which are extended from projections. The value of a voxel on the grid of the reconstructed structure is the sum of the ray values which intersect the voxel. One can select schemes that aim at approximation of the physical reality of the data collection, that is, by weighting the contributions by the areas of the voxels intersected by the ray or by the lengths of the line segments that intersect (Huesman *et al.*, 1977). To reduce the computation time, one usually assumes in electron microscopy that all the mass is located in the center of the voxel, in which case additional accuracy is achieved by application of tri- (or bi-) linear interpolation. The exception is the algebraic reconstruction technique (ART) with blobs algorithm (Marabini *et al.*, 1998) in which the voxels are represented by smooth spherically symmetric volume elements (e.g., Keiser–Bessel function).

In real space, both the projection and backprojection steps can be implemented in two different ways: as voxel driven or as ray driven (Laurette *et al.*, 2000). If we consider a projection, in the voxel-driven approach the volume is scanned voxel by voxel. The nearest projection bin to the projection of each voxel is found, and the values in this bin and three neighboring bins are increased by the corresponding voxel value multiplied by the weights calculated using bilinear interpolation. In the ray-driven approach, the volume is scanned along the projection rays. The value of the projection bin is increased by the values in the volume calculated in equidistant steps along the rays using trilinear interpolation. Because voxel- and ray-driven methods apply interpolation to projections or to voxels, respectively, the interpolation artifacts will be different in each case. Therefore, when calculating reconstructions using iterative algorithms that alternate between projection and backprojection steps, it is important to maintain consistency; that is, to make sure that matrix representing one step is the transpose of the matrix representing the other step. In either case, the computational complexity of each method is $O(K^3)$, which is the same as for the rotation of a 3D volume.

In the reconstruction methods based on the direct Fourier inversion of the 3D ray transform, the interpolation is performed in Fourier space. Regrettably, it is difficult to design an accurate and fast interpolation scheme for the discrete Fourier space, so it is tempting to use interpolation based on Shannon's sampling theorem (Shannon, 1949). It states that a properly sampled, band-limited signal can be fully recovered from its discrete samples. For the signal represented by $K^3$ equispaced Fourier samples $D_{hkl}$, the value at the arbitrary location $D(s_x,s_y,s_z)$ is given by (Crowther *et al.*, 1970a):

$$D\left(s_x, s_y, s_z\right) = \sum_{h=0}^{K-1}\sum_{k=0}^{K-1}\sum_{l=0}^{K-1} D_{hkl} w_h(s_x) w_k\left(s_y\right) w_l(s_z), \tag{1.4}$$

where (Lanzavecchia and Bellon, 1994; Yuen and Fraser, 1979):

$$w_j(s) = \begin{cases} \dfrac{\sin(K\pi(s-j/K))}{\sin(\pi(s-j/K))}, & K\,\text{odd} \\ \dfrac{\sin(K\pi(s-j/K))}{\tan(\pi(s-j/K))}, & K\,\text{even} \end{cases} \quad j = h, k, l. \tag{1.5}$$

For structures with symmetries such as icosahedral structures, the projection data are distributed approximately evenly, in which case Eq. (1.4) can be solved to a good degree of accuracy by performing the interpolation independently along each of the three frequency axes (Crowther *et al.*, 1970b). In this case, the solution to the problem of interpolation in Fourier space becomes a solution to the reconstruction problem. Crowther *et al.* (1970b) solved it as an overdetermined system of linear equations and gave the following least-squares solution:

$$\hat{\mathbf{D}}_{hkl} = \left(\mathbf{W}^{\mathrm{T}}\mathbf{W}\right)^{-1}\mathbf{W}^{\mathrm{T}}\mathbf{D}. \tag{1.6}$$

where $D$ and $D_{hkl}$ are written as 1D arrays $\mathbf{D}$ and $\mathbf{D}_{hkl}$, respectively, and $\mathbf{W}$ denotes the appropriately dimensioned matrix of the interpolants in Eq. (1.5). For general cases the above method is impractical because of the large size of the matrix $\mathbf{W}$. More importantly, the samples in Fourier transforms of projection data are distributed irregularly, so the problem becomes that of interpolation between unevenly sampled data (on 2D planes in polar coordinates) and regularly sampled result (on 3D Cartesian grid) in which case Eq. (1.6) in not applicable. We will discuss relevant algorithms in the section on direct Fourier inversion.

There is a tendency in EM to use simple interpolation schemes both in the design of reconstruction algorithms and in generic image processing operations. Particularly popular is bi- (or tri-) linear interpolation which is easy to implement and computationally very efficient but which also results in noticeable degradation of the image. The effect of interpolation methods is best analyzed in reciprocal space. Thus, nearest-neighbor (NN) interpolation is realized by assigning each sample to the nearest on-grid location on the target grid and effect is that of convolving the image with a rectangular window with a width of one pixel independently along each dimension. In reciprocal space, the effect is that of multiplication by a normalized sinc function:

$$\operatorname{sinc}(s) = \frac{\sin(\pi s)}{\pi s} \tag{1.7}$$

where $s$ is an absolute unit-less frequency bounded above by 0.5: $s_{max} = 0.5$. The bilinear interpolation is realized by assigning each sample to two locations on the target grid with fractional weights obtained as distances to nearest on-grid locations. This is equivalent to the convolution of the image with a triangular function, which in turn can be seen as equivalent to two convolutions with rectangular window. In reciprocal space, the effect is that of multiplication by $\operatorname{sinc}^2(s)$. If interpolation is applied in real space, there is a loss of resolution due to suppression of high frequency, which at Nyquist frequency is 36% for NN interpolation and ~60% for bilinear interpolation (Fig. 1.3). At the same time, the accuracy of interpolation improves with the increased order of interpolation (Pratt, 1992). Higher order interpolation schemes, such as quadratic or B-splines improve the performance further but at significant computational cost.

Application of low-order interpolation schemes in Fourier space, in addition to the introduction of artifacts due to poor accuracy, will result in difficult to account for fall-off of the image densities in real space. The fall-off will be proportional to sinc or $\operatorname{sinc}^2$ in real space depending on the interpolation scheme used. A standard approach to counteracting the adverse effects of interpolation is to oversample the data. As can be seen from Fig. 1.3, the often recommended three times oversampling reduces the suppression of high frequency information (or fall-off in real space) by bilinear interpolation to ~10%. Such extreme oversampling increases the volume of 2D data nine times with a similar increase in computation time. We show in Fig. 1.3 that the so-called "gridding" method (discussed in

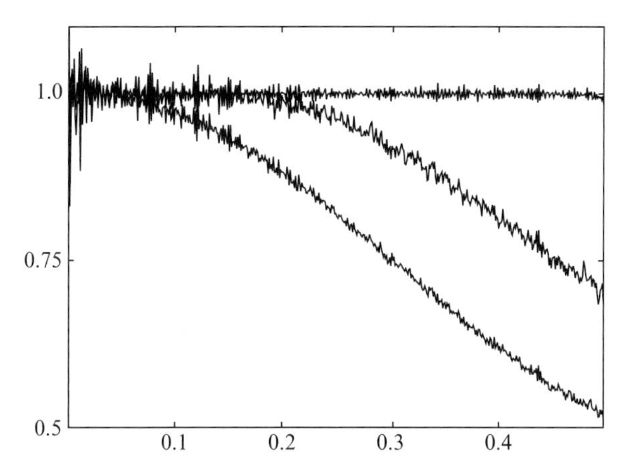


**Figure 1.3** Suppression of high frequency information due to interpolation results in loss of resolution. A 2D $1024^2$ image filled with Gaussian white noise was rotated by 45° and a ratio of rotationally averaged power spectrum of the rotated image to that of the original image was computed: bottom curve–bilinear interpolation; middle curve - quadratic interpolation; top curve–"gridding"-based interpolation (Yang and Penczek, 2008). $x$-axis is given as normalized spatial frequency, that is, 0.5 corresponds to Nyquist frequency.

Section 7) yields nearly perfect results in the entire frequency range when applied to interpolation. As the method all but eliminates the need for oversampling of the scanned data, the relative increase of computational demand is modest (Yang and Penczek, 2008).

## 5. The Algebraic and Iterative Methods

The algebraic methods were derived based on the observation that when the projection equation (1.1) is discretized, it forms a set of linear equations. We begin with forming two vectors: the first, **g**, contains pixels from all available $N$ projections (in an arbitrary order), and the second, **d**, contains the voxels of the 3D object (in an order derived from the order of **g** by algebraic relations):

$$\mathbf{g}_n = \begin{bmatrix} g_n^{k_1 k_2} \\ \vdots \\ g_n^{l_1 l_2} \end{bmatrix}, \quad \mathbf{g} = \begin{bmatrix} \mathbf{g}_1 \\ \mathbf{g}_2 \\ \vdots \\ \mathbf{g}_N \end{bmatrix}, \quad \mathbf{d} = \begin{bmatrix} d^{i_1 i_2 i_3} \\ \vdots \\ d^{j_1 j_2 j_3} \end{bmatrix}. \tag{1.8}$$

The exact sizes of **d** and **g** depend on the choice of support in 2D that can be arbitrary; similarly arbitrary can be the support of the 3D object. This freedom to select the region of interest constitutes a major advantage of iterative algebraic methods. Finally, the operation of projection is defined by the *projection matrix* **P** whose elements are the interpolation weights. The weights are determined by the interpolation scheme used; for the bi- and trilinear interpolations, the weights are between zero and one. The algebraic version of Eq. (1.1) is:

$$\mathbf{g} = \mathbf{Pd}. \tag{1.9}$$

Matrix **P** is rectangular and the system of equation is overdetermined since the number of projections in EM exceeds linear size of the object in pixels. Equation (1.9) can be solved in a least-squares sense:

$$\hat{\mathbf{d}} = \left(\mathbf{P}^T\mathbf{P}\right)^{-1}\mathbf{P}^T\mathbf{g}, \tag{1.10}$$

which yields a unique structure $\hat{\mathbf{d}}$ that minimizes $|\mathbf{Pd} - \mathbf{g}|^2$. Similarly as in the case of direct interpolation in Fourier space Eq. (1.4), the approach introduced by Eq. (1.10) is impractical because of the very large size of the projection matrix. Nevertheless, in the case of the single-axis tilt geometry, the full 3D reconstruction reduces to a series of independent 2D reconstructions, and it becomes possible to solve Eq. (1.10) by using the singular value decomposition (SVD) of the matrix **P**. The advantage of this approach is that for a given geometry the decomposition has to be calculated only once; thus, the method becomes very efficient if the reconstruction has to be performed repeatedly for the same distribution of projections or if additional symmetries, such as helical, are taken into account (Holmes *et al.*, 2003).

In the general 3D case, a least-square solution can be found using one of the iterative approaches that take advantage of the sparsity of the projection matrix. The main idea is that the matrix **P** is not explicitly calculated or stored; instead, its elements are calculated in each iteration as needed. In the following, let $L(\mathbf{d})$ be defined as follows:

$$L(\mathbf{d}) = |\mathbf{Pd} - \mathbf{g}|^2 \tag{1.11}$$

The aim of simultaneous iterative reconstruction technique (SIRT) is to find a structure **d** that minimizes the objective function Eq. (1.11). The algorithm begins by selecting the initial 3D structure $\mathbf{d}^0$ (usually set to zero) and proceeds by iteratively updating the current approximation $\mathbf{d}^{i+1}$ using the gradient of the objective function $\nabla L(\mathbf{d})$:

$$\mathbf{d}^{i+1} = \mathbf{d}^i - \lambda^i\mathbf{P}^T\left(\mathbf{Pd}^i - \mathbf{g}\right) = \mathbf{d}^i - \lambda^i\left(\mathbf{P}^T\mathbf{Pd}^i - \mathbf{P}^T\mathbf{g}\right). \tag{1.12}$$

Setting the relaxation parameter $\lambda^i = \lambda = \text{const}$ yields Richardson's algorithm (Gilbert, 1972).

SIRT is extensively used in SPR (Frank, 2006) because it yields superior results under a wide range of experimental conditions (Penczek *et al.*, 1992); additionally, in the presence of angular gaps in the distribution of projections, SIRT produces the least disturbing artifacts. Furthermore, SIRT offers considerable flexibility in EM applications. First, it is possible to accelerate the convergence by adjusting the relaxation parameters, that is, by setting $\lambda^i = \arg\min_{\lambda \geq 0} L(\mathbf{d}^i - \lambda \nabla L(\mathbf{d}^i))$ we obtain a steepest descent algorithm. Second, even faster convergence (in $\sim$10 iterations) is achieved by solving Eq. (1.12) using the conjugate gradient method; however, this requires the addition of a regularizing term to prevent excessive enhancement of noise. Such a term has the form $|\mathbf{Bd}|^2$, where the matrix $\mathbf{B}$ is a discrete approximation of a Laplacian or higher order derivatives. Third, it is possible to take into account the underfocus settings of the microscope by including the CTF of the microscope and by solving the problem for the structure $\mathbf{d}$ that in effect will be corrected for the CTF:

$$L(\mathbf{d}) = (1 - \eta)|\mathbf{SPd} - \mathbf{g}|^2 + \eta|\mathbf{Bd}|^2. \tag{1.13}$$

We introduced in Eq. (1.13) a Lagrange multiplier $\eta$ whose value determines the smoothness of the solution (Zhu *et al.*, 1997), $\mathbf{S}$ is a matrix containing the algebraic representation of the space-invariant point spread functions (psf, i.e., the inverse Fourier transform of the CTF) of the microscope for all projection images. With this, Eq. (1.13) can be solved using the following iterative scheme:

$$\mathbf{d}^{i+1} = \mathbf{d}^i - \lambda^i\left\{(1 - \eta)\left((\mathbf{SP})^{\mathrm{T}}(\mathbf{SP})\mathbf{d}^i - (\mathbf{SP})^{\mathrm{T}}\mathbf{g}\right) + \eta\mathbf{B}^{\mathrm{T}}\mathbf{Bd}^i\right\} \tag{1.14}$$

Both the psf and regularization terms are normally computed in Fourier space. Algorithms (1.13) and (1.14) are implemented in the SPIDER package (Frank *et al.*, 1996).

The ART predates SIRT; in the context of tomographic reconstructions, it was proposed by Gordon *et al.* (1970) and later it was recognized as a version of Kaczmarz's method for iteratively solving Eq. (1.1) (Kaczmarz, 1993). We first write Eq. (1.9) as a set of systems of equations, each relating individual pixels $g_l$, $l = 1, \ldots, NK^2$ in projections with voxels of the 3D structure:

$$g_l = \mathbf{p}_l^{\mathrm{T}}\mathbf{d}, \quad l = 1, \ldots, NK^2 \tag{1.15}$$

Note Eqs. (1.9) and (1.15) are equivalent because of Eq. (1.8) and $\mathbf{P} = [\mathbf{p}_1\ \mathbf{p}_2\ \ldots\ \mathbf{p}_N]^{\mathrm{T}}$. With this notation and a relaxation parameter $0 < \mu < 2$, ART consists of the following steps:

1. Set $i = 0$ and the initial guess of the structure $\hat{\mathbf{d}}^0$.
2. For $l = 1,\ldots,NK^2$, update

$$\mathbf{d}^{iNK^2+l} = \mathbf{d}^{iNK^2+l-1} - \mu(\mathbf{p}^{\mathrm{T}}\mathbf{d}^{iNK^2+l-1} - g_l)\frac{\mathbf{p}_l}{|\mathbf{p}_l|}. \tag{1.16}$$

3. Set $i \rightarrow i + 1$; go to step 2.

Although the mathematical forms of the update equations in SIRT Eq. (1.12) and in ART Eq. (1.16) appear at first glance to be very similar, there are profound differences between them. In SIRT, all voxels in the structure are corrected simultaneously after projections (**P**) and backprojections ($\mathbf{P}^{\mathrm{T}}$) of the current update of the structure are calculated. In ART, the projection/backprojection in Eq. (1.16) involves only correction with respect to an individual pixel in a single projection immediately followed by the update of the structure. This results in a much faster convergence of ART as compared to SIRT. Further acceleration can be achieved by selecting the order in which pixels enter the correction in Eq. (1.16). It was observed that if a pixel is selected such that its projection direction is perpendicular to the projection direction of the previous pixel, then convergence is achieved faster (Hamaker and Solmon, 1978; Herman and Meyer, 1993). Interestingly, a random order works almost equally well (Natterer and Wübbeling, 2001).

ART has been introduced into SPR as "ART with blobs" (Marabini *et al.*, 1998) and is available in the Xmipp package (Sorzano *et al.*, 2004). In this implementation, the reconstructed structure is represented by a linear combination of spherically symmetric, smooth, spatially limited basis functions, such as Kaiser–Bessel window functions (Lewitt, 1990; Lewitt, 1992; Matej and Lewitt, 1996). Introduction of blobs significantly reduces the number of iterations necessary to reach an acceptable solution (Marabini *et al.*, 1998).

The main advantage of algebraic iterative methods is their applicability to various data collection geometries and to data with uneven distribution of projection directions. Indeed, as the weighting function that would account for particular distribution of projections does not appear in the mathematical formulations of SIRT Eq. (1.12) and ART Eq. (1.16), it would seem that both algorithms can be applied, with minor modifications, to all reconstruction problems encountered in EM. Moreover, a measure of regularization of the solution is naturally achieved by premature termination of the iterative process. However, this is not to say that iterative algebraic reconstruction algorithms do not have shortcomings. For most such methods, the computational requirements are dominated by the backprojection step, so it is safe to assume that the running times of both SIRT and ART will exceed that of other algorithms in proportion to the number of iterations (typically 10–200). Also, applications of SIRT or ART to unevenly distributed set of projection data will result in artifacts in the reconstructed object unless additional provisions are undertaken (Boisset *et al.*, 1998; Sorzano *et al.*, 2001). Finally, since the distribution of projection directions does not appear explicitly in the algebraic formulation of the reconstruction problem (as it does in the Fourier formulation as a weighting function), it is tempting to assume that algebraic iterative methods will yield properly reconstructed objects from very few projections or from data with major gaps in coverage of Fourier space (as in the cases of random conical tilt, and both single- and double-axis tilts in ET). Not only extraordinary claims have been made, but grossly exaggerated results also have been reported. As noted previously, reconstruction artifacts in ART and SIRT tend to be less severe than in other methods in the case of missing projection data; however, neither ART nor SIRT will fill gaps in Fourier space with meaningful information. With more aggressive regularization, as in Maximum Entropy iterative reconstruction, further reduction of artifacts might be possible, but no significant gain in information can be expected. The main reason is that while it is possible to obtain impressive results using simulated data, the limitations of EM data are poorly understood, as are the statistical properties of noise in EM data, CTF effects, and last but not least, errors in orientation parameters. In particular, the latter implies the system of equations (1.9) is inconsistent, which precludes any major gains by using iterative algebraic methods in application to experimental EM data.

In ET, due to slab geometry of the object, the reconstruction problem is compounded by that of truncated views, that is, depending on the tilt angle, each projection contains the projection of a larger or smaller extent of the object. Thus, the system of equations (1.9) becomes inconsistent and strictly speaking algebraic methods are not applicable in this case. So far, the problem has met little attention in the design of ET reconstruction algorithms.

The iterative reconstruction methods can further regularize the solution by taking advantage of *a priori* knowledge, that is, any information about the protein structure that was not initially included in the data processing, and introducing it into the reconstruction process in the form of constraints. Examples of such constraints include similarity to the experimental (measured) data, positivity of the protein mass density (only valid in conjunction with CTF correction), bounded spatial support, etc. Formally, the process of enforcing selected constraints is best described in the framework of the projections onto convex sets (POCS) theory (Stark and Yang, 1998) that was introduced into EM by Carazo and co-workers (Carazo and Carrascosa, 1987; also see Chapter 2).

## 6. Filtered Backprojection

The method of FBP is based on the central section theorem. We will begin with the 2D case, in which the algorithm has a particularly simple form. Let us consider a function $\Upsilon(\mathbf{s})$ describing, in Cartesian coordinates, the 2D Fourier transform of a 1D ray transforms $g$ of a 2D object $d$. The inversion formula is obtained as:

$$\begin{aligned} d(\mathbf{x}) &= \int \Upsilon(\mathbf{s}) \exp(-2\pi i \mathbf{s}\mathbf{x})\, d\mathbf{s} = \int_0^{\pi}\int_{-\infty}^{+\infty} \Upsilon(R,\psi) \exp(-2\pi i \mathbf{x}\tau)|R|\, \mathrm{d}R\, \mathrm{d}\psi \\ &= \int_0^{\pi}\int_{-\infty}^{+\infty} \mathcal{F}\left[g(x_\psi)\right] \exp(-2\pi i \mathbf{x}\tau)|R|dR\, d\psi = \int_0^{\pi} \mathcal{F}^{-1}\left[|R|\mathcal{F}\left[g(x_\psi)\right]\right] \mathrm{d}\psi \\ &= \mathrm{Backprojection}[\mathrm{Filtration}_{|R|}(g)] \end{aligned} \tag{1.17}$$

The 2D FBP algorithm consist of the following steps: (i) compute a 1D Fourier transform of each projection, (ii) multiply the Fourier transform of each projection by the weighting filter $|R|$ (a simple ramp filter), (iii) compute the inverse 1D Fourier transforms of the filtered projections, and (iv) apply backprojection to the processed projections in real space to obtain the 2D reconstructed object. The FBP is applicable to ET single-axis tilt data collection, in which case the 3D reconstruction can be reduced to a series of axial 2D reconstructions (implemented in SPIDER; Frank *et al.*, 1996).

To introduce a 2D FBP for nonuniformly distributed projections, we note Eq. (1.17) includes a Jacobian of a transformation between polar and Cartesian systems of coordinates $|R|\mathrm{d}R\,\mathrm{d}\psi$. When the analytical form of the distribution of projections is known, we can use its discretized form as a weighting function in the filtration step. For an arbitrary distribution a good choice is to select weights $c(|R|,\psi)$ such that the backprojection integral equation (1.17) becomes approximated by a Riemann sum (Penczek *et al.*, 1996):

$$c(|R|,\psi) = |R|dRd\psi \rightarrow c(R_j,\psi_j) = R_j \frac{1}{2\pi}\frac{\psi_{j+1}-\psi_{j-1}}{2} = \frac{R_j \Delta\psi_j}{2\pi}. \tag{1.18}$$

Weights given by Eq. (1.18) yield good reconstruction if there are no major gaps in the distributions of projections, that is, the data is properly sampled by projections. This requires:

$$\forall j \quad \Delta\psi_j \leq r_{\max}^{-1}, \tag{1.19}$$

where $r_{\max}$ is the radius of the object. For distributions which do not fulfill this condition, Bracewell proposed that weighting functions based on an explicitly or implicitly formulated concept of the "local density" of projections might be more appropriate (Bracewell and Riddle, 1967). For a 2D case of nonuniformly distributed projections, Bracewell *et al.* suggested the following heuristic weighting function:

$$c(R_j, \psi_j) = \frac{R_j}{\sum_l \exp\left[-\text{const}(|\psi_i - \psi_l| \bmod \pi)^2\right]} \tag{1.20}$$

where the summation extends over all projection angles.

It is to be noted that it might be tempting to reverse the order of operations in FBP and arrive at the so-called backprojection-filtering reconstruction algorithm (BFP). It is based on the observation that the FT of the 2D object $d$ is related to the FT of the backprojected object by a 2D weighting function $\left(s_x^2 + s_y^2\right)^{1/2} = |\mathbf{s}|$, so the BFP is (Vainshtein and Penczek, 2008):

$$d(\mathbf{x}) = \text{Filtration}_{|\mathbf{s}|}[\text{Backprojection}(g)]. \tag{1.21}$$

One can also arrive at the same conclusion by analyzing the psf of the backprojection step (Radermacher, 1992). While the results are acceptable for uniform distribution of projections, the approach does not yield correct results for nonuniform distributions (thus all interesting 3D cases). The reason is that in the discrete form of FBP Eq. (1.17), the order in which the Jacobian weights given by Eq. (1.18) (or obtained from suitable approximation) are applied to the projection data cannot be reversed. In other words, the sum of weighted projections cannot be replaced by a weighted sum of projections (see also the discussion on direct Fourier inversion algorithms).

For a nontrivial 3D reconstruction from 2D projections, the design of a good weighting function is challenging. First, in 3D there are no uniform distributions of projection directions. So, a reasonable criterion for an appropriate 3D weighting function is to have its backprojection integral approximate the Riemann sum, as in the 2D case (Eq. (1.18)). For example, the weighting function equation (1.20) can be easily extended to 3D, but for a uniform distribution of projections it does not approximate well the analytical form of 3D Jacobian, which we consider optimal.

Radermacher and coworkers (Radermacher, 1992; Radermacher *et al.*, 1986) proposed a *general weighting function* (GWF) derived using a deconvolution kernel calculated for a given (nonuniform) distribution of projections and a finite length of backprojection. GWF is obtained by finding the response of the simple backprojection algorithm to the input composed of delta functions in real space:

$$c_{\mathrm{GWF}}(s_x, s_y, s_z) = \frac{1}{\sum_i 2r_{\max} \operatorname{sinc}\left(2r_{\max}\pi w_\tau^i\right)}, \tag{1.22}$$

where $w_\tau{}^i$ is a variable in Fourier space extending in the direction of the $i$th projection direction $\boldsymbol{\tau}(\varphi_i,\theta_i)$. The GWF is consistent with the analytical requirement of Eq. (1.17) as it can be shown that by assuming infinite support $r_{\max} \to \infty$ and continuous and uniform distribution of projection directions, one obtains in 2D the following:

$$c_{\mathrm{GWF}}(s_x, s_y) = \left(s_x^2 + s_y^2\right) = |\mathbf{s}|.$$

The derivation of Eq. (1.22) is based on the analysis of continuous functions and its direct application to discrete data results in reconstruction artifacts (Radermacher, 2000). In Radermacher (1992), it was proposed (a) to apply the appropriately reformulated GWF 2D projection data *prior* to the back-projection step and (b) to attenuate sinc functions in Eq. (1.22) by Gaussian functions with decay depending on the radius of the structure or simply to replace sinc functions by Gaussian functions. This, however, reduces the concept of the weighting function corresponding to the deconvolution to that of the weighting function representing the "local density" of projections Eq. (1.20). The general weighted backprojection reconstruction is implemented in the SPIDER (using Gaussian weighting functions; Frank *et al.*, 1996) and Xmipp packages (Sorzano *et al.*, 2004).

In Harauz and van Heel (1986), it was proposed that the calculation of the density of projections, and thus the weighting function, be based on the overlap of Fourier transforms of projections in Fourier space. Although the concept is general, its exposition is easier in 2D. If the radius of the object is $r_{\max}$, the width of the Fourier transform of a projection is $1/r_{\max}$, which follows from the central section theorem (1.3). Harauz and van Heel postulated that the weighting should be inversely proportional to the sum of geometrical overlaps between a given central section and the remaining central sections. For a pair of projections $i$ and $l$, this overlap is:

$$o_{il}(R) = T[2r_{\max}R\sin(\psi_i - \psi_l)], \tag{1.23}$$

where $T$ represents the triangle function. Also, due to Friedel symmetry of central sections the angles in Eq. (1.23) are restricted such that

$0 \leq \psi_i - \psi_l \leq (\pi/2)$. In this formulation, the overlap is limited to the maximum frequency:

$$R_{il}^{\max} = \frac{1}{r_{\max} \sin(\psi_i - \psi_l)}, \tag{1.24}$$

in which case the overlap function becomes:

$$o_{il}(R) = \begin{cases} 1 - \dfrac{R}{R_{il}^{\max}} & 0 \leq R \leq R_{il}^{\max}, \\ 0 & R > R_{il}^{\max}, \end{cases} \tag{1.25}$$

and the weighting function, termed an "exact filter" by the authors, is:

$$c(R, \Psi_i) = \frac{1}{1 + \sum\limits_{l \neq i} o_{il}(R)}. \tag{1.26}$$

The weighting function (1.26) easily extends to 3D; however, the calculation of the overlap between central sections in 3D (represented by slabs) is more elaborate (Harauz and van Heel, 1986). The method is conceptually simple and computationally efficient. Regrettably, Eq. (1.26) does not approximate well the correct weighting for a uniform distribution of projections for which it should ideally yield $c(R, \Psi_i) = R$; it is easy to see that this is not the case by integrating Eq. (1.26) over the entire angular range. The exact filter backprojection reconstruction is implemented in the IMAGIC (van Heel *et al.*, 1996) and SPIDER packages (Frank *et al.*, 1996).

The 3D reconstruction methods based on FBP are commonly used in SPR because of their versatility, ease of implementation, and, in comparison with iterative methods, superior computational efficiency. Unlike iterative methods, there are no parameters to adjust, although it has been noted that the results depend on the value of the radius of the structure in all three weighting functions (Eqs. (1.20), (1.22), and (1.26)), so the performance of the reconstruction algorithm can be optimized for a particular data collection geometry by changing the value of $r_{\max}$ (Paul *et al.*, 2004). Because evaluation of the weighting function involves calculation of pair-wise distances between projections, the computational complexity is proportional to the square of the number of projections and for large datasets methods from this class become inefficient. We also note that the weighting functions, both general and exact, remain approximations of the desired weighting function that would approximate the Riemann sum.

## 7. Direct Fourier Inversion

Direct Fourier methods are based on the central section theorem in Fourier space Eq. (1.3). The premise is that the 2D Fourier transforms of projections yield samples of the target 3D object on a nonuniform 3D grid. If the 3D inverse Fourier transform could be realized by means of the 3D inverse FFT, then one would have a very fast reconstruction algorithm. Unfortunately, application of 3D FFT has two nontrivial requirements: (1) accounting for the uneven distribution of samples and (2) recovering the samples on a uniform grid from the available samples on a nonuniform grid.

The most accurate Fourier reconstruction methods are those that employ nonuniform Fourier transforms, particularly the 3D *gridding method* (O'Sullivan, 1985; Schomberg and Timmer, 1995). In EM, a gridding-based fast Fourier summation (FFS) reconstruction algorithm was developed for ET (Sandberg *et al.*, 2003) and implemented in IMOD (Kremer *et al.*, 1996), while the gridding-based direct Fourier reconstruction (GDFR) algorithm was developed specifically for SPR (Penczek *et al.*, 2004). As we showed earlier, the optimum weights accounting for uneven distribution of projections should be such that their use approximates the integral over the distribution of samples through the Riemann sum. Such weights can be obtained as volumes of 3D Voronoi cells obtained from the distribution of sampling points (Okabe *et al.*, 2000). In SPR EM, the number of projections, and thus the number of sampling points in Fourier space, is extremely large. Although the calculation of the gridding weights via the 3D Voronoi diagram of the nonuniformly spaced grid points would lead to an accurate direct Fourier method, the method is very slow and would require excessive computer memory. To circumvent this problem, GDFR introduces an additional step, namely the 2D *reverse gridding* method, to compute the Fourier transform of each 2D projection on a 2D polar grid. By means of the reverse gridding method, the samples of the 3D target object can be obtained on a 3D spherical grid, where the grid points are located both on centered spheres and on straight lines through the origin. Accordingly, it becomes possible to partition the sampled region into suitable sampling cells via the computation of a 2D Voronoi diagram on a unit sphere, rather than a 3D Voronoi diagram in Euclidean space (Fig. 1.4). This significantly reduces the memory requirements and improves the computational efficiency of the method, especially when a fast $O(n \log n)$ algorithm for computing Voronoi diagrams on the unit sphere is employed (Renka, 1997).

GDFR comprises four steps (Fig. 1.4):

1. The reverse gridding method is used to resample 2D input projection images into 2D polar coordinates.

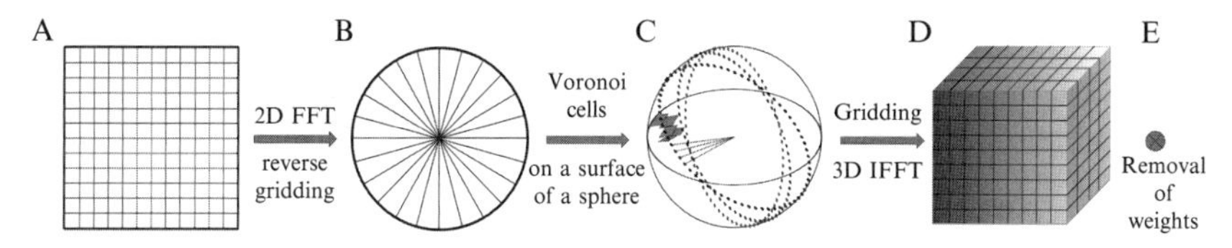


**Figure 1.4** Principle of the gridding-based direct Fourier 3D reconstruction (GDFR) algorithm. (A) 2D FFT of input projection image. (B) The reverse gridding is used to resample 2D Fourier input image into 2D polar Fourier coordinates. (C) "Gridding weights" are computed as cell areas of a 2D Voronoi diagram on a unit sphere (grey polygons) to compensate for the nonuniform distribution of the grid points. (D) Gridding using a convolution kernel with subsequent 3D inverse FFT yields samples on a 3D Cartesian grid. (E) Removal of weights in real space yields the reconstructed 3D object.

2. The "gridding" step involves calculating for the Fourier transform of each projection the convolution

$$\sum_i \mathcal{F}[w] * (c\mathcal{F}[g_i]),$$

where $c$ are "gridding weights" computed as cell areas of a 2D Voronoi diagram on a unit sphere that are designed to compensate for the nonuniform distribution of the grid points and $\mathcal{F}[w]$ is an appropriately chosen convolution kernel. After processing all projections, this step yields samples of $\mathcal{F}[w] \star \mathcal{F}[d]$ on a Cartesian grid.
3. The 3D inverse FFT is used to compute $wd = \mathcal{F}^{-1}[\mathcal{F}[w] \star [d]]$ on a Cartesian grid.
4. The weights are removed using real-space division $d = (wd) / w$.

The reverse gridding in step 1 is obtained by reversing the sequence of steps (2–4) that constitute the proper gridding method:

a. The input image is divided in real space by the weighting function $g/w$.
b. The image is padded with zeros and 2D FFT is used to compute $\mathcal{F}[g / w]$.
c. Gridding is used to compute $\mathcal{F}[w]\star\mathcal{F}[g/w]$ on an arbitrary nonuniform grid. In GDFR, this step yields 2D FTs of input projection images resampled into 2D Fourier polar coordinates.

Note the reverse gridding method does not require explicit gridding weights since the input 2D projection image is given on Cartesian grid, in which case the weights are constant. In addition, as step (c) results in a set of 1D Fourier central lines calculated using constant angular step, upon computing the inverse Fourier transforms they amount to a Radon transform of the projection, and after repeating the process for all available projections, they yield a Radon transform of a 3D object. Thus, GDFR, in addition to being a method of inverting of a 3D ray transform is also a highly accurate method of inverting a 3D Radon transform.

The GDFR method requires a number of parameters. First, we need a *window function* $\mathcal{F}[w]$ whose support in Fourier space is "small." To assure good computational efficiency of the convolution in step 2 of GDFR, this support can be set to six Fourier voxels. In addition, to prevent division by zeroes in steps 1a and 4, the *weighting function* $w$ must be positive within the support of the reconstructed object. A recommended window function is the separable, bell-shaped Kaiser–Bessel window (Jackson *et al.*, 1991; O'Sullivan, 1985; Schomberg and Timmer, 1995) with parameters given in Penczek *et al.* (2004).

The results of a comparison of selected reconstruction algorithms are shown in Penczek *et al.* (2004) and Vainshtein and Penczek (2008). In a series of extensive tests, it was shown that GDFR yields in a noise free case virtually perfect 3D reconstruction within the full frequency range, followed by SIRT and then weighted backprojection algorithms. It was also shown to be ~6 times faster than weighted backprojection methods, which is due to very time consuming calculation of weighting functions in the latter, and 17 times faster than SIRT.

## 8. Implementations of Reconstruction Algorithms in EM Software Packages

The theory of object reconstruction from a set of its line projections is well developed and comprehensive. However, two issues of great concern for EM applications have not received sufficient attention: (1) the problem of reconstruction with transfer function correction or using projection data with nonuniform SSNR and (2) reconstruction using highly nonuniformly distributed or incomplete sets of projections. While some theoretical work have been pursued at early stages of the development of the field (Goncharov *et al.*, 1987; Orlov, 1976a; Orlov, 1976b; Radermacher, 1988), currently used software packages are dominated by *ad hoc* solutions that are aimed at computational efficiency and convenience in data processing. More simply, a number of these *ad hoc* solutions stem from the history of the development of a given package.

The incorporation of the CTF correction of accounting for nonuniform distribution of the SSNR in projection data is relatively easy in iterative algebraic reconstruction algorithms, including both ART and SIRT, as can be seen from Eq. (1.14). Iterative reconstruction *cum* CTF correction remains so far the only theoretically sound approach to the problem. The algorithm is only implemented in SPIDER (Zhu *et al.*, 1997) with the additional requirement that the data be grouped by their respective defocus values in order to make the method computationally efficient; however, it does not appear to be in routine use.

Reconstruction methods based on backprojection and direct Fourier inversion methods require the implementation of a form of Wiener filter, which schematically is written as (see Chapter 2):

$$D = \frac{\sum_n \mathrm{CTF}_n \mathrm{SSNR}_n G_n}{\sum_n \mathrm{CTF}_n^2 \mathrm{SSNR}_n + 1}. \tag{1.27}$$

The summation in the numerator can be realized as a backprojection of the Fourier transforms of $(n-1)$D projections multiplied by their respective CTFs and SSNRs, so the result is $n$D. However, it is far from obvious how the summation in the denominator can be realized such that the result would have the intended meaning after the division is performed.

Cryo-EM software packages circumvent the difficulties listed above with the direct implementation of Eq. (1.27) in various ways. In SPIDER, the routine approach is to group the projection data according to the assigned defocus values (Frank *et al.*, 2000). Depending on the yield of particles, projection images from a single micrograph may constitute a "defocus group," although larger groups are preferable. For each defocus group, an independent 3D reconstruction is computed using any of the algorithms implemented in the package. Finally, a "CTF-corrected" structure is calculated by merging all individual structures using the Wiener filter (Eq. (1.27)). While no doubt successful, this strategy makes it impossible to process data with very low particle yield per micrograph. In the current implementation of SPIDER, the anisotropy of SSNR of individual structures is not accounted for, in which case it is also not clear whether merging partial 3D structures can match the quality of a 3D reconstruction that would include CTF correction. In both IMAGIC and EMAN (including EMAN2), the 3D reconstruction is computed using the so-called class averages, that is, images obtained by merging possibly similar 2D projection images of the molecule using the 2D version of a Wiener filter that includes CTF correction (Eq. (1.27)) or a similar approach. As class averages are corrected for the CTF effects, they can now be used in any 3D reconstruction algorithm. However, such 2D class averages have quite nonuniform distribution of SSNR and properly speaking this unevenness should be accounted for during the 3D reconstruction step. This would again require a reconstruction algorithm based on Eq. (1.27) even though the CTF term is omitted, which is as difficult a problem as the original one.

There are two algorithms that implement Eq. (1.27) directly using a simplified form of direct Fourier inversion and resampling of the nonuniformly distributed projection data onto a 3D Cartesian grid by a form of interpolation. In the FREALIGN package, a modified trilinear interpolation scheme is used to resample the data that leads to Eq. (1.27) with additional weights accounting for interpolation (Grigorieff, 1998). However, it is not clear whether the uneven distribution of projection samples is taken into account. A simpler yet algorithm termed *NN direct* inversion is implemented in SPARX (Zhang *et al.*, 2008). In this algorithm, the 2D input projections are first padded with zeroes to four times the size, 2D Fourier transformed, and samples are accumulated within the target 3D Fourier volume $D$ using simple NN interpolation and a Wiener filter methodology based on Eq. (1.27). After all 2D projections are processed, each element in the 3D Fourier volume is divided by the number of respective per-voxel counts of contributing 2D samples. Finally, a 3D weighting function modeled on Bracewell's "local density" (Bracewell and Riddle, 1967) is constructed and applied to individual voxels of 3D Fourier space to account for possible nonuniform distribution of samples. This function is constructed to fulfill the following requirements:

1. For a given voxel, if all its neighbors are occupied, its weight is set to 1.0 (the minimum weight resulting from the procedure).
2. For a given voxel, if all its neighboring voxels are empty, the assigned weight has the maximum value.
3. Empty voxels located closer to the vacant voxel contribute more to its weight than those located further.

The resulting weighting function is:

$$w(h,k,l) = \frac{1}{1-\alpha \sum_{i,j,m=-n}^{n} l(h+i,k+j,l+m)\exp(-\beta(i+j+m))}, \tag{1.28}$$

where

$$l(h,k,l) = \begin{cases} 1 & \text{if } D_{h,k,l} = 0, \\ 0 & \text{otherwise.} \end{cases} \tag{1.29}$$

$\beta$ and $n$ are constants whose values (0.2 and 3, respectively) were adjusted such that the rotationally averaged power spectrum is preserved upon reconstruction for selected test cases and $\alpha$ is a constant whose value dependence is adjusted such that the first two normalization criteria listed above are fulfilled. As reported in Zhang *et al.* (2008), the inclusion of weights significantly improves the fidelity of the NN direct-inversion reconstruction.

Accounting for uneven distribution of projections and for gaps in angular coverage is the second major problem which is not well resolved in EM reconstruction algorithms. There is strong evidence, both theoretical and experimental, that Riemann sum weights obtained using Voronoi diagrams yield optimum results in both 2D and 3D reconstruction algorithms. However, this approach is valid only if the target space is sufficiently sampled by projection data. In the presence of major gaps, the weights obtained using this methodology tend to be exceedingly large and result in streaking artifacts in the reconstructed object. A similar argument can be applied to various "local density" weighting schemes in FBP algorithms, including "general" and "exact" filters. The problem is particularly difficult to solve for ET reconstructions in which contiguous subsets of projection data is missing. As long as the angular distribution of projections is even, one can apply a simple FBP algorithm with R-weighting (Eq. (1.19)) and ignore the presence of the angular gap. However, for more elaborate angular schemes (Penczek and Frank, 2006) or for double-tilt ET (Penczek *et al.*, 1995) the angular dependence of the weighting function can no longer be ignored. Currently, there are no convincing methods that would yield proper weighting of the projection data within the measured region of Fourier space while avoiding excessive weighting of projections whose Fourier transforms are close to gaps in Fourier space.

The final groups of commonly used EM algorithms are those that use a simplified version of the gridding approach. In the method implemented in EMAN, EMAN2, and in SPIDER, the gridding between projection data samples and a Cartesian target grid is accomplished using a truncated Gaussian convolution kernel. The gridding weights are ignored; instead, fractional Gaussian interpolants are accumulated on an auxiliary Cartesian grid and the final reconstruction is computed as the inverse Fourier transform of the gridded data divided by the accumulated interpolants. The method can be recognized as an implementation of Jackson's original gridding algorithm (Jackson *et al.*, 1991), which was subsequently phased out after the proper order of weighting was introduced (one can see that in Jackson's method, the weights $c$ that appear at the beginning of step 2 in the proper gridding scheme are used only in step 3).

## 9. Conclusions

The field of object reconstruction from a set of its projections underwent rapid development in the past five decades. While it was computerized axial tomography that popularized the technique and drove much of the technical and theoretical progress, applications to studies of the ultrastructure of biological material rivaled these achievements by providing insight into biological processes on a molecular level. EM applications of tomography are distinguished by the predominance of "true" 3D reconstruction from 2D projections, that is, reconstructions that cannot be reduced to a series of 2D axial reconstructions. In addition, EM reconstruction algorithms have to handle problems that remain largely unique to the field, namely, the very large number of projection data, the incorporation of CTF correction, and compensating for the uneven distribution of projection data and gaps in coverage of Fourier space.

There is a great variety of reconstruction algorithms implemented in EM software packages. The first algorithms introduced in the 1980s were generalizations of the FBP approach: general and exact filtrations. The arrival of cryo-preparation in the early 1990s heralded the development of algorithms which can effectively process lower contrast and lower SNR data, and thus the focus shifted to algebraic iterative reconstruction methods: SIRT in SPIDER and ART with blobs in Xmipp. In the late 1990s, the size of datasets increased dramatically. The first cryo-reconstruction of an asymmetric complex (70S ribosome) was done using 303 particle images while most recent work on the same complex processed datasets with sizes exceeding 500,000 2D particle images (Schuette *et al.*, 2009; Seidelt *et al.*, 2009) closely followed by structural work on symmetric complexes. For example, recently published near-atomic structure of a group II chaperonin required ~30,000 particle images, which taking into account D8 point group symmetry of this complex corresponds to ~240,000 images (Zhang *et al.*, 2010). This increase placed new demands on the reconstruction algorithms, mainly affecting the running times and computational efficiency of the algorithms, which resulted in a shift toward very fast direct Fourier inversion algorithms. Currently, virtually all major EM software packages include an implementation of a version of the method (EMAN, EMAN2, FREALIGN, IMOD, SPARX, and SPIDER), with IMAGIC being most likely the only exception. We also note that the large increases in the number of projection images have only been accompanied by modest increases in the volume size (from 64 to ~256 voxels). This implies Fourier space is bound to be oversampled, and consequently, the problem of artifacts due to simple interpolation schemes used in most implementations is not as severe as it would be with a small number of images.

The limited scope of this review did not allow us to discuss some of the advanced subjects that remain of great interest in the EM field. In SPR, one normally assumes that the depth of the field in the microscope is larger than the molecule size. This is not the case for larger biological objects, and certainly not for virus capsids. Efficient algorithms for correcting the defocus spread within the structure have been proposed, but they are not as of yet widely implemented or tested (Philippsen *et al.*, 2007; Wolf *et al.*, 2006).

Similarly, incorporation of *a priori* knowledge into reconstruction algorithms has been proposed early on, but relatively little has been accomplished in terms of the formal treatment of the problem (Sorzano *et al.*, 2008). These and other issues pertinent to the problem of object reconstruction from projections in EM are often addressed in practical implementations using heuristic solutions. In addition, both SPR and ET are still rapidly developing as new ways of extracting information from EM data are being conceived. Methods such as resampling strategies in SPR (Penczek *et al.*, 2006) or averaging of subtomograms in ET (Lucic *et al.*, 2005) call for reevaluation of existing computational methodologies and possible development of new approaches that are more efficient and better suited for the problem at hand. Taken together with the fundamental challenges of the reconstruction problem, we are assured that reconstruction from projections will remain a vibrant research field for decades to come.

## ACKNOWLEDGMENTS

I thank Grant Jensen, Justus Loerke, and Jia Fang for critical reading of the manuscript and for helpful suggestions. This work was supported by grant from the NIH R01 GM 60635 (to PAP).

## REFERENCES

Bellman, S. H., Bender, R., Gordon, R., and Rowe, J. E. (1971). ART is science being a defense of algebraic reconstruction techniques for three-dimensional electron microscopy. *J. Theor. Biol.* **31,** 205–216.

Bender, R., Bellman, S. H., and Gordon, R. (1970). ART and the ribosome: A preliminary report on the three-dimensional structure of individual ribosomes determined by an algebraic reconstruction technique. *J. Theor. Biol.* **29,** 483–487.

Boisset, N., Penczek, P. A., Taveau, J. C., You, V., Dehaas, F., and Lamy, J. (1998). Overabundant single-particle electron microscope views induce a three-dimensional reconstruction artifact. *Ultramicroscopy* **74,** 201–207.

Bracewell, R. N. (1956). Strip integration in radio astronomy. *Aust. J. Phys.* **9,** 198–217.

Bracewell, R. N., and Riddle, A. C. (1967). Inversion of fan-beam scans in radio astronomy. *Astrophys. J.* **150,** 427–434.

Carazo, J. M., and Carrascosa, J. L. (1987). Information recovery in missing angular data cases: An approach by the convex projections method in three dimensions. *J. Microsc.* **145,** 23–43.

Crowther, R. A., Amos, L. A., Finch, J. T., De Rosier, D. J., and Klug, A. (1970a). Three dimensional reconstructions of spherical viruses by Fourier synthesis from electron micrographs. *Nature* **226,** 421–425.

Crowther, R. A., DeRosier, D. J., and Klug, A. (1970b). The reconstruction of a three-dimensional structure from projections and its application to electron microscopy. *Proc. R. Soc. Lond. A* **317,** 319–340.

Crowther, R. A., and Klug, A. (1971). ART and science or conditions for three-dimensional reconstruction from electron microscope images. *J. Theor. Biol.* **32,** 199–203.

DeRosier, D. J., and Klug, A. (1968). Reconstruction of 3-dimensional structures from electron micrographs. *Nature* **217,** 130–134.

Frank, J. (2006). Three-Dimensional Electron Microscopy of Macromolecular Assemblies. Oxford University Press, New York.

Frank, J., Penczek, P. A., Agrawal, R. K., Grassucci, R. A., and Heagle, A. B. (2000). Three-dimensional cryoelectron microscopy of ribosomes. *Methods Enzymol.* **317,** 276–291.

Frank, J., Radermacher, M., Penczek, P., Zhu, J., Li, Y., Ladjadj, M., and Leith, A. (1996). SPIDER and WEB: Processing and visualization of images in 3D electron microscopy and related fields. *J. Struct. Biol.* **116,** 190–199.

Gilbert, H. (1972). Iterative methods for the three-dimensional reconstruction of an object from projections. *J. Theor. Biol.* **36,** 105–117.

Goncharov, A. B., Vainshtein, B. K., Ryskin, A. I., and Vagin, A. A. (1987). Three-dimensional reconstruction of arbitrarily oriented identical particles from their electron photomicrographs. *Sov. Phys. Crystallogr.* **32,** 504–509.

Gordon, R., Bender, R., and Herman, G. T. (1970). Algebraic reconstruction techniques (ART) for three-dimensional electron microscopy and x-ray photography. *J. Theor. Biol.* **29,** 471–481.

Grigorieff, N. (1998). Three-dimensional structure of bovine NADH: Ubiquinone oxido-reductase (complex I) at 22 Å in ice. *J. Mol. Biol.* **277,** 1033–1046.

Hamaker, C., and Solmon, D. C. (1978). Angles between null spaces of X-rays. *J. Math. Anal. Appl.* **62,** 1–23.

Harauz, G., and van Heel, M. (1986). Exact filters for general geometry three dimensional reconstruction. *Optik (Stuttg)* **73,** 146–156.

Herman, G. T. (2009). Fundamentals of Computerized Tomography: Image Reconstruction from Projections. Springer, London.

Herman, G. T., and Meyer, L. B. (1993). Algebraic reconstruction techniques can be made computationally efficient. *IEEE Trans. Med. Imaging* **12,** 600–609.

Holmes, K. C., Angert, I., Kull, F. J., Jahn, W., and Schröder, R. R. (2003). Electron cryo-microscopy shows how strong binding of myosin to actin releases nucleotide. *Nature* **425,** 423–427.

Hoppe, W. (1969). Das Endlichkeitspostulat und das Interpolationstheorem der dreidimensionalen elektronenmikroskopischen Analyse aperiodischer Strukturen. *Optik (Stuttg)* **29,** 617–621.

Huesman, R. H., Gullberg, G. T., Greenberg, W. L., and Budinger, T. F. (1977). RECLBL Library Users Manual–Donner Algorithms for Reconstruction Tomography. University of California, Berkeley.

Jackson, J. I., Meyer, C. H., Nishimura, D. G., and Macovski, A. (1991). Selection of a convolution function for Fourier inversion using gridding. *IEEE Trans. Med. Imaging* **10,** 473–478.

Kaczmarz, S. (1993). Approximate solutions of systems of linear equations. *Int. J. Control* **57,** 1269–1271(Reprint of Kaczmarz, S., (1937). Angenäherte Auflösung von Systemen linearer Gleichungen. Bulletin International de l'Academie Polonaise des Sciences Lett A, 355–357).

Kremer, J. R., Mastronarde, D. N., and Mclntosh, J. R. (1996). Computer visualization of three-dimensional image data using IMOD. *J. Struct. Biol.* **116,** 71–76.

Lanzavecchia, S., and Bellon, P. L. (1994). A moving window Shannon reconstruction for image interpolation. *J. Vis. Commun. Image Represent.* **3,** 255–264.

Laurette, I., Zeng, G. L., Welch, A., Christian, P. E., and Gullberg, G. T. (2000). A three-dimensional ray-driven attenuation, scatter and geometric response correction technique for SPECT in inhomogeneous media. *Phys. Med. Biol.* **45,** 3459–3480.

Lewitt, R. M. (1990). Multidimensional digital image representations using generalized Kaiser–Bessel window functions. *J. Opt. Soc. Am. A.* **7,** 1834–1846.
Lewitt, R. M. (1992). Alternatives to voxels for image representation in iterative reconstruction algorithms. *Phys. Med. Biol.* **37,** 705–716.
Louis, A. K. (1984). Nonuniqueness in inverse Radon problems–the frequency-distribution of the ghosts. *Math. Z.* **185,** 429–440.
Lucic, V., Förster, F., and Baumeister, W. (2005). Structural studies by electron tomography: From cells to molecules. *Annu. Rev. Biochem.* **74,** 833–865.
Maass, P. (1987). The X-ray transform—Singular value decomposition and resolution. *Inverse Probl.* **3,** 729–741.
Marabini, R., Herman, G. T., and Carazo, J. M. (1998). 3D reconstruction in electron microscopy using ART with smooth spherically symmetric volume elements (blobs). *Ultramicroscopy* **72,** 53–65.
Matej, S., and Lewitt, R. M. (1996). Practical considerations for 3-D image reconstruction using spherically symmetric volume elements. *IEEE Trans. Med. Imaging* **15,** 68–78.
Natterer, F. (1986). The Mathematics of Computerized Tomography. Wiley, New York.
Natterer, F., and Wübbeling, F. (2001). Mathematical Methods in Image Reconstruction. SIAM, Philadelphia.
O'Sullivan, J. D. (1985). A fast sinc function gridding algorithm for Fourier inversion in computer tomography. *IEEE Trans. Med. Imaging* **MI 4,** 200–207.
Okabe, A., Boots, B., Sugihara, K., and Chiu, S. N. (2000). Spatial Tessellations: Concepts and Applications of Voronoi Diagrams. Wiley, New York.
Orlov, S. S. (1976a). Theory of three-dimensional reconstruction 1 Conditions for a complete set of projections. *Sov. Phys. Crystallogr.* **20,** 312–314.
Orlov, S. S. (1976b). Theory of three-dimensional reconstruction. 2 The recovery operator. *Sov. Phys. Crystallogr.* **20,** 429–433.
Paul, D., Patwardhan, A., Squire, J. M., and Morris, E. P. (2004). Single particle analysis of filamentous and highly elongated macromolecular assemblies. *J. Struct. Biol.* **148,** 236–250.
Penczek, P., Marko, M., Buttle, K., and Frank, J. (1995). Double-tilt electron tomography. *Ultramicroscopy* **60,** 393–410.
Penczek, P., Radermacher, M., and Frank, J. (1992). Three-dimensional reconstruction of single particles embedded in ice. *Ultramicroscopy* **40,** 33–53.
Penczek, P. A. (2002). Three-dimensional spectral signal-to-noise ratio for a class of reconstruction algorithms. *J. Struct. Biol.* **138,** 34–46.
Penczek, P. A., Chao, Y., Frank, J., and Spahn, C. M. T. (2006). Estimation of variance in single particle reconstruction using the bootstrap technique. *J. Struct. Biol.* **154,** 168–183.
Penczek, P., and Frank, J. (2006). Resolution in electron tomography. *In* "Electron Tomography: Methods for Three-dimensional Visualization of Structures in the Cell," (J. Frank, ed.) pp. 307–330. Springer, Berlin.
Penczek, P. A., Grassucci, R. A., and Frank, J. (1994). The ribosome at improved resolution: New techniques for merging and orientation refinement in 3D cryoelectron microscopy of biological particles. *Ultramicroscopy* **53,** 251–270.
Penczek, P. A., Renka, R., and Schomberg, H. (2004). Gridding-based direct Fourier inversion of the three-dimensional ray transform. *J. Opt. Soc. Am. A* **21,** 499–509.
Penczek, P. A., Zhu, J., and Frank, J. (1996). A common-lines based method for determining orientations for $N > 3$ particle projections simultaneously. *Ultramicroscopy* **63,** 205–218.
Philippsen, A., Engel, H. A., and Engel, A. (2007). The contrast-imaging function for tilted specimens. *Ultramicroscopy* **107,** 202–212.
Pratt, W. K. (1992). Digital Image Processing. Wiley, New York.

Radermacher, M. (1988). Three-dimensional reconstruction of single particles from random and nonrandom tilt series. *J. Electron Microsc. Tech.* **9,** 359–394.

Radermacher, M. (1992). Weighted back-projection methods. *In* "Electron Tomography," (J. Frank, ed.) pp. 91–115. Plenum, New York.

Radermacher, M. (2000). Three-dimensional reconstruction of single particles in electron microscopy. *In* "Image Analysis: Methods and Applications," (D.-P. Häder, ed.), pp. 295–328. CRC Press, Boca Raton, FL.

Radermacher, M., Wagenknecht, T., Verschoor, A., and Frank, J. (1986). A new 3-D reconstruction scheme applied to the 50S ribosomal subunit of *E. coli. J. Microsc.* **141,** RP1–RP2.

Radermacher, M., Wagenknecht, T., Verschoor, A., and Frank, J. (1987). Three-dimensional reconstruction from a single-exposure, random conical tilt series applied to the 50S ribosomal subunit of *Escherichia coli. J. Microsc.* **146,** 113–136.

Radon, J. (1986). On the determination of functions from their integral values along certain manifolds. *IEEE Trans. Med. Imaging* **5,** 170–176 (Reprint of Radon, J. (1917). Über die Bestimmung von Funktionen durch ihre Integralwerte längs gewisser Mannigfaltigkeiten, Ber. Verh. König Sächs. Ges. Wiss., Leipzig, Math.-Phys. Kl. 69, 262–267).

Renka, R. J. (1997). Algorithm 772. STRIPACK: Delaunay triangulation and Voronoi diagram on the surface of a sphere. *ACM Trans. Math. Softw.* **23,** 416–434.

Rieder, A., and Faridani, A. (2003). The semi-discrete filtered backprojection algorithm is optimal for tomographic inversion. *SIAM J. Numer. Anal.* **41,** 869–892.

Sandberg, K., Mastronarde, D. N., and Beylkin, G. (2003). A fast reconstruction algorithm for electron microscope tomography. *J. Struct. Biol.* **144,** 61–72.

Schomberg, H., and Timmer, J. (1995). The gridding method for image reconstruction by Fourier transformation. *IEEE Trans. Med. Imaging.* **14,** 596–607.

Schuette, J. C., Murphy, F. V. T., Kelley, A. C., Weir, J. R., Giesebrecht, J., Connell, S. R., Loerke, J., Mielke, T., Zhang, W., Penczek, P. A., Ramakrishnan, V., and Spahn, C. M. (2009). GTPase activation of elongation factor EF-Tu by the ribosome during decoding. *EMBO J.* **28,** 755–765.

Seidelt, B., Innis, C. A., Wilson, D. N., Gartmann, M., Armache, J. P., Villa, E., Trabuco, L. G., Becker, T., Mielke, T., Schulten, K., Steitz, T. A., and Beckmann, R. (2009). Structural insight into nascent polypeptide chain-mediated translational stalling. *Science* **326,** 1412–1415.

Shannon, C. E. (1949). Communication in the presence of noise. *Proc. Inst. Radio Eng.* **37,** 10–21.

Sorzano, C. O. S., Marabini, R., Boisset, N., Rietzel, E., Schroder, R., Herman, G. T., and Carazo, J. M. (2001). The effect of overabundant projection directions on 3D reconstruction algorithms. *J. Struct. Biol.* **133,** 108–118.

Sorzano, C. O. S., Marabini, R., Velazquez-Muriel, J., Bilbao-Castro, J. R., Scheres, S. H. W., Carazo, J. M., and Pascual-Montano, A. (2004). XMIPP: A new generation of an open-source image processing package for electron microscopy. *J. Struct. Biol.* **148,** 194–204.

Sorzano, C. O. S., Velazquez-Muriel, J. A., Marabini, R., Herman, G. T., and Carazo, J. M. (2008). Volumetric restrictions in single particle 3DEM reconstruction. *Pattern Recognit.* **41,** 616–626.

Stark, H., and Yang, Y. (1998). Vector Space Projections: A Numerical Approach to Signal and Image Processing, Neural Nets, and Optics. Wiley, New York.

Vainshtein, B. K., and Penczek, P. A. (2008). Three-dimensional reconstruction. (U. Shmueli, ed.), Vol. B, pp. 366–375. Springer, New York.

van Heel, M. (1987). Angular reconstitution: *A posteriori* assignment of projection directions for 3D reconstruction. *Ultramicroscopy* **21,** 111–124.

van Heel, M., Harauz, G., and Orlova, E. V. (1996). A new generation of the IMAGIC image processing system. *J. Struct. Biol.* **116,** 17–24.

Wolf, M., DeRosier, D. J., and Grigorieff, N. (2006). Ewald sphere correction for single-particle electron microscopy. *Ultramicroscopy* **106,** 376–382.

Yang, Z., and Penczek, P. A. (2008). Cryo-EM image alignment based on nonuniform fast Fourier transform. *Ultramicroscopy* **108,** 959–969.

Yuen, C. K., and Fraser, D. (1979). *Digital Spectral Analysis* CSIRO Pitman, Adelaide.

Zhang, J., Baker, M. L., Schroder, G. F., Douglas, N. R., Reissmann, S., Jakana, J., Dougherty, M., Fu, C. J., Levitt, M., Ludtke, S. J., Frydman, J., and Chiu, W. (2010). Mechanism of folding chamber closure in a group II chaperonin. *Nature* **463,** 379–383.

Zhang, W., Kimmel, M., Spahn, C. M., and Penczek, P. A. (2008). Heterogeneity of large macromolecular complexes revealed by 3D cryo-EM variance analysis. *Structure* **16,** 1770–1776.

Zhu, J., Penczek, P. A., Schröder, R., and Frank, J. (1997). Three-dimensional reconstruction with contrast transfer function correction from energy-filtered cryoelectron micrographs: Procedure and application to the 70S *Escherichia coli* ribosome. *J. Struct. Biol.* **118,** 197–219.

CHAPTER TWO

# Image Restoration in Cryo-Electron Microscopy

Pawel A. Penczek

## Contents



## Abstract

Image restoration techniques are used to obtain, given experimental measurements, the best possible approximation of the original object within the limits imposed by instrumental conditions and noise level in the data. In molecular electron microscopy (EM), we are mainly interested in linear methods that preserve the respective relationships between mass densities within the

Department of Biochemistry and Molecular Biology, The University of Texas, Houston Medical School, Houston, Texas, USA

*Methods in Enzymology*, Volume 482
ISSN 0076-6879, DOI: 10.1016/S0076-6879(10)82002-6

restored map. Here, we describe the methodology of image restoration in structural EM, and more specifically, we will focus on the problem of the optimum recovery of Fourier amplitudes given electron microscope data collected under various defocus settings. We discuss in detail two classes of commonly used linear methods, the first of which consists of methods based on pseudoinverse restoration, and which is further subdivided into mean-square error, chi-square error, and constrained based restorations, where the methods in the latter two subclasses explicitly incorporates non-white distribution of noise in the data. The second class of methods is based on the Wiener filtration approach. We show that the Wiener filter-based methodology can be used to obtain a solution to the problem of amplitude correction (or "sharpening") of the EM map that makes it visually comparable to maps determined by X-ray crystallography, and thus amenable to comparative interpretation. Finally, we present a semiheuristic Wiener filter-based solution to the problem of image restoration given sets of heterogeneous solutions. We conclude the chapter with a discussion of image restoration protocols implemented in commonly used single particle software packages.

## 1. Introduction

The goal of image restoration is to produce the best possible estimate, within the limits imposed by instrumental conditions, of the original object, given its experimental realization and the noise level. The instrumental limitations are expressed by the *point spread function* (psf) and its Fourier space transform, the *transfer function*, while the noise level, termed *signal-to-noise ratio* (SNR), is conveniently expressed as a ratio of the power of the signal to the power of noise. The distribution of SNR in Fourier space is given as a function of spatial frequency and is called *spectral* SNR (SSNR); regrettably, there is no simple relationship between SNR and SSNR. Relations between an object and its realizations are expressed by *image formation* models that can be quite complicated; they can, for example, depend on localization of the object or be nonlinear. In single-particle electron microscopy (EM) the generally accepted model is linear, with additive noise, and with spatially invariant psf (at least within the area of the imaged object). Given an accurate image formation model and precise knowledge of all its parameters, image restoration methods can inform us about how closely we can recover the original image from its experimental realizations.

The effectiveness of image restoration methods depends on the accuracy of the image formation model, that is, how closely it corresponds to the physical reality of the image formation process, and our ability to precisely establish values of all the parameters of the model as well as statistical properties of random entities included in the model, such as noise.

Fortunately, the linear, weak-phase-object approximation of the image formation process in the EM has been experimentally tested and has been shown to be accurate, and for well-behaved materials, in agreement with experimental results to resolution exceeding 1 Å. For frozen-hydrated biological specimens, dose limitations and beam-induced specimen and media distortions make testing more challenging, but the agreement still appears to be excellent and more accurate image formation models did not result in significantly improved image restoration (Angert *et al.*, 2000). Hence, data processing methods implemented in single-particle software packages employ the standard model with some parameters taken as settings of the microscope (accelerating voltage, spherical aberration constant, and magnification, which together with digitization settings yields effective pixel size), some as generally accepted experimental values (amplitude contrast), and some determined from the data (defocus setting).

A separate category of image formation parameters characterize the random components of the image, such as: the spectral properties of the imaged data and also of noise in the data, and the associated envelope functions that describe suppression of high-frequency information by the microscope and scanning process. The accuracy of these estimates is not very good and the difficulties are compounded by the lack of accepted analytical models for individual components. Finally, we have the suppression of information due to the alignment process necessary to establishing spatial orientations of the collected projection images. For small errors, this loss can be expressed as an additional envelope function (Baldwin and Penczek, 2005; Jensen, 2001).

In cryo-EM the goal of image restoration is to recover the distribution of electron densities within the imaged macromolecular complex, or in the 2D case, its projection image. Since the EM image formation process is linear, it is possible in principle for quantitative EM to yield measurements of atomic densities within the imaged material (Langmore and Smith, 1992). In practice, a more modest but perfectly acceptable goal is to obtain a restored image whose densities are linearly related to those in the original structure. If the linear relationship is perturbed, the interpretation of the final map might be incorrect; hence, image restoration methods should be linear to preserve the desired relationship between the specimen and the restored structure. This essentially excludes from consideration methods based on nonlinear goal functions, such as maximum entropy. Nevertheless, there is room for incorporation of nonlinear constraints that encapsulate *a priori* knowledge about the macromolecular complex, such as nonnegativity of the densities. The appropriate framework is given by the projections onto convex sets (POCS) theory (Stark and Yang, 1998).

The goals of *image restoration* are distinct from those of *image enhancement*. The former is expressed as a well-defined mathematical objective and is aimed at recovery of the original image while keeping distortions due to unavoidable amplification of noise at a minimum. The latter can be broadly described as a set of techniques aimed at modification of the image that facilitate interpretation or visual appreciation of the results. These can range from simply coloring the subunits of a macromolecular complex to make them easily distinguishable, to sophisticated nonlinear image processing methods that are tuned to enhance particular features, for example, components of a given size or shape. Image enhancement methods may be quite elaborate; however, since goals of image enhancement are task-specific and often loosely defined, its methods are generally heuristic and are unrelated to models used in image restoration that are rooted in the physics of image formation.

In this work, we describe the methodology of image restoration in cryo-EM single-particle reconstruction (SPR), and more specifically, we focus on the problem of the optimum recovery of Fourier amplitudes given EM data collected under various defocus settings. The linear theory of image restoration is very well developed (Carazo, 1992; Gonzalez and Woods, 2002; Jain, 1989; Pratt, 1992) and the methods used in SPR are standard. The problem is challenging because of the unique form of the contrast transfer function (CTF) of the EM that has numerous zero crossings. Their number and locations depend on the underfocus setting of the microscope that are difficult to control precisely. Thus, we present the standard criteria, namely the mean-squared error (MSE) and the minimum mean-squared error (MMSE), with an emphasis on the CTF-correction aspects. Finally, we discuss limitations of the image restoration methodology that are mainly caused by the low accuracy of data SSNR estimation and the intrinsic heterogeneity of the processed data set. We conclude the chapter with a description of a semiheuristic approach to image restoration based on a Wiener filter methodology that is meant to alleviate the problems of currently prevalent methods. We note that the step of 3D reconstruction from projections is also part of image restoration methodology in cryo-EM, but because it is a broad subject, we devoted to it an independent contribution which we refer the reader to for details (Chapter 1).

## 2. Image Formation Model in Electron Microscopy

Within the linear, weak-phase-object approximation of the image formation process in the microscope (Wade, 1992), 2D projections represent line integrals of the 3D Coulomb potential of the macromolecule convoluted with the psf of the microscope:

$$g_n(x) = \mathrm{psf}_n(x) * e_n(x) * \left[ \int d(T_n r) dz + m_n^{\mathrm{S}}(x) \right] + m_n^{\mathrm{B}}(x) \qquad (2.1)$$

where $g_n$ denotes the $n$th observed 2D projection image, $e$ is the inverse Fourier transform of the envelope function, $\mathbf{x} = [x\ y]^{\mathrm{T}}$ is a vector of coordinates in the plane of projections, $\mathbf{r} = [r_x\ r_y\ r_z\ \mathbf{1}]^{\mathrm{T}}$ is a vector of coordinates associated with the macromolecule, and $\mathbf{T}$ is the $4 \times 4$ transformation matrix given by

$$T(R, \mathbf{t}) = \begin{bmatrix} R & \mathbf{t} \\ 0 & 1 \end{bmatrix}, \quad \begin{bmatrix} x \\ z \\ 1 \end{bmatrix} = Tr, \tag{2.2}$$

$\mathbf{t} = \left[t_x\, t_y\, 1\right]^{\mathrm{T}}$ being the shift vector denoting translation of the object (and its projection) in the $x$–$y$ plane (translation in $z$ is irrelevant due to the projection operation) and $\mathbf{R}(\psi,\theta,\varphi)$ being the $3 \times 3$ rotation matrix specified by three Eulerian angles (Baldwin and Penczek, 2007). Finally, $m^{\mathrm{B}}$ and $m^{\mathrm{S}}$ denote two additive noises, the first of which is a colored background noise. The second, $m^{\mathrm{S}}$, is attributed to the residual scattering by the solvent and the support carbon film; if used, it is assumed to be white and affected by the transfer function of the microscope in the same way as the imaged macromolecule is. Both types of noise are assumed to be zero mean, mutually uncorrelated, independent between projection images and uncorrelated with the signal:

$$\mathcal{E}\left[m_i^k\right] = 0, \quad k = \mathrm{S}, \mathrm{B}, \tag{2.3}$$

$$\mathcal{E}\left[m_i^k m_j^l\right] = \begin{cases} 0 & \text{if } i \neq j, \\ \sigma_i^{k,l^2} & \text{if } i = j, \end{cases} \quad k, l = \mathrm{S}, \mathrm{B}, \tag{2.4}$$

$$\mathcal{E}\left[d_i m_i^k\right] = 0, \quad k = \mathrm{S}, \mathrm{B}. \tag{2.5}$$

Model (2.1) is in many respects semiempirical. In principle, signal from amorphous ice should not be affected by the CTF; in practice however, the buffer in which the protein is purified contains ions and it is possible to observe CTF effects by imaging frozen buffer alone. It is also debatable whether the envelope function should have the same shape for all components included in the equation. Envelope functions of electron microscopes have been extensively studied and one can identify envelope functions for finite source size, energy spread, drift, specimen charging effects, and multiple inelastic–elastic scattering. Furthermore, one can also identify envelope functions corresponding to the modulation transfer function of the recording medium, which is usually film or CCD camera (for discussion see Zhu *et al.*, 1997). In Fourier space, however, all these functions appear as products of each other, and since they are either Gaussian functions or well approximated by Gaussian functions in the frequency range relevant to single-particle work, and it is impossible in practice to independently retrieve all their parameters from the data, it has become customary in the field to use just one function, namely the Gaussian envelope function ($E$). This function is characterized in Fourier space by the so-called B-factor ($B$):

$$E(s) = \exp\left(-\frac{B}{4}s^2\right) \tag{2.6}$$

where $s$ is the modulus of spatial frequency. While the accepted form simplifies the analysis, the choice of both the terminology and the form adopted is somewhat unfortunate as the unit of the $B$-factor is $\text{Å}^2$ (surface area!). A more useful characterization is a standard deviation of the associated Gaussian function:

$$\sigma_B = \sqrt{\frac{2}{B}}, \tag{2.7}$$

which not only has units of spatial frequency, but is also naturally interpretable as the spatial frequency at which amplitudes of the Fourier transform of the image decrease to ~60% of their original value. It follows from the foregoing statements that at frequency $3\sigma_B$, the amplitudes of the Fourier transform of the image decrease to ~1% of their original value. Thus, $\sigma_B = 0.05\ \text{Å}^{-1}$ (resolution of 20 Å) corresponds to $B = 200\ \text{Å}^2$ in B-factor idiom, while $\sigma_B = 0.04\ \text{Å}^{-1}$ (resolution of 25 Å) to $B = 312\ \text{Å}^2$.

The noise included in Eq. (2.1) is referred to as background noise and in some EM data analysis protocols it has become common to assume that it can be estimated from samples of the background selected from micrographs or simply from areas surrounding windowed particles. In practice, however, the situation is more complicated. If there is no carbon film and if the solvent is appropriately thin, the contribution of solvent to the noise affecting the projection image will be much less than what the estimates obtained from background areas would suggest, as some of the buffer will be displaced by the protein (Fig. 2.1). Regrettably, while the effective height of the particle can be easily estimated, it is all but impossible to obtain reliable measurements of ice thickness. It is therefore advisable to keep in mind that while the model equation (2.1) reflects physical reality of the image formation in cryo-EM to a good extent, its exact form is a compromise between what is theoretically justifiable and what is experimentally and computationally tractable.

In Fourier space, Eq. (2.1) is written by taking advantage of the central section theorem: the Fourier transform of a projection is a Fourier plane (perpendicular to the beam direction) of a rotated Fourier transform of a 3D object:

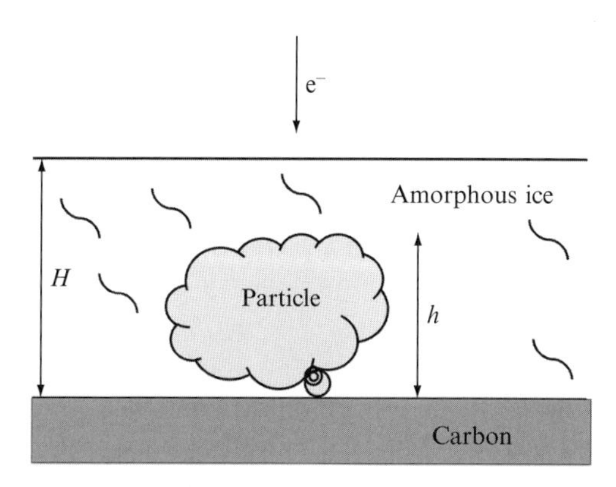


**Figure 2.1** Biological particle suspended in a layer of amorphous ice with an optional thin carbon layer support. Depending on the specimen, the size $h$ of the particle is between 50 and few thousand Ångstroms. The ice thickness $H$ should not exceed the particle size by much.

$$G_n(\mathbf{s}) = \mathrm{CTF}(\mathbf{s}; \Delta z_n) E_n(\mathbf{s}) \Big[ (D(\mathbf{Ts}))_{s_z=0} + M_n^{\mathrm{S}}(\mathbf{s}) \Big] + M_n^{\mathrm{B}}(\mathbf{s}) \tag{2.8}$$

The capital letters denote Fourier transforms of objects appearing in Eq. (2.1) while CTF (a Fourier transform of the psf) depends, among other parameters that are set very accurately (such as the accelerating voltage of the microscope), on the defocus setting $\Delta z_n$ and the amplitude contrast ratio $0 \leq A < 1$ that reflects presence of absorption and thus the amplitude contrast in the images. For the range of frequency considered, $A$ is assumed to be constant and the CTF is written in terms of the phase perturbation function $\gamma$ as:

$$\mathrm{CTF}(\mathbf{s}; \Delta z) = \sqrt{1 - A^2} \sin(\gamma(s; \Delta z)) - A \cos(\gamma(s; \Delta z)) \tag{2.9}$$

where $s = |\mathbf{s}|$ is the magnitude of spatial frequency and where for simplicity we assumed the absence astigmatism. The phase perturbation function is (Scherzer, 1949):

$$\gamma(s; \Delta z) = 2\pi \left( -\frac{1}{2} \Delta z \lambda s^2 + \frac{1}{4} C_s \lambda^3 s^4 \right), \tag{2.10}$$

where $C_s$ is the spherical aberration constant and, ignoring relativistic effects, $\lambda = \sqrt{(h^2)/(2me^- V)}$ is the wavelength of electrons ($h$, Planck's constant; $m$ and $e^-$, electron mass and charge, respectively; $V$, voltage of the microscope).

While what is given here can be written as 2D functions, thus accommodating possible astigmatism, in practice almost all cryo-EM data processed is astigmatism-free. Hence, without loss of generality we write subsequent expressions as 1D functions of the modulus of spatial frequency. For simplicity we may write:

$$\begin{aligned} g_n(\mathbf{x}) &= \mathrm{psf}_n(\mathbf{x}) * e_n(\mathbf{x}) * \int d(\mathbf{T}_n\mathbf{r})dz + m_n(\mathbf{x}) \\ &= \mathrm{psf}_n(\mathbf{x}) * e_n(\mathbf{x}) * f(\mathbf{x}) + m_n(\mathbf{x}) \end{aligned} \tag{2.11}$$

$$\begin{aligned} G_n(\mathbf{s}) &= \mathrm{CTF}(\mathbf{s}; \Delta z_n) E_n(\mathbf{s}) (D(\mathbf{Ts}))|_{s_z=0} + M_n(\mathbf{s}) \\ &= \mathrm{CTF}(\mathbf{s}; \Delta z_n) E_n(\mathbf{s}) F(\mathbf{s}) + M_n(\mathbf{s}) \end{aligned} \tag{2.12}$$

where $m$ is the effective noise containing both the background noise and the CTF-affected contributions from solvent and/or carbon support layer.

Due to linearity of the Fourier transform, assumption stated in equations (2.3)–(2.5) hold for Fourier transforms of the respective entities in Eqs. (2.5) and (2.12). In addition, we assume that in Fourier space noise is approximately uncorrelated between spatial frequencies:

$$\mathcal{E}\{M_n(\mathbf{s})M_n(\mathbf{u})\} = \sigma_n^2(\mathbf{s})\delta(\mathbf{s}-\mathbf{u}) = P_n(\mathbf{s})\delta(\mathbf{s}-\mathbf{u}), \tag{2.13}$$

where based on the assumption that the noise has zero mean, we equated its Fourier space variance with its power spectrum (PW). The rotationally averaged PW of the observed image calculated as the expected value of its squared Fourier intensities (Eq. (2.5)) is given by:

$$\begin{aligned} P_g(s) &= \mathrm{CTF}^2(s)\cdot E^2(s)\cdot\left(P_f(s) + P_m^{\mathrm{S}}(s)\right) + P_m^{\mathrm{B}}(s) \\ &= \mathrm{CTF}^2(s)\cdot E^2(s)\cdot P_f(s) + P_m(s). \end{aligned} \tag{2.14}$$

We assume the PW of the 3D structure to be isotropic, so we can replace it by the (averaged) PW of 2D projections.

Based on Eqs. (2.1), (2.12), and (2.14), the SSNR can be understood either as a ratio of rotationally averaged PWs, and thus an 1D function of modulus of spatial frequency, or as a multidimensional function describing frequency dependence of SSNR in Fourier space (Penczek, 2002). Moreover, there exist in cryo-EM several different understandings of what SSNR represents. A generic one is to define SSNR as a ratio of the power of the "ideal" signal to the power of noise:

$$\mathrm{SSNR}(\mathbf{s}) = \frac{P_f(\mathbf{s})}{P_m(\mathbf{s})}. \tag{2.15}$$

In Eq. (2.15), the details of image formation are excluded. In a definition of SSNR that is commonly used, both the CTF and the envelope function are incorporated into the effective signal, thus making it closely related to the data (Ludtke and Chiu, 2002; Ludtke *et al.*, 1999). We term this definition $\mathrm{SSNR}^{\mathrm{Data}}$ and it is written as:

$$\mathrm{SSNR}^{\mathrm{Data}}(\mathbf{s}) = \frac{\mathrm{CTF}^2(\mathbf{s})E^2(\mathbf{s})P_f(\mathbf{s})}{P_m(\mathbf{s})}. \tag{2.16}$$

Finally, we can consider the SSNR of the average of EM images (or a 3D reconstruction) that will be corrected for the CTF effects, using one of the methods discussed below, but that will not necessarily have the correct distribution of Fourier amplitudes and will also be affected by alignment errors. The latter can be easier considered as a source of noise, or as yet another envelope function (Baldwin and Penczek, 2005). In recognition of alignment as a major source of errors in EM structure determination, we will term the SSNR of the average $\mathrm{SSNR}^{\mathrm{Ali}}$:

$$\mathrm{SSNR}^{\mathrm{Ali}}(\mathbf{s}) = \frac{E^{\mathrm{Ali}^2}(\mathbf{s})P_f(\mathbf{s})}{P_m^{\mathrm{Ali}}(\mathbf{s})}. \tag{2.17}$$

The relationships among the three SSNRs introduced above are not necessarily straightforward, *particularly* in light of the fact that the same SSNR can be estimated using different experimental approaches (see Section 3). Nevertheless, the difference between SSNR (Eq. (2.15)) and $\mathrm{SSNR}^{\mathrm{Data}}$ (Eq. (2.16)) lies chiefly in the inclusion of the CTF and envelope function in the latter, so the choice between them is mainly a matter of convenience or of the particular data processing protocol used. For example, in some approaches one would attempt to directly estimate the effective power of the signal, that is, $\mathrm{CTF}(\mathbf{s})E(\mathbf{s})P_f(\mathbf{s})$, from the data, and in others, estimation of the envelope is separate from estimation of the power of the signal. The latter may be estimated using independent experiments, such as X-ray crystallography or SAXS (Gabashvili *et al.*, 2000; Ludtke *et al.*, 2001).

In what follows we focus on the problem of recovering either $f$ or $d$ given projection images $g$ under the assumption that alignment parameters are approximately known. We also assume that the PW of the 3D structure $d$ is isotropic and that the set of 2D projections $g_n$ samples the entire Fourier space, in which case the rotationally averaged sum of PWs of available 2D projections faithfully represent the rotational PW of the original, albeit initially unknown, 3D structure. Finally, whenever we discuss the restoration of a 3D object from its 2D projections, we largely ignore the difficulties associated with the problem of the reconstruction of an object from a set of its projections; a comprehensive review of reconstruction methods is given in Chapter 1. Instead, we focus on the restoration of the proper distribution of the Fourier amplitudes of the reconstructed object given artifacts introduced by the CTF and fall-off due to the envelope function $E$.

## 3. Estimation of Image Formation Model Characteristics

In this section, we briefly describe the current state of art in estimation of image formation characteristics. In the last decade, there was a surge of publications devoted to the study of estimation of CTF parameters and the resulting computational methods were either incorporated into existing software packages (Huang *et al.*, 2003; Penczek *et al.*, 1997; Saad *et al.*, 2001; Sorzano *et al.*, 2007; van Heel *et al.*, 2000; Velazquez-Muriel *et al.*, 2003) or made available as independent applications (Mallick *et al.*, 2005; Mindell and Grigorieff, 2003; Sander *et al.*, 2003; Yang *et al.*, 2009; Zhou *et al.*, 1996). The CTF is determined from the collected micrographs and the methods used are typically automated with supporting Graphical User Interface (GUI) that allows examination and, if necessary, adjustment of the results. While all methods allow determination of basic image formation parameters such as defocus, sometimes amplitude contrast and astigmatism, others will also yield envelope functions and other more complex characteristics. Overall, a user can always find a CTF-estimation utility that will provide the required functionality; regrettably, however, there is no established standard in the mathematical form of the CTF used by developers. In the following publications (Huang *et al.*, 2003; Penczek *et al.*, 1997; Saad *et al.*, 2001; Sander *et al.*, 2003; Velazquez-Muriel *et al.*, 2003), in addition to the different sign conventions used, amplitude contrast is also defined differently in each case.

The initial steps of CTF parameters estimation are common to all methods. Two parameters of the CTF Eqs. (2.9) and (2.10) are simply taken from the microscope settings and are considered accurate: the accelerating voltage (i.e., the wavelength of electrons) and the spherical aberration constant. The amplitude contrast can in principle be fitted, but very often is assumed to be constant and, for cryo-data, is set in the range of 6–12%. This leaves the defocus and amount and direction of astigmatism, the estimations of which are possible if accurate estimates of PWs of micrographs are available.

Depending on the package, power spectra are computed either from the entire micrograph field (Zhu *et al.*, 1997) or from windowed particles (Saad *et al.*, 2001). The advantages of the first approach are: (1) the larger amount of averaged data and larger window sizes improve the statistical properties and resolution of the estimate and (2) since CTF estimation precedes the particle picking step, it is possible to use CTF information to improve the accuracy of automated particle picking (Huang and Penczek, 2004). The second approach makes it possible to use the information about particle regions and background noise regions to compute additional characteristics of the image formation process. Typically, a 2D estimate of the PW is rotationally averaged, and either a low-order polynomial is fitted to a set of local minima of the resulting 1D PW or a constrained optimization is used to obtain curves bracketing the 1D PW (Huang *et al.*, 2003; Mallick *et al.*, 2005; Yang *et al.*, 2009). The curve "beneath" the PW is associated with the background noise of Eq. (2.11) and is subtracted. The repeated fit yields the curve bracketing the background-subtracted PW from "above," and this curve is used in a direct fit of the analytical form of CTF equation (2.9) to 1D data. The fit is typically restricted to the intermediate range of spatial frequencies within which the 1D PW agrees well with the model. In the low-frequency region, the noise in the data seems to increase rapidly, while in the high-frequency region, the statistical properties of the data seem to change, thus precluding reliance on the simple analytical form of the bracketing curves (Huang and Penczek, 2004). It is possible to use the thus obtained value of defocus as an initial guess in the subsequent analysis of astigmatism in 2D PWs (Huang and Penczek, 2004; Mallick *et al.*, 2005). In the SPARX implementation (Hohn *et al.*, 2007), the initial parameters obtained from estimates based on the entire micrograph are used as initial estimates and are subsequently refined using PWs obtained from windowed particles.

There are two competing approaches both to the initial steps of CTF parameters estimation and the estimation of the envelope function and SSNR in the data. One can either rely on analytical functions to represent the characteristics and fit them to the data, as was initially put forward in a pioneering study by Saad *et al.* (2001), or use the estimated PWs directly to represent image formation characteristics required in subsequent cryo-EM data processing. In the former approach, the selected analytical functions are in general simple, and the choices might either have a theoretical basis, as in using a Gaussian function to represent the envelope function, or be heuristic, as in the use of exponents of low-order polynomials for the envelope function (Huang *et al.*, 2003; Saad *et al.*, 2001). The latter approach is used in EMAN2 (S. Ludtke, personal communication). In both approaches, it is necessary to estimate PWs from micrograph regions that contain projection data ($P_g$ of Eq. (2.11)) and from background noise regions, usually aptly available in micrographs or in areas surrounding windowed particles ($P_m$ of Eq. (2.11)). If there were no errors and if the image formation model (Eq. (2.5)) were to hold exactly, one would have:

$$\mathrm{CTF}^2(s)\cdot E^2(s)\cdot P_f(s) = P_g(s) - P_m(s). \tag{2.18}$$

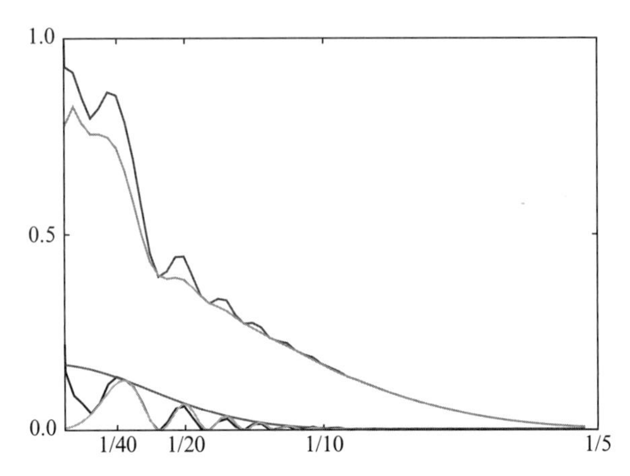


**Figure 2.2** Estimation of image formation model characteristics. The PW (red) was computed using particles windowed from a single micrograph (70S ribosomes on carbon support, accelerating voltage 300 kV). The background noise PW (green) was computed from areas surrounding windowed particles. The defocus (2.95 $\mu$m) was estimated using background-subtracted PW (blue) and the envelope function (magenta) modeled by a Gaussian function. The fitted CTF shows excellent agreement with the experimental PW (light blue). Analysis was done using the e2ctf.py utility of EMAN2. The vertical axis is in arbitrary units, and the horizontal axis corresponds to modulus of spatial frequency [1/Å]. (See Color Insert.)

Since CTF parameters have already been estimated, one would then only need to know the (1D) PW $P_f$ of the molecular complex imaged to estimate the envelope function $E$. There are several ways to carry this out: (1) obtain independent experimental measurements of the 1D PW $P_f$ either from SAXS experiments (Gabashvili *et al.*, 2000; Saad *et al.*, 2001) or from X-ray crystallography (possibly computed from atomic coordinates of the model of the complex), (2) use the 1D PW of a complex of similar size and shape as the resulting differences are negligible in comparison with other uncertainties inherent in the analysis of Eq. (2.18), (3) use an analytical curve fitted to the data from either (1) or (2), and (4) use a simple analytical model that approximates the shape of typical macromolecular complexes well. Next, the background-subtracted 1D PW equation (2.18) is divided by $P_f$, and the fit of a curve bracketing the result from above yields the envelope function $E$ (Fig. 2.2).

## 3.1. Analytical model of 1D rotationally averaged PW of macromolecular complexes in solution

The shape of 1D PWs of proteins was studied by Guinier and Fournet (1955), who noticed that the overall shape of the curve mainly depends on the size of the protein. The authors suggested an approximation by a two-component function where: (1) the low-frequency region is modeled by a Gaussian function whose standard deviation is related to the effective diameter of the protein in real space and (2) the PW in frequencies higher than $\sim(1/8)$ Å$^{-1}$ is assumed to be constant at a level dependent on the number of atoms in the protein. This model does not agree very well with the 1D PWs of pseudoelectron densities obtained from computational conversion of atomic coordinates of selected macromolecular complexes available in the PDB database. On the other hand, it is also debatable whether such a procedure faithfully represents the physical reality and "true" PWs of complexes in solution. One can also dispute the usefulness of SAXS PWs since different buffer conditions or protein concentration may modify the shape of the PW in comparison with that observed in EM. In addition, small angle scattering in physical image formation process is different from that in EM. Reliance on PDB models can be justified by the simplicity of the procedure and by the fact that X-ray models, when available, are considered gold standards in the cryo-EM community. Furthermore, X-ray models are widely used for comparisons and for docking (Volkmann and Hanein, 2003), so macromolecular complexes, even if obtained used different techniques, look "familiar" when they have the same PWs as X-ray atomic models.

The empirical model used in SPARX (Hohn *et al.*, 2007) is heuristic, meets expectations, was thoroughly tested using numerous X-ray crystallographic model, and was found to fit the data very well (Z. Huang and P.A. Penczek, unpublished results):

$$P_f(s) = \exp\left(t_1 + \frac{t_2}{\left(\frac{s}{t_3}+1\right)^2}\right) + \exp\left(t_4 - 0.5\left(\frac{s-t_5}{t_6^2}\right)^2\right), \qquad (2.19)$$

where $t_1$ and $t_4$ are scaling parameters which do not influence the shape of the curve; the first term is responsible for the general shape, and the second, the Gaussian function, represents the $\beta$–$\beta$ spacing which results in a broad peak at $\sim(1/5)$ Å$^{-1}$ (Fig. 2.3). It can be shown that the model effectively depends only on two parameters, namely ($t_2$, $t_3$), and with minimal adjustment can represent a very broad range of macromolecular complexes, virtually irrespective of their shapes.

## 4. Pseudoinverse Restoration—Mean-Square and Chi-Square Error Criteria

### 4.1. Closed form solutions

In MSE pseudoinverse restoration we seek an estimate $\hat{F}$ of the original signal that fits the observed EM images in the mean-square error sense (MSE criterion):

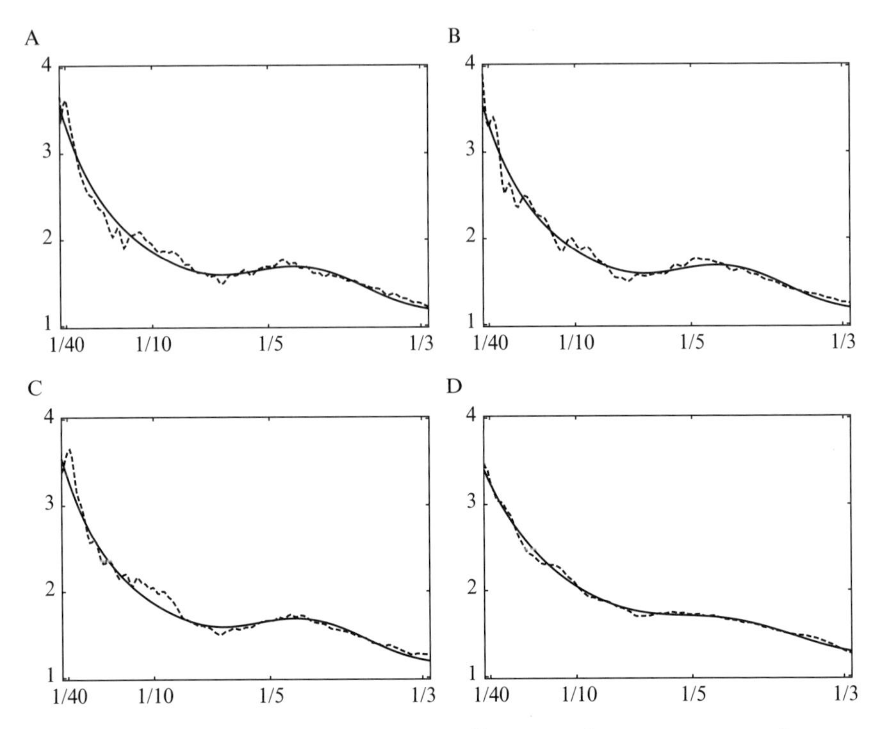


**Figure 2.3** Analytical model of 1D rotationally averaged power spectrum of macromolecular complexes. X-ray crystallographic atomic models were converted to discrete 3D electron density maps using trilinear interpolation and voxel size 1.0 Å. Equation (2.19) was fitted to rotationally averaged 3D power spectra of all maps. The vertical axis is in arbitrary units, and the horizontal axis corresponds to modulus of spatial frequency [1/Å]. (A) 20S proteasome, 630 kDa (PDB:1J2Q), (B) GroEL chaperonin, 715 kDa (PDB:1KP8), (C) calmodulin-dependent protein kinase II (CaMKII), 205 kDa (PDB:1HKX), (D) 70S ribosome, ~2000 kDa (PDB:2AVY). Power spectra of complexes (A–C) were modeled using the same parameters as in Eq. (2.19) while the power spectrum of ribosome (D) required minor adjustments.

$$\min_{\hat{F}} \mathcal{L}_{\mathrm{MSE}}(\hat{F}) = \min_{\hat{F}} \mathcal{E}\left\{ \sum_n \left\| G_n(\mathbf{s}) - \mathrm{CTF}_n(\mathbf{s}) \cdot E_n(\mathbf{s}) \cdot \hat{F}(\mathbf{s}) \right\|^2 \right\} \quad (2.20)$$

The MSE criterion has a number of desirable properties in that: (1) it corresponds to the "total power" of the data, that is, it is equivalent to $\max_{\hat{F}} \mathcal{E}\left\{ \left\| \hat{F} \right\|^2 \right\}$ (Penczek *et al.*, 1992), (2) it is a quadratic form of $\hat{F}$, so in principle a global minimum exists (a unique solution of Eq. (2.20)), and (3) it is differentiable with respect to $\hat{F}$, so when there is no closed form solution, an efficient iterative algorithms for seeking the minimum can be employed.

The most straightforward approach to pseudoinverse restoration is to ignore the presence of noise, in which case, for a single EM image, the solution of Eq. (2.20) has the form:

$$\hat{F}(\mathbf{s}) = \begin{cases} \dfrac{1}{\mathrm{CTF}(\mathbf{s})E(\mathbf{s})} G(\mathbf{s}) & \text{if } \mathrm{CTF}(\mathbf{s}) \neq 0 \\ 0 & \text{otherwise} \end{cases}. \tag{2.21}$$

In practice, a more sensible approach is to select a small constant ε and thus obtain:

$$\hat{F}(s) = \begin{cases} \dfrac{1}{\mathrm{CTF}(\mathbf{s})E(\mathbf{s})} G(\mathbf{s}) & \text{if} \quad \mathrm{abs}(\mathrm{CTF}(\mathbf{s})) > \varepsilon \\ 0 & \text{otherwise} \end{cases}. \tag{2.22}$$

However, even if we were prepared to tolerate an enhancement of noise due to the division by the envelope function, it turns out that strong artifacts in the real-space representation of the "restored" image would result from setting to zero those parts of the image $F$ whose spatial frequencies correspond to small absolute values of the CTF. In addition, for a single image, sections of its Fourier transform are irretrievably lost. Therefore, it is difficult to have a meaningful pseudoinverse-based CTF correction of a single underfocused EM image as the information in some regions of Fourier space cannot be restored (Fig. 2.4A).

A standard approach to overcoming the problem of information loss due to zeros in CTF is to collect EM data using a wide range of defocus settings to obtain images affected by a range of different CTFs. If the defocus settings of the microscope are selected properly, it is possible to collect the data in such a way so that the full range of Fourier space information can be recovered. We also include the presence of frequency-dependent noise that leads to a mean chi-square error (MCE) criterion: in MCE pseudoinverse restoration we seek an estimate $\hat{F}$ of the original signal that fits the observed EM images in the chi-square sense:

$$\min_{\hat{F}} \mathcal{L}_{\mathrm{MCE}}(\hat{F}) = \min_{\hat{F}} \mathcal{E}\left\{ \sum_n \frac{\left\| G_n(\mathbf{s}) - \mathrm{CTF}_n(\mathbf{s}) \cdot E_n(\mathbf{s}) \cdot \hat{F}(\mathbf{s}) \right\|^2}{\sigma^2_{M_n}(\mathbf{s})} \right\}. \tag{2.23}$$

To solve the problem, we express the solution as a linear combination of the input data:

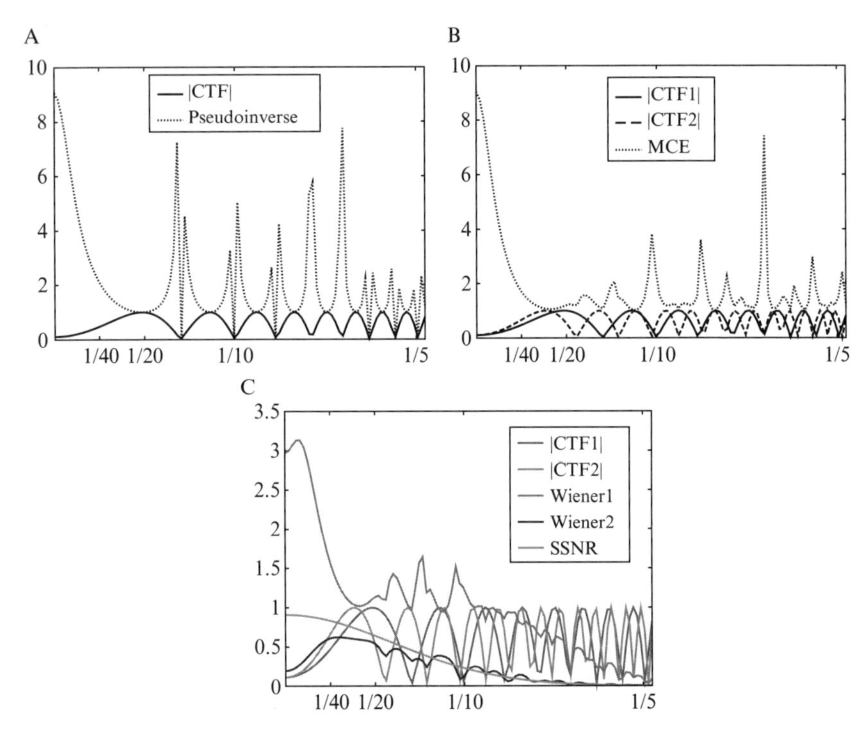


**Figure 2.4** Linear cryo-EM image restoration filters. The horizontal axis corresponds to modulus of spatial frequency [1/Å]. (A) CTF (accelerating voltage 300 kV, amplitude contrast 10%, defocus 1.0 $\mu$m) and the pseudoinverse filter (Eq. (2.22)) with $\varepsilon = 0.02$. (B) Two CTFs, first as in (A), the second with defocus 1.6 $\mu$m, and the mean-square error filter (Eq. (2.29)) with the variance of data omitted. (C) Wiener filters (Eq. (2.45)) plotted assuming the SSNR is the same for each CTF and is given by a Gaussian function. For the first filter (magenta), the maximum SSNR is set to 20 while for the second filter (blue), it is set to 0.9. Note that identical filters can be obtained using the CMCE filter (Eq. (2.34)) with regularization function $U$ set to the inverse of the SSNR of the data. For clarity, we plotted moduli of the CTF function. (See Color Insert.)

$$\hat{F}(\mathbf{s}) = \sum_{n=1}^{N} w_n(\mathbf{s}) G_n(\mathbf{s}) \tag{2.24}$$

and we minimize criterion (2.23) with respect to the weights $w_n(\mathbf{s})$.

In the derivation, since noise in the data, in Fourier space, is uncorrelated with respect to spatial frequencies, we will omit the argument **s**. By combining Eqs. (2.23) and (2.24) we obtain:

$$\mathcal{L}_{\mathrm{MCE}}(\hat{F}) = \mathcal{E}\left\{ \sum_n \frac{\|G_n\|^2}{\sigma^2_{M_n}} - 2Re\left(\left(\sum_n \frac{\mathrm{CTF}_n E_n G_n}{\sigma^2_{M_n}}\right)\left(\sum_n w_n G_n\right)^*\right) + \left(\sum_n \frac{\|\mathrm{CTF}_n E_n\|^2}{\sigma^2_{M_n}}\right)\left\|\sum_n w_n G_n\right\|^2 \right\}. \tag{2.25}$$

To find the minimum, we differentiate with respect to $w_k$, set derivatives to zero, and substitute $G_n$ by its form given by the image formation model equation (2.12):

$$\frac{\partial \mathcal{L}_{\mathrm{MCE}}(\sum_n w_n G_n)}{\partial w_k} = \mathcal{E}\left\{ \left(F\sum_n \frac{|\mathrm{CTF}_n E_n|^2}{\sigma^2_{M_n}} + \sum_n \frac{\mathrm{CTF}_n E_n M_n}{\sigma^2_{M_n}}\right)(\mathrm{CTF}_k E_k F + M_k)^* - \left(\sum_n \frac{|\mathrm{CTF}_n E_n|^2}{\sigma^2_{M_n}}\right)\left(F\sum_n w_n \mathrm{CTF}_n E_n + \sum_n w_n M_n\right)(\mathrm{CTF}_k E_k F + M_k)^* \right\} = 0. \tag{2.26}$$

The expected value of Eq. (2.26) yields a set of $N$ equations:

$$P_f \sum_n \frac{|\mathrm{CTF}_n E_n|^2}{\sigma^2_{M_n}} + 1 = \left(\sum_n \frac{|\mathrm{CTF}_n E_n|^2}{\sigma^2_{M_n}}\right)\left(P_f \sum_n w_n \mathrm{CTF}_n E_n + w_k \frac{\sigma^2_{M_k}}{\mathrm{CTF}_k E_k}\right), \tag{2.27}$$

which can be simultaneously solved by taking advantage of the fact that the term $P_f \sum_n w_n \mathrm{CTF}_n E_n$ is shared. The final form of the weighting term is:

$$w_k = \frac{\mathrm{CTF}_k E_k}{\sigma^2_{M_k}} \Big/ \sum_n \frac{|\mathrm{CTF}_n E_n|^2}{\sigma^2_{M_n}}. \tag{2.28}$$

The image restored from $N$ EM images collected with different defocus setting is thus obtained using the following equation (Ludtke, 2001):

$$\hat{F}(s) = \frac{\sum_{n=1}^{N} \frac{\mathrm{CTF}_n(s) E_n(s) G_n(s)}{\sigma^2_{M_n}(s)}}{\sum_{n=1}^{N} \frac{\mathrm{CTF}_n^2(s) E_n^2(s)}{\sigma^2_{M_n}(s)}}. \tag{2.29}$$

By multiplying the numerator and denominator of Eq. (2.29) by the PW of the original structure $f$, the MCE solution can be written as a function of the SSNR either as that of the data (Eq. (2.16)) (Ludtke, 2001):

$$\hat{F}(s) = \frac{\sum_{n=1}^{N} \frac{1}{\mathrm{CTF}_n(s)E_n(s)} \mathrm{SSNR}_n^{\mathrm{Data}}(s) G_n(s)}{\sum_{n=1}^{N} \mathrm{SSNR}_n^{\mathrm{Data}}(s)}, \tag{2.30}$$

or the ideal SSNR:

$$\hat{F}(\mathbf{s}) = \frac{\sum_{n=1}^{N} \mathrm{CTF}_n(\mathbf{s})E_n(\mathbf{s})\mathrm{SSNR}_n(\mathbf{s})G_n(\mathbf{s})}{\sum_{n=1}^{N} \mathrm{CTF}_n^2(\mathbf{s})E_n^2(\mathbf{s})\mathrm{SSNR}_n(\mathbf{s})}. \tag{2.31}$$

The three forms, Eqs. (2.29)–(2.31), are not mathematically equivalent as Eq. (2.30) is esthetically unappealing due to implied divisions by the zeros of CTFs, or, equally problematic, divisions by the zeroes of $\mathrm{SSNR}_n^{\mathrm{Data}}$ (the two should coincide).

The MCE image restoration given by Eq. (2.31) has the following properties (Fig. 2.4B):

1. For a large number of images collected at approximately uniformly distributed defocus settings, the impact of the filter can be considered separately for two spatial frequency regions consisting of: (a) frequencies lower than the frequency of the first extremum of $\sum_{n=1}^{N} \mathrm{CTF}_n^2(\mathbf{s})$ and (b) frequencies higher than the frequency of the first extremum. In the first region, all CTFs have approximately the same shape irrespective of the defocus settings, and their values for $s \to 0$ are given by the amplitude contrast $A$. Hence, the result of the restoration is an enhancement of the very low frequencies by $A$ times (we assumed that envelope functions are scaled such that $E(0) = 1$); moreover, the degree of this enhancement does not depend on the SSNR. In the second region consisting of the higher frequencies, under the aforementioned assumptions, the denominator will be approximately constant, so the restoration corresponds to the summation of EM images multiplied by their respective CTFs and weighted relative to each other by differences in the envelope function and noise level.
2. For a small number of images or for data sets for which $\sum_{n=1}^{N} \mathrm{CTF}_n^2(\mathbf{s}) \simeq 0$ for some spatial frequencies, the MCS-based restoration will be affected by artifacts shared with the pseudoinverse solution

equation (2.21). In 2D applications, this is unlikely to happen; however, if Eq. (2.31) is used in the context of 3D reconstruction from projections, divisions by very small numbers in the denominator of Eq. (2.31) will result in very unpleasant real-space artifacts. In some applications, this is prevented by inclusion of a small additive heuristic parameter in the denominator (Böttcher and Crowther, 1996; Grigorieff, 1998).

3. The solution does not depend on the form or scaling of the PW of the target, ideal image.
4. The solution does not depend on the absolute scaling of the SSNR but rather on differences between the SSNRs of different images. Since $P_f$ is the same for all images, practically speaking we are only concerned with the PW of the background noise $P_{m_n}$(recall $P_{m_n} = \sigma_{M_n}^{2}$). In other words, if we let $P_m$ denote the average PW of the background noise, then for the PW of the individual images, we can write $P_{m_n} = P_m + \Delta P_n$. It follows that image restoration with Eq. (2.31) depends only mildly on the average distribution of background noise and it is the deviation $\Delta P_n$ that is decisive. Since in cryo-EM PW of the noise is determined from the data and deviations $\Delta P_n$ are small in comparison with $P_m$, one would have to have a sufficiently accurate method of determining $P_{m_n}$ to justify the inclusion of $\mathrm{SSNR}_n$ (or $P_{m_n}$) in the image restoration. Otherwise, errors in the determination of $P_{m_n}$ will propagate into the restored image and result in the deterioration of its SSNR.

## 4.2. Constrained closed form restoration

An elegant way to address the shortcomings of pseudoinverse restorations Eqs. (2.29)–(2.31) is to constrain the desired solution by imposing on it "smoothness." Given a high-pass filter $U(\mathbf{s})$, we seek to minimize the quantity

$$\min_{\hat{F}} \mathcal{L}_{\mathrm{CMCE}}(\hat{F}) = \min_{\hat{F}} \mathcal{E}\{\|U(s)\hat{F}(s)\|^2\}, \tag{2.32}$$

subject to the condition that:

$$\sum_n \frac{\|G_n(s) - \mathrm{CTF}_n(s) \cdot E_n(s) \cdot \hat{F}(s)\|^2}{\sigma_{M_n}^2(s)} = 1. \tag{2.33}$$

As in the previous section, we obtain the solution by introducing the Lagrange multiplier $1/\gamma$ and thus obtain the following constrained mean chi-square error (CMCE) restoration:

$$\hat{F}(\mathbf{s}) = \frac{\sum_{n=1}^{N} \frac{\mathrm{CTF}_n(\mathbf{s}) E_n(\mathbf{s}) G_n(\mathbf{s})}{\sigma^2_{M_n}(\mathbf{s})}}{\sum_{n=1}^{N} \frac{\mathrm{CTF}_n^2(\mathbf{s}) E_n^2(\mathbf{s})}{\sigma^2_{M_n}(\mathbf{s})} + \gamma U^2(\mathbf{s})} \tag{2.34}$$

As expected, the solution reduces to MCE restoration for $\gamma = 0$ and for $\gamma \to \infty$ becomes a sum of input images multiplied by their respective CTFs and envelope functions. It should also be noted that although the mathematical forms of the CMCE restoration and the Wiener filter (which will be discussed in a later chapter) are very similar, the assumptions underlying their derivations are entirely different (Fig. 2.4C). We will return to the apparent similarity of the two solutions later.

The CMCE restoration equation (2.34) has significant advantages over MCE pseudoinverse equation (2.31) in that the possibility of division by zero is eliminated, and artifacts due to divisions by small values in the denominator can be minimized by a proper selection of the constant $\gamma$. In the following, to better understand the behavior of CMCE restoration, we assume that the high-pass filter $U(\mathbf{s})$ is given by an inverse Gaussian function, in which case Eq. (2.34) becomes:

$$\hat{F}(\mathbf{s}) = \frac{\exp\left(-\frac{s^2}{2\sigma^2}\right) \sum_{n=1}^{N} \frac{\mathrm{CTF}_n(\mathbf{s}) E_n(\mathbf{s}) G_n(\mathbf{s})}{\sigma^2_{M_n}(\mathbf{s})}}{\exp\left(-\frac{s^2}{2\sigma^2}\right) \sum_{n=1}^{N} \frac{\mathrm{CTF}_n^2(\mathbf{s}) E_n^2(\mathbf{s})}{\sigma^2_{M_n}(\mathbf{s})} + \gamma}, \tag{2.35}$$

Equation (2.35) is simply a low-passed version of the pseudoinverse restoration equation (2.29) with the difference that with the appropriate choices of $\gamma$ and $\sigma$, any impact of the denominator becomes negligible in high frequencies.

## 4.3. Iterative solution

A standard method of solving least-squares problem (LSQ) of the type given by Eq. (2.20) is by an iterative process which takes advantage of the fact that the goal function is differentiable and employs iterations of the type:

$$\hat{F}^{l+1}(s) = \hat{F}^{l}(s) + \eta^{l} \nabla \mathcal{L}\left(\hat{F}^{l}\right) = \hat{F}^{l}(s) + \eta^{l} \sum_{n} E_n(s) \cdot \mathrm{CTF}_n(s) \left(G_n(s) - \mathrm{CTF}_n(s) \cdot E_n(s) \cdot \hat{F}^{l}(s)\right). \tag{2.36}$$

Note that complex conjugation in Eq. (2.36) is omitted since both the CTF and $E$ are real. The relaxation parameter for the $l$th iteration is denoted by $\eta^l > 0$; by setting $\eta^l = \eta = \text{const} > 0$, we obtain Richardson's algorithm (Richardson, 1910). With properly chosen relaxation parameters, the algorithm will converge to a solution of Eq. (2.20) that is most similar to the prototype image $\hat{F}^0(s)$ in the least-squares sense. The prototype image is usually a blank image filled with zeroes or an acceptable approximation to the solution obtained using one of the fast direct methods.

Termination of the algorithm after a finite number of iterations is a form of regularization of the solution as it will yield an image that roughly corresponds to one that could be obtained from the singular value decomposition (SVD) approach generated by a subset of dominating eigenvectors. The main disadvantages of the iterative method (2.36) include significant computational effort, slow convergence rate, and lack of clear guidelines on what the number of iterations and values of relaxation parameters should be, (which in practice have to be established by trial and error), all of which make the method only moderately attractive. Furthermore, in 2D the iterative method (Eq. (2.36)) yields images of essentially the same quality as closed form solutions, but requires a much larger computational effort. However, the iterative method also has a significant advantage in that it can serve as basis for more sophisticated image restoration algorithms. For example, division by zero is not a problem despite the presence of (overlapping) zeroes of CTFs since Eq. (2.36) does not involve division by CTFs. For a spatial frequency for which all CTFs are equal to zero, the final image can simply inherit the value from the prototype image, and thus make use of the available *a priori* information. When combined with additional regularization provided by early termination of the algorithm, the iterative method yields a solution that has fewer artifacts (particularly if CTFs have overlapping zeroes or very small values) and is "smoother," and hence more agreeable to the eye.

It is possible to dramatically improve the rate of convergence by using more advanced methods for solving Eq. (2.20) while taking advantage of the simple form of the derivative shown in Eq. (2.36). In many applications, it is possible to use steepest decent, conjugate gradients, or limited-memory BFGS optimization methods (Yang *et al.*, 2005). One can also incorporate explicitly rational regularization terms, for example, by constraining the Laplacian of the real-space solution $\hat{F}$ (Zhu *et al.*, 1997), which can also be expressed as a low-pass filtration in Fourier space resulting in an algorithm with iterations of the type:

$$\begin{aligned}\hat{F}^{l+1}(\mathbf{s}) = {} & \hat{F}^l(\mathbf{s}) - \gamma\eta^l U^{-2}(\mathbf{s})\hat{F}^l(\mathbf{s}) + \eta^l \sum_n E_n(\mathbf{s})\cdot \mathrm{CTF}_n(\mathbf{s}) \\ & \left(G_n(\mathbf{s}) - \mathrm{CTF}_n(\mathbf{s})\cdot E_n(\mathbf{s})\cdot \hat{F}^l(\mathbf{s})\right),\end{aligned} \tag{2.37}$$

where the notation has the same meaning as in Eq. (2.34). The improvements outlined above not only accelerate convergence, but also yield solutions less affected by the noise and artifact-inducing inconsistencies in the data.

Finally, in application to 3D reconstruction from projections, iterative methods offer the only mathematically sound approach to reconstruction with CTF correction. For reasons that exceed the scope of this review, there are neither direct Fourier inversion nor filtered backrprojection algorithms that include CTF correction; the only consistent approach is to use a 3D version of Eq. (2.36) and obtain a solution iteratively (Yang *et al.*, 2005; Zhu *et al.*, 1997).

## 4.4. Constrained solutions—incorporation of *a priori* information

The main advantage of iterative methods is that it is straightforward to incorporate *a priori* knowledge about the solution, particularly if it is given as a nonlinear constraint, such as nonnegativity of solution. The theory of POCS offers an elegant and comprehensive paradigm within which various ideas, often articulated in an intuitive, heuristic fashion, can be properly expressed and analyzed (Stark and Yang, 1998). In the scope of this section, we take "projection" to mean the enforcement of certain types of constraints on the solution, and not the integral in the context of the image formation model (Eq. (2.1)). We also note that there are attempts to express constraints as additional sets of linear equations to be solved simultaneously with the 3D reconstruction problem, but they are either designed specifically for structures with icosahedral symmetry (Zheng *et al.*, 2000), or still at initial stages of development (Sorzano *et al.*, 2008).

In POCS, one considers sets whose elements are images, which can be either 2D or 3D. For example, one can consider a set of nonnegative images or a set of frequency-limited images. Given $k$ convex sets of images $C_i$, $i = 1, ..., k$, methodology based on the theory of POCS methodology can be used to find an image that lies in the intersection $C_0$ of all the given sets, and thus simultaneously satisfies all constraints and hence has the desired properties. If we denote by $\mathcal{P}_i$ the projection onto convex set $C_i$, then the sequence of iterations

$$f_{l+1} = \mathcal{P}_1 \mathcal{P}_2 \ldots \mathcal{P}_k f_l \tag{2.38}$$

will converge to an image in the intersection $C_0$ for any initial image $f_0$; if $C_0$ is empty, then the procedure will not terminate. Assuming $C_0$ is nonempty, the properties of the solution will depend on the properties of the initial image $f_0$. The exception is if $C_0$ consists of just a single image, which normally does not happen.

Examples of often used constraints are: nonnegativity (negative elements in the image are set to zero), compact support (image is multiplied by a binary mask), and predefined Fourier amplitudes (the corresponding POCS projection sets the Fourier amplitudes of an image to predefined values while imposing no restrictions on phases). It is easy to see that there are images that could fulfill all the three constraints simultaneously, and for limited resolution data, the number of such images will be very large. However, under favorable conditions, the procedure based on repeated applications of these constraints can be shown to converge in the limit to the so-called *super*-resolution. The method was developed independently by Papoulis (1975) and Gerchberg (1974). While the former was concerned with the extrapolation of a band-limited signal from a part of the original real-space signal, the latter dealt with the problem of reconstruction of the signal from its Fourier spectrum given to the diffraction limit. In X-ray crystallography, the Papillose–Gerchberg (PG) algorithm is known as "solvent flattening" (Wang, 1985). POCS provides a comprehensive framework within which the PG algorithm can be understood, in addition to being a theoretical basis for the development of additional constraints (Sezan, 1992).

In cryo-EM, the initial motivation for exploring POCS-based methods was due to the missing cone problem resulting from application of the Random Conical Tilt method for *ab initio* protein structure determination (Radermacher *et al.*, 1987). Tilt-pairs of electron micrographs are collected and the presence of predominant view on the zero-tilt micrograph leads to a simple and robust way of determining all necessary 3D orientation parameters by using the readout from the microscope goniometer and 2D in-plane alignment of 0°, thus presumably identical, 2D projection views. Regrettably, the method leaves undetermined a conical-shaped region of Fourier space in the 3D reconstructed volume, and the angle of the cone is typically 40°. As a result, the 3D structure is severely distorted and elongated in the direction of the cone. Some effort has been put into POCS-based recovery of the missing region (Carazo and Carrascosa, 1987), and although some reduction of the elongation has been reported, the results were at best modest (Akey and Radermacher, 1993; Radermacher *et al.*, 1994). A related problem in electron tomography is the missing wedge of the single-axis tilt series of EM projection data (20°–30° opening), which is caused by the limited tilting range of the microscope; there have been some attempts to rectify the problem using POCS approaches (Carazo, 1992).

POCS is an attractive approach to the problem of restoration of cryo-EM macromolecular structure; however, the practical application of POCS-based methods is problematic for a number of reasons. Neither the theory of POCS nor the POCS-based methods developed so far account for errors in the 3D results. And while it is easy to demonstrate significant improvements using simulated, error-free data, even modest amounts of noise make the gains doubtful. One possible approach to the aforementioned issues is to consider POCS within a broader context of the structure determination procedure, including refinement of orientation parameters. So far, however, no significant effort in this direction has been reported (for a possible solution, see (Yang *et al.*, 2005)).

# 5. Wiener Filter—the Minimum Mean-Squared Error Criterion

In images restored using pseudoinverse methods, noise can be excessively amplified for some spatial frequencies. This not only produces unpleasant artifacts in real space, but also creates serious problems in iterative alignment procedures, which tend to emphasize artifacts in reference images. A method that prevents excessive enhancement of noise by taking advantage of the estimate of the SSNR distribution in the data is based on minimization of mean-square error (MMSE) between the estimate and the original signal:

$$\min_{\hat{F}} \mathcal{L}_{\mathrm{MMSE}}(\hat{F}) = \min_{\hat{F}} \mathcal{E}\left\{\left\|F(\mathbf{s}) - \hat{F}(\mathbf{s})\right\|^2\right\}. \tag{2.39}$$

## 5.1. Wiener filter in restoration of an image using EM data with varied defocus settings

As previously noted, we seek the solution in a form of linear combination of input data:

$$\hat{F}(s) = \sum_n w_n(s) G_n(s) \tag{2.40}$$

and we minimize Eq. (2.39) with respect to filter coefficients $w_n(\mathbf{s})$. To find the minimum, we differentiate Eq. (2.39) with respect to $w_k$, set derivatives to zero, and obtain a set of $N$ equations:

$$\frac{\partial \mathcal{L}_{MMSE}}{\partial w_k} = \frac{\partial}{\partial w_k} \mathcal{E}\left\{\left\|F - \sum_n w_n G_n\right\|^2\right\} = \mathcal{E}\left\{\left(F - \sum_n w_n G_n\right) G_k^*\right\} = 0, \tag{2.41}$$

where for simplicity we omitted the dependence on spatial frequency. In Eq. (2.41), we substitute $G_n$ by its form given by the image formation model (Eq. (2.11)). The expectation value yields the following set of equations:

$$\mathrm{CTF}_k E_k P_{f_k}\left(1-\left(\sum_n w_n \mathrm{CTF}_n E_n\right)\right) - w_k P_{m_k} = 0, \tag{2.42}$$

which can be solved by noting that the term $\sum_n w_n \mathrm{CTF}_n E_n$ is shared:

$$w_k = \frac{\mathrm{CTF}_k E_k P_f}{P_{m_k}} \Bigg/ \left(\sum_n \frac{|\mathrm{CTF}_n E_n|^2 P_f}{P_{m_k}} + 1\right). \tag{2.43}$$

Weights $w_k$ are coefficients of the Wiener filter, which means that the result is dependent on the ratio of the PW of the signal to the PW of noise:

$$\hat{F}(s) = \frac{\sum_{n=1}^{N} \mathrm{CTF}_n(s) E_n(s) \frac{P_f(s)}{P_{m_n}(s)} G_n(s)}{\sum_{n=1}^{N} \mathrm{CTF}_n^2(s) E_n^2(s) \frac{P_f(s)}{P_{m_n}(s)} + 1}. \tag{2.44}$$

It is customary to express the Wiener filter as a function of SSNR, so, using definition Eq. (2.12), we obtain:

$$\hat{F}(\mathbf{s}) = \frac{\sum_{n=1}^{N} \mathrm{CTF}_n(\mathbf{s}) E_n(\mathbf{s}) \mathrm{SSNR}_n(\mathbf{s}) G_n(\mathbf{s})}{\sum_{n=1}^{N} \mathrm{CTF}_n^2(\mathbf{s}) E_n^2(\mathbf{s}) \mathrm{SSNR}_n(\mathbf{s}) + 1}. \tag{2.45}$$

The difference between the Wiener filter equation (2.45) and the MCE pseudoinverse restoration equation (2.31) appears to be minimal since they differ merely by the addition of one in the denominator. However, this constant has a significant impact on the outcome of the restoration as it lends the Wiener filter the following desired properties: (a) it sets to zero those frequency regions in the input data for which CTF or the envelope function or the SSNR are zero, and (b) it mitigates the enhancement of amplitudes in regions where combined signal $\sum_{n=1}^{N} \mathrm{CTF}_n^2(s) E_n^2(s) \mathrm{SSNR}_n(s)$ is low and depends on the combined value of SSNR in this region.

The mathematical forms of the Wiener filter equation (2.45) and the constrained mean-square restoration CMCE equation (2.34) are all but identical. Nevertheless, the motivations and meanings behind them are entirely different. The Wiener filter will yield a solution similar to the original signal, as permitted by the SSNR level in the data, while CMCE will yield a "smooth" solution that is constrained by its chi-square fit to the data. The main difference is that the Wiener filter does not contain any free parameters, since in principle all its terms should be derived either from the experimental settings or from analysis of the data, whereas in CMCE, both the shape of the high-pass filter $U(\mathbf{s})$ and constant $\gamma$ are entirely arbitrary and depend on the decision of the user. In the context of SPR, it remains to be seen whether it is possible in practice to determine characteristics of the image formation process with an accuracy sufficient to justify usage of the Wiener filter instead of the heuristic CMCE.

The properties of the Wiener filter are best exemplified by the two limiting cases of input images with very high SSNR and those with very low SSNR. In the former case, we obtain:

$$\hat{F}_{\mathrm{SSNR}\to\infty}(s) = \frac{\sum_{n=1}^{N} \mathrm{CTF}_n(s) E_n(s) \mathrm{SSNR}_n(s) G_n(s)}{\sum_{n=1}^{N} \mathrm{CTF}_n^2(s) E_n^2(s) \mathrm{SSNR}_n(s)}. \tag{2.46}$$

Equation (2.45) simply coincides with the MCE pseudoinverse restoration equation (2.31). This reinforces the intuitive notion that unlimited enhancement of amplitudes in regions where products of CTFs and envelope functions are small makes sense only when the power of the noise is negligible relative to the power of the signal, in which case the enhancement of the former will not deteriorate the overall appearance of the image.

For input data that has very low SSNR, that is, the signal is dominated by noise, the Wiener filter equation (2.45) is reduced to:

$$\hat{F}_{\mathrm{SSNR}\to 0}(s) = \sum_{n=1}^{N} \mathrm{CTF}_n(s) E_n(s) \mathrm{SSNR}_n(s) G_n(s). \tag{2.47}$$

Equation (2.47) simply means that under low SSNR conditions there should be no enhancement of amplitudes; to the contrary, all images should be multiplied by their CTFs and envelope functions, which under standard EM imaging conditions corresponds to low-pass filtration of the data. Equation (2.47) has no counterpart in MSE or MCE frameworks and is unique to the MMSE statement of the restoration problem.

In general, the Wiener filter image restoration given by Eq. (2.45) has the following properties (Fig. 2.4C):

1. For a large number of images collected at approximately uniformly distributed defocus settings, the impact of the filter equation (2.45) can be considered separately for two spatial frequency regions consisting of: (a) frequencies lower than the frequency of the first extremum of $\sum_{n=1}^{N}\mathrm{CTF}_n^2(\mathbf{s})$ and (b) frequencies higher than the first extremum. In the first region, all CTFs have approximately the same shape irrespective of the defocus settings, so the enhancement for $s \to 0$ is given by $\left(A^2\sum_{n=1}^{N}\mathrm{SSNR}_n(\mathbf{s})+1\right)/\left(A\sum_{n=1}^{N}\mathrm{SSNR}_n(\mathbf{s})\right)$, which for large SSNRs corresponds to the MCE solution. In the second region consisting of higher frequencies, the denominator will be approximately constant, so the restoration corresponds to the summation of EM images multiplied by their respective CTFs. However, unlike the case in MCE, the variations in the denominator depend on the SSNR values and decrease with decreasing values of the SSNR.
2. For a small number of images or for data sets for which $\sum_{n=1}^{N}\mathrm{CTF}_n^2(\mathbf{s}) \simeq 0$ for some spatial frequencies, the value of the Wiener filter approaches zero, which is markedly different from the MCE solution. This advantage of the Wiener filter makes it the method of choice in direct Fourier inversion 3D reconstruction from projection algorithms in EM (Zhang *et al.*, 2008).
3. The solution depends both on the scaling and frequency dependence of the SSNR. Thus, in protocols where the PW of the signal is obtained from independent experiments or computed from X-ray crystallographic models, while the PW of the background noise is estimated from the data, the additional free variable in the image restoration equation (2.45) is the relative scaling between the two. It is easy to see that the results depend heavily on this parameter and an incorrect scaling cannot be easily accounted for in subsequent data processing steps.

Proper applications of the image restoration methods outlined above vary depending on the dimensionality of the data (2D versus 3D) and the goal (alignment, 3D reconstruction, PW adjustment or "sharpening" of EM map). In 2D applications, the incorrect enhancement of amplitudes in the low-frequencies region can adversely influence the alignment results. Furthermore, the problem is compounded by the need to process heterogeneous data sets, in which case, strictly speaking, neither of the methods reviewed in this section is applicable without modifications. On the other hand, it is the case that in 2D applications each pixel (spatial frequency $\mathbf{s}$) of the restored image has the same number of contributing data images. This, combined with the fact that it would be rare to have a selection of defocus settings such that $\sum_{n=1}^{N}\mathrm{CTF}_n^2(\mathbf{s}) = 0$, implies that restoration in higher frequency regions is less critical, and the practical differences between MCE, the iterative methods, or the Wiener filter become of secondary consideration.

Wiener filter (MMSE) restoration in cryo-EM has many appealing properties, but it also has significant disadvantages. While the solution is better behaved than what one would obtain from MCE pseudoinverse restoration and is less likely to be affected by artifacts, it also depends, to a much higher degree, on the proper estimation of all the image formation characteristics. In particular, the proper scaling and general shape of envelope functions and SSNR distributions are of paramount importance. More importantly, the MMSE criterion cannot be easily used in alignment procedures. Specifically, if we introduce a dependence on the orientation parameters of images $G_n$, then the MSE (Eq. (2.20)) and MCE (Eq. (2.23)) criteria yield to a well-known method of finding the optimum values of these parameters using a cross-correlation function (CCF) (obtained from an inverse Fourier transform of a product in reciprocal space). A similar line of reasoning implies that Eq. (2.39) in combination with Eq. (2.34) would yield a solution for finding the optimum orientations of images $G_n$, under the assumption that an ideal template is given. Regrettably, this reasoning cannot be applied to the MMSE criterion, and has no applicability to EM in general, since such a template is not available. In addition, it can be easily shown that the Wiener filter average (i.e., the restored image) Eq. (2.45) does not minimize the variance of the data set with respect to the alignment parameters.

## 5.2. Wiener filter in image filtration and amplitude correction ("sharpening")

The Wiener image restoration technique described in the previous section is based on the assumption that a reasonable approximation of the SSNR in the data can be established. However, as exemplified by Eq. (2.14), the most comprehensive SSNR characterization of EM results would include accounting for the deterioration due to alignment errors, expressed as an additional envelope function, or additional variance, or both (Baldwin and Penczek, 2005). It would be appropriate to include this information in the restoration of the average by either MCE or Wiener methodology; regrettably, it is more typically the case that information about SSNR of the average becomes available only *after* the average has been calculated. This SSNR of the result is often referred to as a *resolution* of the average reconstruction and is typically estimated using the Fourier shell correlation (FSC) methodology (Chapter 3). Thus, the problem is circular: to properly restore the average, one would have to know its SSNR, but the SSNR can be obtained only after the average has been computed. The issue has never been satisfactorily resolved in the EM data processing methodology, and it is customary to apply a postfiltering of the average, thus breaking the calculation of average into two steps (Penczek, 2008).

The image formation model of a CTF-corrected cryo-EM 2D average or a 3D reconstruction $\nu$ of a macromolecular complex $f$ is given, in Fourier space, as:

$$V(\mathbf{s}) = E(\mathbf{s})E^{\mathrm{Ali}}(\mathbf{s})F(\mathbf{s}) + M(\mathbf{s}), \tag{2.48}$$

where $E$ is the average of the effective envelope functions $E_n$ that appeared in Eq. (2.12), $E^{\mathrm{Ali}}$ is the envelope function due to alignment errors and errors in the image restoration step that resulted in average $V$, and $M$ is the background noise of Eq. (2.12) (albeit reduced due to the averaging/reconstruction step). As before, we assume that the noise is zero mean, uncorrelated between spatial frequencies and uncorrelated with the signal. The Wiener filter derivation for model equation (2.48) follows steps Eqs. (2.40)–(2.43) and yields:

$$\begin{aligned}\hat{F}(s) &= \frac{E(s)E^{\mathrm{Ali}}(s)\frac{P_f(s)}{P_m(s)}}{E^2(s)E^{\mathrm{Ali}^2}(s)\frac{P_f(s)}{P_m(s)}+1}V(s)\\ &= \frac{E(s)E^{\mathrm{Ali}}(s)\mathrm{SSNR}_f(s)}{E^2(s)E^{\mathrm{Ali}^2}(s)\mathrm{SSNR}_f(s)+1}V(s)\\ &= \frac{1}{E(s)E^{\mathrm{Ali}}(s)}\frac{\mathrm{SSNR}_\nu(s)}{\mathrm{SSNR}_\nu(s)+1}V(s).\end{aligned} \tag{2.49}$$

Most properties of the Wiener filter equation (2.45) are shared by the filter given by equation (2.49).

The straightforward application of Eq. (2.49) is in a low-pass filtration of the average/reconstruction emerging from the cryo-EM refinement procedure. In this case, one would neglect correcting for the two envelope functions and take advantage of the simple relationship between the SSNR of the average and the FSC to obtain the appropriate filter (Chapter 3). The full form of Eq. (2.49) provides a solution to the problem of restoration of the rotationally averaged profile of the PW of the average/reconstruction $\nu$; this procedure is sometimes referred to as *sharpening* the result. To derive it, we first note that even though $\mathrm{SSNR}_\nu$ is known, it is all but impossible to estimate the product of two envelope functions. A sensible approach, described in an earlier section, is to independently estimate both the rotational average of the PW of the ideal structure $P_f(s)$ and that of the background noise in the average/reconstruction $P_m(s)$.

Given this and the image formation model in Eq. (2.48), we obtain:

$$E^2(s)E^{\mathrm{Ali}^2}(s)\frac{P_f(s)}{P_m(s)} = \frac{P_v(s)}{P_m(s)} - 1, \tag{2.50}$$

Using Eq. (2.50) we can write the Wiener filter equation (2.49) as:

$$\hat{F}(s) = E(s)E^{\mathrm{Ali}}(s)\frac{P_f(s)}{P_v(s)}V(s). \tag{2.51}$$

The product of the envelope functions is obtained independently using FSC methodology and is simply:

$$E(s)E^{\mathrm{Ali}}(s) = \sqrt{\mathrm{SSNR}_v(s)\frac{P_m(\mathbf{s})}{P_f(s)}}. \tag{2.52}$$

The expression for filtration combined with PW adjustment ("sharpening") of the EM average/reconstruction is:

$$\hat{F}(s) = \begin{cases} \sqrt{\dfrac{\mathrm{SSNR}_v(s)P_m(s)}{P_v(s)}}\sqrt{\dfrac{P_f(s)}{P_v(s)}}V(s) & \text{if } \dfrac{\mathrm{SSNR}_v(s)P_m(s)}{P_v(s)} < 1 \\ \sqrt{\dfrac{P_f(s)}{P_v(s)}}V(s) & \text{otherwise} \end{cases} \tag{2.53}$$

Equation (2.53) has many appealing properties: (1) it contains only entities that can be easily computed from the data, namely, $\mathrm{SSNR}_v$, $\mathrm{FSC}_v$, and PW of the background noise in the average/reconstruction (which can be computed from the area surrounding the object) and (2) the result does not depend on normalizations of PWs of the ideal object and background noise, both of which are difficult to establish. Furthermore, in regions where SSNR is high, $P_m$ should be close to zero, and the additional condition in Eq. (2.53) prevents excessive amplification of amplitudes in $V$.

## 6. Image Restoration for Sets of Heterogeneous Images

The image restoration techniques described in the preceding sections are based on the assumption that the signals of the EM projection images of a macromolecular complex in Eq. (2.1) are identical. This, however, is rarely the case and it might be more appropriate to consider various degrees to which this assumption holds in practice. An EM data set will always contain a certain percentage of nonparticles, that is, objects that passed both automated particle selection and visual scrutiny, but do not represent valid views of the imaged complex ,and are instead artifacts that have strong low-frequency components and that are often found in micrographs. Furthermore, many macromolecular complexes naturally exist as a mixture of various conformations, so the sample cannot be considered homogeneous. Very often, processed sample contains a mixture of complexes in various functional states, for example, some may have ligands bound.

It is a straightforward to conclude that in the case of heterogeneous samples, the images restoration techniques described above are, strictly speaking, not applicable as they are based on the assumption that the signal is identical and images are in the same orientations. Indeed, if a Wiener filter (Eq. (2.45)) were applied to a heterogeneous data set, we would obtain:

$$\hat{F}(s \to 0) = -\frac{A\sum_{n=1}^{N} \mathrm{SSNR}_n^{\mathrm{Data}} G_n}{A^2 \sum_{n=1}^{N} \mathrm{SSNR}_n^{\mathrm{Data}} + 1} = \frac{A\mathrm{SSNR}^{\mathrm{Data}} \sum_{n=1}^{N} G_n}{A^2 \mathrm{SSNR}^{\mathrm{Data}} + {}^{1}/_{N}} \simeq \sum_{n=1}^{N} \frac{1}{A} G_n, \tag{2.54}$$

where we assumed that envelopes and SNRs are approximately the same for all particles in very low-frequency range. It follows from Eq. (2.54) that since all images $G_n$ are different (in the sense that the signal components $F$ are different), the average will be formed from individual images whose low-frequency component was enhanced $A$ times. A typical value of the amplitude contrast ratio ($A$) is 0.1, so the enhancement will be $\sim$10 times, the result of which are very blurry images.

## 6.1. Phase flipping

There are a number of possible heuristic solutions to the problem outlined above. One such solution, adopted in numerous single-particle packages (Ludtke *et al.*, 1999; Tang *et al.*, 2007; van Heel *et al.*, 1996), is to disregard the information about CTF-induced amplitude changes in EM images and correct the data only for the sign of the CTF, and thus obtain the correct signs of phases in Fourier space. Consequently, this heuristic is known as "*phase flipping*":

$$\widetilde{G}_n(\mathbf{s}) = \mathrm{sign}(\mathrm{CTF}_n(\mathbf{s}))G_n(\mathbf{s}). \tag{2.55}$$

The appeals of phase flipping are: (1) signs of phases in images collected under different defocus settings become consistent, so it is possible to apply efficient alignment procedures such as those based on image invariants (Marabini and Carazo, 1996; Schatz and van Heel, 1990, 1992); (2) the problem of excessive low frequencies in the average is circumvented; and (3) the CTF-correction process can be decomposed into a number of steps, thus allowing more elaborate CTF correction to be applied in later stages of data processing. Advantages of using phase-flipped data, even if they are only used in the early stages of data processing, do not outweigh the disadvantages. A most significant one is the omission of CTF amplitude weighting, so the average computed from phase-flipped images will have excessive amount of noise, and thus suboptimal SNR.

## 6.2. The reciprocal space adaptive Wiener filter

We begin by proposing that the desired amplitude-weighted Wiener filter based restoration should be such that in the case of a heterogeneous data set the overall effect of the restoration should result in a PW of the restored image that corresponds to the average of the PWs of the individual images. We note that the aforementioned statement neither implies nor assumes that the PWs of averaged images cannot be changed, as in the phase-flipping approach. To construct such an amplitude-preserving restoration, we impose the additional constraint on the Wiener filter (Eq. (2.45)) requiring that the effective amplitude modification for each spatial frequency should be one:

$$\forall \mathbf{s} : \frac{\sum_{n=1}^{N} q(\mathbf{s})|\mathrm{CTF}_n(\mathbf{s})|\mathrm{SSNR}(\mathbf{s})}{\sum_{n=1}^{N} q(\mathbf{s})\mathrm{CTF}_n^2(\mathbf{s})\mathrm{SSNR}(\mathbf{s}) + 1} = 1, \tag{2.56}$$

where for simplicity we omitted envelope functions and the per image dependence of SSNR. From Eq. (2.56) it follows that the additional normalization factor should be:

$$q(\mathbf{s}) = \frac{1}{\mathrm{SSNR}(\mathbf{s})\left(\sum_{n=1}^{N} |\mathrm{CTF}_n(\mathbf{s})| - \sum_{n=1}^{N} \mathrm{CTF}_n^2(\mathbf{s})\right)}. \tag{2.57}$$

With this the Wiener filter (Eq. (2.45)) is reduced to:

$$\hat{F}(\mathbf{s}) = \frac{\sum_{n=1}^{N} \mathrm{CTF}_n(\mathbf{s}) G_n(\mathbf{s})}{\sum_{n=1}^{N} |\mathrm{CTF}_n(\mathbf{s})|}, \tag{2.58}$$

in which case the restoration becomes a simple CTF-weighted averaging.

To obtain a continuous transition between restoration methods geared for strictly homogeneous (Eq. (2.31)) data sets and strictly heterogeneous (Eq. (2.58)) data sets, we propose the following form of the reciprocal space adaptive Wiener filter:

$$\hat{F}(\mathbf{s}) = \alpha(\mathbf{s}) \cdot \frac{\sum_{n=1}^{N} \mathrm{CTF}_n(\mathbf{s}) \mathrm{SSNR}_n(\mathbf{s}) G_n(\mathbf{s})}{\sum_{n=1}^{N} \mathrm{CTF}_n^2(\mathbf{s}) \mathrm{SSNR}_n(\mathbf{s})}, \tag{2.59}$$

where $\alpha(\mathbf{s})$ depends on the parameter $N_g$, a user specified parameter denoting the number of homogeneous subsets in the data set:

$$\alpha(\mathbf{s}) = \frac{N - N_g}{N - 1} + \frac{N_g - 1}{N - 1} \cdot \frac{\sum_{n=1}^{N} \mathrm{CTF}_n^2(\mathbf{s})}{\sum_{n=1}^{N} |\mathrm{CTF}_n(\mathbf{s})|}. \tag{2.60}$$

The reciprocal space adaptive Wiener filter offers a significant advantage over "phase flipping" as it incorporates proper amplitude weighting of images, thereby maximizing SSNR per each spatial frequency $\mathbf{s}$. It also provides a mechanism for a continuous and intuitive, albeit heuristic, transition between restoration of purely homogeneous and purely heterogeneous EM data sets.

## 7. Discussion

All restoration methods discussed in this review require knowledge of the parameters of the image formation model. These parameters can be derived from analysis of the data, some with a high degree of accuracy (e.g., defocus), while for others (e.g., envelopes, SSNR) there is no consensus on whether their accuracy is sufficient to benefit alignment procedures. Curiously, although elaborate CTF estimation methods have been implemented and used, there are no published studies that would demonstrate that the resolution of the final structure improves if estimates of the SSNR of the data, envelope functions, and background noise are included. Hence, in the absence of conclusive evidence, there are at least three possible approaches to the treatment of the relationship between the image formation model and image restoration: (1) ignore the subtleties, set SSNR of the data to a constant, and adjust the parameters of alignment procedures based on other considerations; (2) model all curves by analytical functions and fit them to the data in the hope that this will reduce uncertainties; and (3) use characteristics such as background noise distribution or SSNR as estimated directly from the data. The latter approach has a significant advantage of being applicable to any data without adherence to any particular model, but at the same time any statistical fluctuations in the characteristics drawn directly from the data are passed on to the alignment procedures unchecked.

All the restoration methods presented here yield solutions that appear very similar, for example, the mathematical forms of the CMCE and Wiener filter are nearly identical. However, the motivations and meanings of the parameters of the respective solutions can be quite different, and the user should be aware which parameters are arbitrary and can be set based on heuristics and alignment results, and which should be derived experimentally from the data. It is also apparent that 2D image restoration is less sensitive to the choice of restoration method since the main difficulty is in the inhomogeneity of the data. Hence we propose a practical solution in terms of a reciprocal space adaptive Wiener filter whose main parameter is the number of homogeneous classes that one assumes are present in the data set. In 3D image restoration, particularly in 3D reconstruction algorithms that incorporate CTF correction, it is necessary to use a form of regularization to avoid artifacts in the resulting 3D structure, in which case, CMCE, iterative methods, and Wiener filters are the only possibilities, while MCE is not applicable. Again, there is relatively little work published on the 3D reconstruction/CTF-correction algorithms and there is no consensus on how to best accommodate large SSNR variations in reciprocal space.

Major single-particle software packages differ significantly in the way that image restoration issues are handled, and published descriptions are minimal for a number of the packages. Hence, a comprehensive comparison among the different packages is a difficult task at best. At the risk of not representing the methods actually used adequately, one can characterize the methodology implemented in the respective packages as follows:

1. In SPIDER (Frank *et al.*, 1996), SSNR is treated as constant and can be set by the user. The 3D reconstruction is done by grouping 2D projection data according to defocus settings, and 3D reconstruction is performed for each group independently. The resulting volumes are

merged using a Wiener filter with constant SSNR. However, since SPIDER is a general purpose image processing package, advanced users tend to implement their own strategies, including phase flipping and others.
2. In IMAGIC (van Heel *et al.*, 1996), data is aligned and clustered in 2D, initially using phased-flipped data; subsequently, only amplitudes are adjusted in the 3D reconstruction step (van Heel *et al.*, 2000).
3. In EMAN1 (Ludtke *et al.*, 1999), the sequence of data processing steps follows that in IMAGIC, for example, the CTF correction is done for 2D projection images, but the major difference is in the incorporation of image formation characteristics using analytical functions fitted to the data.
4. EMAN2 (Tang *et al.*, 2007) inherits the general strategy of EMAN1 but employs a modernized approach to image restoration based on characteristics derived from the data, thus omitting the fitting step and making the process more robust and widely applicable.
5. In SPARX (Hohn *et al.*, 2007), the 3D reconstruction algorithm incorporates CTF correction and both 2D and 3D rely on variations of the reciprocal space adaptive Wiener filters with some image formation characteristics derived from the data. SPARX follows the open source philosophy of SPIDER in allowing users to implement customized strategies by taking advantage of its extensive library of general image processing commands.

Since image restoration methods in single-particle reconstruction software packages are incorporated into the structure determination protocols, it is very difficult to perform comparative testing. Such tests would require repetition of the entire structure determination process, and the results would be influenced by other differences among the packages. Although a consensus is unlikely to be reached soon, there is much that could be done toward carrying out a comprehensive treatment of image formation characteristics. A number of the solutions adopted are heuristic and it should be possible to reach an understanding as to why heuristics are necessary since the physics of image formation in EM is well understood. Similarly, even if the trend is to rely on image formation characteristics directly derived from the data, as implemented in EMAN2, it is possible that more robust and less error-prone approaches can be developed by a reexamination of the theory and the introduction of appropriate constraints to the fitting process.

## ACKNOWLEDGMENTS

I thank Grant Jensen, Justus Loerke, and Jia Fang for critical reading of the manuscript and for helpful suggestions. This work was supported by grant from the NIH R01 GM 60635 (to PAP).

## REFERENCES


Akey, C. W., and Radermacher, M. (1993). Architecture of the *Xenopus* nuclear pore complex revealed by three-dimensional cryo-electron microscopy. *J. Cell Biol.* **122,** 1–19.

Angert, I., Majorovits, E., and Schröder, R. (2000). Zero-loss image formation and modified contrast transfer theory in EFTEM. *Ultramicroscopy* **81,** 203–222.

Baldwin, P. R., and Penczek, P. A. (2005). Estimating alignment errors in sets of 2-D images. *J. Struct. Biol.* **150,** 211–225.

Baldwin, P. R., and Penczek, P. A. (2007). The transform class in SPARX and EMAN2. *J. Struct. Biol.* **157,** 250–261.

Böttcher, B., and Crowther, R. A. (1996). Difference imaging reveals ordered regions of RNA in turnip yellow mosaic virus. *Structure* **4,** 387–394.

Carazo, J. M. (1992). The fidelity of 3D reconstruction from incomplete data and the use of restoration methods. *In* "Electron Tomography," (J. Frank, ed.), pp. 117–166. Plenum, New York.

Carazo, J. M., and Carrascosa, J. L. (1987). Information recovery in missing angular data cases: An approach by the convex projections method in three dimensions. *J. Microsc.* **145,** 23–43.

Frank, J., Radermacher, M., Penczek, P., Zhu, J., Li, Y., Ladjadj, M., and Leith, A. (1996). SPIDER and WEB: Processing and visualization of images in 3D electron microscopy and related fields. *J. Struct. Biol.* **116,** 190–199.

Gabashvili, I. S., Agrawal, R. K., Spahn, C. M., Grassucci, R. A., Svergun, D. I., Frank, J., and Penczek, P. (2000). Solution structure of the *E. coli* 70S ribosome at 11.5 Å resolution. *Cell* **100,** 537–549.

Gerchberg, W. O. (1974). Super resolution through energy reduction. *Opt. Acta* **21,** 709–720.

Gonzalez, R. F., and Woods, R. E. (2002). Digital Image Processing. Prentice Hall, Upper Saddle River, NJ.

Grigorieff, N. (1998). Three-dimensional structure of bovine NADH: Ubiquinone oxidoreductase (complex I) at 22 Å in ice. *J. Mol. Biol.* **277,** 1033–1046.

Guinier, A., and Fournet, G. (1955). *Small-angle Scattering of X-rays* Wiley, New York.

Hohn, M., Tang, G., Goodyear, G., Baldwin, P. R., Huang, Z., Penczek, P. A., Yang, C., Glaeser, R. M., Adams, P. D., and Ludtke, S. J. (2007). SPARX, a new environment for cryo-EM image processing. *J. Struct. Biol.* **157,** 47–55.

Huang, Z., and Penczek, P. A. (2004). Application of template matching technique to particle detection in electron micrographs. *J. Struct. Biol.* **145,** 29–40.

Huang, Z., Baldwin, P. R., Mullapudi, S. R., and Penczek, P. A. (2003). Automated determination of parameters describing power spectra of micrograph images in electron microscopy. *J. Struct. Biol.* **144,** 79–94.

Jain, A. K. (1989). Fundamentals of Digital Image Processing. Prentince Hall, Englewood Cliffs.

Jensen, G. J. (2001). Alignment error envelopes for single particle analysis. *J. Struct. Biol.* **133,** 143–155.

Langmore, J. P., and Smith, M. F. (1992). Quantitative energy-filtered electron microscopy of biological molecules in ice. *Ultramicroscopy* **46,** 349–373.

Ludtke, S. J., Jakana, J., Song, J. L., Chuang, D. T., and Chiu, W. (2001). A 11.5 Å single particle reconstruction of GroEL using EMAN. *J. Mol. Biol.* **314,** 253–262.

Ludtke, S. J., and Chiu, W. (2002). Image restoration in sets of noisy electron micrographs. *In* Proceedings of the IEEE Symposium on Biomedical Imaging, Washington, DC pp. 745–748.

Ludtke, S. J., Baldwin, P. R., and Chiu, W. (1999). EMAN: Semiautomated software for high-resolution single-particle reconstructions. *J. Struct. Biol.* **128,** 82–97.
Ludtke, S. J., Jakana, J., Song, J. L., Chuang, D. T., and Chiu, W. (2001). A 11.5 Å single particle reconstruction of GroEL using EMAN. *J. Mol. Biol.* **314,** 253–262.
Mallick, S. P., Carragher, B., Potter, C. S., and Kriegman, D. J. (2005). ACE: Automated CTF estimation. *Ultramicroscopy* **104,** 8–29.
Marabini, R., and Carazo, J. M. (1996). On a new computationally fast image invariant based on bispectral projections. *Pattern Recognit. Lett.* **17,** 959–967.
Mindell, J. A., and Grigorieff, N. (2003). Accurate determination of local defocus and specimen tilt in electron microscopy. *J. Struct. Biol.* **142,** 334–347.
Papoulis, A. (1975). A new algorithm in spectral analysis and band-limited extrapolation. *IEEE Trans. Circuits Syst.* **CAS-22.**
Penczek, P. A. (2002). Three-dimensional spectral signal-to-noise ratio for a class of reconstruction algorithms. *J. Struct. Biol.* **138,** 34–46.
Penczek, P. A. (2008). Single particle reconstruction. *In* "International Tables for Crystallography," (U. Shmueli, ed.)**B,** pp. 375–388. Springer, New York.
Penczek, P., Radermacher, M., and Frank, J. (1992). Three-dimensional reconstruction of single particles embedded in ice. *Ultramicroscopy* **40,** 33–53.
Penczek, P. A., Zhu, J., Schröder, R., and Frank, J. (1997). Three-dimensional reconstruction with contrast transfer function compensation from defocus series. *Scanning Microsc. Suppl.* **11,** 1–10.
Pratt, W. K. (1992). Digital Image Processing. Wiley, New York.
Radermacher, M., Wagenknecht, T., Verschoor, A., and Frank, J. (1987). Three-dimensional reconstruction from a single-exposure, random conical tilt series applied to the 50S ribosomal subunit of *Escherichia coli*. *J. Microsc.* **146,** 113–136.
Radermacher, M., Rao, V., Grassucci, R., Frank, J., Timerman, A. P., Fleischer, S., and Wagenknecht, T. (1994). Cryo-electron microscopy and three-dimensional reconstruction of the calcium release channel/ryanodine receptor from skeletal muscle. *J. Cell Biol.* **127,** 411–423.
Richardson, L. F. (1910). The approximate arithmetical solution by finite diff erences of physical problems involving diff erential equations with an application to the stresses to a masonry dam. *Philos. Trans. R. Soc. Lond. A.* **210,** 307–357.
Saad, A., Ludtke, S. J., Jakana, J., Rixon, F. J., Tsuruta, H., and Chiu, W. (2001). Fourier amplitude decay of electron cryomicroscopic images of single particles and effects on structure determination. *J. Struct. Biol.* **133,** 32–42.
Sander, B., Golas, M. M., and Stark, H. (2003). Automatic CTF correction for single particles based upon multivariate statistical analysis of individual power spectra. *J. Struct. Biol.* **142,** 392–401.
Schatz, M., and van Heel, M. (1990). Invariant classification of molecular views in electron micrographs. *Ultramicroscopy* **32,** 255–264.
Schatz, M., and van Heel, M. (1992). Invariant recognition of molecular projections in vitreous ice preparations. *Ultramicroscopy* **45,** 15–22.
Scherzer, O. (1949). The theoretical resolution limit of the electron microscope. *J. Appl. Phys.* **20,** 20–29.
Sezan, M. I. (1992). An overview of convex projections theory and its application to image recovery problems. *Ultramicroscopy* **40,** 55–67.
Sorzano, C. O., Jonic, S., Nunez-Ramirez, R., Boisset, N., and Carazo, J. M. (2007). Fast, robust, and accurate determination of transmission electron microscopy contrast transfer function. *J. Struct. Biol.* **160,** 249–262.
Sorzano, C. O. S., Velazquez-Muriel, J. A., Marabini, R., Herman, G. T., and Carazo, J. M. (2008). Volumetric restrictions in single particle 3DEM reconstruction. *Pattern Recognit.* **41,** 616–626.

Stark, H., and Yang, Y. (1998). Vector Space Projections: A Numerical Approach to Signal and Image Processing, Neural Nets, and Optics. Wiley, New York.

Tang, G., Peng, L., Baldwin, P. R., Mann, D. S., Jiang, W., Rees, I., and Ludtke, S. J. (2007). EMAN2: An extensible image processing suite for electron microscopy. *J. Struct. Biol.* **157,** 38–46.

van Heel, M., Harauz, G., and Orlova, E. V. (1996). A new generation of the IMAGIC image processing system. *J. Struct. Biol.* **116,** 17–24.

van Heel, M., Gowen, B., Matadeen, R., Orlova, E. V., Finn, R., Pape, T., Cohen, D., Stark, H., Schmidt, R., Schatz, M., and Patwardhan, A. (2000). Single-particle electron cryo-microscopy: Towards atomic resolution. *Quart. Rev. Biophys.* **33,** 307–369.

Velazquez-Muriel, J. A., Sorzano, C. O., Fernandez, J. J., and Carazo, J. M. (2003). A method for estimating the CTF in electron microscopy based on ARMA models and parameter adjustment. *Ultramicroscopy* **96,** 17–35.

Volkmann, N., and Hanein, D. (2003). Docking of atomic models into reconstructions from electron microscopy. *Methods Enzymol.* **374,** 204–225.

Wade, R. H. (1992). A brief look at imaging and contrast transfer. *Ultramicroscopy* **46,** 145–156.

Wang, B. C. (1985). Resolution of phase ambiguity in macromolecular crystallography. *In* “Methods in Enzymology: Diffraction Methods in Biology, Part B.” (C. Wyckoff, H. W. Hirs, and S. N. Timasheff, eds.), vol. 115, Academic Press, Orlando, FL..

Yang, C., Ng, E. G., and Penczek, P. A. (2005). Unified 3-D structure and projection orientation refinement using quasi-Newton algorithm. *J. Struct. Biol.* **149,** 53–64.

Yang, C., Jiang, W., Chen, D. H., Adiga, U., Ng, E. G., and Chiu, W. (2009). Estimating contrast transfer function and associated parameters by constrained non-linear optimization. *J. Microsc.* **233,** 391–403.

Zhang, W., Kimmel, M., Spahn, C. M., and Penczek, P. A. (2008). Heterogeneity of large macromolecular complexes revealed by 3D cryo-EM variance analysis. *Structure* **16,** 1770–1776.

Zheng, Y. B., Doerschuk, P. C., and Johnson, J. E. (2000). Symmetry-constrained 3-D interpolation of viral X-ray crystallography data. *IEEE Trans. Signal Process.* **48,** 214–222.

Zhou, Z. H., Hardt, S., Wang, B., Sherman, M. B., Jakana, J., and Chiu, W. (1996). CTF determination of images of ice-embedded single particles using a graphics interface. *J. Struct. Biol.* **116,** 216–222.

Zhu, J., Penczek, P. A., Schröder, R., and Frank, J. (1997). Three-dimensional reconstruction with contrast transfer function correction from energy-filtered cryoelectron micrographs: procedure and application to the 70S *Escherichia coli* ribosome. *J. Struct. Biol.* **118,** 197–219.

CHAPTER THREE

# Resolution Measures in Molecular Electron Microscopy

Pawel A. Penczek

## Contents



## Abstract

Resolution measures in molecular electron microscopy provide means to evaluate quality of macromolecular structures computed from sets of their two-dimensional (2D) line projections. When the amount of detail in the computed density map is low there are no external standards by which the resolution of the result can be judged. Instead, resolution measures in molecular electron microscopy evaluate consistency of the results in reciprocal space and present it as a one-dimensional (1D) function of the modulus of spatial frequency. Here we provide description of standard resolution measures commonly used in electron microscopy. We point out that the organizing principle is the relationship between these measures and the spectral signal-to-noise ratio (SSNR) of the computed density map. Within this framework it becomes straightforward to describe the connection between the outcome of resolution evaluations and the quality of electron microscopy maps, in particular, the optimum filtration, in the Wiener sense, of the computed map. We also provide a discussion of practical difficulties of evaluation of resolution in electron microscopy, particularly in terms of its

Department of Biochemistry and Molecular Biology, The University of Texas, Houston Medical School, Houston, Texas, USA

*Methods in Enzymology*, Volume 482
ISSN 0076-6879, DOI: 10.1016/S0076-6879(10)82003-8

**sensitivity to data processing operations used during structure determination process in single particle analysis and in electron tomography (ET).**

# 1. Introduction

Resolution assessment in molecular electron microscopy is of paramount importance both in computational methodology of structure determination and in interpretation of final structural results. In rare cases, when a structure reaches near-atomic resolution, appearance of secondary structure elements, particularly helices, can serve as an independent validation of the correctness of the structure and at least approximate resolution assessment can be made. In most cases, however, both in single-particle reconstruction (SPR) and especially in electron tomography (ET), the amount of detail in the structure is insufficient for such independent evaluation and one has to resort to statistical measures for determining the quality of the results. For historical reasons, they are referred to as resolution measures, although they do not correspond directly to traditional notions of resolution in optics.

The importance of resolution assessment in SPR was recognized early on in the development of the field. The measures were initially introduced for two-dimensional (2D) work, and subsequently extended for three-dimensional (3D) applications. There were a number of competing approaches introduced, such as Q-factor, Fourier ring correlation (FRC), and differential phase residual (DPR). Ultimately, it was the relation of these measures to the spectral signal-to-noise ratio (SSNR) distribution in the resulting structure that provided a unifying framework for the understanding of the resolution issues in molecular EM and their relationship to the optimum filtration of the results.

Resolution assessment of ET reconstruction remains one of the central issues that has resisted a satisfactory solution. The methodologies that have been proposed so far were usually inspired by SPR resolution measures. However, it is not immediately apparent that this is the correct approach since in standard ET there is no averaging of multiple pieces of data, the structure imaged is unique, and no secondary structure elements are resolved. Hence, ideas of self-consistency and reproducibility of the results that are central to validation of SPR results do not seem to be directly transferable to the evaluation of ET results.

In this brief review, we introduce the concept of resolution estimation in EM and what differentiates it from the concept of resolution traditionally used to characterize imaging systems in optics. We point out that it is the self-consistency of the results that is understood as resolution in EM and we show how this assessment is rooted in the availability of multiple data samples, that is, 2D EM projection images of 3D macromolecules. By relating the commonly used Fourier shell correlation (FSC) measure to the SSNR, we show that, if certain statistical assumptions about the data are fulfilled, the concept of resolution is well defined and interpretation of the results straightforward. In the closing section we discuss current approaches to resolution estimation in ET and point the fundamental limitations to their usefulness.

## 2. Optical Resolution Versus Resolution in Electron Microscopy

The resolution of a microscope is defined as the smallest distance between two points on a specimen that can still be distinguished as two separate entities. If we assume that the microscope introduces blurring that is expressed by a Gaussian function, then each of the point sources will be imaged as a bell-shaped object and the closer the two sources are, the worse the separation between the two maxima of the combined bells (Fig. 3.1) will be. It is common

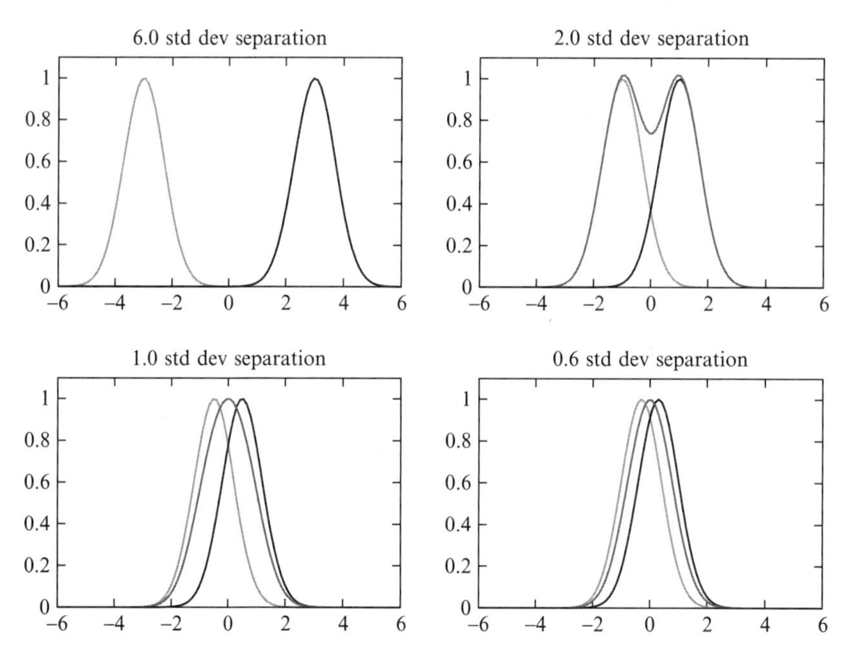


**Figure 3.1** Optical resolution is defined as the smallest distance between two points on a specimen that can be distinguished as two separate entities. Assuming the blur introduced by the microscope to be Gaussian with a known standard deviation, the resolution is defined as a distance between points that equals at least one standard deviation. For distances smaller or equal one standard deviation, the observed pattern, that is, sum of two Gaussian functions (green and blue) has an appearance of a pseudo-Gaussian with one maximum (magenta). (See Color Insert.)

to accept the resolution as the distance that is twice the standard deviation of the Gaussian blur. This concept of resolution is somewhat subjective, as one can presume different values at the minimum as acceptable and thus obtain different values for the resolution of the system. In the traditional definition, there is no accounting for noise in the measurements. Finally, there is a logical inconsistency: if we know that the blurring function is Gaussian, we can simply fit it to the measured data and obtain accuracy of peak separation by far exceeding what the definition would suggest (den Dekker, 1997).

The theoretical resolution of an optical lens is ultimately limited by diffraction and therefore depends on the wavelength of the radiation used to image the object (light for light microscopy, electrons for electron microscopy, X-rays for X-ray crystallography, and so on). In 2D, the Abbe criterion gives the resolution $d$ of a microscope as:

$$d = 0.61 \frac{\lambda}{n \sin \alpha}, \tag{3.1}$$

where $\lambda$ is the wavelength, $n$ is the refractive index of the imaging medium (air, oil, and vacuum), and $\alpha$ is half of the cone angle of light from the specimen plane accepted by the objective (half aperture angle in radians), in which case $n \sin \alpha$ is the numerical aperture. This criterion stems from the requirement for the numerical aperture of the objective lens to be large enough to capture the first-order diffraction pattern produced by the source at the wavelength used. Numerical aperture is $\sim$1 for light microscope and $\sim$0.01 for electron microscope. As the wavelength of light is 400–700 nm, the resolution of a light microscope is $\sim$250–420 nm. For electron microscopes, the electron wavelength $\lambda$ depends on the accelerating voltage $V$. Within the classical approximation, $\lambda = \sqrt{h^2/(2meV)}$, where $h$ is Planck's constant, $m$ is the mass, and $e$ the charge of the electron. For an accelerating voltage of 100 keV, $\lambda = 0.0037$ nm. The theoretical resolution of electron microscopes is 0.23 nm for an accelerating voltage of 100 keV ($\lambda = 0.003701$ nm) and 0.12 nm for 300 keV ($\lambda = 0.001969$ nm), and is mainly limited by the quality of electron optics. For example, spherical aberration correctors can significantly improve the resolution limit of electron microscopes.

While the resolution of electron microscopes is more than sufficient to achieve atomic resolution in macromolecular structural studies, resolution of reconstructed 3D maps of biological specimens is typically worse by an order of magnitude. The resolution-limiting factors in EM include:

1. Wavelength of the electrons.
2. Quality of the electron optics (astigmatism, envelope functions).
3. Underfocus settings. Resolution of the TEM is often defined as the first zero of the contrast transfer function (CTF) at Scherzer (or optimum) defocus.

4. Low contrast of the data. This is due to both the electron microscope being mainly a phase contrast instrument with an amplitude contrast of less than 10% and the similar densities of ice surrounding the molecules and of protein ($\sim$0.97:1.36 g/cm$^3$).
5. Radiation damage of imaged macromolecular complexes. Even if the electron dose is kept very low ($\sim 25e^-/\text{Å}^2$), some degree of damage on the atomic level is unavoidable and regrettably virtually impossible to assess. In addition, exposure to the electron beam is likely to adversely affect the structural integrity of both the medium and the specimen at the microscopic level (shrinkage, local shifts). These artifacts can be reduced in some cases by preirradiation of the sample.
6. Quality of recording devices. The current shift toward collecting EM data on CCD cameras means that there is additional strong suppression of high-frequency signal due to the modulation transfer function of the CCD. The advent of direct detection devices (DDD) should improve the situation (Milazzo *et al.*, 2010).
7. Low signal-to-noise ratio (SNR) level in the data. This stems from the necessity of keeping the electron dose at a minimum to prevent excessive radiation damage. It is generally accepted that the SNR of cryo-data is less than one. The only practical way to increase the SNR of the result is by increasing the number of averaged individual projection images of complexes.
8. Presence of artifactual images in the data set. Due to the very low SNR and despite careful screening of windowed particles there is always a certain percentage of frames that contain objects that appear to be valid projection images, but in reality are artifacts. At present, there are no reliable methods that would allow detection of these artifactual objects or even to assess what might be their share in the sample.
9. Variability of the imaged molecule. This can be caused by natural conformational heterogeneity of the macromolecular assemblies due to fluctuations of the structure around the ground state or due to the presence of different functional states in the sample.
10. Accuracy of the estimation of the image formation model parameters, such as defocus, amplitude contrast, and also SSNR of the data, magnification mismatch and such.
11. Accuracy of alignment of projection images is limited by the very low SNR of the EM data, inhomogeneity of the sample, and limitations of computational methodologies.

Standard definitions of resolution are based on the assumption that it is possible to perform an experiment in which an object with known spacing is imaged in the microscope and the analysis of the result yields a point spread function of the instrument, and thus the resolution. This approach is not applicable to analysis of EM results since the main resolution-limiting factors include instability of the imaged object, high noise in the data, and computational procedures that separate the source data from the result. Thus, even if useful information about resolution of cryo-specimens can be deduced from the analysis of crystalline specimen that do not require orientation adjustment in a computer, practical assessment of resolution in SPR and ET requires more elaborate methodology that has to take into account statistical properties of the aligned and averaged data.

## 3. Principles of Resolution Assessment in EM

Assessment of resolution in EM concerns itself not so much with the information content of individual images or with the resolving power of the microscope as with the consistency of the result, which in 2D is the average of aligned images and in 3D the object reconstructed from its projection images. Moreover, following the example of the *R*-factor in X-ray crystallography, the analysis is cast in Fourier space and the result is presented as a function of the modulus of spatial frequency. *The resolution in EM, if reported as a single number, is defined as a maximum spatial frequency at which the information content can be considered reliable.* Again, the definition is subjective, as it is rather arbitrary what one considers reliable and much of the controversies still present in the EM field revolve around this problem although, as we put forward later, the number by itself is inconsequential.

The resolution measures in EM evaluate the self-consistency of the result. The premise is to take advantage of the information redundancy: each Fourier component of the average/reconstruction is obtained by summing multiple Fourier components of input data at the same spatial frequency. These Fourier components are complex numbers, that is, each comprises a real and an imaginary part and their possible representation is by vectors in an $x$–$y$ system of coordinates, where the $x$-axis is associated with the real part and the $y$-axis with the imaginary part. The length of each vector is called its amplitude, and the angle between the vector and $x$-axis its phase. Thus, formation of an average, that is, the summation of the Fourier transforms (FTs) of images, can be thought of as a summation of such vectors for each spatial frequency. In this representation, assessment of the self-consistency of the result becomes a matter of testing whether vectors that were summed had similar phases, that is, the length of the sum of all vectors is not much smaller than the sum of lengths of individual vectors.

Indeed, if we consider a perfect case in which all images were identical, the vectors would all have the same length and phase. Their sum then would be simply a vector whose length is the same as the sum of the lengths of each vector. The same result is obtained if the vectors have different lengths (amplitudes) but the same direction (phase). Thus, if both sums are the same, we would consider the EM data consistent. Conversely, if there is little consistency, the vectors would point in different directions, in which case the length of their sum would fall short of the sum of their lengths. In EM, the degree to which both sums agree is equated with the consistency of the results (phase agreement), and ultimately with the resolution. Again, this is not the resolution in the optical sense, as it says relatively little about resolving power of the instrument or even about our ability to distinguish between details in real-space representation of the result. It is merely a measure that informs us on the consistency of the result for a particular spatial frequency, without even providing any information as to the cause of the inconsistency. For example, it is easy to see that both noise and structural inhomogeneity of the data will reduce consistency of the result as evaluated using the above recipe.

In what follows, we assume a quite general Fourier space image formation model:

$$G_n = F + M_n, \tag{3.2}$$

where $F$ is the signal (i.e., the projection image of a macromolecular complex), $M_n$ is noise, and $G_n$ is the $n$th observed image. For simplicity, we omit the argument and assume that unless stated otherwise, all entities are functions of spatial frequency and the equation is written for a particular frequency.

There are four major resolution measures that have been introduced into the EM field. For $N$ images, the $Q$-factor is defined as (van Heel and Hollenberg, 1980):

$$Q = \frac{\left|\sum_{n=1}^{N} G_n\right|}{\sum_{n=1}^{N} |G_n|}. \tag{3.3}$$

It is easy to see that $Q$ is zero for pure uncorrelated noise and 1 for a noise-free, aligned signal. We note that Eq. (3.3) is a direct realization of the intuitive notion of consistency expressed as the length of a sum of complex numbers. However, as we shall see later, the ratio of squared sum to the sum of squares of $G_n$'s yields itself better to analysis (Baldwin and Penczek, 2005).

While the $Q$-factor is a particularly simple and straightforward measure, it did not gain much popularity for a number of reasons. First, to compute it for 2D data, one has to compute the FTs of all the individual images, which at least at the time the measure was introduced might have been considered an impediment. Second, while $Q$-factor provides a measure of resolution for each pixel in the image, it needs to be modified to yield resolution per modulus of spatial frequency. Third, since Eq. (3.3) includes moduli of Fourier components, it cannot be easily related to SSNR, which is defined as the ratio of powers (squares) of the respective entities. Finally, it is not clear how to extend Eq. (3.3) to 3D reconstruction from projections, which, as will be described later, would require accounting for the reconstruction process, uneven distribution of data points, and interpolation between polar and Cartesian systems of coordinates.

The DPR (Frank *et al.*, 1981) was introduced to address some of the shortcoming of the Q-factor. With DPR, it is possible to compare FTs of two images, which can be either 2D averages or 3D reconstructions. The weighted squared phase difference between the same frequency Fourier components is summed within a ring (in 2D) or a shell (in 3D) of approximately same spatial frequencies:

$$\mathrm{DPR}(u,v;s) = \min_q \left| \frac{\sum\limits_{\|s_k|-s|\le\varepsilon}^{k_s} [\Delta\varphi_{UV}(s_k)]^2[|U(s_k)| + q|V(s_k)|]}{\sum\limits_{\|s_k|-s|\le\varepsilon}^{k_s} [|U(s_k)| + q|V(s_k)|]} \right|^{1/2}, \quad (3.4)$$

where $\Delta\varphi_{UV}(s_k)$ is the phase difference between Fourier components of $U$ and $V$, $2\varepsilon$ is a preselected ring/shell thickness, $s = |\mathbf{s}_k|$ is the magnitude of the spatial frequency, and $k_s$ is the number of Fourier pixels/voxels in the ring/shell corresponding to frequency $s$. The DPR yields a 1D curve of weighted phase discrepancies as a function of $s$. Regrettably, the DPR is sensitive to relative normalization of $u$ and $v$, so traditionally a minimum value with respect to the multiplicative normalization of $v$ is reported ($q$ in Eq. (3.4)). In order to assess the resolution of a set of images, the data set is randomly split into halves, two averages/reconstructions are computed, and the DPR is evaluated using their FTs. A DPR equal to zero indicates perfect agreement while for two objects containing uncorrelated noise the expectation value of DPR is 103.9° (van Heel, 1987). The accepted cut-off value for resolution limit is 45°. Regrettably, there is no easy way to relate DPR to SSNR or any other resolution measure.

The Fourier Ring Correlation (FRC) (Saxton and Baumeister, 1982) was introduced to provide a measure that would be insensitive to linear transformations of the objects' densities. For historical reasons, in 2D applications the measure is referred to as FRC while in 3D applications as Fourier Shell Correlation (FSC), even though the definition remains the same:

$$\mathrm{FSC}(u,v;s) = \frac{\sum\limits_{\|s_k|-s|\le\varepsilon}^{k_s} U(s_k)V^*(s_k)}{\left\{\left(\sum\limits_{\|s_k|-s|\le\varepsilon}^{k_s} |U(s_k)|^2\right)\left(\sum\limits_{\|s_k|-s|\le\varepsilon}^{k_s} |V(s_k)|^2\right)\right\}^{1/2}}, \quad (3.5)$$

where the notation is the same as in Eq. (3.4). FSC is a 1D function of the modulus of spatial frequency whose values are correlation coefficients computed between the FTs of two images/volumes over rings/shells of approximately equal spatial frequency. An FSC curve that is close to one everywhere reflects a strong similarity between $u$ and $v$ and an FSC curve with values close to zero indicates the lack of similarity between $u$ and $v$. Predominantly negative values of FSC indicate that contrast of one of the images was inverted. Typically, FSC decreases with spatial frequency (although not necessarily monotonically) and various cut-off thresholds have been proposed for serving as indicators of the resolution limit. We postpone their discussion to the Section 4. Finally, we clarify that because $u$ and $v$ are real, their FTs are Friedel symmetric (i.e., $U(\mathbf{s}) = U^{\star}(-\mathbf{s})$), so the result of the summation in the numerator is real. Given that, we can write Eq. (3.5) as:

$$\mathrm{FSC}(u, v; s) = \frac{\sum\limits_{||s_k|-s| \leq \varepsilon}^{k_s/2} |U(s_k)||V(s_k)| \cos(\Delta\varphi_{UV}(s_k))}{\left\{\left(\sum\limits_{||s_k|-s| \leq \varepsilon}^{k_s/2} |U(s_k)|^2\right)\left(\sum\limits_{||s_k|-r| \leq \varepsilon}^{k_s/2} |V(s_k)|^2\right)\right\}^{1/2}}. \tag{3.6}$$

It follows from Eq. (3.6) that FSC is indeed a consistency measure, as it is an amplitude weighted sum of cosines of phase discrepancies between the Fourier components of two images. By comparing Eqs. (3.4) and (3.6), we can also see that the normalization in FSC makes the measure better behaved.

Both DPR and FSC are computed using two maps. These can be either individual images or, more importantly, averages/3D reconstructions obtained from sets of images. The obvious application of these two measures is in the evaluation of resolution in structure determination by SPR. There are two possible approaches that reflect the overall methodology of the SPR. In the first approach, the entire data set is aligned and for the purpose of resolution estimation, it is randomly split into halves. Next, the two averages/reconstructions are computed and compared using DPR or FSC. In the second approach, the data set is first split into halves, each subset is aligned independently, and the two resulting averages/reconstructions are compared using DPR or FSC. The second approach has a distinct advantage in that the problem of "noise alignment" (Grigorieff, 2000; Stewart and Grigorieff, 2004) is avoided. As the orientation parameters of individual images are iteratively refined using the processed data set as a reference, the first approach has the tendency to induce self-consistency of the data set beyond what the level of signal in the data should permit. As a result, all the measures discussed in this chapter tend to report exaggerated resolution as they evaluate the degree of self-consistency of the data set. The problem is avoided by carrying out alignment independently; however, this approach is not without its peril. First, it is all but impossible to achieve "true" independence in SPR; for example, the initial reference is often shared or at least obtained using similar principles. Second, one can argue that alignment of half of the available particles cannot yield results with quality comparable to that obtained by aligning the entire data set, and this is of special concern when the data set is small. Third, two independent alignments may diverge, in which case the reported resolution will be appropriately low, but the problem is clearly with the alignment procedure and not with the resolution as such. Therefore, as long as one is aware that the resolution estimated using the first approach is to some extent exaggerated, there is no significant disadvantage to using it. Finally, we note that Q-factor and SSNR (to be introduced below) evaluate consistency of the entire data set, so they are not applicable to comparisons of independently aligned sets of images.

The SSNR can provide, as Q-factor does, a per-pixel measure of the consistency of the data set, which also distinguishes it from both DPR and FRC that yield measures that are "rotationally averaged" in Fourier space. SSNR was introduced for analyzing sets of 2D images and defined as (Unser *et al.*, 1987):

$$\mathrm{SSNR}(s)\begin{cases} S(s)-1, & \text{if } S(s)>1, \\ 0, & \text{if } S(s)\leq 1, \end{cases} \tag{3.7}$$

where the spectral variance ratio $S$ is:

$$S(s)=\frac{\left|\sum_{n=1}^{N} G_n(s)\right|^2}{\frac{N}{N-1}\sum_{n=1}^{N}\left|G_n(s)-\bar{G}(s)\right|^2}, \tag{3.8}$$

with

$$\bar{G}(s)=\frac{1}{N}\sum_{n=1}^{N} G_n(s). \tag{3.9}$$

Eqs. (3.7)–(3.9) define a per-pixel SSNR. In the original contribution, Unser *et al.* introduced SSNR as a 1D function of the modulus of spatial frequency to maintain its correspondence to FRC:

$$\mathrm{SSNR}(s)\begin{cases} S(s)-1, & \text{if } S(s)>1, \\ 0, & \text{if } S(s)\leq 1, \end{cases} \tag{3.10}$$

with the rotationally averaged spectral variance ratio defined as:

$$S(s)=\frac{\sum_{\|s_k|-s|\leq\varepsilon}^{k_s}\left|\sum_{n=1}^{N} G_n(s_k)\right|^2}{\frac{N}{N-1}\sum_{\|s_k|-s|\leq\varepsilon}^{k_s}\sum_{n=1}^{N}\left|G_n(s_k)-G(s_k)\right|^2}. \tag{3.11}$$

As we demonstrate in Section 4, the SSNR given by Eqs. (3.10) and (3.11) yields results equivalent to FRC in application to 2D data.

Extension of the SSNR to resolution evaluation of a 3D object reconstructed from the set of its 2D projection was only partially successful. The main difficulty lies in accounting for the uneven distribution of projections and the proper inclusion of the interpolation step into Eq. (3.10). To avoid a reliance on a particular reconstruction algorithm, Unser *et al.* (2005) proposed first to estimate the SSNR in 2D by comparing reprojections of the reconstructed structure with the original input projection data, and then averaging the contributions in 3D Fourier space to obtain the 1D dependence of the SSNR on spatial frequency. In addition, the authors proposed to estimate the otherwise-difficult-to-assess degree of averaging of data by the reconstruction process by repeating the calculation of the 3D SSNR for simulated data containing white Gaussian noise. The ratio of the two curves, that is, the one obtained from projection data and the other obtained from simulated noise, yields the desired true SSNR of the reconstructed object. Although the method is appealing in that it can be applied to any (linear) 3D reconstruction algorithm, a serious disadvantage is that the method actually yields a 2D SSNR of the input data, not the 3D SSNR of the reconstruction. This can be seen from the fact that in averaging the 2D contributions to the 3D SSNR, there is no accounting for uneven distributions of projections in Fourier space (Chapter 1).

It is straightforward, however, to extend Eqs. (3.6)–(3.10) to a direct Fourier inversion reconstruction algorithm that is based on nearest-neighbor interpolation; additionally, the method can also explicitly take into account the CTF correction necessary for cryo-EM data (Penczek, 2002; Zhang *et al.*, 2008). It is also possible to introduce SSNR calculation to more sophisticated direct Fourier inversion algorithms that employ the gridding interpolation method, but this is accomplished only at the expense of further approximations and loss of accuracy (Penczek, 2002). As the equations are elaborate and the respective methods rarely used, we refer the reader to the cited literature for details.

The main advantage of evaluating 3D SSNR for 3D reconstruction applications is that the method can yield per-voxel SSNR distributions, and thus provide a measure of anisotropy in the reconstructed object. Given such a measure, it is possible to construct anisotropic Fourier filters that can account for the directionality of Fourier information, and thus potentially improve the performance of 3D structure refinement procedures in cryo-EM single-particle reconstruction technique (Penczek, 2002). Outside of that, the equivalence between FSC and SSNR discussed in the Section 4 limits the motivation for the development of a more robust SSNR measures for the reconstruction problem.

## 4. Fourier Shell Correlation and its Relation to Spectral Signal-to-Noise Ratio

Currently, there is only one resolution measure in widespread use, namely the FSC. Besides the ease of calculation and versatility, the main reason for its popularity is its relation to the SSNR. This relation greatly simplifies the selection of a proper cut-off level for reported resolution, provides bases for the understanding of its relationship to optical resolution, and links resolution estimation to optimum filtration of the resulting average/structure. The SSNR is defined as a ratio of the power of the signal to the power of the noise. We assume zero mean uncorrelated Gaussian noise, that is, $M_n \in N(\mathbf{0}, \sigma^2\mathbf{I})$, then the SSNR for our image formation model equation (3.2) is:

$$\mathrm{SSNR}_{G_n} = \frac{|F|^2}{\sigma^2}. \tag{3.12}$$

The SSNR of an average of $N$ images (Eq. (3.9)) is:

$$\mathrm{SSNR}_{\bar{G}_n} = \frac{|NF|^2}{E\left[\left|\sum_{n=1}^{N} M_n\right|\right]} = \frac{N^2|F|^2}{N\sigma^2} = N\mathrm{SSNR}_{G_n}. \tag{3.13}$$

Thus, the summation of $N$ images that have identical signal and independent Gaussian noise increases the SSNR of the average $N$ times. For 3D reconstruction, the relationship is much more complicated and difficult to compute because of the uneven coverage of 3D Fourier space by projections, and the necessity of the interpolation step between polar and Cartesian coordinates (see Chapter 1).

Given definitions of the FSC($s$) equation (3.5) and SSNR($s$) equations (3.10) and (3.11) relationships between the two resolution measures are (for derivation see Bershad and Rockmore, 1974; Frank and Al-Ali, 1975; Penczek, 2002; Saxton, 1978):

$$\begin{aligned} \mathrm{FSC}(s) &= \frac{\mathrm{SSNR}(s)}{\mathrm{SSNR}(s)+1}, \\ \mathrm{SSNR}(s) &= \frac{\mathrm{FSC}(s)}{1-\mathrm{FSC}(s)}. \end{aligned} \tag{3.14}$$

For cases where the FSC was calculated by splitting the data set into halves, we have that (Unser *et al.*, 1987):

$$\begin{aligned}\mathrm{FSC}(s) &= \frac{\mathrm{SSNR}(s)}{\mathrm{SSNR}(s)+2},\\ \mathrm{SSNR}(s) &= 2\frac{\mathrm{FSC}(s)}{1-\mathrm{FSC}(s)}.\end{aligned} \tag{3.15}$$

Equation (3.15) serves as a basis for establishing a cut-off for reporting resolution as a single number. Since in EM one would typically evaluate the data using FSC methodology, one needs a cut-off level that serves as an indicator of the quality of obtained results. Using the relationship between the FSC and SSNR, the decision can be informed, even though it remains arbitrary. Commonly accepted cut-off levels are: (1) the $3\sigma$ criterion that selects for a cut-off level the point where there is no signal in the results, that is, SSNR falls to zero, in which case FSC $= 0$ (van Heel, 1987); (2) the point at which noise begins to dominate the signal, that is, SSNR $= 1$ or FSC $= 1/3$ (Eq. (3.15)); (3) the midpoint of the FSC curve, that is, SSNR $= 2$ or FSC $= 0.5$, which is also often used for constructing a low-pass filter.

The usage of a particular cut-off threshold requires the construction of a statistical test that would tell us whether the obtained value is significant. Regrettably, the distribution of the correlation coefficient (in our case FSC) is not normal and rather complicated, so the test is constructed using Fisher's $z$-transformation (Bartlett, 1993):

$$z = \frac{1}{2}\log\frac{1+\mathrm{FSC}}{1-\mathrm{FSC}} = \operatorname{arc\,tanh}(\mathrm{FSC}). \tag{3.16}$$

$z$ has a normal distribution:

$$z \in N\left(\zeta = \frac{1}{2}\log\frac{1+\rho}{1-\rho} + \frac{\rho}{k_s}, \frac{1}{\sqrt{k_s-3}}\right), \tag{3.17}$$

where $k_s$ is the number of *independent* sample pairs from which FSC was computed (see Eqs. (3.5) and (3.6)). In the statistical approach outlined here, it is important to use the number of truly independent Fourier components, that is, to account for Friedel symmetry and for point-group symmetry of the complex, if present. Confidence intervals of $z$ at $100(1-\alpha)\%$, where $\alpha$ is the significance level, are:

$$\frac{z^{+}}{-t_{\alpha/2}\frac{1}{\sqrt{k_s-3}}}. \tag{3.18}$$

For example, for $\alpha = 0.05$, $t_{(\alpha/2)} = 1.96$, the confidence limits are $z^{+}/\left(-\left(1.96/\sqrt{k_s-3}\right)\right)$. Once the endpoints for a given value of $z$ are established, they are transformed back to obtain confidence limits for the correlation coefficient (FSC):

$$^{\text{FSC}^+}\big/_{-}\tanh\left(t_{\alpha/2}\frac{1}{\sqrt{k_s-3}}\right) = \frac{\exp\left(2t_{\alpha/2}\frac{1}{\sqrt{k_s-3}}\right)-1}{\exp\left(2t_{\alpha/2}\frac{1}{\sqrt{k_s-3}}\right)+1}. \tag{3.19}$$

Finally, we note that it is much simpler to test the hypothesis that FSC equals zero. Indeed, we first note that for small FSC, $z \simeq$ FSC (Eq. (3.17)) and second for $\rho = 0$, $z \in N\left(0, \left(1/\sqrt{k_s - 3}\right)\right)$. Assuming a significance level $\alpha = 0.026$ corresponding to three standard deviations of a normal distribution, it follows that we can reject the hypothesis FSC $= 0$ when FSC $> 3/\sqrt{k_s - 3}$. The aforementioned is the basis of the $3\sigma$ criterion (van Heel, 1987; Fig. 3.2). We add that the statistical analysis given above is based on the assumption that Fourier components in the map are independent and their number is $k_s$.

The practical meaning of characterizing "resolution" of a result by a single number is unclear. Much has been written on the presumed superiority of one cut-off value over another, but little has been said on the practical difference between reporting resolution of the same result as say 12 Å according to the $3\sigma$ criterion as opposed to 15 Å based on the 0.5 cut-off. It has sometimes been

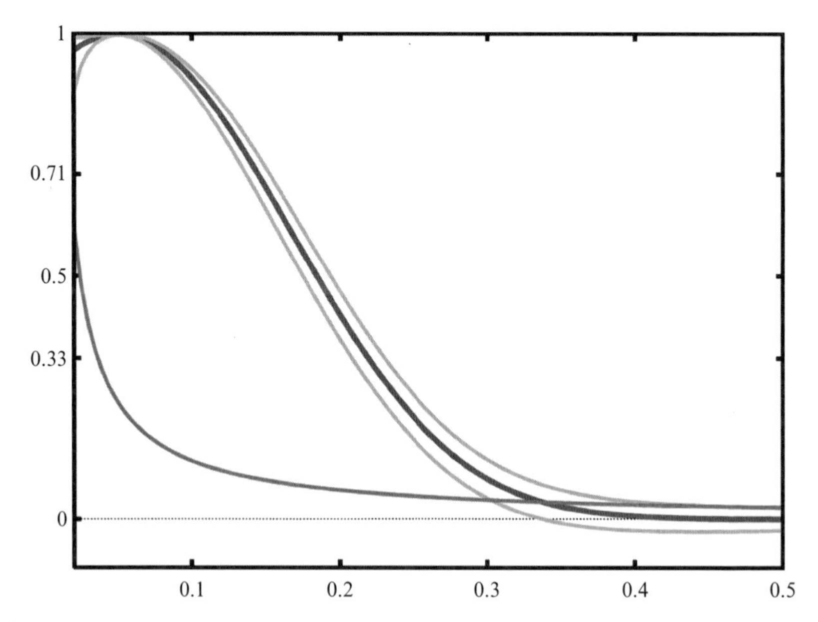


**Figure 3.2** Simulated FSC curve (red) with confidence intervals plotted at $\pm 3\sigma$ (blue) (Eq. (3.19)) and $3\sigma$ criterion curve (magenta) (van Heel, 1987). (See Color Insert.)

claimed that some criteria are more "conservative" than other and supposedly they can lead to Fourier filtration that would suppress interpretable details in the map, and yet no method of filtration is named (for example see exposition in van Heel *et al.* (2000)). The debate is resolved quite simply if we recall the FSC methodology yields a 1D function of the modulus of spatial frequency, which implies that it is the shape of the entire FSC curve that codes the "resolution" and the quality of the results. Finally, the relations between FSC, SSNR, and optimum filtration should put the controversy to rest.

Wiener filtration methodology provides a means for constructing a Fourier linear filter that prevents excessive enhancement of noise and which uses SSNR distribution in the data. The method yields an optimum filter in the sense that the mean-square error between the estimate and the original signal is minimized (see Chapter 2):

$$\hat{F}(s) = \frac{\mathrm{SSNR}(s)}{\mathrm{SSNR}(s) + 1} G(s), \tag{3.20}$$

which, using the relation between SSNR and FSC Eq. (3.14), is simply:

$$\hat{F}(s) = \mathrm{FSC}(s) G(s), \tag{3.21}$$

or, if FSC was computed using a data set split into halves (Eq. (3.15)), is:

$$\hat{F}(s) = 2\frac{\mathrm{FSC}(s)}{\mathrm{FSC}(s) + 1} G(s). \tag{3.22}$$

We conclude that (1) the FSC function gives an optimum low-pass filter; (2) wherever the SSNR is high, the original structure is not modified, and the midpoint, that is, FSC $= 0.5$, corresponds to decrease of the amplitudes two times, while FSC $= 0$ sets respective regions of Fourier space in filtered structure to zero; (3) in practice, FSC oscillates wildly around zero level and generally might be quite irregular, in which case it is preferable to approximate it by an analytical form of a selected digital filter (Gonzalez and Woods, 2002); (4) FSC is not necessarily a monotonically decreasing function of the modulus of spatial frequency, in particular, it may contain imprints of the dominating CTF, in which case an appropriate smooth filter has to be constructed. Finally, if the SSNR of a 3D map reconstructed from projections can be reliably estimated and if there is an indication of anisotropic resolution, a Wiener filter given by Eq. (3.20) is applicable (with spatial frequency replacing its modulus as an argument) either directly or using an appropriate smooth approximation of the SSNR (Penczek, 2002).

The FSC measure has one more very useful application in that it can be used to compare two maps obtained using different experimental techniques. Most often it is used to compare EM maps with X-ray structures of the same complex. In this case, the resulting function is called Fourier cross-resolution (FCR) and its meaning is deduced from the relationship between FRC and SSNR. Usually, one assumes that a target the X-ray structure is error- and noise-free, in which case FRC yields the SSNR of the EM map:

$$\begin{aligned} \mathrm{SSNR}(s) &= \frac{\mathrm{FCR}^2(s)}{1-\mathrm{FCR}^2(s)}, \\ \mathrm{FCR}(s) &= \sqrt{\frac{\mathrm{SSNR}(s)}{\mathrm{SSNR}(s)+1}}. \end{aligned} \tag{3.23}$$

The selected cut-off thresholds are as follows for the FCR: SSNR = 1 corresponds to FCR = 0.71 and SSNR = 2 to FCR = 0.82. Thus, it is important to remember that for a reported resolution of an EM map based on FCR to mean the same as a reported resolution based on FSC, the cut-off thresholds have to be different and higher. Otherwise, the FCR function can be used directly in Eq. (3.21) to obtain an optimum Wiener filtration of the result. Finally, it is important to stress that if there are reasons to believe that the target X-ray models and EM maps represent the same structure, at least within the considered frequency range, then the FCR methodology yields results that are much less marred by the problems prevalent in FSC resolution estimation. In particular, the difficult-to-avoid problem of alignment of noise has no influence on FCR results. Therefore, higher credence should be given to the FCR estimate of SSNR in the result, and hence also the FCR-based resolution.

## 5. Relation Between Optical Resolution, Self-Consistency Measures, and Optimum Filtration of the Map

There is no apparent relationship between optical resolution which is understood as the ability of an imaging instrument to distinguish between closely spaced objects and resolution measures used in EM that are geared toward the evaluation of self-consistency of an aligned data set. Ultimately, the resolution of an EM map will be restricted by the final low-pass filtration based on the resolution curve (Eqs. (3.20)–(3.22)). First we consider two extreme examples: (1) the FSC curve, and thus the filter is Gaussian and (2) the resolution curve is rectangular, that is, it is equal to one up to a certain stop-band frequency and zero in higher frequencies.

For the Gaussian-shaped FSC curve, the traditional definition of resolution applies: using as a cut-off threshold of 0.61, which is the value of a not-normalized Gaussian function at one standard deviation, one obtains that frequency corresponding to the Fourier standard deviation $\sigma_s$ specifies the resolution. Application of the corresponding Fourier Gaussian filter corresponds to a convolution with a real-space Gaussian function with a standard deviation $\sigma_r = 1/(2\pi\sigma_s)$ that effectively limits the resolution to $2\sigma_r = 1/(\pi\sigma_s)$, as per the traditional definition. For example, if $\sigma_s = (1/15)$ Å$^{-1}$, the resolution of the filtered map would be 4.8 Å. The reason it is so high is that the Gaussian filter decreases relatively slowly, which implies the filtered map contains significant amount of frequencies higher than the cut-off of one sigma. On the other hand, a Gaussian filter will also start suppressing amplitudes beginning from very low frequencies, which is why it should not be used for filtration of EM maps.

For a rectangular-shaped FSC curve, one would apply a top-hat low-pass filter that would simply truncate the FT at the stop-band frequency $s_c$ Å$^{-1}$ and thus obtain a map whose resolution is $(1/s_c)$ Å. This is a trivial consequence of the fact that the filtered map would not contain any higher frequencies. It would also appear that this is why it is often incorrectly assumed that a reported "resolution" based on more or less arbitrarily selected cut-off level is equivalent to the actual resolution of the map. In either case, top-hat filters should be avoided as truncation of Fourier series results in real-space artifacts known as "Gibbs oscillations," that is, any step in real-space map will be surrounded by ringing artifacts in the filtered map whose amplitude is ~9% of the step's amplitude (Jerri, 1998). If the original map was nonnegative, Gibbs oscillations will also introduce negative artifacts into the filtered map. Gibbs phenomenon is particularly unwelcome in intermediate resolution EM maps, as ringing artifacts will appear as fine features on the surface of the complex and thus invite spurious interpretations. One has to be aware that all Fourier filters share, to a degree, this problem, particularly if they are steep in the transition regions. Therefore, filtration of the EM map has to be done carefully using dedicated digital filters and one must be aware of the trade-off between desired steepness of the filter in Fourier space with amplitudes of Gibbs oscillations.

A properly designed filter has to approximate well the shape of a given FSC curve and at the same time be sufficiently smooth to minimize ringing artifacts. One possibility is to apply low-pass filtration in real space using convolution with a kernel. For example, if the kernel approximates a Gaussian function, it will have only positive coefficients, in which case both the ringing and negative artifacts are avoided. Regrettably, as discussed above, the practical applicability of Gaussian filtration in EM is limited. Another tempting approach is to use a procedure variously known as "binning" or a "box convolution" in which one replaces a given pixel by an average of neighboring pixels. This procedure is commonly, and incorrectly, often used for decimation of oversampled images (e.g., reduction of large micrographs). The procedure is simply a convolution of the input map with a rectangular function used as kernel. Since all coefficients of the kernel are the same, the resulting algorithm is particularly efficient. In addition, both ringing and negative artifacts are avoided and given that the method is fast, one can apply it repeatedly to approximate convolution with a Gaussian kernel (if repeated three times, it will approximate the Gaussian kernel to within 3%). However, the simple box convolution is deceptive, as the result is not what one would desire. This can be seen from the fact that the FT of a rectangular function is a sinc function, which is not band-limited and decreases rather slowly. Thus, paradoxically, simple box convolution does not suppress high frequencies well and does not have a well-defined stop-band frequency. Finally, if the method is used to decimate image, then it will result in aliasing, that is, frequencies higher than the presumed stop-band frequency will appear in the decimated image as spurious lower frequencies, resulting in real-space artifacts.

An advisable approach to filtration of an EM map is to select a candidate Fourier filter from a family of well-characterized prototypes and fit it to the experimental FSC curve. We note that under the assumption that the SSNR in EM image data is approximately Gaussian, the shape of the FSC curve can be modeled. Indeed, 1D rotationally averaged power spectra of macromolecular complexes in frequencies above 0.05–0.1 $\mathring{A}^{-1}$ are approximately constant at a level dependent on the number of atoms in the protein (Guinier and Fournet, 1955). This leaves SSNR as a ratio between the product of a squared envelope function of the data and the envelope function of misalignment to the power spectrum of the noise in the data. Both envelopes can be approximated, in the interesting frequency region, by Gaussian functions (Baldwin and Penczek, 2005; Saad *et al.*, 2001) and the noise power spectrum by exponent of a low-order polynomial (Saad *et al.*, 2001), so an overall Gaussian function is a good approximation of the SSNR fall-off. Thus, for $N$ averaged images we have: (Chapter 2)

$$\mathrm{SSNR}(s) \simeq N \exp\left(-\frac{B}{4}s^2\right). \tag{3.24}$$

Using Eq. (3.20), we obtain the shape of an optimum Wiener filter as (Fig. 3.3):

$$W(s) = \frac{N \exp\left(-\frac{B}{4}s^2\right)}{N \exp\left(-\frac{B}{4}s^2\right) + \gamma}, \tag{3.25}$$

where $\gamma$ accounts for the unknown normalization in Eq. (3.24). While the filter approximates remarkably well the shape of a typical FSC curve, it remains nonzero in high frequencies, and so is not very well suited for practical applications. Incidentally, Eq. (3.24) yields a simple relationship between resolution and the number of required images. More specifically, by using $\mathrm{SSNR}(s) = 1$ as a resolution cut-off, we obtain that *resolution* $\sim(\ln N)^{-(1/2)}$.

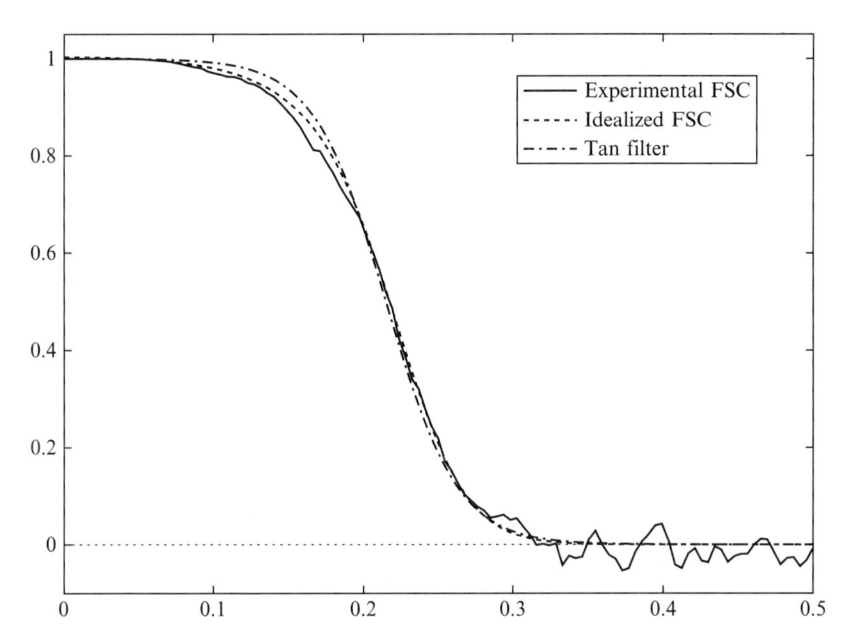


**Figure 3.3** Experimental FSC curve encountered in practice of SPR (solid) plotted as a function of magnitude of spatial frequency with 0.5 corresponding to Nyquist frequency. We also show an idealized FSC curve (Eq. (3.25)) and a hyperbolic tangent filter (Eq. (3.29)) fitted to the experimental FSC.

A commonly used filter is the Butterworth filter:

$$W(s) = \frac{1}{\left(1 + \left(\frac{s}{c}\right)^q\right)^{1/2}} \tag{3.26}$$

where $c$ is the cut-off frequency and $q$ determines the filter fall-off. These two parameters are determined using a pass-band and stop-band frequencies denoted $s_{\text{pass}}$ and $s_{\text{stop}}$, respectively:

$$q = 2\frac{\log\left(\frac{\varepsilon}{\sqrt{a^2-1}}\right)}{\log\left(\frac{s_{\text{pass}}}{s_{\text{stop}}}\right)}, \tag{3.27}$$

$$c = \frac{s_{\text{pass}}}{\varepsilon^{2/q}}, \tag{3.28}$$

where $\varepsilon = 0.882$ and $a = 10.624$. Since the value of the Butterworth filter at $s_{\text{pass}}$ is 0.8 and at $s_{\text{stop}}$ is 0.09, it is straightforward to locate these two points given an FSC curve and then compute the values $c$ and $q$ that parameterize the filter.

The Butterworth filter has a number of desirable properties, but it approaches zero in high frequencies relatively slowly. In addition, it is not characterized by a cut-off point at 0.5 value, so it is not immediately apparent what the resolution of the filtered map is. Therefore, a preference might be given to the hyperbolic tangent filter (tanh) (Basokur, 1998), which is parameterized by the stop-band frequency at 0.5 value, denoted $s_{0.5}$, and width of the fall-off $a$:

$$W(s) = 0.5\,\tanh\left(\frac{\pi(s + s_{0.5})}{2as_{0.5}}\right) - 0.5\,\tanh\left(\frac{\pi(s - s_{0.5})}{2as_{0.5}}\right). \quad (3.29)$$

The shape of the tanh filter is controlled only by one parameter, namely its width. When the width approaches zero, the filter reduces to a top-hat filter (truncation in Fourier space) with the disadvantage of pronounced Gibbs artifacts. Increasing the width reduces the artifacts and results in a better approximation of the shape of the FSC curve. In comparison with the two previous filters, the values of the tanh filter approach zero more rapidly, which makes it better suited for use in refinement procedures in SPR (Fig. 3.3).

## 6. Resolution Assessment in Electron Tomography

The methodologies of data collection and calculation of the 3D reconstruction in ET are dramatically different from those used in SPR and are more similar to those in computed axial tomography. There is only one object—a thin section of a biological specimen—and a tilt projection series is collected in the microscope by tilting the stage holding the specimen. As a consequence, three physical restrictions limit what can be accomplished with the technique: (1) total dose has to be limited to not far above that used in SPR to collect one image, so individual projection images in ET are rather noisy; (2) maximum tilt is limited to $\pm 60°$–$70°$, so the missing Fourier information constitutes an inherent problem; and (3) the specimen should be sufficiently thin to prevent multiple scattering, so the specimen thickness cannot exceed the free mean path of electrons in the substrate for a given energy and is in the range 100–300 nm, and thus the imaged object has to be a slab. The projection images have to be aligned, but in ET the problem is simpler, since the data collection geometry in ET can be controlled (within the limits imposed by the mechanics of the specimen holder), and so the initial orientation parameters are quite accurate.

The possible data collection geometries for ET are (1) single-axis tilting, which results in a missing wedge in coverage of Fourier space; (2) double-axes tilting, that is, two single-axis series, with the second collected after rotating the specimen by 90° around the axis coinciding with the direction of the electron beam (Penczek *et al.*, 1995); and (3) conical tilting, where the specimen tilt angle

(60°) remains constant while the stage is being rotated in equal increments around the axis perpendicular to the specimen plane (Lanzavecchia *et al.*, 2005). Regrettably, in all the three data collection geometries, the Fourier information along the $z$-axis is missing, which all but eliminates from the 3D reconstruction features that are planar in $x$–$y$ planes, thus making it difficult to study objects that are dominated by such features, such as membranes.

In ET, there is only one projection per projection direction, so the evaluation of resolution based on the availability of multiple projections per angular direction, as practiced in SPR, is not applicable. The assessment of resolution in tomography has to include a combination of two key aspects of resolution evaluation in reconstructions of objects from their projections: (1) the distribution of projections should be such that the Fourier space is, as much as possible, evenly covered to the desired maximum spatial frequency; (2) the SSNR in the data should be such that the signal is sufficiently high at the resolution claimed. However, the quality of tomographic reconstructions depends on the maximum tilt angle used and on the data collection geometry, and yet neither of these factors is properly accounted for by resolution measures.

There is no consensus on what should be a general concept of resolution in ET reconstructions, and the resolution measures currently in use in ET are either simply borrowed from SPR (mainly FSC) or are based on slight adjustments of SPR methodologies. In three recently published papers devoted to resolution measures, the authors proposed solutions based on extensions of resolution measures routinely used in SPR. In Penczek (2002), the author proposed application to ET of a 3D SSNR measure developed for a class of 3D reconstruction algorithms that are based on interpolation in Fourier space and described in the earlier section. The 3D SSNR works well for isolated objects and within limited range of spatial frequencies. However, the measure requires calculation of the Fourier space variance, so it will yield correct results only to the maximum frequency limit within which there is sufficient overlap between FTs of projections. Consequently, its appeal for evaluation of ET reconstructions is limited. Similarly, it was suggested that the 3D SSNR measure proposed by Unser *et al.* (2005) and discussed earlier can be made applicable to ET. Regrettably, its application to tomography is doubtful since, as the case with the previous method, it requires sufficient oversampling in Fourier space to yield the correct result.

An interesting approach to resolution estimation was introduced by Cardone *et al.* (2005), who proposed to calculate, for each available projection, two 2D FRC curves: (1) between selected projections and reprojections of the volume reconstructed using the whole set of projections and (2) between selected projections and reprojections of the volume reconstructed with the selected projection omitted. The authors showed that the ratio of these two FRC curves is related to the SSNR in the volume in the Fourier plane perpendicular to the projection direction, as per central section theorem. The authors propose to calculate the SSNR of the whole tomogram by summing the contributions from individual ratios. It is straightforward to note that the method suffers from the same disadvantages as the method by Unser *et al.*: (1) it does not account for the SSNR in the data lying in nonoverlapping regions and (2) it does not yield the proper 3D SSNR because of the omission of reconstruction weights.

In order to address the shortcomings of the method described in Penczek and Frank (2006), the authors propose to take advantage of inherent, even if only directional, oversampling of 3D Fourier space while using standard ET data collection geometries. By exploring these redundancies and by using the standard FSC approach, the authors show that it is possible to: (1) calculate the SSNR in certain regions of Fourier space, (2) calculate the SSNR in individual projections in the entire range of spatial frequencies, and (3) infer/deduce the resolution in nonredundant regions of Fourier space by assuming isotropy of the data. Given the SSNR in projections and known angular step of projections, it becomes possible to calculate the distribution of the SSNR in the reconstructed 3D object. While the approach appears to be sound, results of experimental tests are lacking.

The main challenge of resolution estimation in ET is that imaged objects are not reproducible and they have inherent variability. Thus, unlike the case in crystallography or in single-particle studies, repeated reconstructions of the object from the same category will yield structures that have similar overall features, but are also significantly different. Hence, it is impossible to study resolution of tomographic reconstructions in terms of statistical reproducibility. Moreover, because of the dose limitations there is only one projection for each angular direction, so the standard approach to SSNR estimation based on dividing the data set into halves is not applicable. In order to develop a working approach, one has to consider two aspects of resolution estimation in ET: (1) angular distribution of projections and (2) estimation of the SSNR in the data. A successful and generally accepted approach has not yet emerged.

## 7. Resolution Assessment in Practice

The theoretical foundations of the most commonly used EM resolution measure, the FSC, are very well developed and understood. This includes statistical properties of the FSC (Table 3.1), relation of the FSC to the SSNR, and the relation of FSC to optimum (Wiener) filtration of the results. Based on these, it is straightforward to construct a statistical test that would indicate the significance of the results on the preselected significance level and through the relation of FSC to the SSNR, apply a proper filter to suppress excessive noise. Nevertheless, practical use of resolution measures differs significantly

**Table 3.1** Taxonomy of EM resolution measures

| | Relation to SSNR | Statistical properties | Computed using | Applicable to |
|---|---|---|---|---|
| Q-factor | Remote | Somewhat understood | Individual images | Individual voxels |
| DPR | None | Not understood | Averages | 2D and 3D |
| FSC | Equivalent | Understood | Averages | 2D and 3D |
| SSNR | – | Understood | Individual images | 2D, approximations in 3D |

among software packages and it is advisable to be aware of various factors that influence the results and to also know the details of implementation.

The basic protocol of resolution assessment using FSC is straightforward:

1. Two averages/3D reconstructions are computed using either two "independently" aligned data sets or by splitting the aligned data set randomly into halves.
2. The averages are preprocessed as follows:
   a. A mask is applied to suppress noise surrounding the object.
   b. The averages may be padded with zeroes to *k*x the size (as a result, the FSC curve will have *k*x more sampling points) to obtain a finely sampled resolution curve.
   c. Other occasionally applied operations, particularly nonlinear (e.g., thresholding), will unduly increase the resolution.
3. The FSC curve is computed using Eq. (3.5).
4. Optionally, confidence intervals are computed using Eqs. (3.18) and (3.19) with proper accounting for the reduction of the number of degrees of freedom due to possible point-group symmetry of the complex.
5. An arbitrary cut-off threshold is selected in the range (0,1) with lower values given preference by more advantageous. The decision is supported by a reference to an appropriately chosen methodological paper, "resolution" is read from the FSC curve and solemnly reported as such.
6. A Wiener low-pass filter is constructed using the FSC curve either directly (Eqs. (3.20)–(3.22)) or approximated using one of the candidate analytical filters (Eqs. (3.25)–(3.27)) and applied to the map.
7. Optionally, the power spectrum of the average/reconstruction is adjusted (see Chapter 1) based on the FSC curve and reference power spectrum (obtained from X-ray or SAXS experiments, or simply modeled). This power spectrum-adjusted structure is a proper model for interpretation.

Despite its popularity, the FSC has a number of well-known shortcomings that stem from the violation of underlying assumptions. The FSC is a proper measure of the SSNR in the average under the assumption that the noise in the data is additive and statistically independent from the signal. As in SPR the data has to be aligned in order to bring the signal in images into register, and it is all but impossible to align images without inducing some correlations into the noise component. Even if the data set is split into halves prior to the alignment to eliminate the noise bias, the two resulting maps have to be aligned to calculate the resolution, and this step introduces some correlations and undermines the assumption about noise independence. Second, if the data set contains projection images of different complexes or of different states of the same complex, the signal is no longer identical. Finally, even if the data are actually homogeneous, the alignment procedure may fail and may converge to a local minimum, so the solution will be incorrect but self-consistent and the resolution will appear to be significant.

In step 2 of the basic protocol outlined above, a mask is applied to the averages in real space to remove excessive background noise. However, because this step is equivalent to convolution in Fourier space with the FT of the mask, it results in additional correlations that will falsely "improve" FSC results. In fact, application of the mask can be a source of grave mistakes in resolution assessment by FSC (or any other Fourier space-based similarity measure). First, one has to consider the shape of the mask. It is tempting to use a mask whose shape closely follows the shape of the molecule. However, the design of such a mask is an entirely heuristic endeavor, as presence of noise, influences of envelope function, and filtration of the object make the notion of an ideal shape poorly defined. Many design procedures have been considered (Frank, 2006), but the problem remains without a general solution. As a result, for any particular shape, the mask has to be custom-designed by taking into account the resolution and noise level of the object. Worse yet, some software packages have "automasking" facilities that merely reflect the designer's concept of what a "good" mask is and which contain key parameters not readily accessible to the user. The outcome of such "automasking" is generally unpredictable. The design of an appropriate mask is an important issue because the FSC resolution strongly depends on how close the mask follows the molecule shape and on how elaborate the shape is. More specifically, more elaborate shapes introduce stronger correlations in Fourier space, and thus "improve" the resolution.

Second, the FT of a purely binary mask, that is, a mask whose values are one in the region of interest and zero elsewhere, has high-amplitude ripples extending to the Nyquist frequencies, and will introduce strong correlations in Fourier space. Ideally, one would want to apply as a real-space mask a broad Gaussian function with a standard deviation equal to say half the radius of the structure; however, although such a mask minimizes the artifacts, it does not suppress surrounding noise very well. The best compromise is a mask that has a neutral shape (sphere of ellipsoid), that is, it does not follow the shape of the object closely, and which is equal to 1 in the region of interest and zero outside, and where the two regions joined by a smooth transition region with a fall-off given by Gaussian, cosine, or tangent functions.

Third, if a mask is applied, which is almost always the case, one has to make sure that the tested objects are normalized properly. It is known that if we add a constant to pixel values prior to masking, the resolution will improve. In fact, by adding a sufficiently large number, one can obtain an arbitrarily high resolution using the FSC test. The appropriate approach is to compute the average of the pixel values within a transition region as discussed above (which for a sphere is a shell few pixels wide) and then subtract it before applying the mask.

In general, the masking operation changes the number of independent Fourier pixels, that is the number of degrees of freedom, but it is very difficult to give a precise number. This change adversely affects the results of statistical criteria, particularly in the case of the $3\sigma$ criterion as when FSC approaches zero the curve oscillates widely and too low a threshold can dramatically change the resolution estimation. Nonlinear operations such as thresholding or nonlinear filters such as the median filter will also unduly "improve" the resolution. Finally, various mistakes in the computational EM structure determination process will strongly affect resolution and the overall shape of the FSC curve (Fig. 3.4).

In the study of resolution estimation of maps reconstructed from projection images, it is known that neither FSC nor SSNR yields correct results when the distribution of projection directions is strongly nonuniform. Any major gaps in Fourier space will result in overestimation of resolution and unless 3D SSNR is monitored for anisotropy, there is no simple way to detect the problem. Even if anisotropy is found, the analytical tools that would help to assess its influence on resolution are lacking.

While the resolution measures described here provide specific numerical estimates of resolution, the ultimate assessment of the claimed resolution is always done by examining the appearance of the map. In SPR, when the resolution is in the subnanometer range, features related to secondary structure elements should be identifiable. In the range 5–10 Å, $\alpha$-helices should appear as cylindrical features while $\beta$-sheets as planar objects. At a resolution better than 5 Å, densities approximating the protein backbone trace should be identifiable (Baker *et al.*, 2007). Conversely, presence of small features at a resolution lower than ~8 Å indicates that resolution has been overestimated, that the map has not been low-pass filtered properly, and/or that excessive correction for the envelope function of the microscope has been applied ("sharpening" of the map). In ET, it is also possible to assess the resolution by examining the resolvability of known features of the imaged biological material. These can include spacing in filaments, distances between membranes, or visibility of subcellular structures. It is certainly the case that in both SPR and ET 3D reconstructions are almost

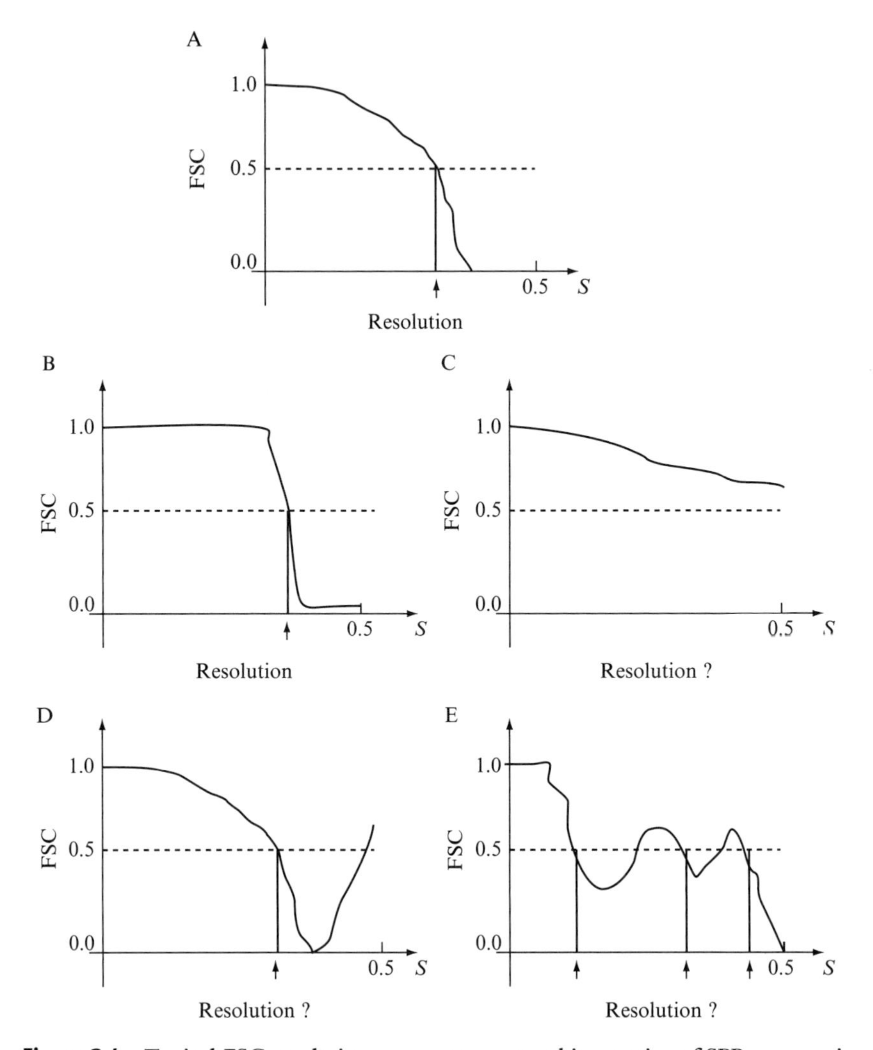


**Figure 3.4** Typical FSC resolution curves encountered in practice of SPR. $s$, magnitude of spatial frequency with 0.5 corresponding to Nyquist frequency. (A) proper FSC curve remains one at low frequencies, which is followed by a semi-Gaussian fall-off (see Eq. (3.25)) and a drop to zero at around 2/3 of Nyquist frequency, in high frequencies oscillates around zero. (B) Artifactual "rectangular" FSC: remains one at low frequencies, followed by a sharp drop, in high frequencies oscillates around zero. Typically it is caused by a combination of alignment of noise and a sharp filtration during the alignment procedure. (C) The FSC never drops to zero in the entire frequency range. Normally, this means that the noise component in the data was aligned, the result is artifactual and the resolution is undetermined. In rare cases, it can also mean that the data was severely undersampled (very large pixel size). (D) After the FSC curve drops to zero, it increases again in high frequencies. This artifact can be caused by the low-pass filtration of the data prior to alignment, errors in the image processing code, mainly in interpolation, by the erroneous correction for the CTF, including errors in estimation of SSNR, and finally, incorrect parameters in 3D reconstruction programs (e.g., iterative 3D reconstruction was terminated too early). It can also mean that all images were rotated by the same angle. (E) FSC oscillates around 0.5. It means that data was dominated by one subset with the same defocus value or there is only one defocus group. The resolution curve is not incorrect per se, but it is unclear what the resolution is. The resulting structure will have strong artifacts in real space.

never performed in the total absence of some *a priori* information about the specimen, for example, the number of monomers and the number of subunits in macromolecular complexes are generally known *a priori*. While this information might be insufficient to assess the resolution to a satisfying degree of accuracy, it can certainly provide sufficient grounds to evaluate the general validity of the results within the resolution limit claimed.

## ACKNOWLEDGMENTS

I thank Grant Jensen, Justus Loerke, and Jia Fang for critical reading of the manuscript and for helpful suggestions. This work was supported by grant from the NIH R01 GM 60635 (to PAP).

## REFERENCES

Baker, M. L., Ju, T., and Chiu, W. (2007). Identification of secondary structure elements in intermediate-resolution density maps. *Structure* **15,** 7–19.

Baldwin, P. R., and Penczek, P. A. (2005). Estimating alignment errors in sets of 2-D images. *J. Struct. Biol.* **150,** 211–225.

Bartlett, R. F. (1993). Linear modelling of Pearson's product moment correlation coefficient: an application of Fisher's z-transformation. *J. R. Statist. Soc. D.* **42,** 45–53.

Basokur, A. T. (1998). Digital filter design using the hyperbolic tangent functions. *J. Balkan Geophys. Soc.* **1,** 14–18.

Bershad, N. J., and Rockmore, A. J. (1974). On estimating signal-to-noise ratio using the sample correlation coefficient. *IEEE Trans. Inf. Theory* **IT20,** 112–113.

Cardone, G., Grünewald, K., and Steven, A. C. (2005). A resolution criterion for electron tomography based on cross-validation. *J. Struct. Biol.* **151,** 117–129.

den Dekker, A. J. (1997). Model-based optical resolution. *IEEE Trans. Instrum. Meas.* **46,** 798–802.

Frank, J. (2006). Three-Dimensional Electron Microscopy of Macromolecular Assemblies. Oxford University Press, New York.

Frank, J., and Al-Ali, L. (1975). Signal-to-noise ratio of electron micrographs obtained by cross correlation. *Nature* **256,** 376–379.

Frank, J., Verschoor, A., and Boublik, M. (1981). Computer averaging of electron micrographs of 40S ribosomal subunits. *Science* **214,** 1353–1355.

Gonzalez, R. F., and Woods, R. E. (2002). Digital Image Processing. Prentice Hall, Upper Saddle River, NJ.

Grigorieff, N. (2000). Resolution measurement in structures derived from single particles. *Acta Crystallogr. D Biol. Crystallogr.* **56,** 1270–1277.

Guinier, A., and Fournet, G. (1955). Small-Angle Scattering of X-rays. Wiley, New York.

Jerri, A. J. (1998). The Gibbs Phenomenon in Fourier Analysis, Splines and Wavelet Applications. Kluwer, Dordrecht, The Netherlands.

Lanzavecchia, S., Cantele, F., Bellon, P., Zampighi, L., Kreman, M., Wright, E., and Zampighi, G. (2005). Conical tomography of freeze-fracture replicas: A method for the study of integral membrane proteins inserted in phospholipid bilayers. *J. Struct. Biol.* **149,** 87–98.

Milazzo, A. C., Moldovan, G., Lanman, J., Jin, L., Bouwer, J. C., Klienfelder, S., Peltier, S. T., Ellisman, M. H., Kirkland, A. I., and Xuong, N. H. (2010). Characterization of a direct detection device imaging camera for transmission electron microscopy. *Ultramicroscopy* **110,** 741–744.

Penczek, P. A. (2002). Three-dimensional spectral signal-to-noise ratio for a class of reconstruction algorithms. *J. Struct. Biol.* **138,** 34–46.

Penczek, P. A., and Frank, J. (2006). Resolution in electron tomography. *In* "Electron Tomography: Methods for Three-dimensional Visualization of Structures in the Cell," (J. Frank, ed.), pp. 307–330. Springer, Berlin.

Penczek, P., Marko, M., Buttle, K., and Frank, J. (1995). Double-tilt electron tomography. *Ultramicroscopy* **60,** 393–410.

Saad, A., Ludtke, S. J., Jakana, J., Rixon, F. J., Tsuruta, H., and Chiu, W. (2001). Fourier amplitude decay of electron cryomicroscopic images of single particles and effects on structure determination. *J. Struct. Biol.* **133,** 32–42.

Saxton, W. O. (1978). Computer techniques for image processing of electron microscopy. Academic Press, New York.

Saxton, W. O., and Baumeister, W. (1982). The correlation averaging of a regularly arranged bacterial envelope protein. *J. Microsc.* **127,** 127–138.

Stewart, A., and Grigorieff, N. (2004). Noise bias in the refinement of structures derived from single particles. *Ultramicroscopy* **102,** 67–84.

Unser, M., Trus, B. L., and Steven, A. C. (1987). A new resolution criterion based on spectral signal-to-noise ratios. *Ultramicroscopy* **23,** 39–51.

Unser, M., Sorzano, C. O., Thevenaz, P., Jonic, S., El-Bez, C., De Carlo, S., Conway, J. F., and Trus, B. L. (2005). Spectral signal-to-noise ratio and resolution assessment of 3D reconstructions. *J. Struct. Biol.* **149,** 243–255.

van Heel, M. (1987). Similarity measures between images. *Ultramicroscopy* **21,** 95–100.

van Heel, M., and Hollenberg, J. (1980). The stretching of distorted images of two-dimensional crystals. *In* "Electron Microscopy at Molecular Dimensions," (W. Baumeister, ed.), pp. 256–260. Springer, Berlin.

van Heel, M., Gowen, B., Matadeen, R., Orlova, E. V., Finn, R., Pape, T., Cohen, D., Stark, H., Schmidt, R., Schatz, M., and Patwardhan, A. (2000). Single-particle electron cryo-microscopy: Towards atomic resolution. *Q. Rev. Biophys.* **33,** 307–369.

Zhang, W., Kimmel, M., Spahn, C. M., and Penczek, P. A. (2008). Heterogeneity of large macromolecular complexes revealed by 3D cryo-EM variance analysis. *Structure* **16,** 1770–1776.

CHAPTER FOUR

# 3D Reconstruction from 2D Crystal Image and Diffraction Data

Andreas D. Schenk,* Daniel Castaño-Díez,† Bryant Gipson,† Marcel Arheit,† Xiangyan Zeng,‡,§ *and* Henning Stahlberg†

## Contents



* Department of Cell Biology, Harvard Medical School, Boston, Massachusetts, USA
† C-CINA, Biozentrum, University of Basel, Basel, Switzerland
‡ Department of Mathematics and Computer Science, Fort Valley State University, Fort Valley, Georgia, USA
§ EON Corporation, California, USA

*Methods in Enzymology*, Volume 482
ISSN 0076-6879, DOI: 10.1016/S0076-6879(10)82004-X

## Abstract

Electron crystallography of 2D protein crystals can determine the structure of membrane embedded proteins at high resolution. Images or electron diffraction patterns are recorded with the electron microscope of the frozen hydrated samples, and the 3D structure of the proteins is then determined by computer data processing. Here we introduce the image-processing algorithms for crystallographic Fourier space based methods using the Medical Research Council (MRC) programs, and illustrate the usage of the software packages 2dx, XDP, and IPLT.

## 1. Introduction to Electron Crystallography Data Processing

Electron crystallography records electron micrographs of thin, sheet-like, two-dimensional (2D) crystals. Such 2D crystals of proteins are sometimes encountered *in vivo*, as for example the purple membrane of *Halobacterium salinarium* (Henderson and Unwin, 1975). Alternatively they can be obtained from detergent-solubilized and purified membrane proteins by reconstitution into a lipid-bilayer at very low lipid-to-protein ratio (Jap *et al.*, 1992; Kühlbrandt, 1992). Images of noncrystalline membrane-reconstituted membrane proteins usually do not provide sufficient signal-to-noise ratio (SNR) to allow detection of the protein location and orientation, so image processing by signal averaging methods fails. However, in 2D crystals the membrane proteins form a regular array in the membrane, so that image processing by Fourier filtering and averaging becomes possible. 2D membrane protein crystals are usually single-layered but in some cases, multilayered 2D crystals can be observed. Multilayered crystals often show a better crystallinity, which can be explained by the additional crystal contacts in between the different layers. If the number of crystal layers is small and defined, for example, 2 or 3, then the structure reconstruction from images of these crystals is possible, and might even benefit from additional symmetry present in the multilayered crystal. However, if the number of layers is too high and/or varying, as would be the case for 2D crystals that have a strong tendency to stack, then the structure determination from such crystals faces strong obstacles with the currently available software tools (Kulik, 2004).

Electron crystallography records images of negatively stained or frozen hydrated 2D crystals, or subjects the 2D crystals to selected area electron diffraction. Such data can be recorded from nontilted crystals, or the crystals can be imaged at a tilt angle. Due to the random orientation of the unit cell vectors in comparison to the tilt axis position, data recording at a fixed set of tilt angles can fully sample Fourier space apart from the missing cone, similar to random conical tilt reconstructions in single-particle cryo-EM. The combination of data from different tilt angles ranging from 0° to 70° then produces the final 3D map of the protein.

Recorded diffraction patterns of 2D crystalline samples show diffraction spots, which can be evaluated to give the amplitudes of the protein structure. This can allow the determination of precise amplitude data up to very high resolution (Gonen *et al.*, 2005). If the 2D crystals were badly ordered, however, then such diffraction pattern would only show low-resolution diffraction orders and would therefore be unsuitable for structure determination by electron diffraction. Recorded real-space images of 2D crystals can be Fourier transformed, and the resulting complex-valued Fourier transforms give access to the amplitudes and *phases* of the protein structure. In case of moderate crystal defects in the 2D crystals, the recorded images can be computer-processed prior to the calculation of the Fourier transformation, to computationally correct ("unbend") the crystal distortions in the images. This process of crystal "unbending" was introduced by Henderson and Unwin (1975) and allows to significantly improve the number and quality of high-resolution spots in the calculated Fourier transformations of the images, thereby giving access to higher resolution data.

Computer data processing for electron crystallography using the "unbending" procedure and evaluation of calculated Fourier transformations or electron diffraction patterns were implemented in the so-called MRC program suite. The MRC image-processing package was written and maintained over the past 30 years by Richard Henderson in the Medical Research Council (MRC) in Cambridge, UK, and others (D. Agard, L. Amos, T. Baker, J. Baldwin, T. Ceska, R. A. Crowther, D. DeRosier, E. Egelman, S. Fuller, T. Horsnell, P. Moore, J. M. Short, G. Vigers, and others; see Crowther *et al.*, 1996). Over many years a large set of programs has been written for processing images of 2D crystals (Henderson *et al.*, 1990; Kühlbrandt *et al.*, 1994; Murata *et al.*, 2000; Unwin and Henderson, 1975). The MRC programs partly make use of functions and routines from the CCP4 program suite (Collaborative Computational Project, 1994), and the MRC programs later became the basis for several other software packages. Parts of the original Fortran MRC software code can be found in several of today's biological image-processing tools, including those for processing noncrystalline specimens (see, e.g., Chapter 16).

SPECTRA from the ICE package was a program that generated Unix Shell scripts that launched an early version of the MRC programs (Hardt *et al.*, 1996; Schmid *et al.*, 1993). SPECTRA contained its own image visualization program, and facilitated the use of the MRC programs. The SPECTRA software was discontinued several years ago and is not available any more. Another similar programming effort is the GRIP system by Wilko Keegstra in the University of Groningen (unpublished). In recent years, other electron crystallography image-processing solutions became available that follow the same unbending idea. Similarly to SPECTRA, 2dx (Gipson *et al.*, 2007b) is a front-end interface for user-friendly interaction with the MRC programs and offers user-guidance and help functions. 2dx in addition offers optional full automation of the 2D and 3D processing and includes tools to organize the project, automate the workflow, merge the extracted data (Gipson *et al.*, 2007a), and apply a single-particle maximum likelihood program to the 2D crystal images (Zeng *et al.*, 2007b). XDP is another front-end software system for the MRC programs, which is mainly focused on the evaluation of electron diffraction patterns (Mitsuoka *et al.*, 1999), and which is complemented by a graphical user interface (GUI) and a set of scripts to simplify the use of the MRC programs. IPLT (Philippsen *et al.*, 2003, 2007b) is a highly modular and adaptable software package developed in C++ and Python that provides a reimplementation of existing algorithms complemented by newly developed algorithms for processing images and diffraction patterns. In addition it provides a GUI to guide the user through the processing and merging of electron diffraction data. CRISP and ELD are commercial software packages for the processing of 2D crystal data (Hovmöller, 1992). XMIPP also offers certain functionality for 2D crystal image processing (Sorzano *et al.*, 2004).

Here we describe the general image-processing algorithm for the image unbending approach, the evaluation of computed Fourier transformations and electron diffraction patterns, and the 3D merging of the measured data to obtain a 3D reconstruction of the protein. We introduce the unbending algorithm as implemented in the MRC programs, and briefly introduce the 2dx, XDP, and IPLT programs and their usage.

## 2. Algorithms for Electron Crystallography

Structural data of 2D protein crystals can either be gained by recording real-space images or electron diffraction patterns in an electron microscope. Both can be recorded on either film or CCD cameras. In practice, however, high-resolution images are best recorded on photographic film and digitized with a scanner, to make use of the better point spread function (PSF) of film compared to that of a CCD for higher voltage electrons (Downing and Hendrickson, 1999). Electron diffraction patterns are best recorded on digital CCD cameras, to make use of the higher bit-depth of the CCD pixels compared to film. In addition recording on a CCD camera offers a wider linear range and an overall increase in data recording speed as no subsequent digitalization of the recorded data is necessary.

Digitized images or diffraction patterns are then processed to extract structural data. Although image and diffraction data share certain features, the different challenges encountered during data extraction mean that two different sets of algorithms have to be applied. The processing of electron crystallography data can roughly be divided into four phases:

1. setup of the processing project and determination of the basic crystal parameters;
2. processing of the individual images and/or diffraction patterns;
3. merging of the data of individual images or patterns into one 3D dataset, refinement, lattice line fitting, and discretization of the dataset; and
4. building of an atomic model and model refinement.

For each of the phases we will outline the steps to be performed in a general fashion—also discussing some of the pitfalls and difficulties inherent to the processing—and then provide introductory protocols for the use of 2dx, XDP, and IPLT.

The general goal of the electron crystallography data processing is to determine the Fourier representation of the protein structure. This then allows calculating the real-space map by a simple Fourier back-transformation. While a 3D crystal would have as its Fourier representation a set of diffraction spots in 3D Fourier space, the 2D crystals have as their Fourier representation a set of vertically arranged "lattice lines." These continuous lines in Fourier space lie on a regularly spaced $h$, $k$ grid, which is defined by the crystal axes $a$, $b$ of the 2D crystal. Because the 2D crystal has in the third dimension, the $z$-axis, only one layer (or a very small number of layers), its Fourier representation is continuous in the vertical $z^\star$ direction, thereby giving rise to the continuous lattice lines. Due to the limited tilt range at which 2D crystal specimens can be imaged (usually $\leq 70°$), a cone of Fourier space cannot be measured, as shown in Fig. 4.1.

It is interesting to note that a 2D crystal sample is limited in the vertical dimension (there is no crystalline structure above and below the 2D crystal sample), but in the horizontal membrane plane direction the crystal is generally continuous, so that no empty boundary wall can be drawn around one crystal unit cell. The task of electron crystallography is to reconstruct the protein structure that populates the volume of the crystal unit cell. This is by definition a periodically continued unit cell in the horizontal two crystal axis directions. The Fourier transformation (calculated by the FFT) also assumes periodic boundaries in the vertical direction for the 2D crystal unit cell, but nevertheless the protein structure is embedded in a larger vertical space that contains empty voxels above and below the protein. In other words, in the vertical direction the crystal structure is masked by some masking function, so that a few nanometers above and below the protein structure the unit cell volume is empty. As the density for the protein structure is free to assume any density distribution, its Fourier representation itself also has no restrictions on its individual Fourier pixels (with the exception of the Friedel symmetry). However, the fact that the real-space volume is multiplied by a mask in the vertical direction means that the Fourier space

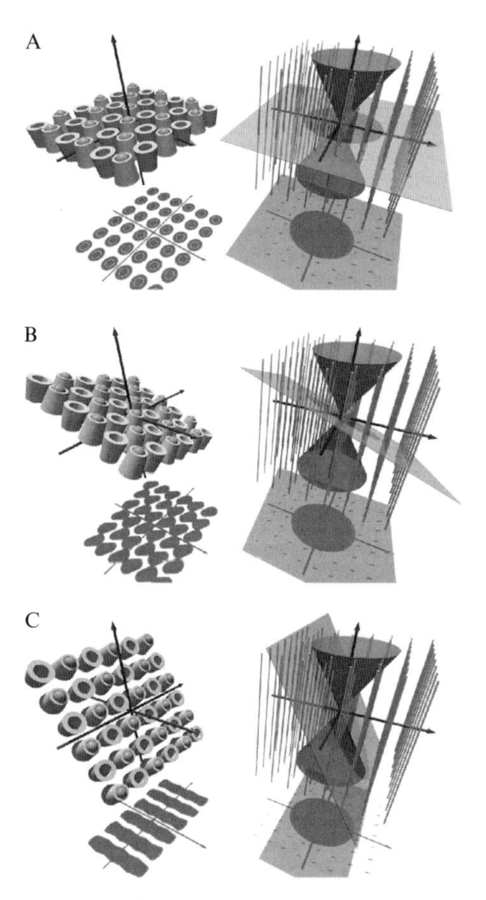


**Figure 4.1** Cartoon depicting the real-space (left) and Fourier space (right) representations of a 2D crystal. Three sample tilt angles are shown: (A) untilted, (B) 30° tilt, (C) 60° tilt. The imaging process in the microscope produces a projection image of the 2D crystal, as symbolized by the shadow under the crystal on the left side. This image is then corrected for lattice distortions and Fourier transformed. The resulting Fourier transformation corresponds to the central section that is indicated on the right as a plane. The values for amplitude and phase on the diffraction peaks in the Fourier transformation are measured and stored along the lattice lines at the position, where the lattice lines intersect with the plane (right). The crystal tilt angle defines the tilt of that plane, thereby defining the vertical $z^{\star}$ height of the measurement on the individual lattice lines. Each recorded image (left) thereby contributes only one (two due to Friedel Symmetry) measurement to a certain lattice line (right), while these measurements can under certain conditions be used for several lattice lines in case of crystal symmetry. Since the 2D crystals cannot be imaged at tilt angles higher than $\pm 70°$, the lattice line values in the indicated missing cone region in Fourier space cannot be experimentally determined, resulting in the so-called "missing cone" problem. (See Color Insert.)

representation of the protein structure is convoluted with the Fourier transformation function of that vertical mask. If the vertical mask in real space is a rectangle, then the convolution function in Fourier space is a Sinc function. (And if the vertical mask is a different shape, then the Fourier space convolution function is *not* necessarily a SINC function, see below.) A convolution of the Fourier representation of the protein structure with a vertical SINC profile means that every Fourier pixel is replaced by the SINC function multiplied by the real and imaginary values of the Fourier pixel. This results in a Fourier space volume, where only in the vertical direction there now is a neighborhood correlation. This fact is the basis for a recently developed software module that will become available in the 2dx software package in the near future, which allows to predict the Fourier space values in the missing cone based on the precisely measured (i.e., oversampled) Fourier pixel values outside of the missing cone (Gipson *et al.*, 2010). For this reason, the Fourier space representation of the crystal's unit cell has in the vertical direction a neighborhood correlation of adjacent Fourier pixels, while in the horizontal direction there is no such neighborhood correlation.

The purpose of processing one image of a tilted 2D crystal is then to obtain correct values for the amplitudes and phases for the points on the lattice lines, where they intersect with the tilted plane in Fig. 4.1. Once a sufficient number of measurements are available for every lattice line, a continuous function can be fitted to these measurements, which is usually a linear combination of SINC functions. These can then be equidistantly sampled, to give an evenly sampled Fourier volume, which is then Fourier back-transformed into real space to produce the final 3D map for the protein.

## 2.1. Initialization of a project

The first step to setup the processing for a new unknown 2D crystal is to determine the unit cell size and the symmetry of the crystal. The easiest way to do this is by recording untilted images of the crystal, either negatively stained or cryo-embedded. Images of negatively stained crystals with low amount of staining allow detecting the lattice also from poorly ordered crystals, due to the higher SNR of such images. In addition, when low amounts of stain are used, the stain preferably occupies the space between the supporting carbon film and the 2D crystal, so that the 2D crystal surface structure on the side toward the carbon film is stronger contrasted than the other crystal surface. Images of such preparations then show uneven surface staining, which allows distinguishing between up- and down-oriented molecules in the unit cells. Analysis of the difference map between maps from evenly stained and unevenly stained samples can allow understanding the 3D orientation of the molecules (Chiu *et al.*, 2007). Recording images of cryo-embedded samples, however, has the advantage of increasing precision and to eliminate the effect of uneven staining, to more reliably recognize screw axis symmetries.

The unit cell parameters of the crystals can be determined from the recorded images by one of several algorithms, as for example described in Zeng *et al.* (2007a). An automated algorithm searches for the reciprocal lattice that has the best fit to the diffraction spots of the Fourier transformation of the image of a nontilted 2D crystal. For a single-layered 2D crystal image, the algorithm generates lattice candidates from the difference vectors between the identified diffraction spots. The strongest linearly independent low-resolution difference vectors are then good candidates for defining the basis vectors of the 2D crystal lattice in reciprocal space. These reciprocal lattice basis vectors $u^\star = (u_1{}^\star, u_2{}^\star)$ and $v^\star = (v_1{}^\star, v_2{}^\star)$ can be translated into their corresponding real-space lattice basis vectors, as

$$u = \left(\frac{v_2^*}{u_1^* v_2^* - u_2^* v_1^*}, \frac{-v_1^*}{u_1^* v_2^* - u_2^* v_1^*}\right) \text{and } v = \left(\frac{-u_2^*}{u_1^* v_2^* - u_2^* v_1^*}, \frac{u_1^*}{u_1^* v_2^* - u_2^* v_1^*}\right) \quad (4.1)$$

The same formula applies to translate real-space vectors into their correspondents in reciprocal space. These real-space unit cell parameters for vector length and included angle are calculated once for the entire 2D crystal project.

The symmetry of an unknown crystal can be determined from the evaluated diffraction spot data for amplitudes and phases, using the MRC program *allspace* (Valpuesta *et al.*, 1994). Certain crystal symmetries can be excluded in advance, depending on the angle between the two unit cell vectors and their relative length. Landsberg and Hankamer (2007) provide a comprehensive overview of the symmetries relevant to 2D crystals. Although prior single-particle studies of the protein in question can give a hint of the symmetry of the building block of a crystal, it should be noted that the crystal nevertheless could exhibit a different overall symmetry (see, e.g., Schenk *et al.*, 2005 or Vonck *et al.*, 2002). Crystals with an apparent three-, four-, or sixfold symmetry might also be affected by twinning (Yeates and Fam, 1999), which can make it impossible to process the data unless the twinning is taken into account. If the symmetry does not exhibit a twofold axis parallel to the membrane plane, the crystals can adsorb to the supporting grid in an up- or down-oriented manner. In this case, the orientation of each 2D crystal on the carbon film support has to be determined correctly, before 3D merging of the data is possible. Some proteins, as for example aquaporins (Gonen *et al.*, 2004; Hiroaki *et al.*, 2006; Kukulski *et al.*, 2005; Schenk *et al.*, 2005), tend to crystallize in double layers. This by itself is not a disadvantage as it often increases crystal quality, but it has to be detected and the $z$ dimension of the crystal has to be adjusted during lattice line fitting and discretization. A double-layered crystal can often be identified—even in low magnification images—by the double fringe at its edges.

## 2.2. Processing of crystal images

Real-space images of 2D crystals can be processed as a collection of densely packed single particles or as a crystalline arrangement using Fourier methods. The single-particle approach offers the possibility to classify particles from the same image into, for example, different conformations of the same protein, and in the case of severe crystal disorder can better localize and align the protein complexes to a common coordinate system than crystal unbending usually does (Zeng *et al.*, 2007b). For this review, we will focus on the unbending approach only.

The unbending of 2D crystal images attempts to position each crystal unit cell into the location that corresponds to the ideal lattice (Fig. 4.2). This

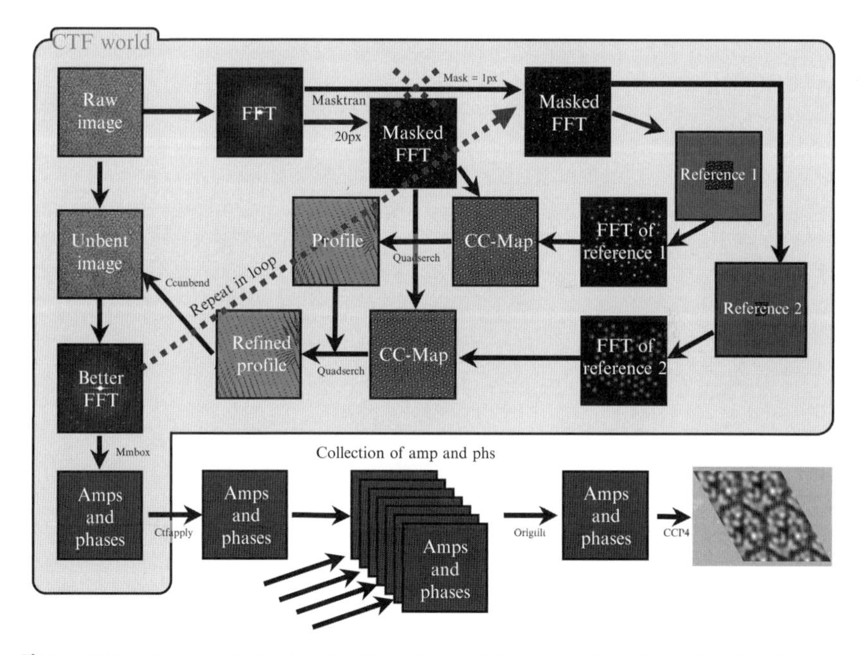


**Figure 4.2** Cartoon depicting the flow chart of the unbending algorithm for the case of a nontilted crystal sample. The raw image is modulated by the microscope's CTF (area on green background), which is typically only corrected after the unbending and evaluation of amplitudes and phases. For a detailed description see text. Figure reproduced from Gipson *et al.* (2007a). (For interpretation of the references to color in this figure legend, the reader is referred to the Web version of this chapter.)

unbending as implemented in the MRC program *ccunbend* typically moves small image blocks into the new location, and for this requires a distortion vector map that describes how each portion of the image has to be translated. This distortion vector information is obtained from a cross-correlation (CC) map between a reference map and a Fourier filtered version of the image. The reference map can be refined in several iterations, during which the reference patches become sharper and smaller, thereby allowing to trace at increasing confidence also the finer distortions in the crystal. *ccunbend* then uses this information to perform a block-wise unbending, whereby it shifts segments of the image, but does not rotate them. The alternative would be a smooth image unbending, using a spline-based pixel-wise interpolation as for example implemented in *2dx_ccunbendh.for*, or as also available in IPLT, to produce unbent images without fractures. In case of images of materials sciences specimens that show atoms at high SNRs, such smooth unbending would appear more beautiful to the human eye, as it does not risk cutting projections of individual atoms into fragments (Morgan *et al.*, 2009). Another advantage of the pixel-wise unbending is that it compensates for rotation of parts of the image. Nevertheless, for biological 2D crystal samples, the advantage of a smooth interpolation method is not clear. With a well-aligned TEM, even in badly distorted 2D crystal images, the individual protein complexes should be imaged without any distortions within themselves. In some image cases, a spline-based pixel-wise unbending might distort the shape of the individual proteins, after which their high-resolution structure would be disturbed.

The unbending approach bears the danger to alter the high-resolution ripples of the extended PSF of the microscope, so that this could result in a loss of high-resolution information (Philippsen *et al.*, 2007a). In that case, a solution would be to first correct for the effect of the contrast transfer function (CTF), and then correct the lattice distortions (Mariani, 2009).

This is easily implemented for images of nontilted samples, but becomes difficult for images of tilted samples due to the complicated nature of the CTF-analog for tilted specimens. Due to the defocus gradient across the image, the diffraction spots in the Fourier transform of the image are split into two subpeaks separated in the direction perpendicular to the tilt axis. The distance of the two subpeaks increases with the distance from the tilt axis. For small lattice distortions the original amplitude and phase information can be extracted using the tilted transfer function (TTF) algorithm introduced by Henderson *et al.* (1986) that convolutes the local area around each peak with a corresponding peak profile to restore the original diffraction peak. The TTF algorithm is implemented in the set of *tt**** programs, as for example *ttbox*, *ttmask*, and *ttrefine*. This algorithm, however, does not account for the asymmetry of the split subpeaks, which gets more prominent with increasing resolution as described by the tilted contrast transfer imaging function (TCIF) in Philippsen *et al.* (2007a). In addition the TTF algorithm cannot correct the full image but only the diffraction peaks. Therefore, for the correction of images of noncrystalline samples (e.g., in thin-section tomography), or of badly distorted crystalline samples, where sample information outside the diffraction peak itself is desirable for optimal distortion correction, or for the correction of crystalline samples at very high resolution, where the nonsymmetric nature of the split peaks plays a role, a more general algorithm for correction of the imaging effects encountered in tilted samples would be of great value. Unfortunately, even though the work of Philippsen *et al.* provides the mathematical basis for this TCIF, a direct inversion and therefore correction of the TCIF is beyond the current computing power of today's limited computational resources.

The description of the lattice distortions is typically obtained by the MRC program *quadserch* from a peak search within a CC map between the raw image and a reference map. The latter can be obtained by Fourier filtering the raw image and can be later refined by Fourier filtering unbent versions of the raw image. CC of this reference with the raw image then gives a CC map that contains peaks at the positions of the crystal unit cells. A peak search algorithm as implemented in the MRC program *quadserch* tries to predict the approximate position of a peak in that CC map, based on the neighboring lattice nodes, and then locates the actual CC peak in the immediate vicinity of the predicted position only. This process can be iteratively refined, to produce a precise map of the crystal unit cells even with images of extremely low SNRs.

This list of unit cell coordinates can then be used to crop single-particle windows with the individual crystal unit cells out of the crystal image, as is offered by the 2dx software (Gipson *et al.*, 2007b). These unit cell particles can then be used for a single-particle processing, using a maximum likelihood algorithm (Zeng *et al.*, 2007b). The MRC software instead uses these coordinates to unbend the raw image, with the MRC program *ccunbend*. The Fourier transformation of the unbent image is then evaluated for the amplitude and phase values of the diffraction spots with the MRC program *mmbox* (respectively *ttbox*), and each measurement is accompanied by a value for the average amplitude in the vicinity of the spot as background information, which then is translated into a figure of merit for the measurement. The amplitude for a spot is measured as the amplitude above the background value. The MRC software classifies the evaluated spots according to their SNR with so-called "IQ values": An IQ = 1 spot is at least seven times stronger than its background, an IQ = 7 spot is equally strong as the background, an IQ = 8 is weaker than the background, and an IQ = 9 spot has a negative amplitude, meaning that the Fourier transformation at that location has an intensity that is below the local average noise level. Spots of IQ values 1–3 are typically related to real measurements. However, since during the merging process the Fourier data are individually weighted according to their SNR, even spots of IQ values 4–6 might contribute valid information to the merging process, while spot measurements of IQ 7–9 can usually be discarded.

## 2.3. Processing of electron diffraction patterns

The first step in processing individual diffraction patterns encompasses the determination of the beamstop position and shape and takes care of the masking of the beamstop and the center of the diffraction pattern.

The second step deals with the subtraction of the background of inelastically scattered electrons, which are dominating at lower scattering angles, and of the background from electrons that were elastically scattered from noncrystalline parts of the sample (e.g., the carbon film), which may appear also at higher scattering angles. This can be done with the MRC program *backauto*.

The next step concerns the determination of the reciprocal lattice vectors. Lattice determination for diffraction patterns is somewhat more involved than lattice determination for Fourier transforms of images, as the origin of the lattice is not in the center of the image and the lattice origin itself together with the lowest resolution diffraction spots is not visible because it is blocked by the beamstop. Depending on the software used for analyzing the diffraction data, the lattice can either be determined automatically, as offered by the MRC program *autoindex*, or as also possible with IPLT, or the user has to index the lattice manually by providing the position and index of several diffraction spots.

Collapsed vesicles—a quite common form of 2D crystals—often give rise to a set of two epitaxial, twinned lattices. Since the unit cell vectors of the two lattices do not coincide, the two lattices can be separated. Unfortunately, there is often a fair number of diffraction peaks overlapping. These spots have to be excluded from further processing. In the case of electron diffraction pattern processing, it is often easier to discard these patterns as a whole, if enough single-layered patterns are available. The indexing of the lattice also defines the tilt geometry. Due to the lack of a defocus gradient in comparison to images, the lattice distortion is the only source to define the tilt geometry for that crystal sample.

Once the lattice and the tilt geometry are determined, the diffraction intensity of all the peaks can be calculated by integrating over all the pixels contained in the individual peaks, as done by the MRC program *pickauto*. The size of the box for integration has to be optimized depending on the sharpness of the peak, which varies not only based on the sample itself but also based on the recording conditions, that is, the relation of CCD pixel size and camera length. A too large box size leads in general to an increase in noise in the data, whereas a too small box size leads to an underestimation of the peak intensity, as parts of the peak might get excluded from integration.

The peak intensities are corrected for background contribution, which most commonly is measured by calculating a background average from the area surrounding the peak.

## 2.4. Merging and 3D reconstruction

Merging of the individual measurements for image data (amplitudes and phases) or diffraction data (amplitudes only) into one merged dataset comprises the following steps: (1) symmetrization, (2) scaling, (3) merging, (4) tilt geometry refinement, (5) lattice line fitting, and (6) discretization. The first two steps 1–2, symmetrization and scaling, deal with each image or diffraction pattern individually. The third step, merging, fuses these into one combined dataset. The last three steps 4–6 benefit from the combined dataset to refine and evaluate the data.

Here we will give a general overview of the different steps. The specific protocols for merging data can be found in the descriptions of the individual software systems below.

Amplitude and phase data are present as a function of their $h$ and $k$ Miller indices. These, together with the information about the tilt geometry, allow to assign for each reflection also a $z^{\star}$ height, which describes the vertical position of these measurements along the corresponding $h$, $k$ lattice line.

### 2.4.1. Symmetrization

During symmetrization the diffraction peak measurements are moved into the unique subarea of Fourier space (called the "asymmetric unit"), according to the earlier determined symmetry. The measured peak intensities do not have to be adjusted, as they are invariant in all symmetry operations. The symmetrization reduces the number of lattice lines to be fitted and increases the number of data points per fitted lattice line. In addition, comparison of symmetry related points gives an estimate of the quality of the dataset. If the symmetry of the crystal contains a screw axis, then the Fourier space representation of the data will have systematic absences. In this case, those Fourier pixels can then be set to zero to enforce those absences.

### 2.4.2. Scaling

In contrast to X-ray crystallography, where the whole dataset is recorded from one crystal, every 2D crystal image or electron diffraction pattern is recorded from a different crystal, which is usually embedded and/or imaged slightly differently. This leads to a varying absolute scale of the amplitudes, which means that the individual image data or diffraction patterns have to be scaled against each other before merging. For this, an image dataset or diffraction pattern can be compared to the reference in the area with a similar $z^{\star}$ range (common line scaling) or the patterns can be scaled according to their weighted resolution binned average intensity, similar to the well-known Wilson plot (Wilson, 1942) in X-ray crystallography. As this method does not depend on comparing individual reflections in a narrow $z^\star$ range, it avoids the necessity to collect a large fraction of diffraction data at low tilt angle to be able to scale the dataset. Potential errors in determination of the scale factors will negatively impact the lattice line fitting procedure, thereby contributing to the error in the amplitudes during lattice line fitting.

### 2.4.3. Merging

Once the individual datasets are scaled, they can be merged into one big dataset. This can be done with the MRC program *origtilt* for images and *mergediff* for diffraction patterns.

### 2.4.4. Tilt geometry refinement

An important step in merging a dataset is the refinement of the tilt geometry. For data from only slightly tilted specimens, small errors in the measured lattice vectors can falsely be interpreted as strong changes to the tilt geometry. For image data and specimen tilts $\leq 25°$, the tilt geometry is therefore more reliably be determined from the defocus gradient across the image. For electron diffraction data of specimens at low tilt, the tilt geometry determined from the lattice vectors cannot be considered reliable. At higher tilt angles $\geq 30°$, the tilt geometry assignment for an individual dataset can be refined from the lattice distortion. After the merging step, the available 3D dataset can also be used to refine the tilt geometry for each individual image or diffraction pattern, by comparison of the data with the 3D reference dataset (MRC program *origtilt*). Potential errors in the tilt geometry lead to errors in the $z^\star$ direction later on during the lattice line fitting and discretization.

### 2.4.5. Lattice line fitting by SINC interpolation

Before the dataset can be discretized, the lattice lines have to be fitted along $z^\star$. Most programs use a sum of *SINC* functions in one variation or another to fit the lattice lines (MRC program *latline*). The parameterized lattice line functions are then discretized into evenly sampled Fourier pixel values, to produce an evenly spaced Fourier volume with $h$, $k$, and $l$ dimensions. The sampling interval for the discretization is calculated from the thickness of the crystal. This Fourier volume is then back-transformed into real space to produce a 3D map for the protein structure.

### 2.4.6. Lattice line fitting by matrix transformation and SVD

An alternative approach introduced by Gipson *et al.* (2010) does not make use of the *SINC* interpolation approach. Instead, Gipson *et al.* attempt for each lattice line to find a linear transformation matrix that most closely can translate the regularly sampled pixels of the real-space form of that lattice line into the irregularly sampled measured data in Fourier space along that one lattice line, as depicted in Fig. 4.3. This matrix is determined by taking advantage of the finite thickness of the 2D crystal sample. The real-space form of that lattice line is thereby still placed in a reciprocal space concerning its $h$, $k$ coordinates, so that this representation is in a mixed space, where the horizontal $h$ and $k$ dimensions are reciprocal, while the vertical $z$ dimension is real. The required transformation matrix for that lattice line then transforms from this regularly sampled hybrid space into the irregularly sampled native Fourier space that contains the experimental measurements. This transformation is defined as

$$\mathfrak{I}_{z,z^*} x_z = \hat{x}_{z^*} \tag{4.2}$$

where $z$ is referring to the real-space pixel numbers along the real-space version of that lattice line, $z^\star$ is the index referring to the (unevenly sampled) experimental lattice line measurement, $\mathfrak{I}_{z,\ z^\star}$ is the discrete to

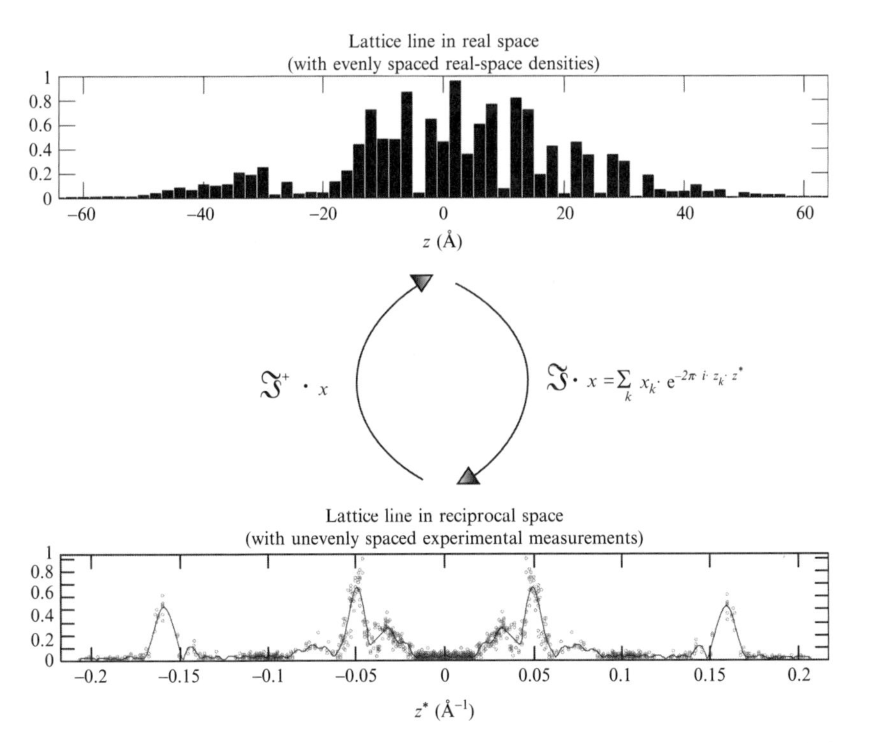


**Figure 4.3** The matrix $\mathfrak{I}$ accomplishes the linear transform between the regularly spaced real-space representation of a lattice line and the experimentally measured, unevenly spaced lattice line values in reciprocal space.

$z^\star$-sampled Fourier transform matrix, $x_z$ is an unknown regularly spaced real-space solution, and $\hat{x}_{z*}$ is the vector containing the experimentally measured, unevenly sampled Fourier space lattice line values sampled at $z^\star$.

Once in this form, the standard tools from statistics and linear analysis become available for solving this problem. The standard method currently used to find the least squares solution for this system involves truncated singular value decomposition (SVD) (Hansen, 1987). This process decomposes the Fourier matrix $\mathfrak{I}_{z,\,z^\star}$ into a quasi-diagonalized form

$$\mathfrak{I}_{z,z*} = U \cdot S \cdot V^{\mathrm{T}} \tag{4.3}$$

with $S$ being a diagonal matrix of "singular values" closely related to Eigenvalues, and $U$ and $V^{\mathrm{T}}$ being matrices to accommodate the remainder of the linear transformation.

The least squares solution to this problem can then be found by transposing the system, then inverting the nonzero values of $S$.

$$\mathfrak{I}^{+}_{z,z^*} = V \cdot S^{+} \cdot U^{\mathrm{T}} \tag{4.4}$$

This "pseudoinverse" matrix $\mathfrak{I}_{z,\,z^\star}{}^{+}$ then gives the least squares solution to Eq. (4.2), when applied to the list of spots for a given lattice line.

$$x_z = \mathfrak{I}^{+}_{z,z^*} \cdot \hat{x}_{z^*} \tag{4.5}$$

The threshold for the smallest singular value not assumed to be zero gives this method a large degree of control over noise sensitivity, resulting in a dynamically adjustable, noise-robust method for 3D structure determination.

Besides finding an optimal solution to this problem, this method provides the "real-space" terms directly, without the need for a sum of SINC functions fit and resampling of the fitted curves. These regularly spaced real-space lattice line values can then either be directly used for further processing or can be Fourier transformed, so that a regularly spaced *h*,*k*,*l* conventional Fourier space dataset is created.

## 2.5. Molecular replacement and model building

For well-ordered large ($>1\ \mu$m) 2D crystals, electron diffraction data collection can rapidly produce a dataset of high resolution that does not contain phase information. In this case, and if a homology model is available for the structure to be determined, it is possible to apply the technique of molecular replacement (Vagin and Teplyakov, 1997) that is commonly used for structure determination of 3D crystals in X-ray crystallography. Because the measurement of the unit cell is considered to be of limited precision due to the uncertainty in the magnification calibration and the flatness of the specimen grid, the unit cell size might have to be slightly adjusted to fit the reference structure. In addition—if the target resolution for the density map is beyond 3 Å—the molecular replacement profits from replacing the atomic scattering factors for X-rays by atomic scattering factors for electrons (Grigorieff *et al.*, 1996). If no suitable model for molecular replacement is available, the missing phase data have to be taken from the real-space image data. After processing and discretization of the image data, these can be combined with the discretized diffraction data to yield a high-resolution density map (Henderson *et al.*, 1990).

## 3. Image Processing with 2dx

In 2dx, a project is created and managed through the program 2dx_merge (Gipson *et al.*, 2007a). The recorded real-space images are aggregated by the program and grouped in directories according to their tilt angles. The program 2dx_image is used for the processing of an individual image.

### 3.1. Project initialization in 2dx

When starting 2dx_merge the user is requested to select a project directory. Just creating a new folder starts a new project. 2dx_merge will then create the "2dx_master.cfg" configuration file and the "merge" directory with its files in the project folder. After the initialization 2dx_merge will open a GUI window that shows different panels to display data, or provide specific functions and routines.

2dx_merge manages the recorded real-space images of 2D crystals and displays them in the central panel of the GUI. At first, images have to be imported via the "File" pull-down menu option "Import Images...", which lets the user add a selection of images. Image files can be in MRC or TIFF format. Information embedded in the file names of the image selection can be extracted in the import dialog by a regular expression in the "Translation" field. The regular expression is used to fill the following fields: the name of the protein, the tilt angle category, the micrograph number, and its subimage identifier. Each field is depicted by parenthesis in the regular expression. The result of the file name extraction is listed on the fly in a table of the dialog window, in which the first column can be used to set default values. The specified naming scheme is then used by 2dx_merge to organize the image files into separate directories for each tilt category and subdirectories for each image in the project folder. Upon the import the initialization script of 2dx_image is run on every single image to prepare it for processing. The unit cell size of the unknown crystal is determined by processing the first individual image and can then be passed from the individual image configuration file to the "2dx_master.cfg" configuration file.

## 3.2. Processing of an individual image in 2dx

Individual processing of a real-space image is done by 2dx_image (Gipson *et al.*, 2007b). 2dx_image is called by 2dx_merge when double-clicking a listed image in the central table of the GUI. 2dx_image can also be used as a stand-alone program. The general approach is to first process one image in 2dx_image, to determine the lattice and other project-wide parameters, and to search for optimal processing parameters for this type of images. With these, 2dx_image then should be able to process the majority of the following images automatically. After merging in 2dx_merge, the obtained merged 2D or 3D dataset can be used to refine the processing with 2dx_image, as depicted in Fig. 4.4.

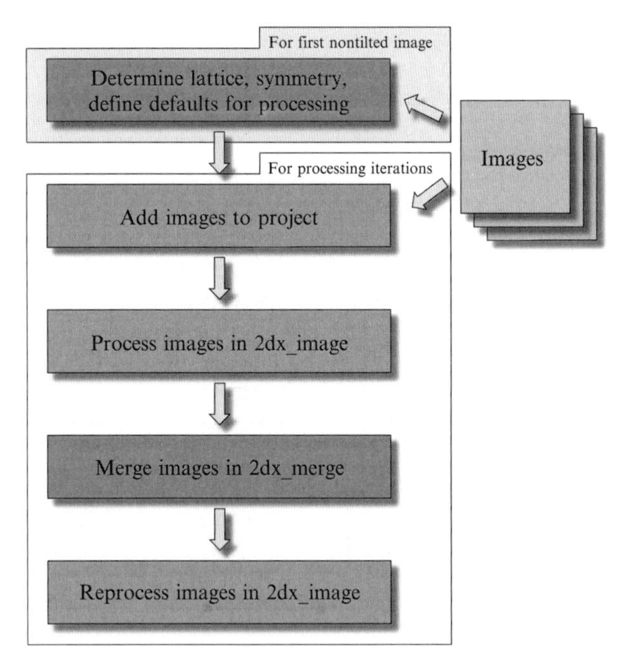


**Figure 4.4** Flow chart of the image processing in 2dx. The manual steps are highlighted in light gray. The first image is processed manually, during which project-wide parameters (e.g., lattice dimensions, symmetry) and optimal processing settings are determined. 2dx can then process the remaining images largely automatically. Merging remains a manual task, after which the reconstruction can iteratively be used to refine the individual image processing.

The GUI of 2dx_image consists of different panels displaying data and parameters as well as the image-processing routines. In the "Processing Data" panel parameters can be set, as for example the magnification of the microscope, but most values are calculated through the scripts held in the "Standard Scripts." Each script represents an individual processing routine, which not only calls the corresponding MRC and other stand-alone programs but also determines the parameters and data that are displayed in the GUI. The scripts order as listed in the panel defines the complete workflow of 2D crystal image processing: initialization, Fourier transformation, defocus and astigmatism calculation, tilt geometry determination, automatic lattice determination, lattice refinement, spot list determination, unbending, CTF correction, and projection map generation. For each script a description can be displayed by pressing the "Help" button. An image can be processed automatically by selecting all the standard scripts and pressing the "Run" button.

Most of the scripts call corresponding MRC programs for the actual image processing, and some additional functions were added. The defocus, astigmatism, and tilt geometry are calculated using ctffind2 (Grigorieff, 1998). The reciprocal crystal lattice is determined automatically: for images of tilted crystals, the approximate tilt geometry is first determined from the defocus gradient across the image. A hypothetical test lattice is then created from the known real-space dimensions of the crystal, which is then distorted according to the obtained approximate tilt geometry information, and translated into the reciprocal space. These test lattices are then rotated around the $z$ direction in the image by 360°, while each time taking the tilt geometry distortions into account, and while slightly varying lattice parameters, tilt geometry, and magnification, to search for the best fit with the measured lattice peaks in the calculated Fourier transformation of the image of the tilted crystals, until the best fitting lattice is found (Zeng *et al.*, 2007a). For the determination of several lattices in the case of multiple overlaying 2D crystals, the above algorithm is used to first find the strongest lattice. The diffraction peaks that correspond to that lattice are then removed from the list of diffraction spots, and the algorithm is iteratively reapplied to the remaining diffraction spots, until all possible lattices are found.

The lattice is refined using the MRC program mmlatref, but can also be refined manually by selecting diffraction spots in the Fourier transform displayed by the integrated full-screen image browser. The user is assisted in the spot selection by a peak search in the surroundings of a mouse double-click. The unbending of the 2D crystal image is split into multiple rounds. A first reference is created by using a one-pixel diameter Fourier mask. The following unbending rounds with wider Fourier masks then retrieve the structure's underlying signal. Iterative refinement of the reference from unbent images allows refining the unbending protocol, while each time the raw original image is unbent only once. Thanks to predictive lattice node tracking implemented in the MRC program team *quadserch* and *ccunbend*, unit cells can be localized with good precision, and this information can be used to extract the unit cell image stacks for the single-particle maximum likelihood algorithm that is available in 2dx (Zeng *et al.*, 2007b).

### 3.3. Merging in 2dx

2dx_merge not only assists in the logistics of managing the processing workflow but also performs the actual 3D merging of the image data. For this, 2dx_merge offers a 2D mode, where evaluated image data (amplitudes and phases) from nontilted crystals are merged into a final 2D projection map. 2dx_merge also offers a 3D mode, where data from tilted images can be included. This is done by first merging data from crystals at low tilt angles to the nontilted dataset, and later successively adding data from higher tilt angles. While 2dx_image allows optionally fully automated image processing, 2dx_merge so far only assists in the manual merging process, following the MRC program philosophy described above. Once a 3D reconstruction is produced, this can then be used to calculate synthetic reference maps for a certain tilt geometry, to be used by 2dx_image as ideal reference (using the MRC program *maketran*) to perform an optimized unbending. This iteratively can lead to a refined 3D reconstruction, which in turn can benefit a better unbending. In its current state (as of summer 2010), 2dx does not assist in the processing of electron diffraction patterns. 2dx is freely available (Gnu public license) at http://2dx.org.

## 4. Electron Diffraction Processing with XDP

### 4.1. Project initialization in XDP

The directory setup for a XDP project is flexible and can be chosen as the user wishes. Nevertheless it is advantageous to group diffraction patterns with identical nominal tilt angles into separate directories, as it reduces cluttering of the main project directory and simplifies the organization during the merging steps. Before using diffraction patterns in XDP, they have to be converted to data only format using em2em or a similar tool and renamed. XDP uses as image name a two-letter prefix to identify a project followed by a six-digit number. The first two digits are defined by the nominal tilt angle and the remaining four digits can be freely chosen. Running the script padd generates a parameter file for the image, calculates a histogram, prepares the diffraction pattern to be used in XDP, and converts it to the mrc file format. At this stage all the necessary files are created for the diffraction to be processed within the XDP GUI.

## 4.2. Processing of an individual electron diffraction pattern in XDP

The general workflow to be used in XDP is illustrated in Fig. 4.5. To process the diffraction data, XDP is started with the -c option followed by the two-letter project prefix. Diffraction patterns can be loaded using the screen menu; the starting image being the cut file. In XDP, determination of the beamstop is done manually by selecting four points defining a quadrangle covering the stem of the beamstop (Menu Box). The part of the beamstop covering the center of the beam is ignored as the center region of the diffraction pattern is excluded from processing later on during processing anyway to avoid over saturated areas of the CCD.

The lattice is determined in XDP by manually picking and indexing one point of the lattice in each quadrant of the diffraction pattern (Menu index). The user has to take care that the overall sum of $h$ and $k$ indexes is not equal to 0, because otherwise the lattice determination in the subsequent step will fail.

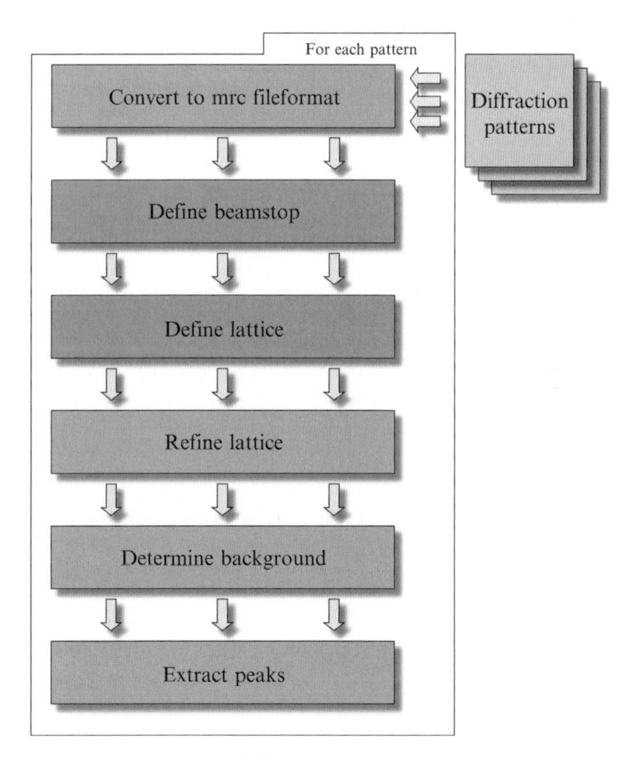


**Figure 4.5** Flow chart of the diffraction processing in XDP. The manual steps performed in the image viewer are highlighted in light gray. Each of the processing steps has to be run for every diffraction pattern individually (indicated by multiple arrows between the processing steps).

In the next step running diffint refines the user-defined lattice. At this point the lattice information is still saved in a separate file. To add it to the extra header section of the mrc image file, tomx is run. Running back generates the mean radial density profile, which is used for a global background correction of the diffraction pattern. As soon as the global background correction is done, the background corrected file (cmpct file) is available in the screen menu.

pick performs the diffraction peak integration and the correction of the diffraction intensities for the local background contribution. The average background for a peak is determined in the area around the peak.

## 4.3. Merging in XDP

In XDP, diffraction patterns are merged starting with the best, untilted diffraction pattern as reference. The other patterns are compared to the reference, scaled, and gradually merged into the combined dataset going from low tilt angles to high tilt angles. Diffraction patterns above 45° should only merged into the dataset after the initial merge is completely refined. Initially the data should only be merged at a relatively low resolution (~5 Å). The resolution can then be gradually increased during the refinement cycles. XDP uses the two programs `mergediff` and `syncfit` of the MRC program collection to perform the merging. `mergediff` is responsible for symmetrization, scaling, merging, and refinement and `syncfit` performs the actual lattice line fitting using a sum of *SINC* funtions and discretization of the fitted lattice line. To simplify the use of these two programs, XDP provides several command line scripts acting as frontends to `mergediff` and `syncfit`.

The merging is performed in a separate directory, which is subdivided into directories for the individual tilt angles. For every dataset to be merged, the files containing the lattice vectors, the parameters, and the integrated data points have to be copied to the corresponding merge subdirectory. The script `mergint` scans the directory of an individual tilt angle and generates a list with the patterns to be merged. `mergset` can then be used to set some basic merging parameters like the maximal resolution. The following steps in processing need some template files, which can be fetched with `mgsetdif_XX` (`mgsetdif_XXT` for tilted patterns) where `XX` is the project prefix. For a new project the template files have to be adjusted as they contain project specific parameters as for example the unit cell size. `mgsetdif_XX/T` provides additional sets of templates for tilt geometry and anisotropic scaling refinement. Merging and scaling within the directory of one tilt angle can be performed using `difcomp`. This script also allows refining the tilt geometry for nominally untilted images. Once all the tilt angle directories are processed, the overall merging can be performed by `difinit`. This is followed by `syncfit-1st.com`, which fits the lattice lines based on the merged data and discretizes the fit based on the entered sample thickness. The discretized data, which is stored in text format, can be converted to the mtz format by `f2mtz` and be used for molecular replacement and further processing in CCP4 (Collaborative Computational Project, 1994).

# 5. Electron Diffraction Processing in IPLT

In IPLT the project handling is done in the IPLT Diff Manager `giplt_diff_manager`. It serves as a platform for creation of new projects, adding and processing new diffraction data, verifying processed data, and organizing the data merging. Complementary to the Diff Manager, there is a command line interface called `iplt_diff`, which can be used to process data without having to start a GUI.

## 5.1. Project Initialization in IPLT

The general workflow for diffraction pattern processing in IPLT is depicted in Fig. 4.6. Starting the Diff Manager and selecting "New Project" creates a new project. The user has to enter the main directory for the project, a project prefix, the file format of the recorded diffraction patterns, and the unit cell and spacegroup parameters. The Diff Manager will create the project directory, the main settings file (project.xml), and open the central processing GUI. New diffraction patterns are added using the GUI and will be placed into separate directories within the project directory. New diffraction patterns can be encoded in any of the many formats readable by IPLT, but the dm3, tiff, and mrc formats are probably the most commonly used.

From one of the diffraction patterns the general beamstop shape has to be determined using the "Define Beamstop Mask" tool within the Diff Manager. This information is used later on for the automatic determination of the beamstop position.

## 5.2. Processing of an individual electron diffraction pattern in IPLT

The first step in the processing of individual diffraction patterns is the determination of the beamstop position and lattice for all diffraction patterns. In case that the diffraction pattern is distorted—for example, due to lens distortions encountered with an off-axis positioned CCD—IPLT provides also an option to fit the parameters for a barrel or spiral distortion of the lattice. The automatic lattice determination also includes a global fit of the background produced by inelastically scattered electrons. This fit then allows to extrapolate the position of the highest background intensity, which coincides with the origin of the lattice. This fit, however, is not used for the background correction for the diffraction peak intensities that

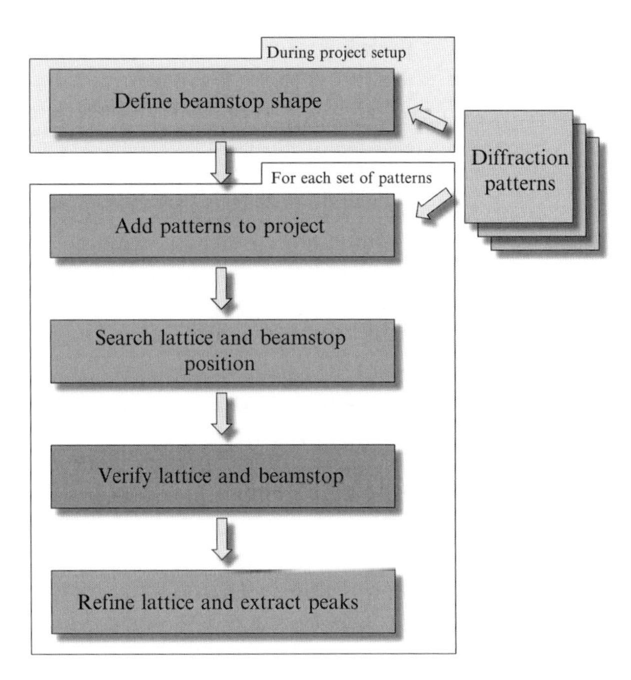


**Figure 4.6** Flow chart of the diffraction processing in IPLT. The manual steps performed in the image viewer are highlighted in light gray. IPLT can process a whole set of patterns in one go (indicated by single arrows between steps). The beamstop shape has to be determined only once during the project setup.

are later evaluated, because the model for the background is too coarse to account for the exact background variations, which are influenced by many effects like sample contamination, uneven illumination conditions, lens distortions, and electron scattering of the amorphous ice.

Once the lattice and the beamstop position are determined for a pattern, they can be visually checked and if necessary corrected in the Diff Manager using the "Verify Beamstops and Lattices" option. The rest of the processing steps do not need any user interaction and can be performed by selecting the "Refine Lattices and Extract Data" option. The automatic steps consist of a refinement of the lattice vectors, the peak extraction and correction for background contribution, and the determination of the tilt geometry.

## 5.3. Merging in IPLT

Automatic merging using the default parameters and directory setup can be done by selecting the "Merge" option in the Project tab. More customized merging can be performed by running the individual steps for merging manually in the merging tab of the Diffraction Manager.

Alternatively—if no GUI is desired—the identical merging steps can also be performed using the `iplt_diff_merge` command line executable.

Before the datasets are merged, they are symmetrized according to the symmetry defined during project setup. Following symmetrization the datasets are preliminary merged to provide a first reference for scaling. They are then scaled according to their weighted resolution binned average intensity to bring them to a common relative level. The symmetrization, premerging, and scaling steps can be performed by selecting "Scale" in the merging tab of the Diffraction Manager.

Once the datasets are scaled, they can be subjected to a refinement of the tilt geometry by selecting "Refine" in the merging tab. The tilt geometry is refined by varying tilt angle and tilt axis position slightly from the determined values and minimizing the difference between the dataset of the currently refined image and the overall merged dataset, which serves as reference. The search range for the tilt geometry is defined for its normal vector on the crystal plane. If parameterized in tilt angle and tilt axis, the search range for the tilt axis is a function of the tilt angle: for a pattern from an untilted specimen the tilt axis position is searched in a range of 180°, whereas for a pattern from a highly tilted sample only a small deviation of the determined tilt axis position is allowed.

After refinement of the tilt geometry, lattice lines can be fitted and a discretized dataset can be generated by sampling the lattice line fit at equally spaced intervals. Selecting the "Lattice Line Fit" option in the merging tab runs both steps together. IPLT uses a sum of squared SINC functions to fit the intensity data. The interval spacing for discretization is determined by the thickness of the original sample. Whereas the original datapoints can be negative if the background intensity is higher than the peak intensity, the fitted line will always have a positive or zero value. To improve the reliability of the fitting errors, IPLT provides a bootstrapping mode where subsets of the data are fitted several times and the fitting errors are calculated from the difference of the individual fits. Using the lattice line fitted data as reference, the sequence of scaling, tilt geometry refinement, and lattice line fitting can repeated in an iterative way to further refine the merged dataset.

Data files generated by IPLT can directly be used for molecular replacement and further processing in CCP4. No conversion is necessary.

## 6. Conclusions

The last decade has seen the consolidation of electron crystallography as a reliable alternative to X-ray crystallography and NMR for structural studies on membrane proteins, as it provides unique insights into their molecular arrangement in their lipid environment.

In practice, the realization of this potential advantage is not a trivial matter. Besides the experimental complexity, final success in structure determination is critically dependent on the image processing of the collected data. The involved image processing builds from concepts stemming from X-ray crystallography and electron microscopy, but new algorithms are required to account for the particular geometry, distortion treatment, and the integration of amplitude and phase information.

Earliest efforts to develop these specific software tools have resulted in the MRC software suite for electron crystallography. Initially conceived as a multitude of Fortran programs linked by a common file format, they were the first specific tool to be used by the community. The feedback provided by this extensive testing led to a constant improvement and eventually to the consolidation of a reliable processing pipeline.

The growing acceptance achieved by electron crystallography has driven the development of new software resources that make these already well-established algorithms available to a wider spectrum of users. Recently developed and freely distributed software packages like 2dx and IPLT envision the integration of all the different aspects of the data processing, offering the user a unified interface and reducing the requirement for computational expertise for the users. As these packages explicitly aim at attending the needs of a constantly increasing community of users, developments like the creation of graphical interfaces and production of exhaustive documentation and tutorials have been accompanied with an increase in direct contact with the users. The method of electron crystallography and image processing by the several programs including MRC, 2dx, and IPLT is now taught in the biannual series of workshops in electron crystallography (see http://2dx.org/workshop), and further support is ensured by the IPLT.org and 2dx.org web sites maintained at the University of Basel.

This general orientation toward an increased user-friendliness of already established computational techniques has not prevented a parallel effort in developing new algorithms that push the range of approachable problems beyond the current limitations. The maximum likelihood single-particle approach has been introduced as a powerful alternative to deal with images of 2D crystals, which is particularly strong in the case of severe crystal distortions. The recently developed method of Projective Constraint Optimization can compensate for the inherent geometrical constraints that used to limit the achievable vertical resolution of the technique, by filling data into the missing cone (Gipson *et al.*, 2010).

A further development direction is the cross-linking of the different software solutions. Experience has already identified some particular aspects of the image processing that are better resolved by some programs than others. Even if the attained file format compatibility already allows for simultaneous use of different approaches, a complete integration of the different software approaches will be advantageous. Nevertheless, the available electron crystallography software resources already constitute a useful toolbox for the full spectrum of potential users. The workflow guidance provided to users by software systems like 2dx gives them a good level of operational confidence in a short learning period. On the same time, experienced users of 2dx have the possibility of accessing the full range of thoroughly tested and optimized algorithms in the MRC software, and can adapt them to their particular needs. Finally, developers have an extended repertoire of tools that supports the creation of new features and conceptual advances. In case of 2dx, this is possible through the simple integration of independently compiled stand-alone programs that are called through the flexible 2dx interface. In case of IPLT, users can contribute to the software at several levels, ranging from high-level python scripts to low-level C++ routines. These new features can in turn get easily integrated into the existing general software frameworks that provide an efficient channel for distribution to the whole community.

In summary, electron crystallography is a mature method (Glaeser *et al.*, 2007) that provides important contributions to our understanding of the structure and function of membrane proteins. Electron crystallography is a powerful tool to study at intermediate resolution the conformational changes that membrane proteins can undergo in the membrane embedded state and to determine at the level of atomic resolution the structure of membrane proteins and their interaction with the lipid bilayers.

## ACKNOWLEDGMENT

This work was partially supported by the Swiss National Science Foundation and by the NIH grant GM084921.

## REFERENCES

Chiu, P. L., Pagel, M. D., Evans, J., Chou, H. T., Zeng, X., Gipson, B., Stahlberg, H., and Nimigean, C. M. (2007). The structure of the prokaryotic cyclic nucleotide-modulated potassium channel MloK1 at 16 Å resolution. *Structure* **15,** 1053–1064.

Collaborative Computational Project (1994). The CCP4 Suite: Programs for protein crystallography. *Acta Crystallogr.* **50,** 760–763.

Crowther, R. A., Henderson, R., and Smith, J. M. (1996). MRC image processing programs. *J. Struct. Biol.* **116,** 9–16.

Downing, K. H., and Hendrickson, F. M. (1999). Performance of a 2k CCD camera designed for electron crystallography at 400 kV. *Ultramicroscopy* **75,** 215–233.

Gipson, B., Zeng, X., and Stahlberg, H. (2007a). 2dx_merge: Data management and merging for 2D crystal images. *J. Struct. Biol.* **160,** 375–384.

Gipson, B., Zeng, X., Zhang, Z., and Stahlberg, H. (2007b). 2dx–User-friendly image processing for 2D crystals. *J. Struct. Biol.* **157,** 64–72.

Gipson, B., Masiel, D. J., Browing, N. D., Spence, J., Mitsuoka, K., and Stahlberg, H. (2010). Automatic recovery of missing amplitudes and phases in tilt-limited electron crystallography of 2D crystals (manuscript in preparation).

Glaeser, R., Downing, K., DeRosier, D., Chiu, W., and Frank, J. (2007). *Electron Crystallography of Biological Macromolecules*. Oxford University Press, USA.

Gonen, T., Sliz, P., Kistler, J., Cheng, Y., and Walz, T. (2004). Aquaporin-0 membrane junctions reveal the structure of a closed water pore. *Nature* **429,** 193–197.

Gonen, T., Cheng, Y., Sliz, P., Hiroaki, Y., Fujiyoshi, Y., Harrison, S. C., and Walz, T. (2005). Lipid-protein interactions in double-layered two-dimensional AQP0 crystals. *Nature* **438,** 633–638.

Grigorieff, N. (1998). Three-dimensional structure of bovine NADH:ubiquinone oxidoreductase (complex I) at 22 Å in ice. *J. Mol. Biol.* **277,** 1033–1046.

Grigorieff, N., Ceska, T. A., Downing, K. H., Baldwin, J. M., and Henderson, R. (1996). Electron-crystallographic refinement of the structure of bacteriorhodopsin. *J. Mol. Biol.* **259,** 393–421.

Hansen, P. C. (1987). The truncated SVD as a method for regularization. *BIT Arch.* **27,** 534–553.

Hardt, S., Wang, B., and Schmid, M. F. (1996). A brief description of I.C.E.: The integrated crystallographic environment. *J. Struct. Biol.* **116,** 68–70.

Henderson, R., and Unwin, P. N. (1975). Three-dimensional model of purple membrane obtained by electron microscopy. *Nature* **257,** 28–32.

Henderson, R., Baldwin, J. M., Downing, K. H., Lepault, J., and Zemlin, F. (1986). Structure of purple membrane from *Halobacterium halobium*: Recording, measurement and evaluation of electron micrographs at 3.5 Å resolution. *Ultramicroscopy* **19,** 147–178.

Henderson, R., Baldwin, J. M., Ceska, T. A., Zemlin, F., Beckmann, E., and Downing, K. H. (1990). Model for the structure of bacteriorhodopsin based on high-resolution electron cryo-microscopy. *J. Mol. Biol.* **213,** 899–929.

Hiroaki, Y., Tani, K., Kamegawa, A., Gyobu, N., Nishikawa, K., Suzuki, H., Walz, T., Sasaki, S., Mitsuoka, K., Kimura, K., Mizoguchi, A., and Fujiyoshi, Y. (2006). Implications of the aquaporin-4 structure on array formation and cell adhesion. *J. Mol. Biol.* **355,** 628–639.

Hovmöller, S. (1992). CRISP: Crystallographic image processing on a personal computer. *Ultramicroscopy* **41,** 121–135.

Jap, B. K., Zulauf, M., Scheybani, T., Hefti, A., Baumeister, W., Aebi, U., and Engel, A. (1992). 2D crystallization: From art to science. *Ultramicroscopy* **46,** 45–84.

Kühlbrandt, W. (1992). Two-dimensional crystallization of membrane proteins. *Q. Rev. Biophys.* **25,** 1–49.

Kühlbrandt, W., Wang, D. N., and Fujiyoshi, Y. (1994). Atomic model of plant light-harvesting complex by electron crystallography. *Nature* **367,** 614–621.

Kukulski, W., Schenk, A. D., Johanson, U., Braun, T., de Groot, B. L., Fotiadis, D., Kjellbom, P., and Engel, A. (2005). The 5 Å structure of heterologously expressed plant aquaporin SoPIP2;1. *J. Mol. Biol.* **350,** 611–616.

Kulik, V. (2004). Structure of bovine liver catalase solved by electron diffraction on multilayered crystals. Department of Physics, University of Osnabrueck, Osnabrueck, pp. 159.

Landsberg, M. J., and Hankamer, B. (2007). Symmetry: A guide to its application in 2D electron crystallography. *J. Struct. Biol.* **160,** 332–343.

Mariani, V. (2009). Transfer of tilted sample information in transmission electron microscopy. Biozentrum, University of Basel, Basel.

Mitsuoka, K., Hirai, T., Murata, K., Miyazawa, A., Kidera, A., Kimura, Y., and Fujiyoshi, Y. (1999). The structure of bacteriorhodopsin at 3.0 Å resolution based on electron crystallography: Implication of the charge distribution. *J. Mol. Biol.* **286,** 861–882.

Morgan, D. G., Ramasse, Q. M., and Browning, N. D. (2009). Application of two-dimensional crystallography and image processing to atomic resolution Z-contrast images. *J. Electron Microsc. (Tokyo)* **58,** 223–244.
Murata, K., Mitsuoka, K., Hirai, T., Walz, T., Agre, P., Heymann, J. B., Engel, A., and Fujiyoshi, Y. (2000). Structural determinants of water permeation through aquaporin-1. *Nature* **407,** 599–605.
Philippsen, A., Schenk, A. D., Stahlberg, H., and Engel, A. (2003). IPLT-image processing library and toolkit for the electron microscopy community. *J. Struct. Biol.* **144,** 4–12.
Philippsen, A., Engel, H. A., and Engel, A. (2007a). The contrast-imaging function for tilted specimens. *Ultramicroscopy* **107,** 202–212.
Philippsen, A., Schenk, A. D., Signorell, G. A., Mariani, V., Berneche, S., and Engel, A. (2007b). Collaborative EM image processing with the IPLT image processing library and toolbox. *J. Struct. Biol.* **157,** 28–37.
Schenk, A. D., Werten, P. J., Scheuring, S., de Groot, B. L., Müller, S. A., Stahlberg, H., Philippsen, A., and Engel, A. (2005). The 4.5 Å structure of human AQP2. *J. Mol. Biol.* **350,** 278–289.
Schmid, M. F., Dargahi, R., and Tam, M. W. (1993). SPECTRA: A system for processing electron images of crystals. *Ultramicroscopy* **48,** 251–264.
Sorzano, C. O., Marabini, R., Velazquez-Muriel, J., Bilbao-Castro, J. R., Scheres, S. H., Carazo, J. M., and Pascual-Montano, A. (2004). XMIPP: A new generation of an open-source image processing package for electron microscopy. *J. Struct. Biol.* **148,** 194–204.
Unwin, P. N., and Henderson, R. (1975). Molecular structure determination by electron microscopy of unstained crystalline specimens. *J. Mol. Biol.* **94,** 425–440.
Vagin, A., and Teplyakov, A. (1997). MOLREP: An automated program for molecular replacement. *J. Appl. Cryst.* **30,** 1022–1025.
Valpuesta, J. M., Carrascosa, J. L., and Henderson, R. (1994). Analysis of electron microscope images and electron diffraction patterns of thin crystals of phi 29 connectors in ice. *J. Mol. Biol.* **240,** 281–287.
Vonck, J., von Nidda, T. K., Meier, T., Matthey, U., Mills, D. J., Kühlbrandt, W., and Dimroth, P. (2002). Molecular architecture of the undecameric rotor of a bacterial Na(+)-ATP synthase. *J. Mol. Biol.* **321,** 307–316.
Wilson, A. (1942). Determination of absolute from relative X-ray intensity. *Nature* **150,** 152.
Yeates, T. O., and Fam, B. C. (1999). Protein crystals and their evil twins. *Structure* **7,** R25–R29.
Zeng, X., Gipson, B., Zheng, Z. Y., Renault, L., and Stahlberg, H. (2007a). Automatic lattice determination for two-dimensional crystal images. *J. Struct. Biol.* **160,** 353–361.
Zeng, X., Stahlberg, H., and Grigorieff, N. (2007b). A maximum-likelihood approach to two-dimensional crystals. *J. Struct. Biol.* **160,** 362–374.

CHAPTER FIVE

# Fourier–Bessel Reconstruction of Helical Assemblies

Ruben Diaz,[*] William, J. Rice,[*] *and* David L. Stokes[*,†]

## Contents



[*] Cryo-electron Microscopy Facility, New York Structural Biology Center, New York, USA
[†] Skirball Institute, Department of Cell Biology, New York University School of Medicine, New York, USA

*Methods in Enzymology*, Volume 482
ISSN 0076-6879, DOI: 10.1016/S0076-6879(10)82005-1

## Abstract

Helical symmetry is commonly used for building macromolecular assemblies. Helical symmetry is naturally present in viruses and cytoskeletal filaments and also occurs during crystallization of isolated proteins, such as Ca-ATPase and the nicotinic acetyl choline receptor. Structure determination of helical assemblies by electron microscopy has a long history dating back to the original work on three-dimensional (3D) reconstruction. A helix offers distinct advantages for structure determination. Not only can one improve resolution by averaging across the constituent subunits, but each helical assembly provides multiple views of these subunits and thus provides a complete 3D data set. This review focuses on Fourier methods of helical reconstruction, covering the theoretical background, a step-by-step guide to the process, and a practical example based on previous work with Ca-ATPase. Given recent results from helical reconstructions at atomic resolution and the development of graphical user interfaces to aid in the process, these methods are likely to continue to make an important contribution to the field of structural biology.

## 1. Introduction

The helix is a very utilitarian shape found in all walks of life. Everyday examples include a spring, a circular staircase, and a barber's pole. Likewise, biological organisms have adopted helices for a wide variety of tasks, ranging from a basic building block for macromolecular structures to higher-order assemblies such as the placement of scales on a pine cone. The double-helical arrangement of bases in DNA is perhaps the most famous example, followed by the ubiquitous $\alpha$-helix used as a structural element of many proteins. Filaments composing the cytoskeleton are composed of globular proteins that are themselves organized on a helical lattice, such as actin, microtubules, and myosin. Similarly, extracellular fibers such as flagella and pili on the surface of bacteria and collagen in the extracellular matrix are all based on the helical assemblies of their constituent proteins. These examples demonstrate one of the most important properties of the helix, which is the construction of an extended, fibrous element from a single type of building block. The elastic properties of the fiber are determined by the geometry of the helix and the local interactions between the building blocks. Due to the ubiquity of this molecular design, a great deal of work has gone into understanding the basic geometrical properties of the helix and into developing algorithms for reconstructing the structure of relevant biological assemblies.

# 2. Basic Principles

## 2.1. Mathematical description of a helix

In Cartesian coordinates, a continuous helix can be defined by a set of three equations, $x = r\cos(2\pi z/P)$, $y = r\sin(2\pi z/P)$, and $z = z$; these describe a circle in the $x$–$y$ plane that gradually rises along the $z$ axis (Fig. 5.1). The diagnostic parameters of the helix include the radius ($r$) and the repeat distance along the $z$ axis, or pitch ($P$). In cylindrical coordinates, which are the most convenient way to describe a helix, these equations become $r = r$, $\phi = 2\pi z/P$, and $z = z$. These equations describe a continuous helix, such as the continuous wire path of a spring (Fig. 5.1A), but biological assemblies generally involve a discontinuous helix, built with individual building blocks, or subunits, positioned at regular intervals along the helical path. These assemblies are characterized by the angular and axial interval between the subunits, $\Delta\phi$ and $\Delta z$, or alternatively are characterized by the number of subunits per turn of the helix. Although the helix in Fig. 5.1B has an integral number of subunits/turn (8), this parameter does not have to be integral, for example, there are 3.6 amino acids per turn of an $\alpha$-helix. Generally speaking, the structure repeats at some distance along $z$

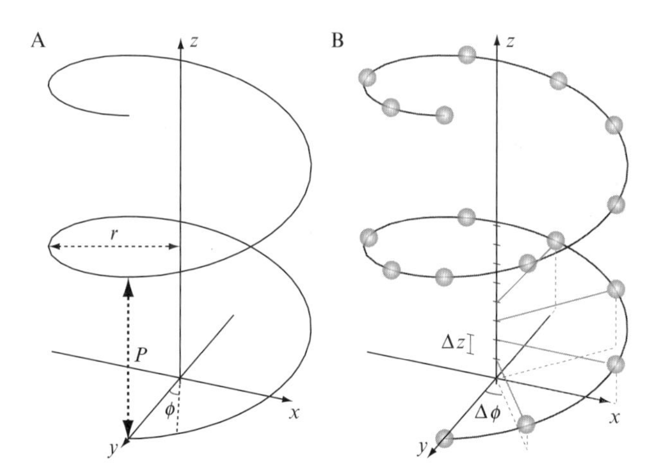


**Figure 5.1** Diagrams depicting the geometry of a helix. (A) A continuous helix is characterized by the pitch ($P$) and the radius ($r$) adopted by the spiral. Either a Cartesian coordinate system ($x$,$y$,$z$) or cylindrical coordinate system ($r$,$\varphi$,$z$) can be used. In either case, the $z$ axis corresponds to the helical axis. (B) Helical assemblies are generally composed of identical subunits arranged along the path of a continuous helix. This requires additional parameters, $\Delta\varphi$ and $\Delta z$, which describe the incremental translation and rotation between the subunits. This helix contains an exact repeat of eight subunits in one turn, thus, $\Delta\varphi = 45°$.

and the repeating element can be described as having $u$ subunits in $t$ turns, for example, 18 subunits in 5 turns for the $\alpha$-helix. From these values, the repeat distance ($c$) can be calculated as $c = u\Delta z = tP$.

The relationship between a helical assembly and a two-dimensional (2D) lattice is useful for understanding the process of 3D reconstruction. To construct a helical assembly, one starts by plotting a planar 2D lattice of points on a sheet of paper (Fig. 5.2A). To convert this planar lattice into a helix, a vector connecting two lattice points is designated as the "circumferential vector"; that is, a vector in the 2D plane that will be converted into an equator in the helical assembly (Fig. 5.2B). The sheet of paper is then rolled into a cylinder such that the lattice points at either end of the circumferential vector are superimposed. In 2D crystallography, unit cell vectors are arbitrarily chosen for the (1,0) and the (0,1) directions of the lattice. Based on these c, lines drawn through any series of lattice points can be assigned a so-called Miller index ($h$,$k$). The corresponding lattice planes produce discrete diffraction spots in the 2D Fourier transform of the lattice. After rolling the sheet into a cylinder, these lines become a family of helices spiraling around the helical axis at a constant radius. Because of the finite

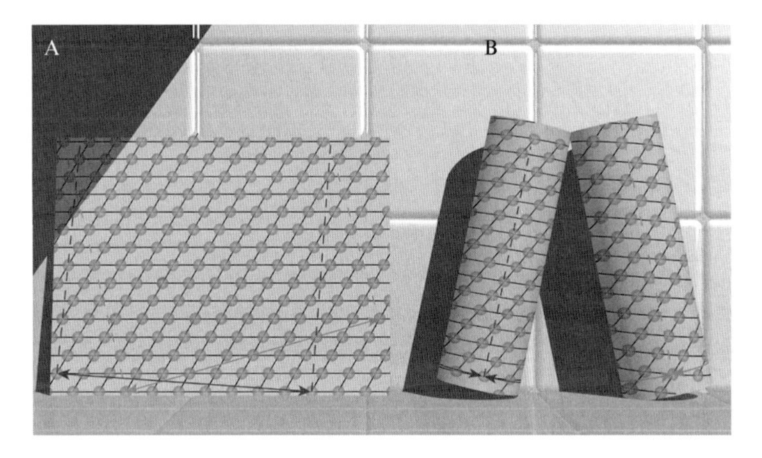


**Figure 5.2** Relationship between a planar 2D lattice and a helical assembly. (A) The 2D lattice is characterized by a regular array of points. An infinite variety of lines can be drawn through these points and each set of lines can be assigned a Miller index ($h$,$k$). For example, the black lines shown here could be assigned to the (1,0) and (0,1) directions. Two circumferential vectors are shown in green and red and these can be used to generate two unique helical structures shown in panel B. The dashed red and green lines are parallel to the $z$ axis in the resulting helical structures. (B) Helical lattices result from superimposing lattice points on either end of the circumferential vectors shown in panel A. Each set of lines through the 2D lattice are transformed into a family of helices. The start number ($n$) of each helix corresponds to the number of lines that cross the circumferential vector. The red circumferential vector produces helices with $n = 1$ and $n = 10$. The green circumferential vector produces helices with $n = -4$ and $n = 8$. For a left-handed helix, $n < 0$. (See Color Insert.)

lateral extent of the helix, each family of helices crosses the equator (circumferential vector) an integral number of times, and this parameter is defined as the start number ($n$) of the helical family. DNA is called a double helix because the start number of its fundamental helix is two. The peptide bond of an $\alpha$-helix generates a one-start helix. Generally, wider cylinders require larger values of $n$, for example, a microtubule has 13 protofilaments arranged in a 13-start helix. However, just as many different lines can be drawn through a 2D lattice, many different helical families are defined by a helical assembly. This situation is shown in Fig. 5.2, where the two sets of black lines drawn on the planar lattice can be assigned to the (1,0) and (0,1) lattice planes. Depending on the circumferential vector chosen, the (1,0) lattice planes give rise to 1-start or 4-start helices; the (0,1) lattice planes give rise to 8-start or 10-start helices. Note that the hand of the (1,0) helices is opposite in the two helical assemblies. By convention, left-handed helices have $n < 0$ and, in this case, $n = -4$.

## 2.2. Fourier transform of a helix

The mathematical derivation for the Fourier transform of a helix was initially described by Cochran *et al.* (1952) for understanding diffraction patterns from DNA and by Klug *et al.* (1958) for understanding fiber diffraction from tobacco mosaic virus. Application of these mathematics to electron microscopy was outlined by DeRosier and colleagues (DeRosier and Klug, 1968; DeRosier and Moore, 1970; Chapter Historical Perspective) and the relevant formulas are presented in the Appendix. The derivation explains why the Fourier transform of a helical assembly (e.g., Fig. 5.3A) is characterized by a series of horizontal "layer lines." These layer lines are analogous to the discrete reflections that characterize the Fourier transform of a 2D lattice, in that each layer line corresponds to one of the helical families discussed above, which in turn correspond to one set of lattice planes through the 2D lattice. Thus, each layer line can be assigned a Miller index ($h$,$k$). However, the $\delta$ function used to describe each reflection from an infinite 2D crystal is convoluted with a Bessel function ($J_n(R)$) to produce the corresponding layer line in the Fourier transform of a helix. This convolution is a direct consequence of the cylindrical shape of the helix; indeed, Bessel functions are also known as cylinder functions or cylindrical harmonics (Lebedev, 1972). The order, $n$, of the Bessel function appearing on a given layer line corresponds to the start number for the corresponding helical family. Thus, if the (0,1) family of helices are 10-start, then the $J_{10}(R)$ Bessel function appears on the corresponding layer line, where $R$ represents the radial coordinate of Fourier space.

Bessel functions are oscillating functions that appear in the integral used to express the Fourier transform in cylindrical coordinates (see the Appendix, Fig. 5.3C). Bessel functions generally have a value of zero at

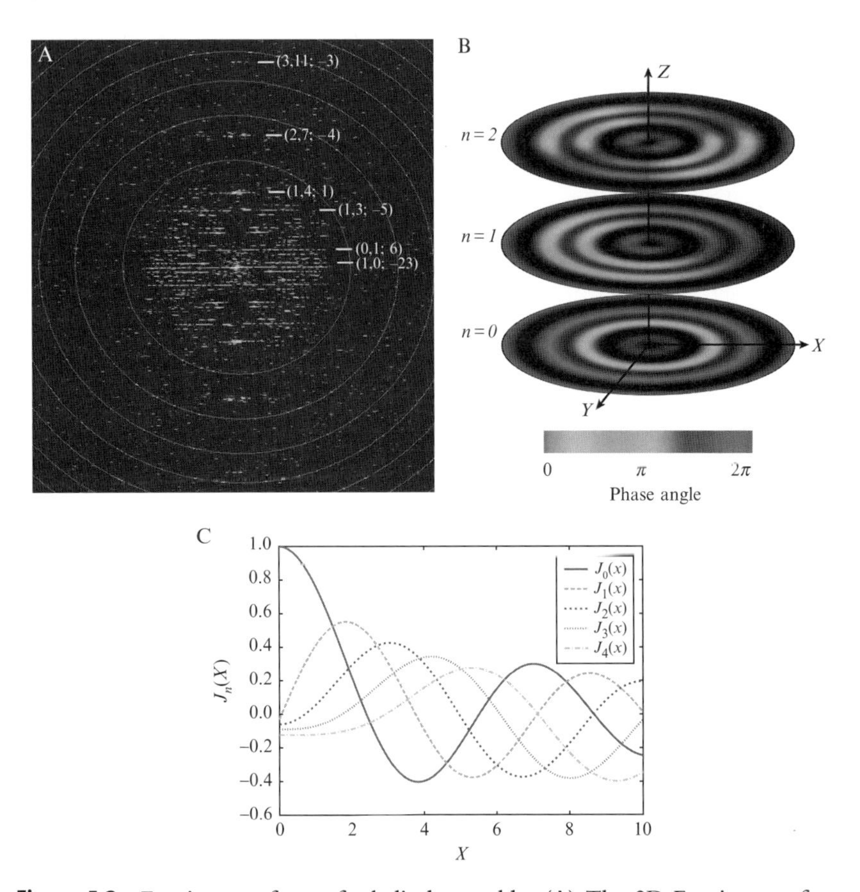


**Figure 5.3** Fourier transform of a helical assembly. (A) The 2D Fourier transform from a Ca-ATPase helical tube (e.g., Fig. 5.5A) is characterized by discrete layer lines that run horizontally across the transform. Each layer line corresponds to a helical family (cf., Fig. 5.2) and can be assigned a Miller index ($h$,$k$). The layer line running through the origin is called the equator and has a Miller index of (0,0). The vertical axis is called the meridian and the transform has mirror symmetry across the meridian. The start number of each helix ($n$) is shown next to each Miller index ($h$,$k$; $n$), and this start number determines the order of the Bessel function appearing on that layer line. The red circles indicate the zeros of the contrast transfer function and the highest layer line (3,11) corresponds to 10 Å resolution. (B) 3D distribution of three layer lines from a hypothetical helical assembly with Bessel orders of 0, 1, and 2, as indicated. The $Z$ axis corresponds to the meridian, the $X$ axis corresponds to the equator, and the $Y$ axis is the imaging direction. Thus, the $X$–$Z$ plane would be obtained by Fourier transformation of a projection image (e.g., panel A). The amplitude of the 3D Fourier transform is cylindrically symmetric about the meridian, but the phase (depicted by the color table at the bottom of (B)) oscillates azimuthally, depending on the Bessel order. Thus, the phase along the $n = 0$ layer line (equator) remains constant; the phase along the $n = 1$ layer line sweeps through one period, and the phase along the $n = 2$ layer line sweeps through two periods. (C) Amplitudes of Bessel functions with orders $n = 0$–4. Note that as $n$ increases, the position of the first maximum moves away from the origin. (See Color Insert.)

the origin and rise to a maximum at a distance corresponding to $\sim n + 2$. The zero-order Bessel function, which typifies the equator of all helices, is an exception in having its maximum at the origin. Thus, the order of the Bessel function appearing on a given layer line can be estimated from the radial position of its first maximum. This estimate is not precise because the values along the layer line are influenced by the Fourier transform of the molecules composing the helix, but it is a reasonable approximation and allows one to "index" the helical diffraction pattern. From a mathematical point of view, data along the layer lines is known as $G_{n,\ell}(R)$, where $\ell$ is the height of the layer line along the $Z$ axis of the Fourier transform.

## 2.3. Advantages of helical symmetry for 3D reconstruction

In order to calculate a 3D structure, one must index the Fourier transform by assigning the correct values of $n$ to all of the layer lines composing the Fourier transform. Once accomplished, these indices allow one to apply an inverse Fourier–Bessel transform to the layer line data. This calculation is a two-step process. In the first step, the Fourier–Bessel transform is applied individually to data from each layer line ($G_{n,\ell}(R)$) to produce a set of $g_{n,\ell}(r)$, where $r$ is the radial component in a real space cylindrical coordinate system (see Appendix). The $g_{n,\ell}(r)$ are real-space functions that characterize the density wave along the direction of the corresponding helical families. By analogy, the Fourier transform of a single reflection from a planar, 2D lattice would characterize a plane wave along the corresponding direction in real-space. Unlike the situation with planar crystals where density remains in the plane, the helical crystals have cylindrical symmetry. Thus, the $g_{n,\ell}(r)$ spiral around the helical axis and by summing them all up, one is able to obtain a 3D structure: $\rho(r,\phi,z)$. Unlike 2D crystals that must be tilted to produce a 3D structure, full recovery of the 3D Fourier data set is possible from a single image of a helical assembly. This is a major advantage, as the tilting of planar structures inevitably leaves a missing cone of information and, furthermore, the logistics of imaging becomes increasingly difficult as the tilt angle rises.

The availability of 3D information from a single projection image of a helical assembly can be appreciated by applying cylindrical symmetry to the layer lines that make up the 2D Fourier transform. Basically, the 3D Fourier transform can be generated by rotating the 2D transform around the helical axis. As the layer lines sweep around this axis, known as the meridian, the amplitude remains constant. This azimuthal invariance of the amplitude imparts mirror symmetry to the Fourier transform, which is an important characteristic in recognizing objects with helical symmetry. On the other hand, the phase varies as a function of the azimuthal angle. Specifically, the phase oscillates through an integral number of periods as the azimuthal angle in the Fourier transform ($\psi$) varies from 0° to 360°, with the number of periods traversed by a given layer line being equal to its Bessel order,

$n$ (Fig. 5.3B). This behavior explains why it is so important to correctly index the helical symmetry, as an incorrect assignment will completely scramble the phases of the 3D Fourier data. This phase behavior also places an important constraint that aids indexing the Fourier transform of a helical assembly with unknown symmetry. In a projection image, the phases along a layer line on opposite sides of the meridian are constrained to differ by either 0° or 180°. Specifically, if $n$ is even, the phase will have traversed an integral number of periods as it moves half-way around the meridian to the other side of the 2D Fourier transform (i.e., $\psi = 180°$). On the other hand, if $n$ is odd, this phase will be half a period out of sync and will therefore differ by 180°.

## 3. Step-by-step Procedure for Helical Reconstruction

As with all methods of 3D reconstruction, there are many individual steps required for helical reconstruction and a variety of software packages have been developed to manipulate the images and perform the relevant calculations. In particular, the classic paper by DeRosier and Moore (1970) provided a theoretical background and outlined the basic steps provided by the software developed at the Medical Research Council Laboratory of Molecular Biology (the MRC package, Crowther *et al.*, 1996). This software has been updated and customized by a number of groups, including the DeRosier laboratory (Brandeis package, Owen *et al.*, 1996), the Scripps Research Institute (Phoelix package, Whittaker *et al.*, 1995), the Toyoshima laboratory (Yonekura *et al.*, 2003b), the Kikkawa laboratory (Ruby-Helix package, Metlagel *et al.*, 2007), the Unwin laboratory (Beroukhim and Unwin, 1997) among others. The steps described below represent Unwin's approach and have been implemented in our own Python-based graphical user interface called EMIP, which also provides a platform for a wide variety of other image processing tasks, including 2D crystallography and tomography (Fig. 5.4).

### 3.1. Scanning the film image

Conventionally, film has been the medium of choice for recording images of helical assemblies. This is because their filamentous shape has made it difficult to encompass an entire assembly with a sufficiently fine sampling interval within the frame of a digital camera. In particular, the membranous tubes produced by ordered arrays of membrane proteins can be very long (5–10 μm) and it is not possible to select a straight, well-ordered portion during low-dose imaging. If recorded on film, a relatively low magnification image can be screened by optical diffraction and an optimal area can

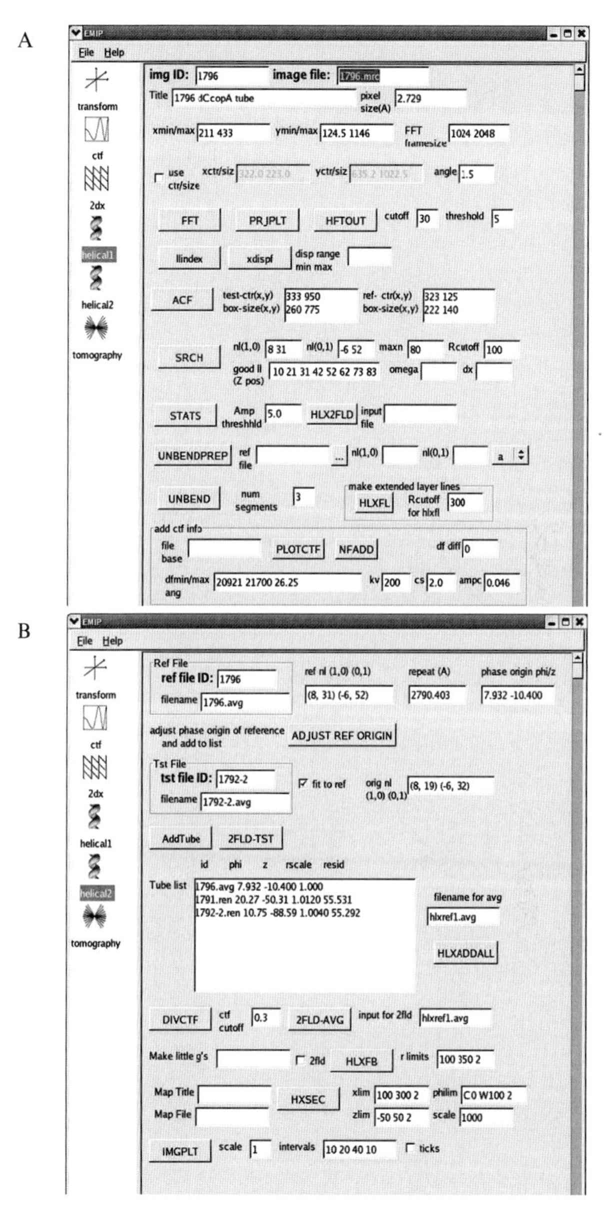


**Figure 5.4** Graphical user interface for helical reconstruction. This program (EMIP) collects information from the user and guides him/her through the various steps required for 3D reconstruction. Popup text provides information about each of the steps and a right-click on each button displays relevant log files. (A) Steps in processing individual tubes include masking, Fourier transformation, finding the repeat distance, searching for out-of-plane tilt, unbending, and addition of CTF parameters. (B) Steps in averaging Fourier data together and calculation of the 3D map. This user interface was written in Python using the wxPython library for creation of graphical widgets and is available upon request.

then be digitized with a high precision film scanner at a small sampling interval (e.g., 7 $\mu$m) with minimal degradation of the resolution (Henderson *et al.*, 2007). In contrast, current CCD cameras not only have limited numbers of pixels, but the modulation transfer function is significantly worse than film (Sherman *et al.*, 1996; Zhang *et al.*, 2003). As large-format, direct detection digital cameras with better modulation transfer functions become available (Faruqi, 2009), they should also be suitable for imaging. In the meantime, the film should be rotated on the scanner such that the long axis of the helix is parallel to the scan axis (Fig. 5.5A).

## 3.2. Orient, mask, and center the helix

Prior to calculating a Fourier transform, the boundaries of the helical assembly are defined in order to mask off peripheral material that would otherwise contribute only noise. The width of the mask should be slightly larger than the apparent width of the helical assembly; the length can be arbitrarily chosen, for example, 1024 pixels. The masked edge should be apodized, and the average density at the edge of the mask should be subtracted from the image in order to minimize Fourier artifacts from this potentially high contrast feature. For the Fourier transform, the box is padded to 512 pixels in the horizontal direction, producing an image with dimensions of 512 $\times$ 1024. The helical assembly must be centered and two alternate methods can be used. The first method involves calculating a projection of the helical assembly along the $y$ axis and comparing the right and left sides of the resulting 1D projection plot; if properly centered, the two sides will show minimal deviations from the average profile (Fig. 5.5B). This plot is also useful in measuring the radius of the tube, which will be important later. The second method involves examining phases along the equator of the Fourier transform. If properly centered, the equator should be real (Fig. 5.5C). If a phase gradient exists, the center should be shifted until it is minimized. Finally, the alignment of the helical axis with the scan raster of the image should be refined. This alignment involves an in-plane rotation, which can be evaluated both from the layer lines in the Fourier transform and from the projection plot. The equator and other layer lines that extend to a high radius should be inspected and the tube rotated until they remain aligned with the $x$ axis of the Fourier transform. Also, the relief of the projection plot will be maximal when the helical axis is rotationally aligned.

Some software packages provide a facility for straightening curved helical assemblies (Metlagel *et al.*, 2007). This is typically done by marking points along the helical axis and then using a polynomial fit to model the path of the axis. The image densities are then reinterpolated to straighten the helical assembly. For higher resolution analysis, bent assemblies are generally discarded as this bending most likely produces distortions in the subunits.

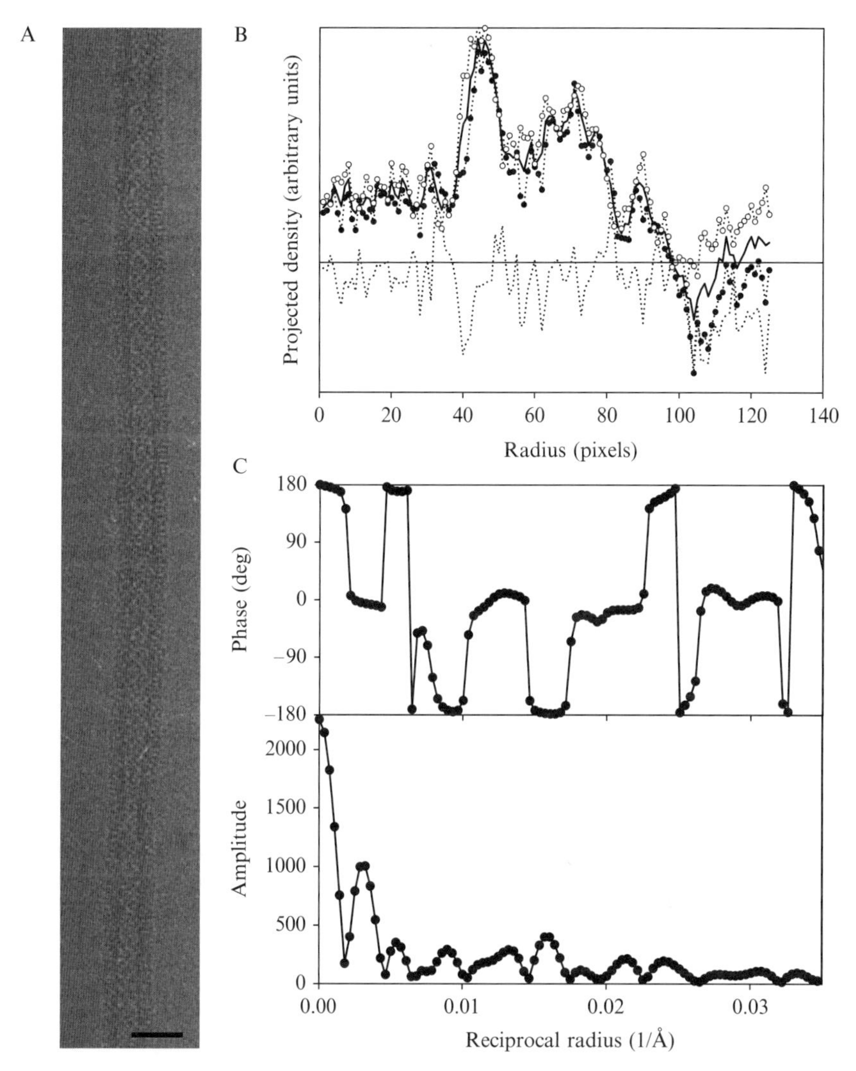


**Figure 5.5** Masking and centering of an individual helical assembly. (A) An image of a helical tube of Ca-ATPase. Only the straightest part of the assembly would be used for reconstruction, that is, the upper half. Scale bar corresponds to 60 nm. (B) Plot of density after projecting the image along the helical axis. The origin of the plot corresponds to the center of the tube and density from right (○) and left (●) sides have been plotted together with their average (solid line). The difference between the two sides is plotted as a broken line. The outer radius of this tube is ~115 pixels, which falls just outside the negative density ripple caused by the contrast transfer function. (C) Amplitude and phase data from the equator. The fact that phases are close to either 0° or 180° indicates that the tube is well centered.

However, if the radius of curvature is large relative to the repeat distance, then an "unbending" approach, discussed later, can be used to improve resolution.

### 3.3. Determine repeat distance and interpolate the image

In early work, helical assemblies were relatively short and produced layer lines that spanned multiple pixels in the axial direction of the Fourier transform. Therefore, a 1D plot of the layer line data ($G_{n,\ell}(R)$) was interpolated from the neighboring pixels in the Fourier transform (Bullitt *et al.*, 1988). For longer, better ordered helical assemblies such as those from the nicotinic acetylcholine receptor (Toyoshima and Unwin, 1990) and Ca-ATPase (Toyoshima *et al.*, 1993a), these layer lines were very narrow and, if they fell between the pixels in Fourier space, the observed amplitude was greatly diminished. The alternative of interpolating image densities was therefore developed, not only to align the helical axis with the sampling lattice, but also to ensure that an integral number of repeats was included in the image, thus producing layer lines lying precisely on a pixel in Fourier space. This repeat distance is determined by autocorrelation of a small segment of the helix at one end with the rest of the assembly. This autocorrelation inevitably produces many peaks, indicating the presence of approximate repeats at various sites along the length of the helix. This result reflects the fact that the helix frequently produces an approximate repeat when the azimuthal angle returns to the starting position. Depending on the angular interval of the subunits, some of these repeats are closer than others and an "exact" repeat may never be found within a reasonable length of a given assembly. However, interpolation of the image based on one of these approximate repeats is generally sufficient to place the layer lines very close to integral pixel values in the Fourier transform.

A problem can occur if the helical assembly has a very short repeat distance, such that multiple repeats (say 5) are included in a length of $\sim$1024 pixels. In this case, layer lines are constrained to exist only on multiples of 5 pixels in the transform, creating a situation where layer lines with different Bessel orders fall at the same axial position in the transform. This situation should be avoided, because the corresponding Bessel functions will overlap and produce phase interference. Often, a different repeat distance can be chosen that will resolve this redundancy. From a practical point of view, once a repeat distance has been chosen, the image is stretched along the axial direction such that an integral number of repeats are contained within the fixed box size of the image (typically 2048 pixels). The precise positions of the layer lines depend on the particular repeat distance that has been selected and, because each assembly may be stretched differently, will not necessarily correlate from one image to the next.

## 3.4. Indexing of layer lines

Indexing is perhaps the most vexing of the steps because of the lack of objective measures for verifying the correct answer. This process starts by assigning arbitrary Miller indices to the layer lines in order to define the unit cell, similar to the indexing of a planar, 2D crystal. However, the mirror symmetry in the Fourier transform of a helix creates an ambiguity in this assignment. This mirror symmetry can be thought of as representing two overlapping lattices from the near and far sides of the cylindrical structure. Thus, only one side of the layer line should be considered in the assignment of these Miller indices. In the example in Fig. 5.6, only the left side of the (1,0) and the right side of the (0,1) are relevant to the indexing. All the other layer lines will correspond to linear combinations of these primary layer lines, for example, (1,1), (2,0), and (1,2). If either the (1,0) or (0,1) has been assigned to the wrong side of the meridian, then there will be inconsistency in the locations of the higher-order layer lines. Note that although the axial positions of the layer lines fall exactly on the lattice, the radial locations are less exact due to the convolution of 2D lattice points with Bessel functions that extends these points in the radial direction.

After assigning Miller indices, the Bessel order for each of the layer lines must be determined. For this process, it is essential to understand the argument of the Bessel function that characterizes the distribution of amplitude: $J_n(2\pi rR)$, where $r$ is the radius of the helical assembly in real-space and $R$ is the radius along the layer line in Fourier-space. The location of the first maximum of $J_n(X)$ is well established (Fig. 5.3C) and by noting the value of $R$ for this maximum along a given layer line, one can use a measured value of $r$ to estimate the corresponding value of $n$. For this estimate, it is wise to use the maximum real-space radius of the helical assembly determined from the projection plot (i.e., beyond the ripple from the contrast transfer function (CTF), Fig. 5.5) and the Fourier-space radius at which the layer line amplitude first starts to rise toward the first maximum. By creating a table containing the measured radii from several layer lines and the Bessel orders calculated for each Miller index (e.g., the Bessel order for the (1,1) layer line is equal to the sum of Bessel orders from the (1,0) and (0,1) layer lines, Fig. 5.6B), one can strive to achieve an indexing scheme that produces the observed real-space radius of the assembly (Table 5.1, which is consistent with the Fourier transform in Fig. 5.6, but slightly different from that in Fig. 5.3). Higher-order layer lines that lie close to the meridian represent important constraints to this procedure, because although the difference between $n = -21$ and $n = -23$ for a low-order (e.g., 1,0) layer line may not be distinguishable, the consequent differences to the higher-order layer line (e.g., $n = 0$ or $n = -2$ for the (1,3)) may be definitive. EMIP aids in the construction of this table or, alternatively, a graphical software tool has been described by Ward *et al.* (2003) that directs the indexing process.

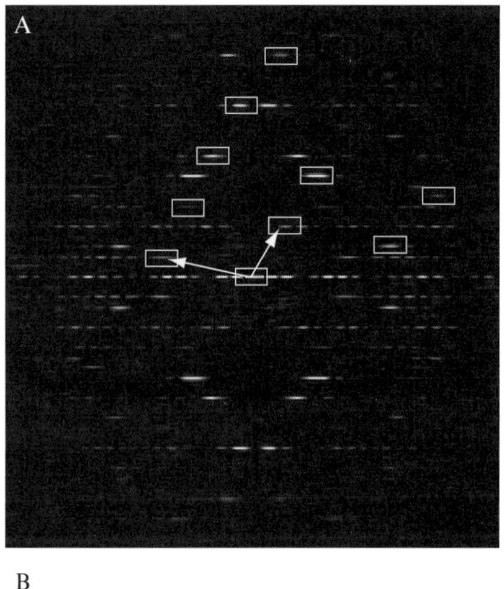


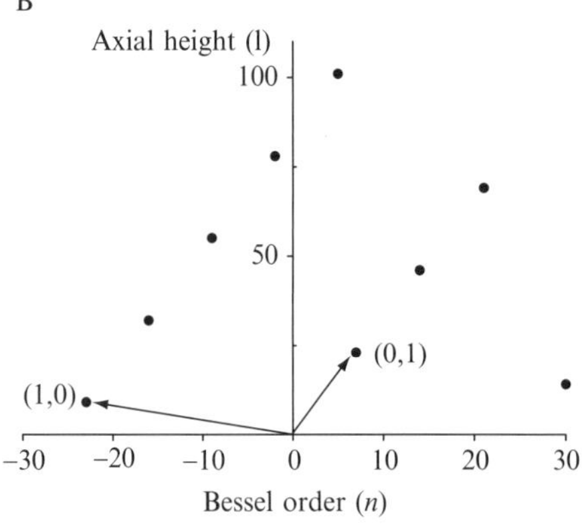


**Figure 5.6** Indexing of layer lines in the Fourier transform of a helical assembly. (A) Overlay of the near-side lattice on the Fourier transform of Ca-ATPase. (B) Corresponding plot of Bessel order ($n$) versus layer line height ($\ell$). Assignment of (1,0) and (0,1) layer lines is arbitrary, but once chosen then all of the other visible layer lines should be either a linear combination of these two, or a consequence of *mm* symmetry in the transform. The radial positions of the layer lines are distorted relative to a planar 2D lattice due to the behavior of Bessel functions, which have a nonlinear relationship between the radial position of their first maximum and their order, $n$. Nevertheless, the axial positions of the layer lines should be accurate.

This indexing procedure leaves an ambiguity with regards to the hand of the structure. A positive Bessel order indicates that the corresponding family of helices is right-handed and the mirror symmetry in the Fourier transform is consistent with the opposite conclusion. In order to resolve this ambiguity, the most straight forward approach is to use rotary shadowing to view the top side of the helical assembly. Alternatively, one can tilt the sample in

**Table 5.1** Indexing layer lines

| Miller index (*h*,*k*) | Bessel order (*n*) | Measured Fourier-space radius (pixels$^{-1}$) | Calculated real-space radius (pixels) |
|---|---|---|---|
| 1,0 | −23 | 36 | 118.61 |
| 0,1 | 7 | 12 | 116.80 |
| 1,1 | 27 | 27 | 111.67 |
| 0,2 | 14 | 22 | 120.75 |
| 1,2 | −9 | 14 | 125.72 |
| 0,3 | 21 | 34 | 115.04 |
| 1,3 | −2 | 4 | 126.31 |
| 1,4 | 5 | 9 | 115.89 |

the electron microscope about an axis normal to the helical axis. At certain angles that depend on the pitch of the helix, this tilt will produce a scalloped appearance along one side of the helix, and a smooth appearance along the other side. As described by Finch (1972), the sense of the tilt and the hand of the helix determine whether the scalloped edge appears on the right or left side of the helical assembly.

## 3.5. Refine center and determine out-of-plane tilt

The helical axis does not necessarily lie precisely parallel to the imaging plane and the resulting out-of-plane tilt creates systematic shifts to layer line phases that must be taken into account. This situation is illustrated in Fig. 5.7, which shows how this tilt results in a sampling of the layer line at azimuthal angles ($\psi$) other than 0° and 180°. As shown in Fig. 5.3B, the azimuthal dependence of the phase depends on the Bessel function, so once the angle of out-of-plane tilt ($\Omega$) is determined, a correction can be applied to each point along the layer line to recover the phase corresponding to $\Omega = 0$. The out-of-plane tilt is determined by comparing phases between the two sides of each layer line for a range of values for $\Omega$. Because this phase relationship is also dependent on the centering of the helical assembly, both $\Omega$ and an $x$-shift ($\Delta x$) are refined together. This procedure relies heavily on the correct indexing of the helical symmetry and the magnitude of the phase residual can therefore be used to distinguish between two alternative indexing schemes. This distinction works best when there is a considerable amount of out-of-plane tilt (e.g., a few degrees), thus producing substantial differences in the phase correction on the higher-order layer lines.

## 3.6. Extract layer line data and calculate phase statistics

After establishing the helical symmetry and the amount of out-of-plane tilt, amplitude and phase data is extracted along each of the layer lines, producing a list of $G_{n,\ell}(R)$. Two redundant sets of data are produced, which are

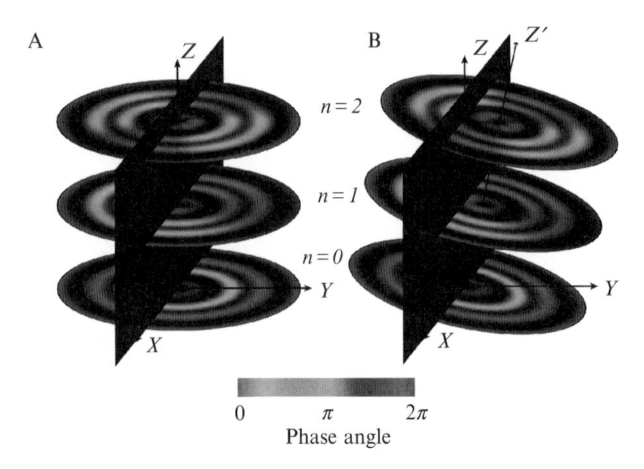


**Figure 5.7** Out-of-plane tilt. These diagrams illustrate the relationship between the 3D Fourier transform and the central section that results from the projection along the viewing direction (*Y*). Due to this projection, layer lines are sampled where they intersect the *X*–*Z* plane (black). (A) Untilted helical assembly where the helical axis is coincident with the *Z* axis of the transform and layer lines are sampled at azimuthal angles ($\psi$) equal to 0° and 180°. (B) Helical assembly that is tilted away from the viewing direction, causing sampling of layer lines at $\psi \neq 0°$ and 180°. $Z'$ corresponds to the helical axis and the angle between $Z'$ and $Z$ corresponds to the out-of-plane tilt, $\Omega$. This tilt produces systematic phase shifts that are dependent on the order of the Bessel function along each layer line. Phases are represented by the color table shown at the bottom. (See Color Insert.)

referred to as the near and the far side of the helical assembly and which reflect the two alternative 2D lattices identified during indexing. Phase statistics from this dataset provide a useful measure of the quality of the data. If no symmetry exists, then a simple comparison of near and far data can produce a resolution-dependent phase residual. Symmetry is common, in which case the near and far data sets are averaged and the twofold related phase error is plotted as a function of resolution (Fig. 5.8).

## 3.7. Determine defocus and apply the contrast transfer function

The CTF of the electron microscope effectively multiplies the Fourier transform by an oscillating function that depends on defocus and the spherical aberration coefficient of the objective lens (Erickson and Klug, 1971). This effect must be corrected, not only because of the aberrations that it produces in the 3D reconstruction, but also because different levels of defocus will alter the period of oscillation and produce destructive phase interference if uncorrected images are averaged together. Indeed, the accuracy of the correction is critical for recovering high resolution information,

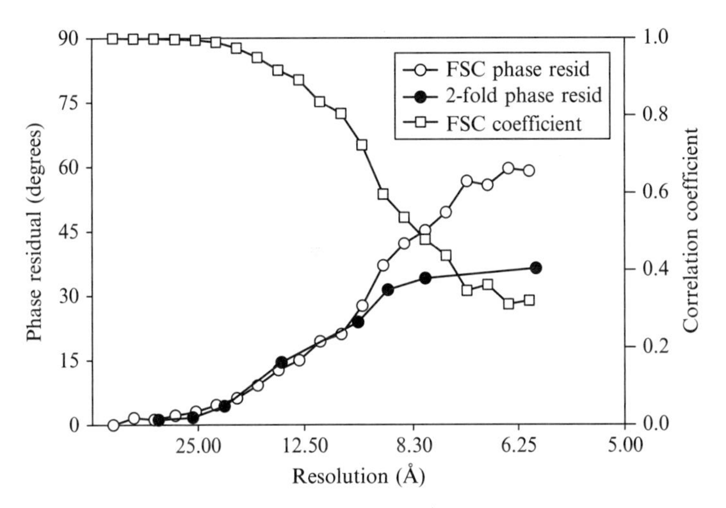


**Figure 5.8** Evaluation of the resolution of a helical reconstruction. Both the Fourier shell coefficient (FSC) and the Fourier shell phase residual result from comparing masked and aligned molecules obtained from independent halves of the data set. The twofold phase residual is calculated from averaged $G_{n,\ell}(R)$ derived from the entire data set. A twofold phase residual of 45° is random, whereas a Fourier Shell phase residual of 90° is random. Data reproduced from Xu et al. (2002)

because the frequency of the oscillations increases with resolution. A number of excellent programs are available for determining the defocus and astigmatism of an image, such as CTFFIND3 (Mindell and Grigorieff, 2003), PLTCTFX (Tani *et al.*, 1996), and many others (Fernando and Fuller, 2007; Huang *et al.*, 2003; Mallick *et al.*, 2005; Velazquez-Muriel *et al.*, 2003). Once values for the defocus and astigmatism have been determined, the phases along the layer lines are adjusted to eliminate the reversals that characterize the CTF and the amplitudes are multiplied by the magnitude of the CTF to differentially weight data from different images during averaging (Chapters 1–3). These CTF corrections are applied independently to the near and far data sets in order to compensate for astigmatism in the image. Once the CTF corrections are applied, the near and far data sets are averaged to create the final data set for combination with other images.

## 3.8. Averaging data from multiple images

In order to average data from multiple images in Fourier space, all of the helical assemblies must have the same symmetry. This means that the Bessel orders from all of the corresponding layer lines must be the same. However, layer lines from the various images do not necessarily lie at the same axial positions, because variations in the axial repeat distance cause the layer lines to fall at different axial heights. Ideally, these differences simply reflect a linear scale factor along the helical axis due to the interpolation of images during masking, but in reality, slight distortions in the unit cell cause the layer lines to move slightly with respect to one another. A similar situation occurs with planar 2D crystals, where the unit cell dimensions and included angle are somewhat variable. As long as this variability is small, the data can be averaged together without significantly reducing the resolution of the reconstruction (Toyoshima and Unwin, 1990).

Prior to averaging, the axial coordinates ($\ell$) of each layer line are simply renumbered to match the reference data set and a common phase origin must be found. If twofold symmetry is present, then the phase origin of a reference image (i.e., the best of the group) is constrained to one of the twofold axes and the phase origins of the other images are adjusted to match the reference. A sum of the CTF-weighted data from all the images can then be produced, together with the sum of the CTF weights. After dividing these averaged amplitudes by the summed CTF weights, this data is suitable for calculating a 3D reconstruction (see below).

## 3.9. Correcting distortions in the helical assembly (unbending)

As mentioned, one approach to straightening helical assemblies has been to fit a polynomial curve to the helical axis and then to reinterpolate image densities to produce a straighter structure prior to calculating the initial Fourier transform. When combined with cross-correlation to determine the position of the helical axis (Metlagel *et al.*, 2007), this method is analogous to the unbending procedure used for planar 2D crystals (Henderson *et al.*, 1986). Both of these methods compensate for in-plane translation of image densities away from their ideal positions and thus significantly improve the resolution of Fourier data that are then used for 3D reconstruction. However, neither of these approaches considers out-of-plane bending nor more subtle distortions such as stretching in the axial or radial directions. A more sophisticated approach was developed by Beroukhim and Unwin (1997) for their high resolution studies of the helical tubes of acetylcholine receptor. For this method, in-plane and out-of-plane tilt, translation, and radial and axial scale factors (stretch) are determined for short segments of the tube. Some of these parameters are determined solely from phase relationships along the layer lines. Other parameters involve comparison of data from the individual segments with reference data produced by averaging together a number of images. The fundamental limitation to the length of the segments is the signal-to-noise ratio that they produce, which must be high enough to allow reliable alignment. Helical assemblies of acetylcholine receptor are ~700 Å in diameter with a wall thickness of ~100 Å, similar to assemblies of Ca-ATPase. In these cases, segments of ~600 Å in length represent the minimum that can be used for this unbending procedure. This places a limit on the range of disorder that can be compensated and continues to require selection of extremely straight or only very slightly bent assemblies. Also, during unbending, data is cut off at the first maximum of the CTF in order to assure maximal reliability of the orientational data, specifically by maximizing signal-to-noise ratio and eliminating the nodes of the CTF. Even though this resolution cutoff limits the precision of the alignments, this precision far exceeds the nominal resolution of the data used for the fits (e.g., 15 Å). This unbending involves many iterative steps of alignment, reboxing, and realignment, which have been tied together by a series of scripts, all of which are controlled by the EMIP user interface. Using this strategy on a large number of images, Unwin was able to extend the resolution of the acetylcholine receptor to 4 Å resolution (Miyazawa *et al.*, 2003); Yonekura *et al.* (2003a) achieved a similar resolution for bacterial flagella and a Ca-ATPase structure was determined at 6.5 Å resolution (Xu *et al.*, 2002).

## 3.10. Create final averaged data set and correct CTF

The unbending procedure produces data sets for the near and far sides of the individual images that have been weighted by the CTF. Similar to the averaging step above, these near and far data sets (and their respective weights) should be added together to produce an average for each tube, and these should be added together to produce a final averaged data set. The CTF is then corrected by dividing each point along each layer line by the combined sum of the $CTF^2$: the first CTF factor coming from the image formation by the microscope and the second factor from the weight that was applied during the image processing. However, care should be taken in applying this correction to the equatorial data. In particular, the magnitude of the CTF approaches zero at the origin, thus generating very large correction factors for the low resolution amplitudes along the equator. The proportion of amplitude contrast produced by the specimen will greatly influence this correction near the origin, and although Toyoshima has determined the appropriate factor for the acetylcholine receptor (Toyoshima and Unwin, 1988; Toyoshima *et al.*, 1993b), some ambiguity exists for other samples. In most cases, it is necessary to carefully consider the correction applied to the low resolution values along the equator. This is because these values have a large influence over the radial distribution of mass in the final 3D reconstruction. Incorrect handling of the equatorial correction can produce a distorted molecule where, for example, the transmembrane domain is exaggerated relative to the cytoplasmic domain, or *vice versa* (Fig. 5.9). One approach is to limit the maximum correction that can be applied along this equator, or to prevent correction to the very

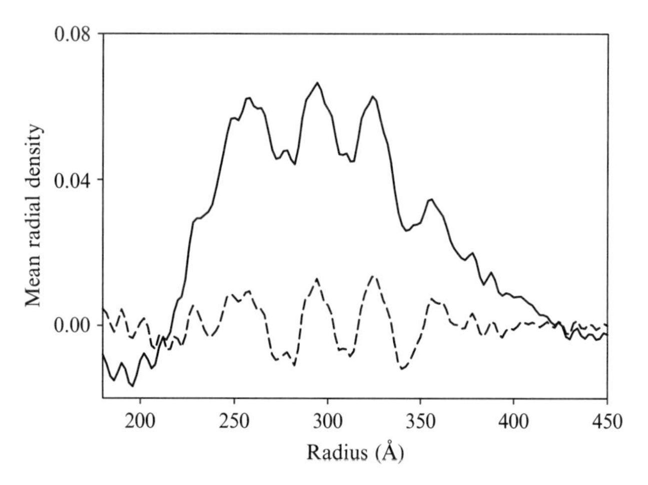


**Figure 5.9** Mean radial density distribution derived from the equator of an averaged data set. The solid line corresponds to data that has been appropriately corrected for the CTF, thus producing positive density at radii between 225 and 400 Å. For the dashed line, the CTF correction was limited along the equator. Although the structure at any given radius is unchanged, the overall distribution of mass is dramatically affected, making it impossible to render the molecular surface based on a single density threshold (cf., Fig. 5.10c).

lowest resolution terms on the equator. This issue is best handled by trial and error at the very last stages of map calculation.

## 3.11. Map calculation and display

Once an averaged and CTF-corrected data set has been produced, calculation of the 3D map is a two-step operation. The first step involves a Fourier–Bessel transform that converts the Fourier-space layer line data ($G_{n,\ell}(R)$) into real-space density waves ($g_{n,\ell}(r)$). The second step involves summing up these $g_{n,\ell}(R)$ functions over the range of $r$, $\varphi$, and $z$ required for the final density map (see Appendix). Conventionally, the 3D maps were displayed as a stack of contour plots and programs exist for cutting sections both parallel to and perpendicular to the helical axis. Such contour plots can still be useful in understanding the composition of the asymmetric unit and in delineating individual molecules (Fig. 5.10). In particular, an individual molecule, or multimer, can be masked from the helical lattice by using IMOD (Kremer *et al.*, 1996). Ultimately, the masked molecules should be rendered as an isodensity surface using a program like Chimera (Pettersen *et al.*, 2004). Such programs also facilitate fitting atomic models to the density, which is a great aid in interpretation and presentation of the structure.

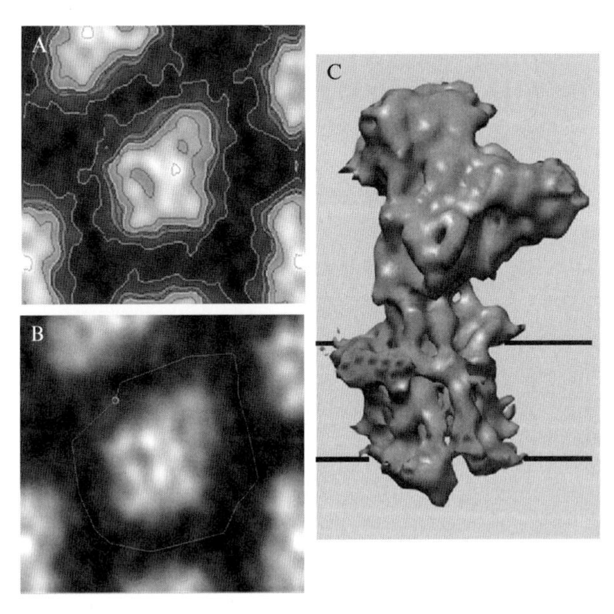


**Figure 5.10** Fourier–Bessel reconstruction of Ca-ATPase. (A) Section from the reconstruction with contours superimposed on the densities. Evaluation of contour maps can be useful in delineating the individual molecules composing the structure. (B) Masking of a single molecule from the map, which is useful both for real-space averaging and for display. (C) Surface representation of a single molecule of Ca-ATPase defined by density threshold. The black, horizontal lines correspond to the boundaries of the membrane. In this case, this density threshold corresponds to a volume recovery of ~75% of the expected molecular mass. IMOD (Kremer *et al.*, 1996) was used for panels A and B, and Chimera (Pettersen *et al.*, 2004) was used for panel C.

## 4. Case Study of Helical Reconstruction of Ca-ATPase

Ca-ATPase is an ATP-dependent $Ca^{2+}$ pump found in the sarcoplasmic reticulum and endoplasmic reticulum. Due to its abundance in striated muscle, it has been extensively characterized by biochemical and biophysical methods (Moller *et al.*, 2005). Ca-ATPase in skeletal muscle has a propensity to form 2D crystals within the native membrane (Franzini-Armstrong *et al.*, 1986) and these crystals were studied extensively by electron microscopy in the 1980s and 1990s (Castellani *et al.*, 1985; Dux and Martonosi, 1983; Taylor *et al.*, 1986; Toyoshima *et al.*, 1993a; Zhang *et al.*, 1998). In particular, small vesicles from isolated sarcoplasmic reticulum or from purified, reconstituted Ca-ATPase preparations display 2D lattices predominantly at their highly curved edges. Under the right conditions, Ca-ATPase forms long helical tubes with a diameter of 600–800 Å

(Fig. 5.5, see also Chapter 7, Vol. 483). These lattices are composed of Ca-ATPase dimers that are oriented asymmetrically in the membrane and the intermolecular distance between the cytoplasmic domains on one side of the membrane is greater than the corresponding distance between the luminal domains on the other side of the membrane, thus inducing curvature (Young *et al.*, 1997).

## 4.1. Variable helical symmetry

Although the diameter of a given tube is generally constant along its length, this diameter is variable from one tube to the next; this variability reflects differences in the helical symmetry adopted by different tubes. For 3D reconstruction, the thinnest tubes were chosen in order to constrain the range of helical symmetries to be analyzed. Even from the thinnest tubes, a series of related, but different symmetries were obtained, which can be characterized by the Bessel orders of the (1,0) and (0,1) layer lines (e.g., −23,6 in Fig. 5.3 and −23,7 in Fig. 5.6). Specifically, tubes with symmetries ranging from (−19,6) to (−27,9) have been used for various 3D reconstructions (Xu *et al.*, 2002; Young *et al.*, 2001a, b; Zhang *et al.*, 1998). The underlying unit cell appears to be relatively well conserved in all of these cases and the differing symmetry simply involves use of a different circumferential vector in generating the helical assembly. Inspection of Fig. 5.2 indicates that the longer circumferential vectors required to generate the larger Bessel orders are strictly correlated with larger diameters of the resulting tubes. In fact, once the basic helical symmetry is understood, simple measurement of the tube diameter using, for example, the projection plot in Fig. 5.5 gives a good estimate of the helical symmetry. Any remaining ambiguities can be resolved by checking the even/odd character of the Bessel orders on key layer lines.

## 4.2. Real-space averaging of different helical symmetries

Well ordered tubes frequently produced visible layer lines at 15 Å resolution and, in exceptional cases a layer line was visible at 10 Å resolution (Fig. 5.3). Averaging is essential not only for filling in the nodes of the CTF, but also for improving the signal-to-noise ratio at high resolution. By averaging together as few as 12 individual tubes from a single symmetry group, it has been possible to obtain better than 8 Å resolution (Stokes *et al.*, 2005). For higher resolution, structures from multiple symmetry groups were averaged together in real space. For this process, software developed by Toyoshima and colleagues (Yonekura *et al.*, 1997) was used to mask off a single unit cell and then to divide this unit cell in half, which effectively separated the two monomers composing the asymmetric unit. The mean radial density distribution was used to adjust the magnification of the reconstruction and cross-correlation was then used to independently align each of the two monomers from reconstructions of several different symmetry groups (e.g., the −19,6; −20,6; −21,6; −22,6; and −23,6 symmetries in Xu *et al.* (2002)). After averaging the masked monomers, a hybrid unit cell was created. An inverse Fourier–Bessel transform was then calculated to produce a set of layer line data ($G_{n,\ell}(R)$) conforming to the symmetry of the reference data set. Examination of the twofold related phase residuals revealed a substantial improvement in resolution to 6.5 Å from a total of 70 tubes (Xu *et al.*, 2002). Unwin and colleagues took a similar approach in averaging tubes of acetylcholine receptor with various symmetries (Miyazawa 2003). In this case, 359 tubes falling into four different symmetry groups (−16,6; −17,5; −15,7; and −18;6) were averaged together. The cylindrical shape of the acetylcholine molecule made it relatively easy to construct a mask around a single pentameric molecule, which was then used for alignment and averaging. The final resolution of 4 Å was verified by dividing the images into two equal halves and calculating an amplitude-weighted phase difference as well as a Fourier shell correlation coefficient.

## 4.3. Difference maps

A similar strategy of comparing masked molecules in real space has also been used for calculating difference maps from Ca-ATPase crystallized in the presence of various ligands. In particular, the location of the inhibitor thapsigargin (Young *et al.*, 2001a) and of the regulatory protein phospholamban (Young *et al.*, 2001b) was investigated in this way. In addition to aligning the two structures in real space, it was important to optimize the density scaling to minimize the background level of density differences. In order to evaluate the statistical significance of the differences, it is possible to conduct a pixel-by-pixel Student's t-test (Milligan and Flicker, 1987; Yonekura *et al.*, 1997). This involves calculating maps for each of the individual tubes used for the two reconstructions. From these individual maps, a standard deviation can be calculated for each voxel in the averaged reconstruction, which can then be used in calculation of the $T$ statistic. This procedure becomes more complicated when using CTF weighting and real-space averaging. Therefore, in more recent studies, an empirical $\sigma$ value was used to characterize the significance of the differences, based solely on the Gaussian distribution of densities in the difference map.

## 4.4. Alternative real-space averaging of $g_{n,\ell}(r)$ of different helical symmetries

An alternative method for averaging data from different symmetry groups involves the averaging of $g_{n,\ell}(r)$ functions rather than the density maps themselves (DeRosier *et al.*, 1999). Although not widely adopted, this elegant method is simpler and more analytically correct than the real space method described above, because it does not require the subjective step of masking individual molecules from the maps. The method is based on the fundamental concept that $g_{n,\ell}(r)$ functions are derived from individual layer lines and that they correspond directly to $F_{h,k}(r)$, which are the Fourier coefficients derived from the underlying, planar 2D lattice (Fig. 5.2). In particular, both $g_{n,\ell}(r)$ and $F_{h,k}(r)$ come from a specific set of planes through this lattice. As long as the molecular packing and geometry of the unit cell is preserved in the different helical symmetries, the different $g_{n,\ell}(r)$ derived from particular $(h,k)$ planes can be directly averaged together. Thereafter, the averaged $g_{n,\ell}(r)$ can be summed to create a 3D map. There are two alignments necessary before averaging $g_{n,\ell}(r)$; first is a shift in radius, which reflects the fact that the different symmetries produce helical tubes of different radii. Second, the $g_{n,\ell}(r)$ must be adjusted to a common phase origin (e.g., a specific twofold symmetry axis). These methods were tested on Ca-ATPase tubes from three different symmetries and resolution-dependent phase residuals indicate that the improvements in resolution for $g_{n,\ell}(r)$ averaging were almost identical to those for real-space averaging (Fig. 5.11).

## 4.5. Comparison with the iterative helical real-space reconstruction method

An alternative to Fourier–Bessel reconstruction of helical assemblies has been developed by Egelman and colleagues and termed the iterative helical real-space reconstruction (IHRSR) method (Egelman, 2000, Chapter 6). Initially developed for flexible assemblies of DNA binding proteins (Yu *et al.*, 2001), IHRSR has been used with success for a broad range of structures, including bacterial pili (Mu *et al.*, 2002), actin (Galkin *et al.*, 2003), myosin (Woodhead *et al.*, 2005), and dynamin (Chen *et al.*, 2004). The method is based on matching short segments from the image of a helical assembly to a series of projections from a model using the SPIDER software suite (Frank *et al.*, 1996) in a manner analogous to the single particle analysis of isolated macromolecular complexes. The segments are typically much shorter than for Unwin's unbending methods discussed above and IHRSR is therefore able to compensate for shorter-range disorder. Furthermore, indexing of the layer lines in the Fourier transform is unnecessary, though knowledge of $\Delta\phi$ and $\Delta z$ for the smallest pitch helix is generally required. Specifically, after using projection matching to determine the relative orientations of all the individual segments along the helical assembly, a 3D structure is generated by backprojection. The helical symmetry of this 3D structure is then determined empirically by examining autocorrelation coefficients after systematically rotating and translating the structure about its helical axis. Once helical parameters ($\Delta\phi$ and $\Delta z$) are determined, the

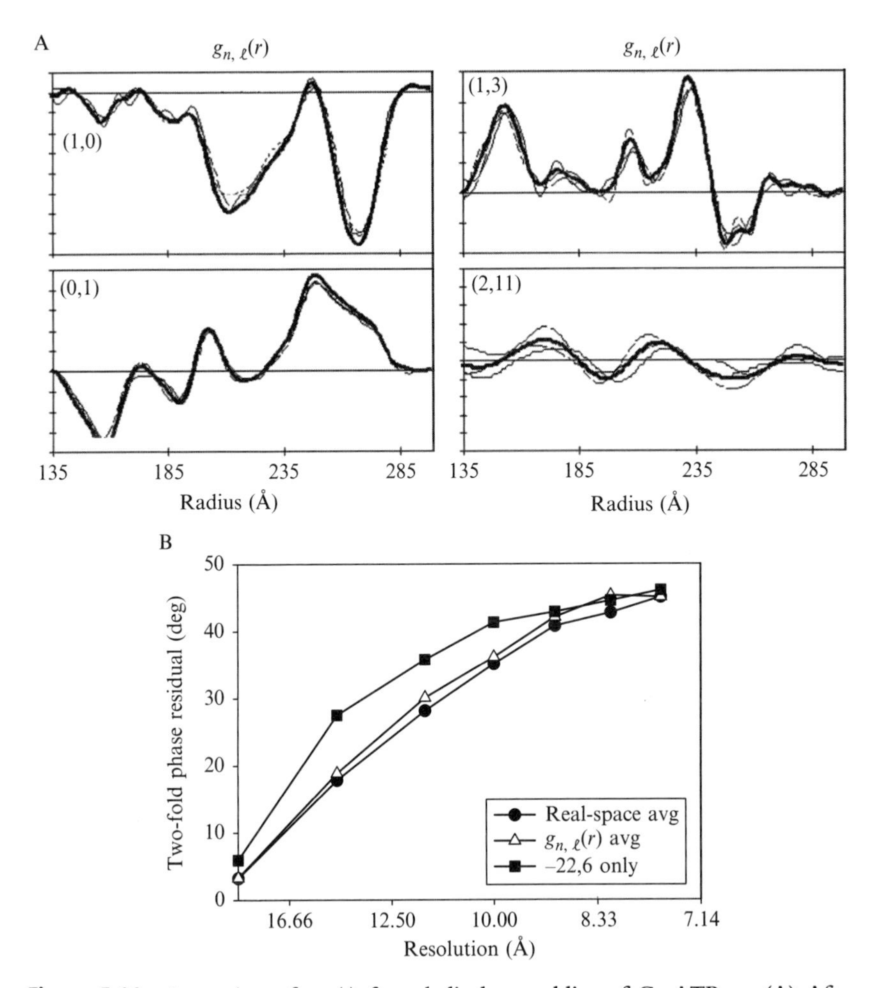


**Figure 5.11** Averaging of $g_{n,\ell}(r)$ from helical assemblies of Ca-ATPase. (A) After alignment, amplitudes from three different symmetry groups are shown (thin, dashed lines), together with the average (thick solid line). (B) Twofold phase residuals are compared for the Fourier-space average from a single symmetry group (−22,6, ■), for the real-space average of the three symmetry groups (●), and for the average of $g_{n,\ell}(r)$ from these same three symmetry groups (Δ). The improvements obtained by averaging $g_{n,\ell}(r)$ are comparable to those obtained by real-space averaging. Data reproduced from DeRosier *et al.* (1999).

structure is symmetrized and used for the next round of alignment and projection matching.

When applied to images of Ca-ATPase and the related Na,K-ATPase, IHRSR generated a plausible 3D structure, but the resolution-dependent phase residuals were not as good as Fourier–Bessel reconstructions (Pomfret *et al.*, 2007). This was attributed to the fact that Fourier–Bessel reconstruction excludes data that lie between the layer lines, thus filtering out a large amount of noise from the reconstruction and allowing more accurate alignments. On the other hand, IHRSR successfully generated a 3D structure of Ca-ATPase from scallop muscle when Fourier–Bessel methods failed, because these helical tubes were too poorly ordered to successfully complete the necessary alignments based on layer line data. Related real-space methods have been used for very high resolution reconstructions of tobacco mosaic virus, illustrating that if the helical symmetry is well determined and the helical assembly is well ordered, the real-space methods represent a viable alternative to the more conventional Fourier–Bessel analysis (Sachse *et al.*, 2007). However, bacterial flagella are another well ordered helical assembly and it is notable that a Fourier–Bessel reconstruction at 4 Å resolution was reported from only 102 images, comprising ~41,000 molecules of the building block, flagellin (Yonekura *et al.*, 2003a). This is a remarkable achievement especially considering the small size of the flagellin molecule (<500 residues). The authors attributed this success to a number of factors, including the high quality of the images, the strict order of molecules in the flagellar filament, and innovative software techniques such as solvent flattening to remove noise from the maps during alignment of individual filaments (Yonekura *et al.*, 2005). Another important factor was the use of optical diffraction to objectively pick the very best helical assemblies for image processing. Specifically, the presence of sharp layer lines extending to high resolution represents a relatively easy basis for selecting the best 60 images from a pool of >1000; the fact that these images came from one of the world's best electron microscopes also improved the odds for a high resolution structure.

# 5. Conclusions

3D reconstruction of helical assemblies continues to be an important application for electron microscopists. Not only are helical assemblies ubiquitous in nature, but these methods offer advantages for producing a 3D structure. First, the ability to form a regular structure means that individual molecules are likely to adopt a fixed conformation, thus greatly reducing the structural heterogeneity that plagues single particle approaches. Second, the cylindrical symmetry of the helix ensures that all the necessary views of the molecule are present in a single image, thus eliminating the missing cone of information that produces anisotropic resolution in electron crystallography of planar 2D crystals. These advantages have been illustrated by high resolution structures of acetylcholine receptor and bacterial flagella, as well as a medium resolution structure of Ca-ATPase. Additional considerations include the large computational resources that are required for single particle approaches and the potential for combining amplitudes from fiber diffraction with phases derived by the helical reconstruction methods described here. Historically, there has been great variety in the software approaches to Fourier–Bessel reconstructions, making it difficult for a novice to undertake the process of structure determination. However, graphical user interfaces have been recently developed to guide the user through the many complex steps of helical reconstruction, making this process more accessible (Yonekura *et al.*, 2003b). Thus, development of biochemical methods to generate ordered helical assemblies, for example during the 2D crystallization of membrane proteins, may be an important area of future development (Chapter 2, Vol. 481).

## *ACKNOWLEDGEMENTS*

The authors would like to acknowledge David DeRosier and Don Caspar, who were pioneers in analyzing structures with helical symmetry. Their publications and teachings laid the foundation for the current review.

# Appendix A. Mathematical Foundations of Fourier-Based Helical Reconstruction

## A1. Definition of a Fourier transform

For an integrable, complex valued function of a real variable ($f \in \mathbf{C}^{\mathbf{R}}$), the Fourier transform is defined as:

$$\hat{f}(t) = F[f(x)](t) = \frac{1}{\sqrt{2\pi}} \int_{-\infty}^{\infty} f(x) e^{2\pi i x t} dx.$$

In three dimensions, the scalar arguments of the functions become vectors in Euclidean space. As a result, the argument of the exponent involves a scalar product of real and reciprocal space variables and is written as,

$$\hat{f}(X) = F[f(\mathrm{x})](X) = \frac{1}{(2\pi)^{(3/2)}} \iiint_{R^3} f(\mathrm{x}) e^{2\pi i \mathrm{x} \cdot X} d^3 x. \qquad \text{(A1)}$$

## A2. Fourier transform in cylindrical coordinates

A helix is a curve in three-dimensional space, defined in parametric form as a function $f$: $(\mathbf{R}^3)^{\mathbf{R}}$, such that $f(t) = (r\cos(t), r\sin(t), t)$. Since any point in this helix is equidistant from the $z$-axis, it is natural to use cylindrical coordinates to take advantage of the symmetries of helices.

Using the cylindrical coordinates defined as $x = r\cos(\phi)$; $y = r\sin(\phi)$; $z = z$, and $X = R\cos(\Phi)$; $Y = R\sin(\Phi)$; $Z = Z$, the scalar product in the exponential term of Eq. (A1) can be written as,

$$\begin{aligned} \mathrm{x}\cdot X &= xX + yY + zZ \\ &= rR\cos\Phi\cos\Phi + rR\sin\Phi\sin\Phi + zZ \\ &= rR\cos(\Phi - \phi) + zZ. \end{aligned} \tag{A2}$$

Substituting for the volume element $\mathrm{d}^3x = r\mathrm{d}rd\phi dz$, the Fourier transform in cylindrical coordinates adopts the form

$$\hat{f}(R,\Phi,Z) = \frac{1}{(2\pi)^{\frac{3}{2}}}\iiint_{R^3} f(r,\Phi,z)e^{2\pi i(rR\cos(\Phi-\phi)+zZ)}rdrd\phi dz, \tag{A3}$$

with the integration taking place over $0 \leq r < \infty$, $0 \leq \phi < 2\pi$, and $-\infty < z < \infty$.

To simplify this expression, we can substitute $\cos(\Phi - \phi) = \sin(\Phi - \phi + \pi/2)$ and use Euler's formula ($\mathrm{e}^{i\phi} = \cos(\phi) + i\sin(\phi)$ ) to express the sine as a sum of complex exponentials ($\sin(\theta) = (1/2i)(\mathrm{e}^{i\theta} - \mathrm{e}^{-i\theta})$ ). We then have

$$e^{2\pi i(rR\cos(\Phi-\phi))} = e^{2\pi(rR(1/2)(u-u^{-1}))} \tag{A4}$$

where $u = \mathrm{e}^{i(\Phi-\phi+\pi/2)}$.

The exponential term can be reexpressed using the formula for generating Bessel functions (Lebedev, 1972),

$$e^{(1/2)x(t-t^{-1})} = \sum_{n=-\infty}^{\infty} J_n(x)t^n, \tag{A5}$$

where $J_n(x)$ are the Bessel functions of the first kind. Thus the exponential term in Eq. (A3) can be written as

$$e^{2\pi(rR(1/2)(u-u^{-1}))}e^{2\pi izZ} = \sum_{n=-\infty}^{\infty} J_n(2\pi rR)e^{in(\Phi-\phi+\pi/2)}e^{2\pi izZ}.$$

Plugging this into Eq. (A3), we get

$$\hat{f}(R,\Phi,Z)=\frac{1}{(2\pi)^{(3/2)}}\iiint_{R^3} f(r,\phi,z)\sum_{n=-\infty}^{\infty} J_n(2\pi rR)e^{in(\Phi-\phi+\pi/2)}e^{2\pi izZ}rdrd\phi dz$$
$$=\frac{1}{(2\pi)^{(3/2)}}\sum_{n=-\infty}^{\infty} e^{in(\Phi+\pi/2)}\iiint_{R^3} f(r,\phi,z)J_n(2\pi rR)e^{-in\phi}e^{2\pi izZ}rdrd\phi dz.$$

A convenient way to write this expression is achieved by defining a "helical structure factor" $g_n(r, Z)$ as

$$g_n(r,Z)=\frac{1}{(2\pi)^{(3/2)}}\int_0^{2\pi}\int_{-\infty}^{\infty} f(r,\phi,z)e^{-in\phi}e^{2\pi izZ}d\phi dz, \tag{A6}$$

and thus Eq. (A3) can be written as

$$\hat{f}(R,\Phi,Z)=\sum_{n=-\infty}^{\infty} e^{in(\Phi+\pi/2)}\int_0^{\infty} g_n(r,Z)J_n(2\pi rR)rdr. \tag{A7}$$

Lastly, if we define

$$\begin{aligned}G_n(R,Z)&=\int_0^{\infty} g_n(r,Z)J_n(2\pi rR)rdr\\ &=\iiint_{R^3} f(r,\phi,z)J_n(2\pi rR)e^{-in\phi}e^{2\pi izZ}rdrd\phi dz,\end{aligned} \tag{A8}$$

the Fourier transform in cylindrical coordinates becomes

$$\hat{f}(R,\Phi,Z)=\sum_{n=-\infty}^{\infty} e^{in(\Phi+\pi/2)}G_n(R,Z). \tag{A9}$$

Notice that up to this point, all we have done is to write the Fourier transform for a function $f(r, \phi, z)$ in cylindrical coordinates. The reason to introduce functions known as "big $G$'s" and "little $g$'s" is that they correspond to data from individual layer lines lying at discrete values of $Z$ in the Fourier transforms of helical particles.

## A3. Helical symmetry

In the case, that $f(r, \phi, z)$ represents a density function with helical symmetry, we have a density function that is invariant after a twist of $\Delta\phi$ and an axial translation $\Delta z$, that is,

$$f(r,\phi,z) = f(r,\phi+\Delta\phi,z+\Delta z). \tag{A10}$$

Similarly, the Fourier transform of these two functions are equal.

Making the transformation $\phi \rightarrow (\phi + \Delta\phi)$; $z \rightarrow (z + \Delta z)$ in Eq. (A8), we have

$$\begin{aligned}
G'_n(R,Z) &= \frac{1}{(2\pi)^{3/2}}\iiint_{R^3} f(r,\phi+\Delta\phi,z+\Delta z)J_n(2\pi rR)e^{-in(\phi+\Delta\phi)}e^{2\pi i(z+\Delta z)Z}rdrd\phi dz \\
&= \frac{1}{(2\pi)^{3/2}}\iiint_{R^3} f(r,\phi,z)J_n(2\pi rR)e^{-in(\phi+\Delta\phi)}e^{2\pi i(z+\Delta z)Z}rdrd\phi dz \\
&= \frac{1}{(2\pi)^{3/2}}\iiint_{R^3} f(r,\phi,z)J_n(2\pi rR)e^{-in\phi}e^{2\pi izZ}e^{-in\Delta\phi}e^{2\pi i\Delta zZ}rdrd\phi dz.
\end{aligned}$$

In order to satisfy $G'_n(R, Z) = G_n(R, Z)$,

$$e^{-in\Delta\phi+2\pi i\Delta zZ} = 1. \tag{A11}$$

This expression implies that for a density function with helical symmetry,

$$-n\Delta\phi + 2\pi\Delta zZ = 2m\pi, \tag{A12}$$

where $m$ is an arbitrary integer.

The pitch of the helix is defined as the axial displacement of the helix after one full turn. This can be written in terms of the azimuthal and axial displacements between subunits ($\Delta\phi$ and $\Delta z$) as $\mathcal{P} = (2\pi\Delta z)/(\Delta\phi)$. Substituting for $\Delta\phi$ in Eq. (A12),

$$Z = \frac{m}{\Delta z} + \frac{n}{\mathcal{P}}. \tag{A13}$$

In general, a helical structure will not repeat after a single turn, but will require an axial displacement ($c$) that depends on $\Delta\phi$ and $\Delta z$ and that includes several turns. Again, the structure and its transform are invariant after a translation $z = c$. Applying this fact to Eq. (A8), we obtain that $e^{2\pi icZ} = 1$, which implies that $2\pi cZ = 2\ell\pi$, where $\ell$ is any integer. From this equation, we have that

$$Z = \ell/c. \tag{A14}$$

This equation implies that the signal in the three-dimensional Fourier transform of an object with helical symmetry will be nonzero only in planes perpendicular to the $Z$-axis, which are separated by a distance of $1/c$.

## A4. Selection rule

Substituting the fact that $Z = \ell/c$ in Eq. (A13), we get that

$$\ell = c\frac{m}{\Delta z} + c\frac{n}{\mathcal{P}}.$$

Defining $u = c/\Delta z$ as the number of subunits contained in a single repeat and $t = c/\mathrm{P}$ as the number of helical turns in a repeat, this expression becomes

$$\ell = um + tn, \tag{A15}$$

which is known as the helical selection rule, based on its resemblance to the selection rules in some quantum mechanical systems.

If the assembly has a cyclic periodicity, that is, if the helical axis also corresponds to a $k$-fold symmetry axis, we have the additional requirement that $f(r, \phi + 2\pi/k, z) = f(r, \phi, z)$. This will add the requirement $e^{-2\pi in/k} = 1$, which implies that $n$ has to be a multiple of $k$, in addition to the requirements imposed in Eq. (A15).

## A5. Example: tobacco mosaic virus

Tobacco mosaic virus consists of a coat protein arranged around a single strand of RNA, with an axial repeat distance of 69 Å. In this repeat, the virus has 49 subunits, and the basic helix makes three turns in a right-handed fashion. Thus, we have that $t = 3$ and $u = 49$. Then the selection rule becomes $\ell = 49m + 3n$. The equator ($\ell = 0$) can always be satisfied with $m = n = 0$ and, as a result, $J_0(2\pi rR)$ will always be found on the equator. However, in this case the equator is also consistent with $m = 1$; $n = -49$. Therefore, $J_{-49}(2\pi rR)$ also contributes to the equator, although only at rather high spatial frequencies (starting at $R \approx 11$ Å, since $J_{-49}$ has its first maximum when $2\pi rR \approx 51$ and $r = 90$ Å).

For the first layer line, $\ell = 1 = 49m + 3n$, which is consistent with $m = 1$ and $n = -16$. The second layer line generates the equation $\ell = 2 = 49m + 3n$, which can be satisfied by $m = -1$ and $n = 17$. Thus, the first and second layer lines are generated by $J_{-16}$ and $J_{17}$, respectively, which explains their locations relatively far from the meridian in the diffraction pattern. The third layer line generates the equation $\ell = 3 = 49m + 3n$, which is satisfied by $m = 0$ and $n = 1$. Therefore the third layer line is characterized by $J_1(2\pi rR)$, which will have its maximum near the meridian.

It is worth noting that the equation for the first layer line, $1 = 49m + 3n$ has $m = 1$; $n = -16$ as solutions, as well as $m = -2$; $n = 33$. That is, this layer line will contain both $J_{-16}$ *and* $J_{33}$. Although $J_{33}$ starts contributing at a considerably higher spatial frequencies (~16 Å), this overlap cannot be ignored, especially if we aim to produce a high resolution three-dimensional reconstruction. In cases like this, there is no established method to deconvolve the Bessel functions as would be required for Fourier–Bessel reconstruction.

## REFERENCES


Beroukhim, R., and Unwin, N. (1997). Distortion correction of tubular crystals: Improvements in the acetylcholine receptor structure. *Ultramicroscopy* **70,** 57–81.

Bullitt, E. S., DeRosier, D. J., Coluccio, L. M., and Tilney, L. G. (1988). Three-dimensional reconstruction of an actin bundle. *J. Cell Biol.* **107,** 597–611.

Castellani, L., Hardwicke, P. M., and Vibert, P. (1985). Dimer ribbons in the three-dimensional structure of sarcoplasmic reticulum. *J. Mol. Biol.* **185,** 579–594.

Chen, Y. J., Zhang, P., Egelman, E. H., and Hinshaw, J. E. (2004). The stalk region of dynamin drives the constriction of dynamin tubes. *Nat. Struct. Mol. Biol.* **11,** 574–575.

Cochran, W., Crick, F. H. C., and Vand, V. (1952). The structure of synthetic polypeptides I. The transform of atoms on a helix. *Acta Crystallogr.* **5,** 581–586.

Crowther, R. A., Henderson, R., and Smith, J. M. (1996). MRC image processing programs. *J. Struct. Biol.* **116,** 9–16.

DeRosier, D. J., and Klug, A. (1968). Reconstruction of three dimensional structures from electron micrographs. *Nature* **217,** 130–134.

DeRosier, D. J., and Moore, P. B. (1970). Reconstruction of three-dimensional images from electron micrographs of structures with helical symmetry. *J. Mol. Biol.* **52,** 355–369.

DeRosier, D., Stokes, D. L., and Darst, S. A. (1999). Averaging data derived from images of helical structures with different symmetries. *J. Mol. Biol.* **289,** 159–165.

Dux, L., and Martonosi, A. (1983). Two-dimensional arrays of proteins in sarcoplasmic reticulum and purified $Ca^{2+}$-ATPase vesicles treated with vanadate. *J. Biol. Chem.* **258,** 2599–2603.

Egelman, E. H. (2000). A robust algorithm for the reconstruction of helical filaments using single-particle methods. *Ultramicroscopy* **85,** 225–234.

Erickson, H. P., and Klug, A. (1971). Measurement and compensation of defocusing and aberrations by Fourier processing of electron micrographs. *Philos. Trans. R. Soc. Lond.* **261,** 105–118.

Faruqi, A. R. (2009). Principles and prospects of direct high resolution electron image acquisition with CMOS detectors at low energies. *J. Phys. Condens. Matter* **21,** 314004–314013.

Fernando, K. V., and Fuller, S. D. (2007). Determination of astigmatism in TEM images. *J. Struct. Biol.* **157,** 189–200.

Finch, J. T. (1972). The hand of the helix of tobacco virus. *J. Mol. Biol.* **66,** 291–294.

Frank, J., Radermacher, M., Penczek, P., Zhu, J., Li, Y., Ladjadj, M., and Leith, A. (1996). SPIDER and WEB: Processing and visualization of images in 3D electron microscopy and related fields. *J. Struct. Biol.* **116,** 190–199.

Franzini-Armstrong, C., Ferguson, D. G., Castellani, L., and Kenney, L. (1986). The density and disposition of Ca-ATPase in *in situ* and isolated sarcoplasmic reticulum. *Ann. N. Y. Acad. Sci.* **483,** 44–56.

Galkin, V. E., Orlova, A., VanLoock, M. S., and Egelman, E. H. (2003). Do the utrophin tandem calponin homology domains bind F-actin in a compact or extended conformation? *J. Mol. Biol.* **331,** 967–972.

Henderson, R., Baldwin, J. M., Downing, K. H., Lepault, J., and Zemlin, F. (1986). Structure of purple membrane from Halobacterium halobium: recording, measurement and evaluation of electron microgaphs at 3.5 Å resolution. *Ultramicroscopy* **19,** 147–178.

Henderson, R., Cattermole, D., McMullan, G., Scotcher, S., Fordham, M., Amos, W. B., and Faruqi, A. R. (2007). Digitisation of electron microscope films: six useful tests applied to three film scanners. *Ultramicroscopy* **107,** 73–80.

Huang, Z., Baldwin, P. R., Mullapudi, S., and Penczek, P. A. (2003). Automated determination of parameters describing power spectra of micrograph images in electron microscopy. *J. Struct. Biol.* **144,** 79–94.

Klug, A., Crick, F. H. C., and Wyckoff, H. W. (1958). Diffraction by helical structures. *Acta Crystallogr.* **11,** 199–213.

Kremer, J. R., Mastronarde, D. N., and McIntosh, J. R. (1996). Computer visualization of three-dimensional image data using IMOD. *J. Struct. Biol.* **116,** 71–76.

Lebedev, N. N. (1972). Special functions and their applications. Dover, Mineola, NY.

Mallick, S. P., Carragher, B., Potter, C. S., and Kriegman, D. J. (2005). ACE: Automated CTF estimation. *Ultramicroscopy* **104,** 8–29.

Metlagel, Z., Kikkawa, Y. S., and Kikkawa, M. (2007). Ruby-Helix: An implementation of helical image processing based on object-oriented scripting language. *J. Struct. Biol.* **157,** 95–105.

Milligan, R. A., and Flicker, P. F. (1987). Structural relationships of actin, myosin, and tropomyosin revealed by cryo-electron microscopy. *J. Cell Biol.* **105,** 29–39.

Mindell, J. A., and Grigorieff, N. (2003). Accurate determination of local defocus and specimen tilt in electron microscopy. *J. Struct. Biol.* **142,** 334–347.

Miyazawa, A., Fujiyoshi, Y., and Unwin, N. (2003). Structure and gating mechanism of the acetylcholine receptor pore. *Nature* **423,** 949–955.

Moller, J. V., Olesen, C., Jensen, A. M., and Nissen, P. (2005). The structural basis for coupling of Ca2+ transport to ATP hydrolysis by the sarcoplasmic reticulum $Ca^{2+}$-ATPase. *J. Bioenerg. Biomembr.* **37,** 359–364.

Mu, X. Q., Egelman, E. H., and Bullitt, E. (2002). Structure and function of Hib pili from Haemophilus influenzae type b. *J. Bacteriol.* **184,** 4868–4874.

Owen, C. H., Morgan, D. G., and DeRosier, D. J. (1996). Image analysis of helical objects: The Brandeis helical package. *J. Struct. Biol.* **116,** 167–175.

Pettersen, E. F., Goddard, T. D., Huang, C. C., Couch, G. S., Greenblatt, D. M., Meng, E. C., and Ferrin, T. E. (2004). UCSF Chimera–a visualization system for exploratory research and analysis. *J. Comput. Chem.* **25,** 1605–1612.

Pomfret, A. J., Rice, W. J., and Stokes, D. L. (2007). Application of the iterative helical real-space reconstruction method to large membranous tubular crystals of P-type ATPases. *J. Struct. Biol.* **157,** 106–116.

Sachse, C., Chen, J. Z., Coureux, P. D., Stroupe, M. E., Fandrich, M., and Grigorieff, N. (2007). High-resolution electron microscopy of helical specimens: a fresh look at tobacco mosaic virus. *J. Mol. Biol.* **371,** 812–835.

Sherman, M. B., Brink, J., and Chiu, W. (1996). Performance of a slow-scan CCD camera for macromolecular imaging in a 400 kV electron cryomicroscope. *Micron* **27,** 129–139.

Stokes, D. L., Delavoie, F., Rice, W. J., Champeil, P., McIntosh, D. B., and Lacapere, J. J. (2005). Structural studies of a stabilized phosphoenzyme intermediate of $Ca^{2+}$-ATPase. *J. Biol. Chem.* **280,** 18063–18072.

Tani, K., Sasabe, H., and Toyoshima, C. (1996). A set of computer programs for determining defocus and astigmatism in electron images. *Ultramicroscopy* **65,** 31–44.

Taylor, K. A., Dux, L., and Martonosi, A. (1986). Three-dimensional reconstruction of negatively stained crystals of the $Ca^{++}$-ATPase from muscle sarcoplasmic reticulum. *J. Mol. Biol.* **187,** 417–427.

Toyoshima, C., and Unwin, N. (1988). Contrast transfer for frozen-hydrated specimens: determination from pairs of defocused images. *Ultramicroscopy* **25,** 279–292.

Toyoshima, C., and Unwin, N. (1990). Three-dimensional structure of the acetylcholine receptor by cryoelectron microscopy and helical image reconstruction. *J. Cell Biol.* **111,** 2623–2635.

Toyoshima, C., Sasabe, H., and Stokes, D. L. (1993a). Three-dimensional cryo-electron microscopy of the calcium ion pump in the sarcoplasmic reticulum membrane. *Nature* **362,** 469–471.

Toyoshima, C., Yonekura, K., and Sasabe, H. (1993b). Contrast transfer for frozen-hydrated specimens: II Amplitude contrast at very low frequencies. *Ultramicroscopy* **48,** 165–176.

Velazquez-Muriel, J. A., Sorzano, C. O., Fernandez, J. J., and Carazo, J. M. (2003). A method for estimating the CTF in electron microscopy based on ARMA models and parameter adjustment. *Ultramicroscopy* **96,** 17–35.

Ward, A., Moody, M. F., Sheehan, B., Milligan, R. A., and Carragher, B. (2003). Windex: A toolset for indexing helices. *J. Struct. Biol.* **144,** 172–183.

Whittaker, M., Carragher, B. O., and Milligan, R. A. (1995). PHOELIX: A package for semi-automated helical reconstruction. *Ultramicroscopy* **58,** 245–259.

Woodhead, J. L., Zhao, F. Q., Craig, R., Egelman, E. H., Alamo, L., and Padron, R. (2005). Atomic model of a myosin filament in the relaxed state. *Nature* **436,** 1195–1199.

Xu, C., Rice, W. J., He, W., and Stokes, D. L. (2002). A structural model for the catalytic cycle of $Ca^{2+}$-ATPase. *J. Mol. Biol.* **316,** 201–211.

Yonekura, K., Stokes, D. L., Sasabe, H., and Toyoshima, C. (1997). The ATP-binding site of $Ca^{2+}$-ATPase revealed by electron image analysis. *Biophys. J.* **72,** 997–1005.

Yonekura, K., Maki-Yonekura, S., and Namba, K. (2003a). Complete atomic model of the bacterial flagellar filament by electron cryomicroscopy. *Nature* **424,** 643–650.

Yonekura, K., Toyoshima, C., Maki-Yonekura, S., and Namba, K. (2003b). GUI programs for processing individual images in early stages of helical image reconstruction–for high-resolution structure analysis. *J. Struct. Biol.* **144,** 184–194.

Yonekura, K., Maki-Yonekura, S., and Namba, K. (2005). Building the atomic model for the bacterial flagellar filament by electron cryomicroscopy and image analysis. *Structure* **13,** 407–412.

Young, H. S., Rigaud, J.-L., Lacapere, J.-J., Reddy, L. G., and Stokes, D. L. (1997). How to make tubular crystals by reconstitution of detergent-solubilized $Ca^{2+}$-ATPase. *Biophys. J.* **72,** 2545–2558.

Young, H., Xu, C., Zhang, P., and Stokes, D. (2001a). Locating the thapsigargin binding site on $Ca^{2+}$-ATPase by cryoelectron microscopy. *J. Mol. Biol.* **308,** 231–240.

Young, H. S., Jones, L. R., and Stokes, D. L. (2001b). Locating phospholamban in co-crystals with $Ca^{2+}$-ATPase by cryoelectron microscopy. *Biophys. J.* **81,** 884–894.

Yu, X., Jacobs, S. A., West, S. C., Ogawa, T., and Egelman, E. H. (2001). Domain structure and dynamics in the helical filaments formed by RecA and Rad51 on DNA. *Proc. Nat. Acad. Sci. USA* **98,** 8419–8424.

Zhang, P., Toyoshima, C., Yonekura, K., Green, N. M., and Stokes, D. L. (1998). Structure of the calcium pump from sarcoplasmic reticulum at 8 Å resolution. *Nature* **392,** 835–839.

Zhang, P., Borgnia, M. J., Mooney, P., Shi, D., Pan, M., O'Herron, P., Mao, A., Brogan, D., Milne, J. L., and Subramaniam, S. (2003). Automated image acquisition and processing using a new generation of 4K x 4K CCD cameras for cryo electron microscopic studies of macromolecular assemblies. *J. Struct. Biol.* **143,** 135–144.

CHAPTER SIX

# Reconstruction of Helical Filaments and Tubes

Edward H. Egelman

## Contents



## Abstract

While Fourier–Bessel methods gave rise to the first three-dimensional reconstruction of an object from electron microscopic images, and these methods have dominated three-dimensional reconstruction of helical filaments and tubes for 30 years, single-particle approaches to helical reconstruction have emerged within the past 10 years that are now the main method being used. The Iterative Helical Real Space Reconstruction (IHRSR) approach has been the main methodology, and it surmounts many of the problems posed by real polymers that are flexible, display less than crystalline order, or are weakly scattering. The main difficulty in applying this method, or even Fourier–Bessel methods, is in determining the approximate helical symmetry. This chapter focuses on some of the intrinsic ambiguities that are present when trying to determine the helical symmetry from power spectra of images and argues that complementary techniques or some form of prior knowledge about the subunit may be needed to have confidence in the solution that is found.

Much of the protein in bacterial, archaeal, and eukaryotic cells exists in the form of helical polymers. Whether the result of some design (either moronic or intelligent) or not (Egelman, 2010), the abundance of helices can be best explained by the fact that helical polymers are the consequence of the simplest bonding rule that can be created between any two copies of an asymmetric unit, such as a protein (Egelman, 2003). In general, such a bonding

Department of Biochemistry and Molecular Genetics, University of Virginia, Charlottesville, Virginia, USA

*Methods in Enzymology*, Volume 482
ISSN 0076-6879, DOI: 10.1016/S0076-6879(10)82006-3

rule can be repeated many times to generate a helical lattice. Large numbers of helical polymers, from F-actin (Moore *et al.*, 1970) to tubes containing the dynamin molecule (Chen *et al.*, 2004), have been studied by electron microscopy (EM), reflecting the central role of such structures in all aspects of cell and molecular biology. But helical polymers have also been ideal objects for EM study and three-dimensional reconstruction. In many cases, a single image of a helical polymer provides all of the views necessary to generate a three-dimensional reconstruction. This is because a single asymmetric unit is repeated many times in a helical polymer, with each copy of this asymmetric unit providing a different projection due to a different orientation. The apparent simplicity of helical reconstruction is why the first three-dimensional reconstruction from EM images was of a helical polymer (DeRosier and Klug, 1968).

Because of the utility of helical arrays, and the elimination of the need to obtain tilts when working with two-dimensional crystals, some effort has actually been placed in forming synthetic helical tubes as a means to reconstruct a protein or complex (Dang *et al.*, 2005; Toyoshima and Unwin, 1990; Zhang *et al.*, 1998). On the other hand, some viral capsid proteins have been observed to form helical tubes (or polyheads), rather than closed shells, under the appropriate conditions (Li *et al.*, 2000; Steven *et al.*, 1976). Most other helical assemblies that have been studied (and there are many!) occur naturally. I will not distinguish in this chapter between artificially created helical tubes and naturally occurring helical filaments, as the methods and problems are quite similar.

For more than 30 years after DeRosier and Klug (1968), almost all helical reconstructions involved the Fourier–Bessel formalism (Klug *et al.*, 1958). In this approach, the Fourier transform of a helical object is described in terms of Bessel functions. Since a helical polymer has axial periodicity, the three-dimensional Fourier transform is nonzero only on layer planes. The images recorded in an EM correspond (ideally) to the projection of a three-dimensional density distribution onto a two-dimensional image, and the Fourier transform of the projected image is the central section of the three-dimensional Fourier transform (see chapter 1 in this volume). Layer planes are thus intersected by the central section to become the layer lines seen in a two-dimensional Fourier transform. On each layer line, the amplitudes can be described very simply by a Bessel function whose argument involves the radius of a point scatterer giving rise to the diffraction. Since any atomic density can be thought of as a sum of many point-like atoms, the sum of all of these Bessel terms generates the observed amplitudes. To reconstruct in three-dimensions, one must simply "index" the diffraction pattern in terms of assigning an order $n$ to each Bessel function, and then do a Fourier–Bessel inversion. There are a number of excellent reviews that discuss the Fourier–Bessel approach in some depth (Stewart, 1988).

Over the past 10 years, however, it has become apparent that single-particle type approaches to helical reconstruction offer many advantages

(Egelman, 2000, 2007a,b; Sachse *et al.*, 2007, 2008). In the limit of nearly crystalline helical structures (Yonekura *et al.*, 2003) one may do as well with Fourier–Bessel methods as with the single-particle type approach, but such highly ordered samples are quite rare. One of the most highly ordered helical polymers, tobacco mosaic virus (TMV), has now been solved by a single-particle approach at better than 5 Å resolution (Sachse *et al.*, 2007), so even in this rare limit there appear to be few reasons to use the Fourier–Bessel method. The advantages of the single-particle approach are many and include: (1) elimination of the problem of Bessel overlap when the symmetry generates more than one Bessel function on a layer line (Crowther *et al.*, 1985; Woodhead *et al.*, 2005); (2) elimination of the need to computationally straighten images of flexible filaments (Egelman, 1986); (3) ability to solve weakly diffracting filaments, where Fourier transforms from individual filaments do not show the layer lines needed for Fourier–Bessel approaches (Craig *et al.*, 2006; Fujii *et al.*, 2009); and (4) ability to deal with variability in structure (Egelman *et al.*, 1982), a property of most helical filaments, where long-range order is not maintained. In addition, because the method is based upon using large data sets in a single-particle approach (Frank *et al.*, 1981), many different strategies may be employed for sorting and classification to achieve more homogeneous subsets of images.

# 1. The Iterative Helical Real Space Reconstruction Approach

A general method for reconstructing helical polymers (Egelman, 2000) was proposed that involves an iterative determination and imposition of helical symmetry upon objects that have been reconstructed without any helical symmetry imposed. The method works very well, has been applied extensively (in more than 100 publications), and surmounts many of the problems mentioned above that are inherent in Fourier–Bessel approaches. As discussed below, however, the method does not eliminate the need to understand helical symmetry and helical diffraction. A schematic diagram (Fig. 6.1) shows the iterative helical real space reconstruction (IHRSR) cycle. A reference volume (top) is used to generate reference projections, where each involves a different azimuthal rotation of the reference volume. The actual angular increment between projections depends upon the diameter of the object ($D$) and the expected resolution ($d$), and should be $360° \times d/(2\pi D)$. The number of reference projections is therefore simply $2\pi D/d$. In the case of a point group symmetry ($C_n$) in addition to the helical screw symmetry, the number of reference projections is reduced to $2\pi D/nd$. The reference projections are cross-correlated against the thousands or tens

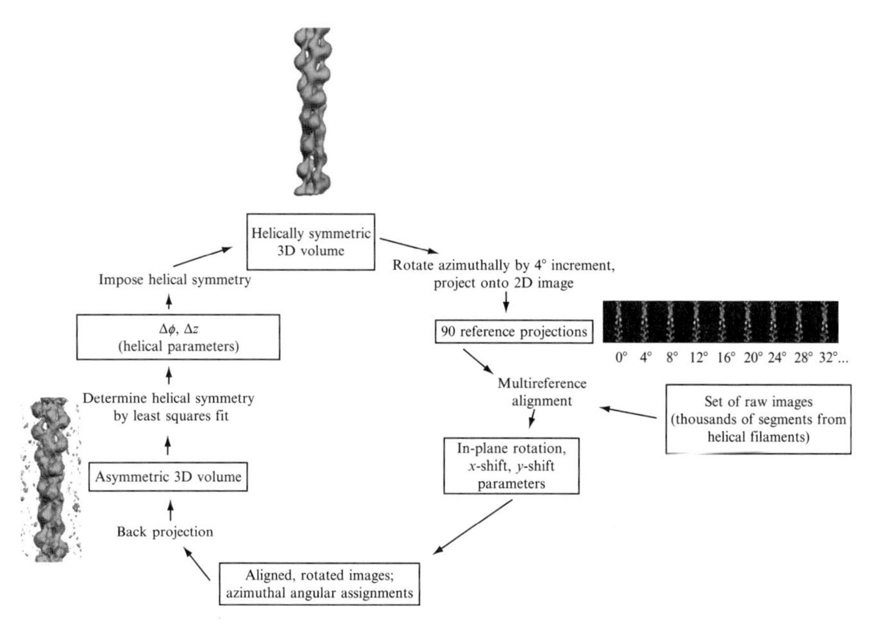


**Figure 6.1** The IHRSR cycle, reproduced from Egelman (2007b). A reference volume (top) is used to generate 90 reference projections, where each involves a 4° azimuthal rotation of the reference volume. The actual angular increment (4° in this example) depends upon the diameter of the object ($D$) and the expected resolution ($d$). The reference projections (90 in this case) are cross-correlated against the actual image segments. The highest correlation determines the azimuthal orientation of the image in question, as well as providing the in-plane rotation and translation needed to bring it into register with the reference projection. The aligned images are then used in a back-projection to generate a three-dimensional reconstruction (lower left corner). This volume is searched for the helical screw operator (the coupled rotation and axial translation) that minimizes the variance between the actual volume and a symmetrized version of the volume. This screw operator is then imposed on the reconstruction to generate a helically symmetric volume used as a new reference (top) for the next cycle of the procedure.

of thousands of actual image segments. The highest correlation yields the azimuthal orientation of the image in question, as well as the in-plane rotation and translation needed to bring it into register with the reference projection. The aligned images are then used in a back-projection algorithm to generate a three-dimensional reconstruction (Fig. 6.1, lower left corner). This volume is searched for the helical screw operator (the coupled rotation and axial translation) that minimizes the variance between the actual volume and a symmetrized version of the volume. This screw operator is then imposed on the reconstruction to generate a helically symmetric volume used as a new reference (Fig. 6.1, top) for the next cycle of the procedure.

Since each cycle involves the use of a reference volume (Fig. 6.1, top), a question is obviously raised as to how one can begin with no prior knowledge of what the structure looks like. Numerous applications have shown that no prior knowledge of the appearance of the structure is needed, as the method always works when starting with a solid cylinder as an initial reference (Egelman, 2007b). Images are cross-correlated each cycle with different projections of the reference volume to assign Euler angles to each image. Since all projections of a solid cylinder are the same when the azimuthal angle is changed, the effect of using a solid cylinder as an initial reference is to simply assign random azimuthal angles to each image. In fact, the solid cylinder can be eliminated, and random azimuthal angles can be assigned to generate the starting volume. In this sense, the IHRSR method does not require any starting reference volume. The method *does* require some estimate of the helical symmetry, and this is potentially problematic. The problems arise not from limitations of the IHRSR procedure, but from intrinsic ambiguities that exist at limited resolution, and this is discussed in much more detail later.

## 2. Using IHRSR

With the hope that the caveats discussed below about helical ambiguities will be fully appreciated, IHRSR works very simply when one starts with an estimate of the helical symmetry. This estimate can come from indexing power spectra obtained from the images, from knowledge of the mass per unit length, or from trial and error. At this point, IHRSR scripts only exist for SPIDER (Frank *et al.*, 1996), but that will be changing and IHRSR approaches should be available in the future for packages such as EMAN (Ludtke *et al.*, 1999), EMAN2 (Tang *et al.*, 2007), and SPARX (Hohn *et al.*, 2007). A graphical user interface (Fig. 6.2), called *generator*, exists to write the SPIDER script and create the initial symmetry file. This symmetry file is updated in each cycle (Fig. 6.3) as the iterations proceed.

There are many practical questions that arise when using IHRSR, and one of the first is the optimal length of the segments to be used and the optimal overlap (or shift) between adjacent boxes. In conventional single-particle reconstruction, the box size is simply determined by the size of the object. For helical filaments, the minimum width of the box is determined by the object size, but the length is arbitrary. A "gedanken" experiment is shown in Fig. 6.4 for the resolution as a function of box length. For a perfectly ordered specimen (thin dotted line), the longer the box length the better the resolution that one can achieve. This can be understood due to the fact that the shorter the box length, the greater the error in alignment of segments against reference projections. The main limitation on resolution would thus be the errors in alignment. For most real specimens, however,

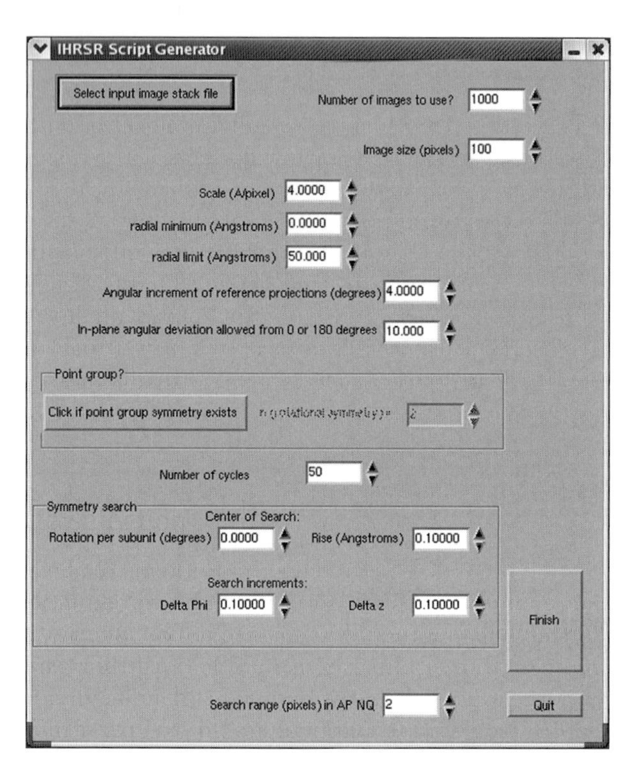


**Figure 6.2** A graphical interface is shown for writing the IHRSR Spider script. All of the needed parameters are entered through this interface, including the initial estimate of the helical symmetry (the rotation and axial translation that operates on one subunit to generate another one). For objects with a rotational point group symmetry $C_n$ in addition to a helical symmetry, the value of $n$ can be entered.

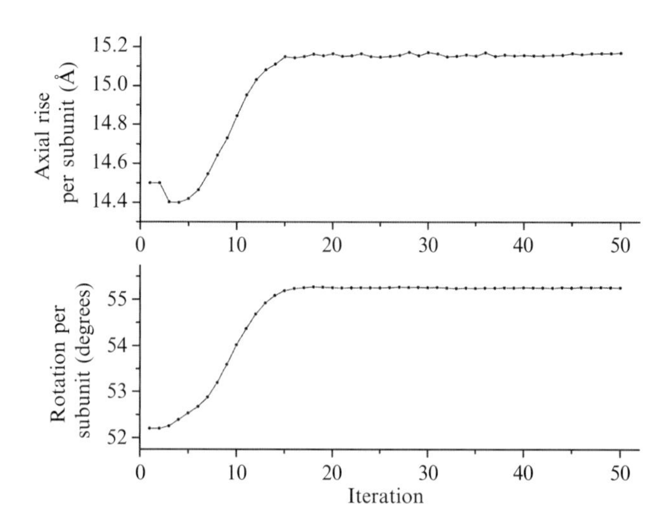


**Figure 6.3** The helical screw symmetry (axial rise and rotation per subunit) is determined each iteration of IHRSR. The figure shows the convergence of both parameters from initial values for filaments of the human Dmc1 protein formed on DNA (Sheridan *et al.*, 2008).

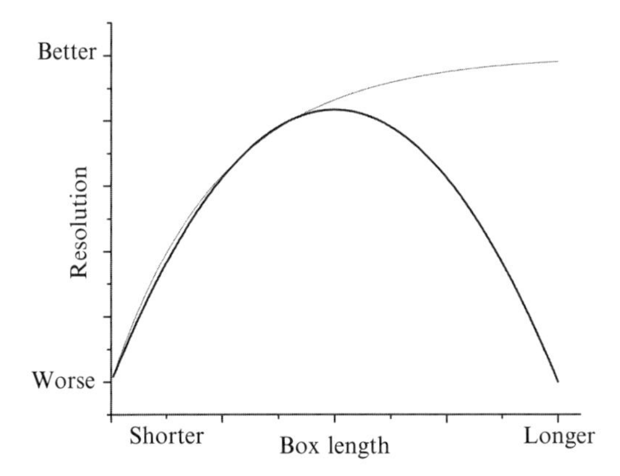


**Figure 6.4** A "gedanken" experiment showing resolution as a function of box length. For a perfectly ordered ideal helical specimen (dotted line) the resolution will improve as the box becomes longer, due to the fact that there will be less alignment error the longer the box. For most real specimens (solid curve) there is an optimal box length that can only be found empirically.

the resolution as a function of box length would more likely follow the solid curve. When one uses boxes longer than optimal, the resolution suffers due to the fact that the "coherence" of the polymer (Egelman and DeRosier, 1982) is shorter than the box length, and one is blurring the reconstruction by adding in portions that are not in helical register. Similarly, the flexibility and curvature of the polymer will be a factor in determining the optimum box length. While algorithms have existed to computationally straighten images of curved polymers (Egelman, 1986), these are intrinsically artifactual, as they assume that the polymer is undergoing a normal mode of bending. While such an assumption is reasonable for a homogeneous solid, it fails for a helical polymer that has grooves which may be compressed more readily than subunits. The IHRSR approach relies upon using a piece-wise linear approximation of the continuum, so a gently curving polymer is broken up into short lengths that are each treated as a straight segment. The radius of curvature will obviously be important in determining the optimal box length. If the box length is too long, curvature *within* a segment will degrade the resolution, so as the box becomes longer than optimal (Fig. 6.4) the resolution will decline. Unfortunately, the optimal box length may only be determined empirically for each specimen, will depend upon the signal-to-noise ratio in the images as well as the properties of the polymer, and there are no simple formulas describing what is optimal.

There is absolutely no relation between the box length chosen and the classical notion of a helical repeat. A helical repeat is defined as the distance that a subunit must be translated along the axis to be in register with another subunit. This helical repeat must be the product of an integer multiplied by the axial rise per subunit. I have argued (Egelman, 2007b) that this description of helical symmetry is a poor one, as it is ill-conditioned. Consider an actin filament with the simplest possible repeat, 13 subunits in 6 turns of the left-handed 1-start helix, with an axial rise per subunit of 27.3 Å. This filament will have a repeat of 13 × 27.3 Å = 355 Å. The angle between adjacent subunits is 360° × 6/13 = 166.15°. If this angle is changed by 0.03°, the best approximation of the repeat might be 938 subunits in 433 turns, with a repeat distance of 25,607 Å. Thus, a nearly infinitesimal change in the twist gives rise to a catastrophic change in the repeat distance. On the other hand, one would have a very difficult time experimentally distinguishing between an actin helix that repeats every 938 subunits and one that repeats every 13 subunits, which is another reason why the use of the term "repeat" is a poor one. The length of the box chosen must be at least three or four times the axial rise per subunit (27.3 Å in the case of F-actin) for the helical search and imposition to work, but there is no relationship between this box length and whether F-actin repeats in 13, 28, or 938 subunits.

The optimal overlap between adjacent boxes also depends upon the axial rise per subunit. The search range that is used in SPIDER in the multi-reference alignment sets a limit for how far images are translated when seeking the best alignment against reference projections. This search range must be at least half the axial rise per subunit. That is the maximal axial translation that is needed to bring any given image into register with the best-matching projection. If the shift between adjacent overlapping segments is only twice the search range, then one image may be translated down by the search range to align it with a reference projection, and the subsequent image may be translated up by the search range so that they are both in register with the same reference projection. These two images will therefore be used in the back-projection as the same view of the object being reconstructed. One will waste time in both the alignment phase and the reconstruction phase without including any new information. On the other hand, if there is little or no overlap, one is essentially throwing out useful data, as an overlapping image shifted by several subunits provides a completely different view of the helical structure. So one needs to introduce as much overlap of adjacent boxes as possible, as long as two adjacent boxes are not being translated in the alignment and reconstruction procedure to provide the same view of the helix.

## 3. Intrinsic Ambiguities in Helical Symmetry

To understand these intrinsic ambiguities, let us look at the amplitudes along a layer line of order $n$ generated by a helical arrangement of atoms, all at a radius $r = r_0$. As shown (Klug *et al.*, 1958), the layer line amplitude $F(R)$

in the Fourier transform (where $R$ is the radius in reciprocal space from the meridian, the imaginary vertical line that runs through the center of the transform) will be proportional to $J_n(2\pi R r_o)$, with $J_n$ being an ordinary Bessel function of order $n$. Let us set $r_o = 24$ Å for a three-dimensional helical array of atoms, and an image, corresponding to the projection of the three-dimensional array onto a plane, is shown in Fig. 6.5A. The power spectrum of this image is shown in Fig. 6.5B. Indexing this power spectrum involves assigning Bessel orders to each layer line. Let us look at the layer line marked as "ll = 97", where we can measure $R \approx 0.043$ Å$^{-1}$ for the position of the first maximum. Since EM image analysis of helical polymers (as opposed to electron or X-ray diffraction) begins with images, one can easily measure the maximum diameter of this array of atoms, which would yield an estimate of $r_o \approx 24$ Å. We can therefore see that $2\pi(0.043)(24) = 6.5$, and $|n|$ must be 5 on this layer line, since the function $J_5(x)$ has its first maximum when the argument $x$ is equal to 6.4 (the difference between 6.4 and 6.5 is experimental error). We only need to do this for two independent layer lines (such as the two marked by the basis vectors shown in Fig. 6.5C), and we can then index all layer lines as shown in the $n$, $l$ plot. The only ambiguity that exists is with the helical hand. In the $n$, $l$ plot shown in Fig. 6.5C, it is assumed that the 1-start helix (which passes through every subunit) is right-handed, hence it has been assigned $n = +1$. But from the information shown, this could just as well be left-handed (generating the same projection seen in Fig. 6.5A), and the $n$, $l$ plot would be mirrored. In this case, there would be $n = +5$ on ll = 97, rather than $n = -5$ as shown. Information would need to be obtained, such as by tilting the sample in the EM (Finch, 1972) or by metal shadowing (Woodward *et al.*, 2008), to determine the hand of the helix. Because of the projection theorem (relating the image to the projection of the three-dimensional density distribution onto two-dimensions), any information about hand has been simply lost, whether one is using cryo-EM of unstained, frozen-hydrated samples, or conventional EM of negatively stained samples (Egelman and Amos, 2009).

The point of this exercise has been to show that the helical diffraction pattern of an array of points (or atoms) can be simply indexed when these points are all at a single radius. The problem with EM of helical protein polymers is that subunits are rarely approximated well by a single point scatterer. Let us look at the example of TMV, which has been a model system in EM (Jeng *et al.*, 1989; Sachse *et al.*, 2007; Zhu *et al.*, 2001). Figure 6.6A shows the 5 Å resolution reconstruction of TMV (Sachse *et al.*, 2007) when filtered to 10 Å resolution. Projections were made from this volume at random azimuthal angles, generating 1000 images. These images were reconstructed using the IHRSR algorithm shown in Fig. 6.1, starting with a symmetry of 16.4 subunits per turn of a 23 Å pitch helix, and with a solid cylinder as an initial reference volume. These parameters are quite

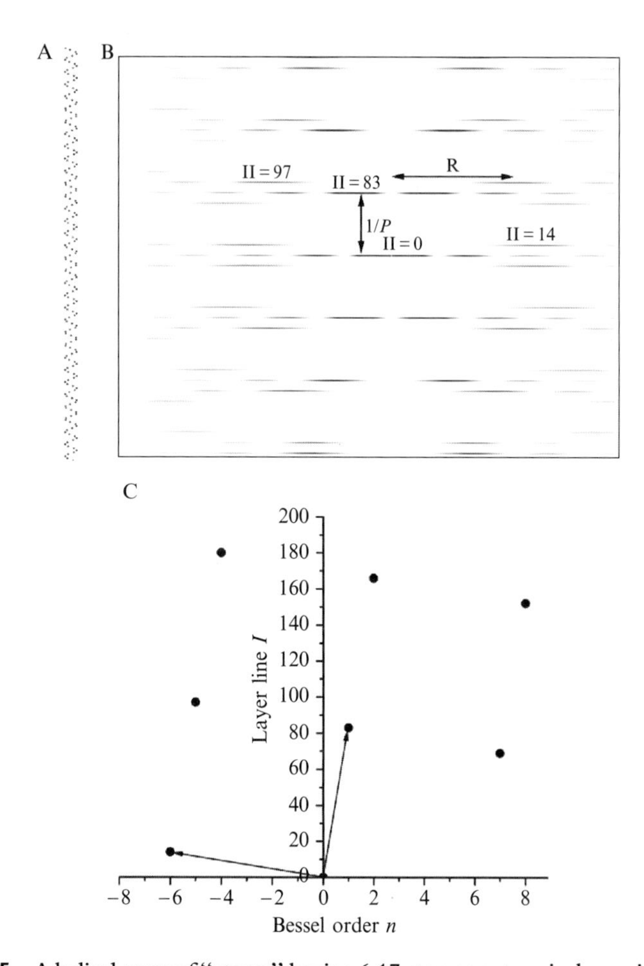


**Figure 6.5** A helical array of "atoms" having 6.17 atoms per turn is shown in (A). The power spectrum of this array (the squared modulus of the Fourier transform) is shown in (B). The $n$, $l$ plot, showing which Bessel function $n$ is allowed on each layer line $l$, is shown in (C). The $n = 1$ layer line, labeled ll $= 83$ in (B), corresponds to a 1-start helix which passes through every subunit. The distance from this layer line to the equator is the reciprocal of the pitch of this 1-start helix, and this distance is labeled $1/P$ in (B). The layer line nearest the equator, ll $= 14$, can be seen from (C) to have $n = -6$, while ll $= 97$ can be seen to have $n = -5$. Two arbitrary basis vectors have been drawn in (C), and every other point in the $n$, $l$ plot arises from a linear combination of these two vectors. Other basis vectors might have been chosen, such as ($n = -5$, ll $= 97$) and ($n = 7$, ll $= 69$), as long as these vectors are independent and not simply different orders of the same vector (such as ($n = 1$, ll $= 83$) and ($n = 2$, ll $= 166$)). The $n$, $l$ plot corresponds to a right-handed 1-start helix, but this information cannot be determined from the projection in (A), as a left-handed 1-start helix would give rise to the same projection in (A) and power spectrum in (B).

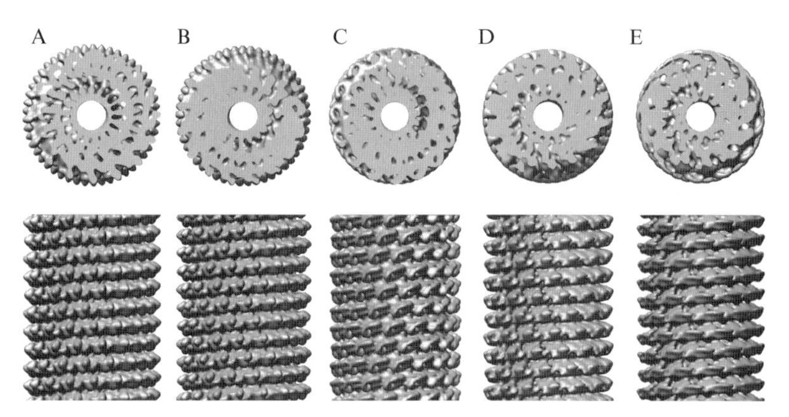


**Figure 6.6** A model for TMV has been constructed (A) using the EM structure (Sachse *et al.*, 2007), but filtered to 10 Å resolution. An axial view is shown at the top and the side view at the bottom. Random azimuthal projections (120 pixel long, with 2 Å per pixel sampling) have been generated from this model ($n = 1000$), and these have been used for different IHRSR reconstructions (B–E). The reconstruction in (B) has the correct symmetry, 16.33 subunits per turn of a 23 Å pitch helix, and is nearly indistinguishable from the starting model (A). Different incorrect symmetries are shown in (C–E), which look extremely different from the starting model (A), but which provide stable solutions to the IHRSR algorithm. In (C), the symmetry is 15.67 subunits per turn, in (D) it is 14.33 subunits per turn and in (E) it is 12.33 subunits per turn.

close to the actual symmetry of TMV, with ~16.33 subunits per turn of a 23 Å pitch helix. The IHRSR method rapidly converges to the correct structure, and a volume is shown after 10 cycles (Fig. 6.6B). Next, the IHRSR algorithm was started with a twist and axial rise corresponding to 15.67 subunits per turn of a 23 Å pitch helix. This also converges rapidly, but to this wrong solution (Fig. 6.6C). Similarly, convergence occurs to 14.33 units per turn (Fig. 6.6D) and 12.33 units per turn (Fig. 6.6E) when the IHRSR algorithm is started near these values.

To understand what is taking place here one needs to recognize the intrinsic ambiguities in helical indexing, the procedure where one assigns a Bessel function to a layer line. The power spectrum from the images used in this simulation is shown in Fig. 6.7A, and the layer lines are labeled 1–6. Since the three-dimensional volume was filtered to 10 Å resolution, the power spectrum does not extend beyond 1/(10 Å). This is an important point, as the ambiguities that I discuss disappear in this case when higher resolution is available. The $n$, $l$ plot for this power spectrum is shown in Fig. 6.7B and was generated since we *know* the symmetry of the model! The $n$, $l$ plot shows us that there is an $n = -16$ on ll = 1, and an $n = 17$ on ll = 2. What can be determined unambiguously from the power spectrum, since one knows the outer diameter of the particles used to generate the power spectrum

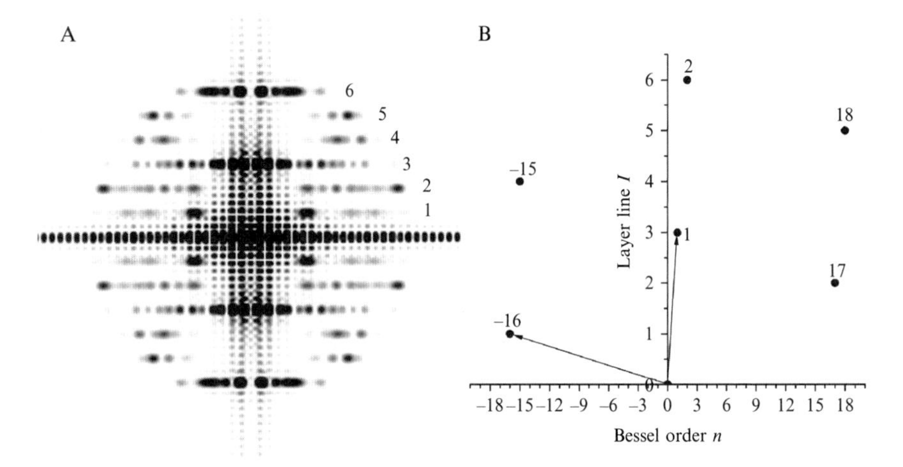


**Figure 6.7** The averaged power spectrum (A) from the 120 px (240 Å) long projections of the TMV model in Fig. 6.6A. The layer lines are numbered 1–6, where layer line 3 is at a spacing of 1/(23 Å) and arises from the 23 Å pitch 1-start helix of TMV. The $n$, $l$ plot for TMV (B) shows that layer line 1 contains $n = -16$, layer line 2 contains $n = 17$, etc.

(∼180 Å), is that there must be an $n = 1$ on ll = 3 and an $n = 2$ on ll = 6. This follows from the fact that the outer radius of the particle sets a *maximum limit* on the Bessel function generating a peak at a radius $R$ in the power spectrum. Determining the Bessel orders on the other layer lines is problematic, in contrast to what has been shown in Fig. 6.5 for an array of atoms at a single radius, since the outer diameter of the particle does not set a *minimum* allowable Bessel function. The problem that arises is in conflict with statements in the literature suggesting that the indexing of such power spectra from real structures (and not an array of atoms at a single radius) is simple.

The solution in Fig. 6.6C had 15.67 units per turn, rather than the correct 16.33 units per turn. The power spectrum from this reconstruction (Fig. 6.8B) has an $n = 16$ (rather than $n = -16$) on the first layer line, and an $n = -17$ (rather than an $n = 17$) on the second layer line. The peaks of these layer lines appear in exactly the same position as do the peaks from the correct reconstruction in Fig. 6.8A. But whereas the correct solution (Fig. 6.8A) has $n = -15$ and $n = 18$ on layer lines 4 and 5, respectively, the incorrect solution in Fig. 6.8B has $n = 17$ and $n = -14$ on layer lines 4 and 5, respectively. What can be seen is that the peaks actually occur in the same positions (in $R$) in Fig. 6.8A and B, even though there are very different Bessel orders involved. This is due to the fact that the radius ($r$) at which contrast is being generated differs between these two reconstructions. Similarly, the power spectra from the two other incorrect solutions (Fig. 6.6D, E) are shown (Fig. 6.8C, D, respectively) and one can see that

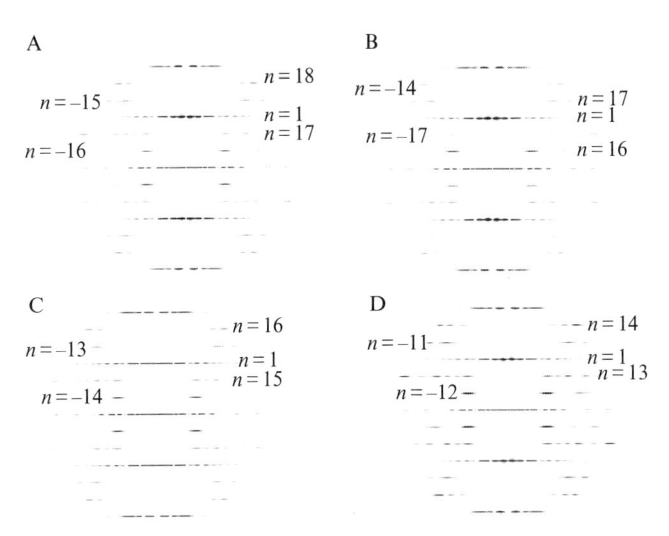


**Figure 6.8** The degeneracy of solutions in Fig. 6.6 can be best understood by comparing the power spectra from the different reconstructions. The power spectra in (A–D) are from the reconstructions in Fig. 6.6B–E, respectively. The power spectrum in (A) is from the correct solution, and the layer lines are indexed exactly as shown in the $n$, $l$ plot in Fig. 6.7B. The power spectrum in (B) comes from a three-dimensional volume with 15.67 units per turn (rather than 16.33), and has an $n = 16$ (rather than an $n = -16$) on the first layer line, and an $n = -17$ (rather than an $n = +17$) on the second layer line. The power spectrum in (C) comes from a three-dimensional volume with 14.33 units per turn (rather than 16.33), and has an $n = -14$ (rather than an $n = -16$) on the first layer line and an $n = 15$ (rather than an $n = 17$) on the second layer line. Similarly, the power spectrum in (D) comes from a three-dimensional volume with 12.33 units per turn (rather than 16.33) and has an $n = -12$ (rather than an $n = -16$) on the first layer line, and an $n = 13$ (rather than an $n = 17$) on the second layer line. Despite having different Bessel orders on layer lines 1, 2, 4, and 5, the power spectra all have peaks at identical positions due to the fact that the diffraction is coming from different radii.

these are also indistinguishable from the correct power spectrum (Fig. 6.8A). What this means is that starting with a power spectrum from the actual images (Fig. 6.7A), at a limiting resolution of 10 Å in this example, there are multiple ways to index the diffraction, but only one is correct, and the correct solution cannot be distinguished from the incorrect ones solely from the power spectrum.

In comparing power spectra from reconstructions (Fig. 6.8) with averaged power spectra from images (Fig. 6.7A) we are throwing out potentially valuable phase information. This phase information might be useful in discriminating between a correct and an incorrect solution. However, I previously provided an example (Egelman, 2007b) where the projections of a correct and an incorrect reconstructions were indistinguishable, demonstrating that the two-dimensional Fourier transforms (containing both phase and amplitude information) of the correct and incorrect reconstructions must also be indistinguishable. If we look at the present example, one way to use the phase information would be to do cross-correlations between the projections of the different reconstructions and the images. These cross-correlations would be sensitive to both the phases and the amplitudes, while power spectra involve throwing out the phase information and only looking at the squared amplitudes.

Figure 6.9 shows the frequency of the highest cross-correlation between the images and projections of the five reconstructions shown in Fig. 6.6. Strikingly, the correct solution (Fig. 6.6B) does not achieve the highest cross-correlation. If one had only to choose between a symmetry of 16.33 units per turn (Fig. 6.6B) and a wrong symmetry of 15.67 units per turn (Fig. 6.6C), then this approach would be useful as it would show a higher cross-correlation for the correct solution. However, the incorrect solution of Fig. 6.6D with 14.33 units per turn actually has a higher cross-correlation with the images than the correct solution. One must also keep in mind that the present example has been done with perfect "images" in the absence of any noise. In reality, observed power spectra are rarely as perfect as the one in Fig. 6.7A, and this introduces further ambiguities.

So how can one ever solve a helical structure given these potential problems? In some cases, the available resolution allows one to distinguish between different solutions. For example, one solution may predict an $n = 0$ and another solution predicts an $n = 2$ on the same layer line.

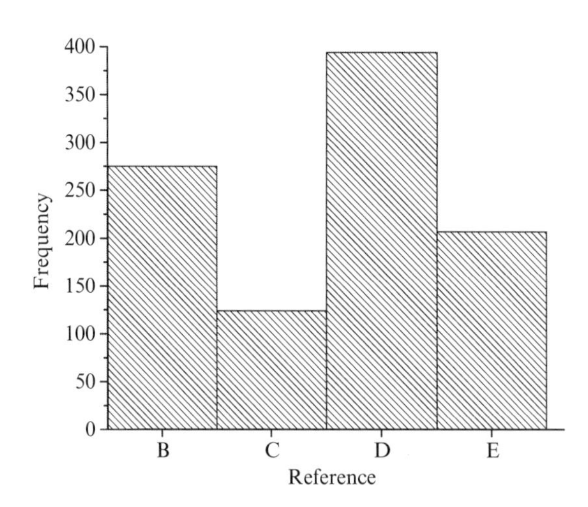


**Figure 6.9** The reconstructions in Fig. 6.6 can be used for a cross-correlation based sorting of the images. The histogram shows that of the three incorrect solutions (C–E), two (C, E) have a lower frequency than the correct solution (B). However, solution (D), which is wrong, has a higher frequency than the correct solution!

Observing intensity on the meridian at this layer line requires that there be an $n = 0$, establishing that it cannot be $n = 2$. In other cases one may have prior knowledge of the structure of a subunit (but not of the polymer). As can be seen in Fig. 6.6, different solutions generate very different appearances for the subunit, so a solution that provides an excellent match with the prior structure provides a strong argument for the correctness of the solution. In many cases one can use additional information to discriminate. The possible solutions of 16.33, 14.33, and 12.33 units per turn make different predictions about the mass per unit length given a knowledge of the subunit molecular weight. Scanning transmission electron microscopy (STEM) is a very valuable technique for measuring mass per unit length (Wall and Hainfeld, 1986) and could discriminate among these solutions. The different solutions also make different predictions about the hand of surface helices. Techniques such as quick-freeze/deep-etch EM (Heuser, 1981), metal shadowing (Woodward *et al.*, 2008), or AFM (Frederix *et al.*, 2009) provide surface information that can also distinguish among solutions. Similarly, one can actually do tilts in the EM (Finch, 1972) to provide information that is lost when one is simply collecting projections. A recent paper (Yu and Egelman, 2010) serves as an example of how such helical ambiguities can arise and can be resolved.

## 4. Conclusion

This chapter has not been intended to serve as a comprehensive overview of helical symmetry, helical diffraction, and three-dimensional reconstruction from helical polymers. Rather, it is meant to provide an introduction to the IHRSR approach and some guidance and caution when dealing with the most important problem when reconstructing from images of helical polymers, deducing the helical symmetry. I show that when resolution is limited there are intrinsic ambiguities that cannot be resolved with the images alone. One must rely upon prior knowledge of the subunit appearance, or use data from complementary techniques such as metal shadowing, STEM, or AFM, to distinguish between a correct solution and ones that are wrong.

## *Acknowledgments*

This work was supported by NIH EB001567. I thank the members of my laboratory for most of the applications of IHRSR that have appeared.

## REFERENCES


Chen, Y. J., Zhang, P., Egelman, E. H., and Hinshaw, J. E. (2004). The stalk region of dynamin drives the constriction of dynamin tubes. *Nat. Struct. Mol. Biol.* **11,** 574–575.

Craig, L., Volkmann, N., Arvai, A. S., Pique, M. E., Yeager, M., Egelman, E. H., and Tainer, J. A. (2006). Type IV pilus structure by cryo-electron microscopy and crystallography: Implications for pilus assembly and functions. *Mol. Cell* **23,** 651–662.

Crowther, R. A., Padron, R., and Craig, R. (1985). Arrangement of the heads of myosin in relaxed thick filaments from tarantula muscle. *J. Mol. Biol.* **184,** 429–439.

Dang, T. X., Farah, S. J., Gast, A., Robertson, C., Carragher, B., Egelman, E., and Wilson-Kubalek, E. M. (2005). Helical crystallization on lipid nanotubes: Streptavidin as a model protein. *J. Struct. Biol.* **150,** 90–99.

DeRosier, D. J., and Klug, A. (1968). Reconstruction of three-dimensional structures from electron micrographs. *Nature* **217,** 130–134.

Egelman, E. H. (1986). An algorithm for straightening images of curved filamentous structures. *Ultramicroscopy* **19,** 367–373.

Egelman, E. H. (2000). A robust algorithm for the reconstruction of helical filaments using single-particle methods. *Ultramicroscopy* **85,** 225–234.

Egelman, E. H. (2003). A tale of two polymers: New insights into helical filaments. *Nat. Rev. Mol. Cell Biol.* **4,** 621–630.

Egelman, E. H. (2007a). Single-particle reconstruction from EM images of helical filaments. *Curr. Opin. Struct. Biol.* **17,** 556–561.

Egelman, E. H. (2007b). The iterative helical real space reconstruction method: Surmounting the problems posed by real polymers. *J. Struct. Biol.* **157,** 83–94.

Egelman, E. H. (2010). Reducing irreducible complexity: Divergence of quaternary structure and function in macromolecular assemblies. *Curr. Opin. Cell Biol.* **22,** 68–74.

Egelman, E. H., and Amos, L. A. (2009). Electron microscopy of helical filaments: Rediscovering buried treasures in negative stain. *Bioessays* **31,** 909–911.

Egelman, E. H., and DeRosier, D. J. (1982). The Fourier transform of actin and other helical systems with cumulative random angular disorder. *Acta Crystallogr.* **A38,** 796–799.

Egelman, E. H., Francis, N., and DeRosier, D. J. (1982). F-actin is a helix with a random variable twist. *Nature* **298,** 131–135.

Finch, J. T. (1972). The hand of the helix of tobacco virus. *J. Mol. Biol.* **66,** 291–294.

Frank, J., Verschoor, A., and Boublik, M. (1981). Computer averaging of electron micrographs of 40S ribosomal subunits. *Science* **214,** 1353–1355.

Frank, J., Radermacher, M., Penczek, P., Zhu, J., Li, Y., Ladjadj, M., and Leith, A. (1996). SPIDER and WEB: Processing and visualization of images in 3D electron microscopy and related fields. *J. Struct. Biol.* **116,** 190–199.

Frederix, P. L., Bosshart, P. D., and Engel, A. (2009). Atomic force microscopy of biological membranes. *Biophys. J.* **96,** 329–338.

Fujii, T., Kato, T., and Namba, K. (2009). Specific arrangement of alpha-helical coiled coils in the core domain of the bacterial flagellar hook for the universal joint function. *Structure* **17,** 1485–1493.

Heuser, J. (1981). Preparing biological samples for stereomicroscopy by the quick-freeze, deep-etch, rotary-replication technique. *Methods Cell Biol.* **22,** 97–122.

Hohn, M., Tang, G., Goodyear, G., Baldwin, P. R., Huang, Z., Penczek, P. A., Yang, C., Glaeser, R. M., Adams, P. D., and Ludtke, S. J. (2007). SPARX, a new environment for cryo-EM image processing. *J. Struct. Biol.* **157,** 47–55.

Jeng, T. W., Crowther, R. A., Stubbs, G., and Chiu, W. (1989). Visualization of alpha-helices in tobacco mosaic virus by cryo-electron microscopy. *J. Mol. Biol.* **205,** 251–257.

Klug, A., Crick, F. H., and Wyckoff, H. W. (1958). Diffraction by helical structures. *Acta Crystallogr.* **11,** 199–213.

Li, S., Hill, C. P., Sundquist, W. I., and Finch, J. T. (2000). Image reconstructions of helical assemblies of the HIV-1 CA protein. *Nature* **407,** 409–413.

Ludtke, S. J., Baldwin, P. R., and Chiu, W. (1999). EMAN: Semiautomated software for high-resolution single-particle reconstructions. *J. Struct. Biol.* **128,** 82–97.

Moore, P. B., Huxley, H. E., and DeRosier, D. J. (1970). Three-dimensional reconstruction of F-actin, thin filaments, and decorated thin filaments. *J. Mol. Biol.* **50,** 279–295.

Sachse, C., Chen, J. Z., Coureux, P. D., Stroupe, M. E., Fandrich, M., and Grigorieff, N. (2007). High-resolution electron microscopy of helical specimens: A fresh look at tobacco mosaic virus. *J. Mol. Biol.* **371,** 812–835.

Sachse, C., Fandrich, M., and Grigorieff, N. (2008). Paired beta-sheet structure of an Abeta (1-40) amyloid fibril revealed by electron microscopy. *Proc. Natl. Acad. Sci. USA* **105,** 7462–7466.

Sheridan, S. D., Yu, X., Roth, R., Heuser, J. E., Sehorn, M. G., Sung, P., Egelman, E. H., and Bishop, D. K. (2008). A comparative analysis of Dmc1 and Rad51 nucleoprotein filaments. *Nucleic Acids Res.* **36,** 4057–4066.

Steven, A. C., Couture, E., Aebi, U., and Showe, M. K. (1976). Structure of T4 polyheads. II. A pathway of polyhead transformation as a model for T4 capsid maturation. *J. Mol. Biol.* **106,** 187–221.

Stewart, M. (1988). Computer image processing of electron micrographs of biological structures with helical symmetry. *J. Electron Microsc. Tech.* **9,** 325–358.

Tang, G., Peng, L., Baldwin, P. R., Mann, D. S., Jiang, W., Rees, I., and Ludtke, S. J. (2007). EMAN2: An extensible image processing suite for electron microscopy. *J. Struct. Biol.* **157,** 38–46.

Toyoshima, C., and Unwin, N. (1990). Three-dimensional structure of the acetylcholine receptor by cryoelectron microscopy and helical image reconstruction. *J. Cell Biol.* **111,** 2623–2635.

Wall, J. S., and Hainfeld, J. F. (1986). Mass mapping with the scanning transmission electron microscope. *Annu. Rev. Biophys. Biophys. Chem.* **15,** 355–376.

Woodhead, J. L., Zhao, F. Q., Craig, R., Egelman, E. H., Alamo, L., and Padron, R. (2005). Atomic model of a myosin filament in the relaxed state. *Nature* **436,** 1195–1199.

Woodward, J. D., Weber, B. W., Scheffer, M. P., Benedik, M. J., Hoenger, A., and Sewell, B. T. (2008). Helical structure of unidirectionally shadowed metal replicas of cyanide hydratase from *Gloeocercospora sorghi*. *J. Struct. Biol.* **161,** 111–119.

Yonekura, K., Maki-Yonekura, S., and Namba, K. (2003). Complete atomic model of the bacterial flagellar filament by electron cryomicroscopy. *Nature* **424,** 643–650.

Yu, X., and Egelman, E. H. (2010). Helical filaments of human Dmc1 protein on single-stranded DNA: a cautionary tale. *J. Mol. Biol.* in press.

Zhang, P., Toyoshima, C., Yonekura, K., Green, N. M., and Stokes, D. L. (1998). Structure of the calcium pump from sarcoplasmic reticulum at 8-Å resolution. *Nature* **392,** 835–839.

Zhu, Y., Carragher, B., Kriegman, D. J., Milligan, R. A., and Potter, C. S. (2001). Automated identification of filaments in cryoelectron microscopy images. *J. Struct. Biol.* **135,** 302–312.

CHAPTER SEVEN

# Three-Dimensional Asymmetric Reconstruction of Tailed Bacteriophage

Jinghua Tang,* Robert S. Sinkovits,* *and* Timothy S. Baker*,†

## Contents



* Department of Chemistry & Biochemistry, University of California, San Diego, La Jolla, California, USA
† Division of Biological Sciences, University of California, San Diego, La Jolla, California, USA

*Methods in Enzymology*, Volume 482
ISSN 0076-6879, DOI: 10.1016/S0076-6879(10)82008-7

## Abstract

A universal goal in studying the structures of macromolecules and macromolecular complexes by means of electron cryo-microscopy (cryo-TEM) and three-dimensional (3D) image reconstruction is the derivation of a reliable atomic or pseudoatomic model. Such a model provides the foundation for exploring in detail the mechanisms by which biomolecules function. Though a variety of highly ordered, symmetric specimens such as 2D crystals, helices, and icosahedral virus capsids have been studied by these methods at near-atomic resolution, until recently, numerous challenges have made it difficult to achieve sub-nanometer resolution with large ($\geq\sim$500 Å), asymmetric molecules such as the tailed bacteriophages.

After briefly reviewing some of the history behind the development of asymmetric virus reconstructions, we use recent structural studies of the prolate phage $\phi$29 as an example to illustrate the step-by-step procedures used to compute an asymmetric reconstruction at sub-nanometer resolution. In contrast to methods that have been employed to study other asymmetric complexes, we demonstrate how symmetries in the head and tail components of the phage can be exploited to obtain the structure of the entire phage in an expedited, stepwise process. Prospects for future enhancements to the procedures currently employed are noted in the concluding section.

# 1. Introduction: 3D Asymmetric Reconstruction of Tailed Bacteriophage

Electron microscopy and three-dimensional (3D) image reconstruction have been the preferred tools for more than 40 years for studying large macromolecular structures that resist crystallization. Even for those viruses that can be crystallized, an advantage of electron microscopy is that it can be used to capture more transient, intermediate stages in the viral life cycle (e.g., Steven *et al.*, 2005) or to visualize the virus complexed with antibodies, receptors, or other molecules and ligands (e.g., Smith, 2003; Stewart *et al.*, 2003). These techniques were first applied to images of negatively stained samples of the helical, contractile tail of bacteriophage T4 (DeRosier and Klug, 1968; see DeRosier's Personal Account in this volume) and to the icosahedral tomato bushy stunt and human papilloma viruses (Crowther *et al.*, 1970), and they have since been used to solve the structures of a rapidly expanding universe of macromolecules and macromolecular complexes imaged by means of electron cryo-microscopy (cryo-TEM) (Cheng and Walz, 2009; Jonic *et al.*, 2008; also, see other chapters in this volume).

Whenever possible, advantage is taken of the inherent symmetry of the particles since this generally allows one to reach higher resolutions and maximize the signal-to-noise ratio in the final reconstructed map. One of the best examples of exploiting symmetry is found in the study of icosahedral viruses, where each image contains information from 60 equivalent views (Baker *et al.*, 1999; Crowther *et al.*, 1970). Recent advances in both the microscopy and image processing methods have enabled cryo-reconstructions of icosahedral viruses to approach atomic resolution (Chen *et al.*, 2009; Wolf *et al.*, 2010; Zhang *et al.*, 2008, 2010; Zhou, 2008).

## 1.1. Symmetry mismatch in phage structure

Many bacteriophage contain a multi-subunit, multicomponent tail connected to a unique vertex of the capsid, and it is through this vertex that the genome is packaged during assembly and released during infection. The heads of tailed phages have either a prolate (e.g., T4 and $\phi$29) or an isometric (e.g., P22, T7, $\lambda$, and $\varepsilon$15) morphology, but all heads possess a fivefold rotational symmetry about an axis that passes through the tail and opposing vertex. The tail complex generally has 6- or 12-fold redundancy in the constituent viral proteins, which results in a symmetry mismatch at the junction between the head and tail (Jiang *et al.*, 2006; Lander *et al.*, 2006; Xiang *et al.*, 2006). The DNA genome is a molecule with a unique sequence and hence cannot adopt a structure that matches any of the local symmetries in the capsid or tail components. Given that there is just one copy of the genome and at least one prominent, symmetry mismatch between the capsid and tail, the phage as a whole must be asymmetric.

Symmetry averaging is often used as part of the reconstruction process, and whether applied to viruses with full icosahedral symmetry or to tailed phages with fivefold symmetric heads, comes at a cost. Only those components that possess the imposed symmetry will be accurately represented in the averaged final map. Other features of the virus structure, such as the packaged genome or unique tail, will be smeared out since they do not share the imposed symmetry. Reconstructions of these viruses could, in principle, be carried out from start to finish without applying any symmetry using techniques similar to those used to study ribosomes (Frank, 2009), in which no symmetry is imposed. But the tailed phages provide a unique set of challenges since intrinsic local symmetries in parts of the phage can complicate the reconstruction process. For example, an attempt to process images of P22 phage, in which only $C_1$ symmetry was assumed from the start, failed to lead directly to a valid reconstruction (Chang *et al.*, 2006). Thus, it can be beneficial to exploit the inherent local symmetry in components of the phage when designing a reconstruction strategy that will yield a reliable density map from images of particles whose global structure is asymmetric.

### 1.2. History of asymmetric virus reconstructions by single-particle cryo-TEM

The widespread success with icosahedral virus reconstructions has also stimulated the development of techniques to study nonicosahedral viruses (Johnson and Chiu, 2007). After generalizing our polar Fourier transform method (Baker and Cheng, 1996) to handle cyclic symmetries, we obtained the first reconstruction of a tailed bacteriophage with a prolate head (Tao *et al.*, 1998). In that study of $\phi$29, fivefold symmetry was imposed to enhance features in the head structure. However, since the $\phi$29 tail is not fivefold symmetric, its structure was smeared out in the reconstruction. Later, a two-step processing scheme was developed to preserve symmetry mismatched components in the entire phage (Morais *et al.*, 2001). In step one, a fivefold-averaged reconstruction was computed in which the head but not tail was symmetrized. Then, the fivefold symmetry constraint of the head was removed and each particle image was compared to five separate, related projections of the model. The projection with the highest correlation to each particle image was used to assign the view orientation for that particle. This new set of particle orientations provided the necessary information needed to combine the images and compute a new reconstruction in which no symmetry was enforced. The second step of the procedure was repeated and led to the first asymmetric reconstruction of the complete $\phi$29 phage at 33-Å resolution (Morais *et al.*, 2001).

Since this initial $\phi$29 asymmetric reconstruction, several other 3D density maps of entire tailed phages have been determined at progressively higher resolutions (Table 7.1). In all these studies, even those at lower resolutions (~20 Å), the head and tail structures were clearly resolved. Our studies of $\phi$29 have led to two, sub-nanometer resolution, asymmetric reconstructions that have made it possible to resolve features corresponding to helices in the head–tail connector as well as a highly condensed, toroid-like DNA structure embedded within a cavity at the connector–tail junction (Tang *et al.*, 2008b).

### 1.3. Alternative strategies to determine the structures of asymmetric viruses

Not all phages or asymmetric complexes can be readily solved using the strategy we have outlined above. For example, T4 phage with its long, contractile tail, remains a significant challenge for single particle, asymmetric reconstruction methods. For T4, a concerted, "divide and conquer" approach was used to solve the head (Fokine *et al.*, 2004) and tail

**Table 7.1** Chronological history of asymmetric cryo-reconstructions of entire tailed bacteriophage

| Phage | Head shape | Software[a] | Resolution (Å) | Reference |
|---|---|---|---|---|
| $\phi$29 | Prolate | PFT | 33 | Morais *et al.*, 2001 |
| T7 | Isometric | XMIPP | 24 | Agirrezabala, 2005 |
| ε15 | Isometric | EMAN | 20 | Jiang *et al.*, 2006 |
| P22 | Isometric | EMAN | 20 | Chang *et al.*, 2006 |
| P22 | Isometric | SPIDER | 17 | Lander *et al.*, 2006 |
| $\phi$29 | Prolate | EMAN | 16 | Xiang *et al.*, 2006 |
| $\phi$29 | Prolate | EMAN, FREALIGN, AUTO3DEM | 7.8 | Tang *et al.*, 2008b |
| N4 | Isometric | EMAN | 29 | Choi *et al.*, 2008 |

[a] Software indicates main program(s) used to perform reconstruction. It is assumed that, in all or most cases, additional scripts were required for image preprocessing, data manipulation, file format conversions, and other tasks.

(Kostyuchenko *et al.*, 2005) structures separately. Asymmetric reconstruction techniques have also been used to examine virus–host interactions at limited resolution. Studies of polio and Semliki Forest virus attachment to liposomes were aided by manually adding high intensity dots in the images at points in the membrane where virus particles attached (Bubeck *et al.*, 2008). These dots served as fiducial markers to help in determining the relative orientations of individual virus particles and to compute from those images a reconstruction that indicated a unique vertex is involved in the delivery of the genome in both types of virus.

## 1.4. Cryo-TEM of $\phi$29

The sample preparation and microscopy steps required to produce a set of images from which the reconstructed structure of an asymmetric virus can be obtained are identical to those used to study icosahedral particles (e.g., Baker *et al.*, 1999). We recently computed asymmetric reconstructions of two different $\phi$29 particles (Tang *et al.*, 2008b). These included fiberless (gp8.5-) virions and fiberless "ghosts," which are particles formed by inducing virions *in vitro* to lose their dsDNA genome and the two molecules of viral gene product 3 (gp3) that are covalently linked to the ends of the linear genome. The same procedures were used to determine the 3D structures of both types of particles (to 7.8 and 9.3 Å, respectively), and we limit our discussion here to the procedures used to study the virion. Briefly, the microscopy involved first taking purified samples of fiberless $\phi$29 virions and vitrifing them over holey, carbon-coated grids (Chapter 3, Vol. 481). Images of these samples were then recorded on Kodak SO163 electron image film at a nominal magnification of 38,000× and an electron dose of $\sim 20e^{-}/\text{Å}^2$ in an FEI CM200 FEG microscope operated at 200 keV. Micrographs that exhibited minimal astigmatism and specimen drift, and with the objective lens under-focused by 1–5 $\mu$m, were digitized at 7 $\mu$m intervals on a Zeiss PHODIS scanner and bin-averaged to yield an effective pixel size of 3.68 Å. A total of 12,682 particle images were selected from 74 micrographs for further processing.

## 1.5. Strategy for determining the $\phi$29 structure

The general image reconstruction scheme we used followed a model-based refinement procedure. This entailed aligning raw particle images relative to a series of projections of an existing 3D model to estimate the origin and orientation of each particle. The particle origin is defined by the $(x,y)$ pixel coordinates of the position of the center of the particle in the image, and the particle orientation is defined by three angles ($\theta$, $\phi$, $\omega$) that specify the direction from which the particle is viewed in the image (See Baker et al., 1999 for definition of these angles.). With this set of five particle parameters, a new 3D reconstruction can be computed and used as the model for the next iteration of the process.

The specific reconstruction strategy that we adopted in our $\phi$29 study involved two major branches as shown schematically in Fig. 7.1. This strategy took into account that fact that $\phi$29 has a prolate head ($\sim$480 × 600 Å) and a tail of comparable length ($\sim$400 Å). The first branch of the processing scheme constructs a reliable starting model and achieves optimal particle boxing. This includes calculating separate head and tail models and then combining them into a single, hybrid model. The second branch takes this hybrid model, computes projections, and compares these to each raw image to assign to it an origin and orientation. The images are used to compute a new reconstruction and several cycles of alignment and image screening (to weed out "bad" particles) are carried out until no further improvement in resolution is achieved.

## 1.6. Image reconstruction software

Owing to the complexity of performing asymmetric reconstructions, we found it necessary to use different image reconstruction packages at various stages of the process. In the $\phi$29 project and as described below, we used EMAN (Ludtke *et al.*, 1999), FREALIGN (Grigorieff, 2007), AUTO3-DEM (Yan *et al.*, 2007b), and BSOFT (Heymann, 2001) to ultimately reach sub-nanometer resolution. It is important to stress that our choice of programs more reflects our familiarity with the capabilities of the software

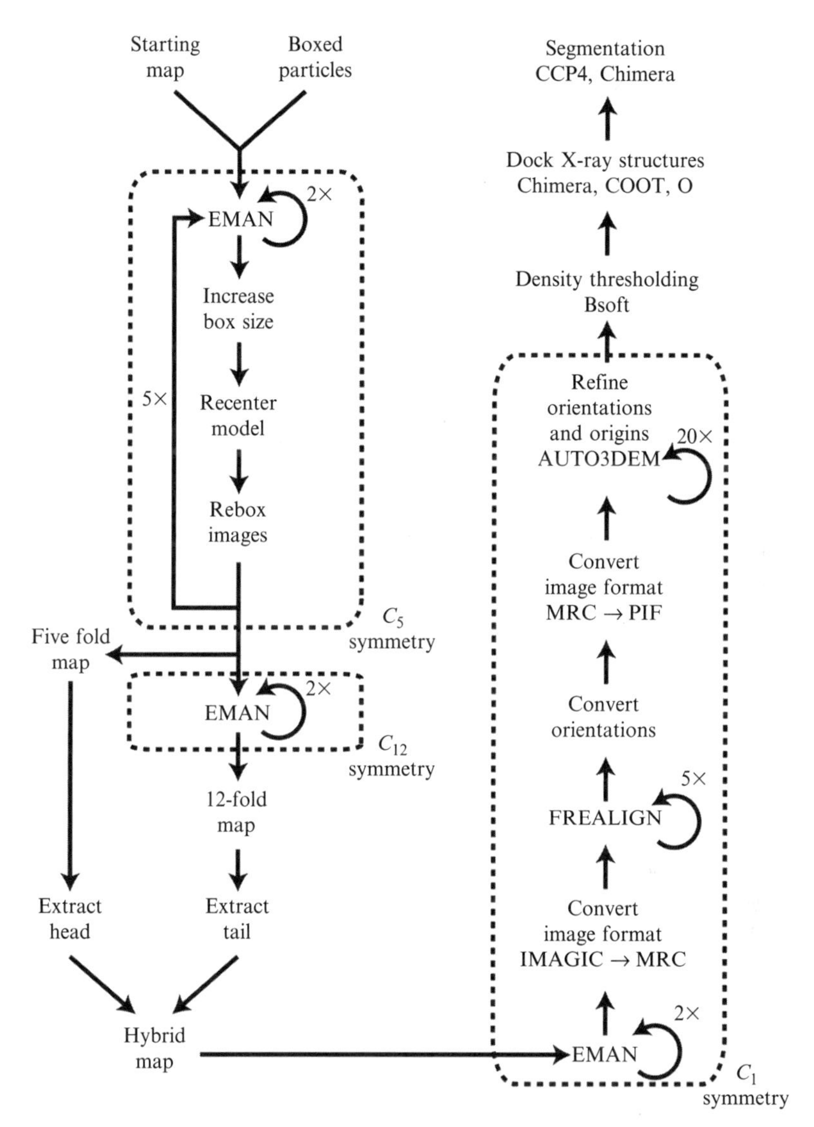


**Figure 7.1** $\phi$29 image reconstruction flowchart. Left hand side of flowchart shows steps taken to generate hybrid-starting model built from fivefold symmetric phage head and 12-fold symmetric tail. Right hand side illustrates steps in fully asymmetric reconstruction. Numbers next to loops indicate the number of iterations used in stages of the $\phi$29 reconstruction and will likely vary for other phage species.

than it does with providing an optimized strategy. These packages and others such as SPIDER (Frank *et al.*, 1996) and XMIPP (Sorzano *et al.*, 2004) contain many useful features that require significant expertise in order to use to their full potential (see also Chapter 15 in this volume).

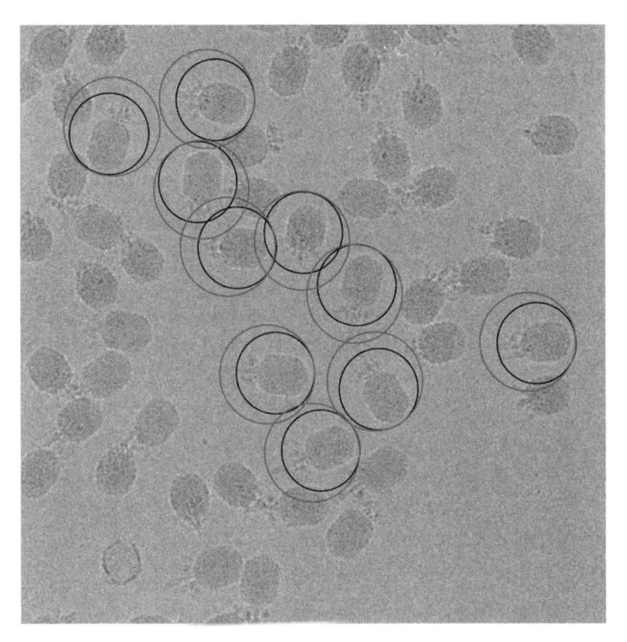

**Figure 7.2** Progressive boxing of phage particles. The small blue circles indicate the initial boxing of the particles, with box size chosen to capture head and proximal portion of tail. As the reconstruction progresses, the box size is gradually expanded to include more of the tail (large red circles). Note that the red and blue circles are not concentric and that the top of the phage head remains a constant distance from the edge of the circle. (See Color Insert.)

## 2. Particle Boxing

The purpose of boxing is to window out individual particles from their surroundings and to minimize the contribution of extraneous noise in the images to the final 3D reconstruction. Ideally, the boxing procedure centers each particle in its own box and excludes all neighboring particles. The defocus used in generating phase contrast during microscopy causes particle details to spread outside the particle boundary (Rosenthal and Henderson, 2003), and hence the pixel dimensions of the box need to extend well enough beyond this boundary to preserve structural information. The choice of the box size depends on resolution and amount of defocus. For the $\phi$29 image data, we chose a box size approximately 50% larger than the width of the prolate head (Fig. 7.2, red circles).

The type of boxing employed depends on the shape of the phage head and the length of the tail. If the head is isometric and the tail is short relative to the size of the capsid, which is true for some phage like P22 (Lander *et al.*, 2006), the particle can simply be boxed so that the head center coincides with the center of the box. Alternatively, if the tail dimension is comparable to that of the head, then the center of the box should coincide with the center of the whole particle to avoid having to make the box size excessively large.

## 2.1. Initial tight boxing of the phage head

Given that we did not have a starting model of the entire $\phi$29 virion, we decided to adopt a systematic, stepwise approach to solving its structure. This involved first constructing a model just for the head from a subset of ~1000 images masked tightly to include primarily the phage heads. Subsequently, we incrementally increased the size of the box to include more and more of the tail with the eventual goal of including the entire image of each phage particle in the final reconstruction as described below (Section 3). Hence, we started by boxing out just the particle heads and thereby excluded most of the tail and most of the neighboring particles (Fig. 7.2, blue circles). This tight boxing strategy helped to enhance the reliability with which initial particle origin and orientation parameters could be determined. Such a procedure is unnecessary with icosahedral particles since the origin of spherically symmetric objects is generally fairly easy to define quite accurately. The procedure does limit the resolution of the initial 3D reconstruction, but the overriding concern at this point is to obtain a reliable starting model. All boxing was carried out with the program RobEM (http://cryoem.ucsd.edu/programs.shtml) and images were converted from PIF to IMAGIC format using BSOFT (Heymann, 2001) before the next step of processing was performed using EMAN (Ludtke *et al.*, 1999).

## 2.2. Reboxing to center the particle

Whether particle boxing is performed manually or automatically, it is difficult to assure that each particle will be centered to 1 pixel accuracy in its box. Experience shows that reconstruction quality can be improved by periodically reboxing the particles using the latest set of origin positions obtained during the iterative refinement process (Gurda *et al.*, 2010). More significantly, with asymmetric particles it is important to make sure that the origin of each boxed particle correlates with the defined origin of the most current reconstructed model. Hence, after each cycle of EMAN as described later (Section 3.1), the identified origin of the particle within the box is used to reextract without interpolation the particle image from the micrograph such that this origin lies within 1 pixel of the center of the new box.

## 2.3. Expanding the box to include the tail

After a reliable head map was obtained (i.e., one in which pentameric and hexameric gp8 capsomers were clearly resolved), all particles were boxed anew from the raw micrographs using a circular mask, typically larger by 10–15 pixels in radius, to include more of the tail structure in each image. This necessitated that the density map of the head just calculated be padded with voxels set to the background density to match the size of the newly boxed particles and used as the model to align the images. Padding entailed adding voxels to expand all three dimensions of the cubic density map by equal amounts. The long axis of the phage was kept centered in the map but voxels in the axial direction were primarily added toward the tail side. This procedure assured that the particle (head + currently included portion of tail) was centered within the box containing the reconstructed 3D map.

The padded density map was used to generate a new set of projected images from which the origin and orientation parameters for each image could be redetermined. These were then used to compute a new reconstruction and to recenter and rebox the particles. At this point the mask was expanded, the map repadded with the background density, and the origins and orientations redetermined. This cycle of steps was repeated four additional times, at the end of which the entire phage particle was included in the density map.

### 2.4. Additional considerations about boxing

Typically, the early stages of the reconstruction process are accelerated significantly by making use of images that are two- or fourfold, bin-averaged. This is feasible because the primary goal at this stage is to obtain a low-resolution starting model whose size and shape are approximately correct. Hence, it is not necessary to use data at full pixel resolution. It is worth noting that, though automatic boxing routines can be used to speed up the boxing process, it was helpful with $\phi$29 to take the extra time to manually screen the entire data set of images and select the best ones to include in the processing. Ultimately, even the most sophisticated alignment algorithms cannot compensate for poor particle selection. Finally, though our reboxing strategy added additional steps to the entire image processing procedure, it proved to be quite effective in assuring that the particles were optimally centered in the boxes.

## 3. Generating a Starting Model of the Complete $\phi$29 Phage

Considerable effort is often required at the onset of a new project to generate a reliable, nonbiased model for initiating refinement of particle origin and orientation parameters. If the head essentially has icosahedral symmetry (i.e., neglecting the presence of the tail), the random model method (Yan *et al.*, 2007a) provides a relatively straightforward means to obtain a suitable starting model. To this then, one can either graft to one vertex of the icosahedral reconstruction an available reconstructed density map of the tail (Lander *et al.*, 2006) or a very simple, cylindrically averaged 3D model constructed from the image of a single, clearly visible tail (Jiang *et al.*, 2006). Even though the tail in the latter instance would not have the correct rotational symmetry, it is good enough to jump start refinement.

Because the $\phi$29 head has a prolate rather than isometric shape, the random model method could not be used to generate a starting model for the reconstruction process. However, we were able to use the $\phi$29 prohead structure (Morais *et al.*, 2005) as a starting model for the head of the mature phage since the two are similar in size and shape and both have fivefold axial symmetry (Fig. 7.3A).

## 3.1. Fivefold, symmetrized model of $\phi$29 head

The first 3D reconstruction of the head and truncated tail was obtained by combining all the tightly boxed particle images (Section 2.1) and imposing fivefold axial symmetry in EMAN. Then, as described in Sections 2.1 and 2.2, we obtained a series of four phage reconstructions that progressively encompassed more of the tail (Fig. 7.3B) and ultimately included the entire phage (Fig. 7.4A).

## 3.2. Hybrid head–tail model of $\phi$29

Given that the capsid portion of the $\phi$29 head has fivefold rotational symmetry, its structure was preserved in all of the fivefold averaged reconstructions generated as just described (Section 3.1). There is ample evidence

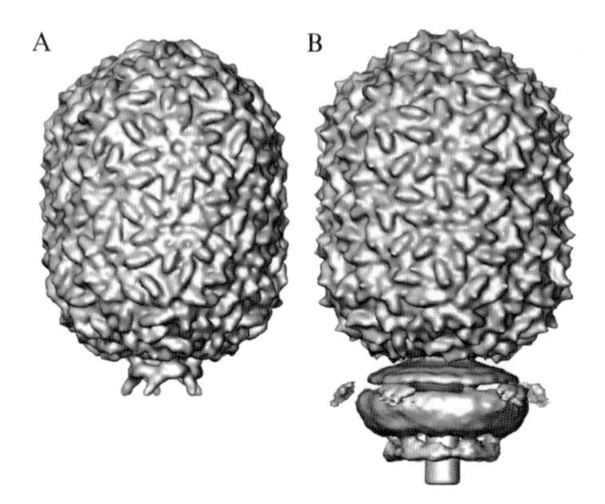


**Figure 7.3** Starting model for phage head and connector. (A) Shaded surface representation of $\phi$29 prohead map used as the starting point for the reconstruction of the complete phage. (B) Initial, fivefold averaged phage reconstruction obtained with tightly boxed particle images (blue circles, Fig. 7.2). Fivefold averaging yields clearly defined features in the head but features of the connector and tail are smeared out. (For interpretation of the references to color in this figure legend, the reader is referred to the Web version of this chapter.)

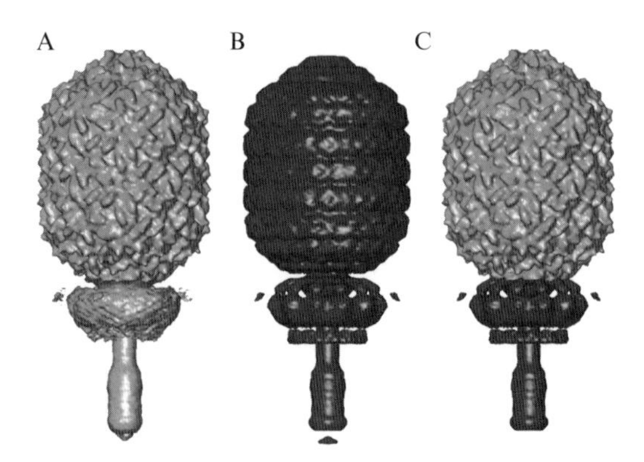


**Figure 7.4** Construction of asymmetric, hybrid model. (A) Shaded surface representation of complete phage reconstruction with fivefold symmetry enforced during processing. (B) Same as (A) for phage with 12-fold symmetry. Since the head and tail do not share the same symmetry, these reconstructions smear out the tail and head densities in panels (A) and (B), respectively. (C) Hybrid model obtained by combining head from fivefold reconstruction and tail from 12-fold reconstruction. Colors in the hybrid map highlight contributions from the two symmetrized maps. At this point the symmetry mismatch between the head and tail was unknown and no effort was made to impose a particular rotational alignment between the two segments. (See Color Insert.)

that the connectors and tail portions of all tailed bacteriophage, including $\phi$29, do not possess fivefold symmetry like the capsids (Jiang *et al.*, 2006; Lander *et al.*, 2006; Xiang *et al.*, 2006). The presence of a head–tail symmetry mismatch was also evident in the fivefold averaged $\phi$29 reconstruction since features in the head were more distinct and not smeared as in the tail (Fig. 7.4A). The entire neck and tail of $\phi$29 was shown to have quasi-six-fold symmetry (Peterson *et al.*, 2001) and the existence of 12 distinct appendages was revealed in some of the very earliest electron micrographs of negatively stained phage (Anderson *et al.*, 1966). Hence, to establish the rotational symmetry, if any, in the structure of the $\phi$29 tail, we used EMAN to recompute the phage reconstruction with imposed three-, six-, and 12-fold symmetries instead of fivefold. Inspection of these three reconstructions showed, as expected, a smeared capsid structure but tails with features more distinct than in the fivefold averaged map. This strategy works since the head portions of the images have an equally poor agreement with all projections of the model and the assignment of particle orientations becomes influenced most by the tail structure. In addition, the ring of appendages near the head–tail junction appeared most distinct in the 12-fold symmetrized map (Fig. 7.4B).

At this point we constructed a hybrid model in the following way. We used the *subregion selection* tool in Chimera (Pettersen *et al.*, 2004) to segment out the fivefold symmetrized head portion of the first reconstruction (Fig. 7.4A) and the 12-fold symmetrized tail portion of the second reconstruction (Fig. 7.4B). These separately segmented volumes were then combined into a single, hybrid density map (Fig. 7.4C). No attempt was made to impose any particular rotational alignment of the tail and head segmented maps in constructing the hybrid model. Regardless, the exact same hybrid model was used to produce the final virion and ghost $\phi$29 reconstructions, which have tail structures that differ in several ways including the asymmetric arrangements of appendage conformations that break the 12-fold axial symmetry present in the hybrid model (Tang *et al.*, 2008b).

# 4. Asymmetric Reconstruction of the Entire $\phi$29 Phage

A hybrid density map generated from full resolution image data was used as input to the second branch of the reconstruction scheme (Fig. 7.1), which began with two cycles of refinement carried out in EMAN. This produced an asymmetric reconstruction of the virion at ~30-Å resolution. At this point, the use of unbinned data and the abandonment of symmetry averaging led to no further improvement in resolution and resulted in significant computational overhead. This failure to reach higher resolutions does not imply there are limitations in the capabilities of EMAN, but more likely reflects our own lack of expertise with the software.

Our group has developed a set of programs as part of AUTO3DEM that can be run efficiently on computer clusters and other parallel computers (Yan *et al.*, 2007b). Most of the underlying programs were designed to handle cyclic, dihedral, and cubic point group symmetries, but the global search procedure implemented in PPFT is strictly only applicable to spherical particles since it requires particle images and projections of the maps to be reinterpolated into a series of concentric, circular annuli. We then turned to FREALIGN (Grigorieff, 2007) to initiate a refinement process that could be transitioned relatively easily to AUTO3DEM.

## 4.1. Origin and orientation determination and refinement with FREALIGN

The version of FREALIGN available at the time this work on $\phi$29 was being performed required that the particle images be combined in a single stack file. Hence, particle images were first converted from IMAGIC to MRC format using the EMAN *proc2d* command. Also, to accommodate FREALIGN requirements, the MRC image file header was converted with the MRC program *image_convert.exe*.

The final map generated from EMAN was used as input to FREALIGN, which provides several modes of operation. We used its global search procedure (Mode 3) to assign initial origin and orientation parameters for each particle image. Also, the *matching projection* function was enabled to produce side-by-side comparisons of particle images and corresponding projections of the current reconstruction. This was carried out for the entire data set of images and, though somewhat tedious, gave a useful way to validate the assigned orientation parameters and to detect and screen out potentially "bad" particles that failed to align properly to the model.

The accuracy of the particle origins and orientations was further verified because the resultant cryo-reconstruction computed from the particle images with these assigned parameters had distinct, easily interpretable features. This was followed by standard refinement and map calculation in FREALIGN (Mode 1) for several more cycles (Fig. 7.1). The asymmetric reconstruction of the entire phage that emerged from this procedure reached a resolution of ~25 Å. Here, we had ample confidence in the current set of assigned origin and orientation parameters to carry out further processing steps with AUTO3DEM.

## 4.2. Origin and orientation refinement in AUTO3DEM

After obtaining estimates of the origins and orientations of all particles and a reliable 3D model of the entire phage with FREALIGN, we switched to AUTO3DEM for subsequent processing. AUTO3DEM runs in either serial or distributed-memory parallel mode and hence, access to a computer cluster can lead to a dramatic improvement in algorithm performance and significantly reduced computation time. The program $PO^2R$ (Ji *et al.*, 2006) in AUTO3DEM carries out the Fourier-based origin and orientation refinement process, and program P3DR (Marinescu and Ji, 2003) computes the 3D density map from a selected ("best") set of particle images. Both programs can be run with or without symmetry constraints and hence are suitable for analysis of asymmetric particles like $\phi$29.

FREALIGN uses an Euler angle convention ($\theta_e$, $\phi_e$, $\psi_e$) to specify particle orientation, whereas AUTO3DEM uses the ($\theta_a$, $\phi_a$, $\omega_a$) convention first described by Finch and Klug (Finch and Klug, 1965) and implemented in the original icosahedral processing programs developed by Crowther (Crowther *et al.*, 1970). We employed the following relationships to convert orientation parameters from the FREALIGN convention to that used in AUTO3DEM:

$$\begin{aligned} \theta_a &= 180 - \theta_e \\ \phi_a &= \varphi_e - 180 \\ \omega_a &= \psi_e - 90 \end{aligned}$$

Once the above conversions were made, we employed the full capabilities of AUTO3DEM refinement to help improve the asymmetric reconstruction of $\phi$29 as much as possible with the set of images that were available. AUTO3DEM works to progressively improve the resolution of any reconstruction by automatically optimizing numerous refinement parameters. For example, the images are typically band-pass filtered in Fourier space to limit the data used in comparing images to model projections. The upper Fourier limit (i.e., highest spatial frequency) included in the next cycle of refinement calculations is slowly increased as long as the resulting reconstruction shows improvement over the previous one. Improvement can be monitored in a variety of ways, but typically includes conventional Fourier shell correlation procedures (van Heel and Schatz, 2005). Other important parameters adjusted automatically by AUTO3DEM or manually by the user include the step sizes used to define the range of origin and orientation parameters to be tested. Typically, the origin and orientation intervals start out at about 1.0 pixel and 1–2°, respectively, and these are generally reduced when refinement stalls (i.e., no longer yields improvement in reconstruction resolution). At the end of refinement the origin and orientation step sizes might drop to 0.1 pixel and 0.1°, respectively. At even smaller step sizes, the process can lead to over-refinement and unreliable resolution estimates, as the noise in the data can drive refinement (Stewart and Grigorieff, 2004).

Another empirical parameter that we employed with some success is use of the inverse temperature factor (Fernandez *et al.*, 2008; Havelka *et al.*, 1995) to enhance high spatial frequency details in the reconstruction and during particle refinement (Tang *et al.*, 2008a). We generally did not make use of this strategy until the reconstructed density map reached about 10–12 Å resolution, at which point an initial inverse temperature factor of 1/100 $Å^{-2}$ would be employed during refinement. As refinement progressed and the resolution improved, the sharpening factor would be increased progressively to about 1/400 $Å^{-2}$. It proved important, if not essential, to carefully inspect the 3D density map calculated at the end of each refinement cycle to ensure that the signal-to-noise of reliably represented features (e.g., tubes of density ascribed to $\alpha$-helices) did not decrease as this would signify the refinement was being driven by the dominant, high frequency noise in the data. A carefully monitored, trial and error approach and inverse temperature factors even as low as 1/1200 $Å^{-2}$ were used during the $\phi$29 asymmetric refinement.

An additional strategy we used in the $\phi$29 asymmetric reconstruction study was to mask out the genome density in the reconstructed density map, which is a method that typically improves the refinement of icosahedral viruses (Chen *et al.*, 2009; Yan *et al.*, 2007b; Zhang *et al.*, 2008). This procedure leads to enhancement of reliably represented features in the projected images of the model and improves the accuracy in determining particle parameters. In a similar manner, we used with some success a density threshold procedure to remove some of the more obvious, random noise in the reconstructed $\phi$29 density map. In density maps of icosahedral viruses computed from thousands or more particle images, the noise level outside the particle is generally quite low owing to the benefits of the 60-fold symmetry averaging that is an inherent part of the 3D reconstruction process. Hence, in an asymmetric reconstruction computed from a comparable number of particle images, the average noise level is significantly higher throughout the map. We tried a number of different masking and threshold procedures to zero noise outside the outer envelope of the phage particle. These are akin to solvent flattening used during the phase extension step in X-ray crystallographic studies (Wang, 1985). The technique that worked best in this instance involved the following steps. A second map was first recalculated at a lower resolution, typically about 20 Å. The two maps were then read into Chimera, which could take the low-resolution map and define a surface that enclosed the final, sub-nanometer resolution map. The Chimera *mask* tool was used to set values of all voxels that lie outside the defined surface to zero. This provided an additional enhancement of the signal-to-noise in the projections of the model used to correlate with the raw particle images for refinement of their parameters.

## 5. Analysis and Interpretation of $\phi$29 Reconstruction

Careful analysis and interpretation of a reconstruction are critical for understanding the biology of the viral system, and these also help guide the refinement process by distinguishing genuine structural features from the noise. The availability of X-ray crystallographic data for individual components is particularly valuable as it helps in validating the reliability of the cryo-reconstruction, determining the significance of various structural features, and defining the boundaries between individual viral components.

### 5.1. Model docking

Fitting X-ray crystal structures into cryo-TEM reconstructions to produce pseudoatomic models of macromolecular complexes has become a powerful tool in the arsenal of analysis procedures ever since it was first introduced in studies of viruses (Stewart *et al.*, 1993), virus–antibody (Smith *et al.*, 1993; Wang *et al.*, 1992) and virus–receptor (Olson *et al.*, 1993) complexes, and acto-myosin filaments (Rayment *et al.*, 1993). While useful for a wide variety of macromolecular systems, the combination of cryo-TEM and X-ray crystallographic structures can have a particularly profound impact on the study of tailed phages for two main reasons. First, owing to their shape and size, intact tailed phages resist crystallization and this will likely continue to thwart attempts at obtaining diffraction quality single crystals of a complete asymmetric phage. Second, the head–tail symmetry mismatch precludes the imposition of any symmetry and hence limits the resolutions that can be achieved. The docking of X-ray models of individual phage components into cryo-reconstructions currently offers the best means for obtaining a pseudoatomic model of the complete phage.

If an X-ray structure of a viral component is available, it is generally a relatively straightforward procedure to dock the atomic model manually into the reconstructed density map using a variety of interactive programs such as O (Jones *et al.*, 1991), COOT (Emsley and Cowtan, 2004), and Chimera (Pettersen *et al.*, 2004). After obtaining a reasonable fit of model to density via manual procedures, one can then quantitatively refine the fit by translating and rotating the atomic model as a rigid body until the correlation coefficient between the model and density map is maximized. This can be accomplished in numerous programs such as RSREF (Chapman, 1995), SITUS (Wriggers and Birmanns, 2001), EMfit (Rossmann *et al.*, 2001), CoAn (Volkmann and Hanein, 2003), and Chimera (Pettersen *et al.*, 2004).

Rigid body docking of the $\phi$29 connector (gp10) crystal structure (Guasch *et al.*, 2002) into reconstructions of the prohead, ghost, and mature virion shows that the top of the connector fits well into the cryo-TEM density in all three cases, but the lower portion of the connector only fits well into the prohead (Fig. 7.5). Ghosts and virions both contain additional proteins that were added sequentially during assembly onto the bottom of the connector to create a functional tail, and it is clear from the rigid body fits that the attachment of the tail and packaging of the genome induce conformational changes in the connector. At the time of this study (Tang *et al.*, 2008b), the existing software was not capable of modifying the crystal structure to better fit the density.

Flexible fitting methods have been developed to permit models to be modified in various ways to achieve better fits to the cryo-TEM density.

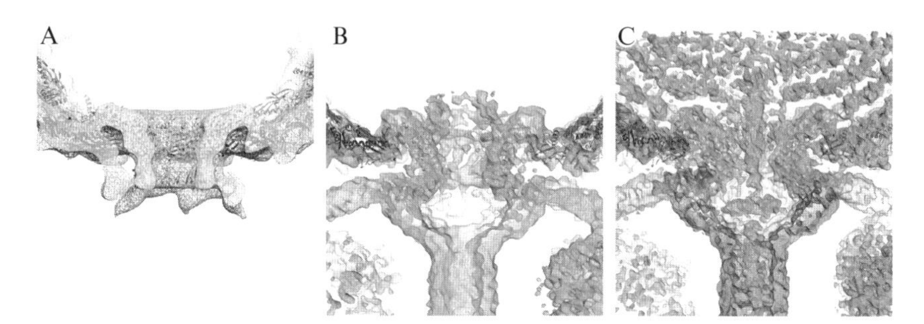


**Figure 7.5** Rigid body fit of gp10 connector crystal structure (magenta ribbon model) and gp8 capsid subunit homology model (red), into $\phi$29 density maps (gray). The top portion of the connector fits well into the prohead (A), ghost (B), and virion (C) reconstructions, whereas the lower portion only fits well into the prohead. (See Color Insert.)

Several software packages such as NMFF (normal mode flexible fitting; Tama *et al.*, 2004), DireX (Schroder *et al.*, 2007), and Flex-EM (Topf *et al.*, 2008) sample the conformational space of the crystal structure to improve the fit. Molecular dynamics based packages such as MDFF (molecular dynamics flexible fitting; Trabuco *et al.*, 2008) adjust the crystal structure in a physically reasonable way and even show a potential trajectory between conformations.

All fitting methods are particularly powerful when models are available for the entire reconstructed density, but become problematic when the densities from multiple components overlap to form a larger, more complex shape. This situation occurs in $\phi$29 where gp10 interacts with the capsid protein (gp8), the lower collar (gp11), and the appendages (gp12*) as well as the genome.

## 5.2. Map segmentation

An important step in correctly interpreting a reconstructed density map is to identify the individual components and understand their interactions in the context of the macromolecular complex (Chapter 1, Vol. 483). Segmentation is the process by which the boundaries between components are defined (Chapter 2, Vol. 483) and success here directly depends on the quality and resolution of the map. Substructures with large, solvent accessible surfaces, and hence clear boundaries, are relatively easy to segment, whereas components with extensive interactions present a greater challenge. As described below, we employed three different techniques to isolate major components of the phage. These included (1) using X-ray structures to mask out regions of the density map, (2) using difference map calculations to identify particular viral components, and (3) manually segmenting out portions of the density map.

In our $\phi$29 study, the connector crystal structure was first docked into the phage reconstruction and the Chimera *mask* operation was then used to segment out the connector portion of the density map. A crystal structure of the $\phi$29 capsid protein was not available, but a pseudoatomic homology model derived from the X-ray structure of the capsid protein of HK97 phage (Wikoff *et al.*, 2000) was used to segment the head density.

In instances where no molecular model is available, reliable segmentation of the components can sometimes be achieved by difference map analysis. For example, by subtracting the $\phi$29 ghost density from the virion density we achieved a relatively unambiguous segmentation of the packaged DNA genome. Indeed, the difference map revealed the bulk DNA in the phage head and also highly organized, linear stretches of DNA above and inside the connector complex and inside the lower collar, and a unique, toroid-like DNA structure inside the cavity between the connector and collar (Tang *et al.*, 2008b; Fig. 7.6). It is important to note that this segmentation technique only works if the two structures being compared by means of difference map analysis are essentially identical except for the

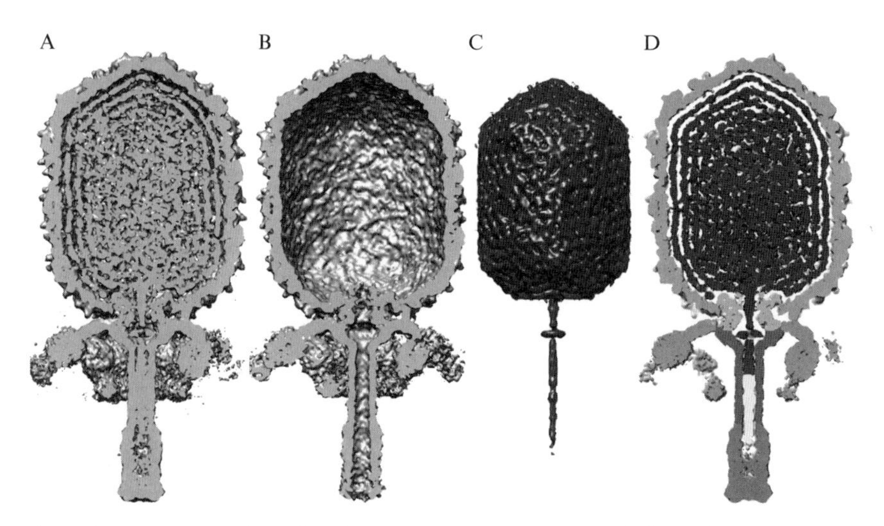


**Figure 7.6** Segmentation of viral components in reconstructed phage density map. (A) Shaded surface representation of virion reconstruction with front half of map removed to reveal internal structure. (B) Same as (A) for ghost reconstruction. (C) Difference map obtained by subtracting ghost from virion contains density (red) attributed to dsDNA genome plus the terminal protein, gp3. (D) Segmented map of $\phi$29 virion with components distinguished by color: capsid (gp8), light blue; connector (gp10), yellow; lower collar and tube (gp11), green; knob (gp9), cyan; appendages (gp12*), magenta; terminal protein (gp3) covalently attached to right end of DNA, white; and DNA, red. (See Color Insert.)

absence or presence of the component of interest. If any conformational changes to the capsid had occurred as a consequence of genome release, these would also have appeared in the difference map and may have complicated the identification of the genome boundaries.

With the aid of an X-ray model for gp10 and a homology model for gp8, we were able to isolate the connector and capsid, and using difference maps we could clearly identify the bulk genome density. However, segmentation of other components proved to be a more challenging task. Computer algorithms have been developed to aid in the identification of secondary structural elements and automate feature-guided segmentation (Baker *et al.*, 2006), but we were not successful in applying them to $\phi$29. Certainly, limited resolution, noise, and the close proximity of multiple components could individually or collectively thwart attempts to segment the entire map in a definitive manner.

Based on knowledge of the $\phi$29 tail structure from previous studies (Xiang *et al.*, 2006), we were able to easily identify and manually segment those portions belonging to the 12 appendages (magenta in Fig. 7.6D). This was accomplished using the Chimera *volume eraser* tool to remove density defined by the user to be part of other components. After subtracting density corresponding to all of the components already identified and using an estimate of the molecular volume of the lower collar (gp11), we were able to define a plausible boundary between gp11 and the gp9 tail knob (green and cyan, respectively in Fig. 7.6D). Similarly, we defined an approximate boundary between the right end of the dsDNA and terminal protein (gp3) (red and white, respectively in Fig. 7.6D). Our application of three separate segmentation techniques as described above provided a complete picture of the overall organization of the phage components.

## 6. Summary and Future Prospects

The asymmetric tailed phages present a number of unique challenges for 3D image reconstruction. The reconstruction process is not nearly as straightforward as that employed for solving the structures of totally asymmetric particles or of highly symmetric particles. Although the entire phage lacks global symmetry, strategies that exploit the symmetries of the head and tail at an early stage of the reconstruction process, before transitioning to a fully asymmetric reconstruction, can improve the likelihood of success, expedite the process, and lead to a higher resolution final map. In our studies of $\phi$29, the lower symmetry (fivefold) and nonspherical shape of the prolate head made it crucial that extra care be taken both in construction of a reliable head model and boxing of the particles from the micrographs. In our processing strategy, boxing was performed not just at the start of the project, but rather became an integral part of the reconstruction process and the particle images and phage map were gradually expanded and improved. Owing to the low signal-to-noise ratio from the lack of symmetry averaging, the particle refinement process became more intensive and the interpretation of the reconstruction also required more attention.

### 6.1. Streamlined processing

Our approach to solving the asymmetric structure at sub-nanometer resolution necessitated the use of multiple software packages. The choice of program at each step was based on our level of understanding of the various packages and their relative strengths and weaknesses. We found EMAN to be very good at starting a new project when faced with limited knowledge of the final structure. FREALIGN contains a valuable feature to validate the initial assignment of particle orientations. The computational speed of AUTO3DEM enabled us to rapidly test a large number of reconstruction parameters and finish the refinement in a timely fashion. The drawback in employing multiple software packages was that extreme care had to be taken when transitioning between programs, particularly with respect to file formats and image orientation conventions.

AUTO3DEM is currently being enhanced so it can be used for all steps of the asymmetric reconstruction process. Recent releases of this software have been modified so that a global orientation search can be performed entirely in reciprocal space, but this feature has been used primarily to improve icosahedral reconstructions and remains untested for analysis of images of asymmetric particles. We have also added the capability to compare the 60 equivalent orientations related by icosahedral symmetry. This functionality was originally implemented to study nonstoichiometric binding of antibodies to icosahedral capsids, but could also be used to identify the unique vertices of otherwise isometric particles such as P22 (Lander *et al.*, 2006) and PBCV-1 (Cherrier *et al.*, 2009). Our study of $\phi$29 illustrated the importance of using an inverse temperature factor during the refinement process. Currently the choice of value is empirical and the resultant map must be visually inspected to determine whether the quality of the map increased or decreased. In the future, we hope to develop automated techniques for assessing map quality and allow AUTO3DEM to choose the optimal inverse temperature factor at each step of the reconstruction process.

## 6.2. Development of smart masking

Given the elongated shape of $\phi$29, boxing out particles using a circular mask results in the inclusion of much unwanted background and often portions of neighboring particles. Elliptical or ovoid masks that more tightly follow the particle boundaries would help remedy some of the shortcomings of circular masks. Also, the technique used in EMAN, in which masks are individually tailored to the particles, seems a promising approach to adopt. In this scheme, initial boxing would still be performed with a circular mask, but particle images would then be reboxed using masks derived from the projections of the models that agree most closely with the images. Regardless of which advanced boxing method is employed, tighter masking should help enhance the signal-to-noise ratio that is inherently poor in asymmetric reconstructions. This also might, in turn, lead to higher resolution.

## 6.3. Future prospects

Tremendous progress has been made in the cryo-reconstruction of tailed phages in recent years. Indeed, the first report of a 3D reconstruction of a complete phage at sub-nanometer resolution followed less than a decade after the original one at 33-Å resolution was published (Table 7.1). Notwithstanding that considerably higher resolutions are now routinely achieved for icosahedral viruses (Zhou, 2008), the progress made in reconstructing the 3D asymmetric structures of whole phage like $\phi$29 sets the stage for even greater achievements in the near future. These of course will enable mechanistic questions concerning phage structure to be answered that currently cannot be addressed at moderate resolutions in the 7–10 Å realm.

An obvious bottleneck is the large amount of data needed for asymmetric reconstructions. Assuming everything else is equivalent (e.g., specimen stability, microscope conditions, accuracy in determining particle parameters, etc.), at least 60 times more particle images are needed to achieve a resolution comparable to that obtained with images of icosahedral particles. Fortunately, tools for automatic data collection such as LEGINON (Suloway *et al.*, 2005) and automatic data processing such as APPION (Lander *et al.*, 2009) are being developed to record and handle large numbers of particle images and to determine microscopy conditions such as objective lens defocus and astigmatism. In addition, advances in microscopy and image processing that have recently led to near-atomic resolution icosahedral reconstructions will undoubtedly help pave the way to achieve higher resolutions in asymmetric virus reconstructions.

In this chapter, we described a set of techniques that we used to solve the structures of fiberless $\phi$29 virions and ghosts to sub-nanometer resolutions (Tang *et al.*, 2008b). These methods should be readily applicable to other tailed phages and, with relatively minor modifications, other asymmetric macromolecules that possess high degrees of local symmetry. This marks just one step in the quest to achieve atomic or near-atomic resolution cryo-reconstructions of asymmetric particles through a combination of improvements in sample preparation and microscopy, and in image processing, particle selection, interpretation, and visualization algorithms.

## ACKNOWLEDGMENTS

We thank D. Anderson, S. Grimes, and P. Jardine for their sustained enthusiasm about structure-function studies of $\phi$29. We also thank numerous other colleagues for their inspiration and perspiration in helping to unravel the 3D structure of $\phi$29 (N. Olson, W. Xu, W. Grochulski, Y. Tao, A. Simpson, M. Rossmann, M. Morais, and M. Sherman). The work reported in this chapter was supported in part by NIH grant R37 GM-033050, NSF shared instrument grant BIR-9112921, the University of California, San Diego, and a gift from the Agouron foundation (all to T.S.B.).

## REFERENCES

Agirrezabala, X., Martín-Benito, J., Castón, J. R., Miranda, R., Valpuesta, J. M., and Carrascosa, J. L. (2005). Maturation of phage T7 involves structural modification of both shell and inner core components. *EMBO J.* **24,** 3820–3829.

Anderson, D. L., Hickman, D. D., and Reilly, B. E. (1966). Structure of *Bacillus subtilis* bacteriophage $\phi$29 and the length of $\phi$29 deoxyribonucleic acid. *J. Bacteriol.* **91,** 2081–2089.

Baker, M. L., Yu, Z., Chiu, W., and Bajaj, C. (2006). Automated segmentation of molecular subunits in electron cryomicroscopy density maps. *J. Struct. Biol.* **156,** 432–441.

Baker, T. S., and Cheng, R. H. (1996). A model-based approach for determining orientations of biological macromolecules imaged by cryoelectron microscopy. *J. Struct. Biol.* **116,** 120–130.

Baker, T. S., Olson, N. H., and Fuller, S. D. (1999). Adding the third dimension to virus life cycles: Three-dimensional reconstruction of icosahedral viruses from cryo-electron micrographs. *Microbiol. Mol. Biol. Rev.* **63,** 862–922.

Bubeck, D., Filman, D. J., Kuzmin, M., Fuller, S. D., and Hogle, J. M. (2008). Post-imaging fiducial markers aid in the orientation determination of complexes with mixed or unknown symmetry. *J. Struct. Biol.* **162,** 480–490.

Chang, J., Weigele, P., King, J., Chiu, W., and Jiang, W. (2006). Cryo-EM asymmetric reconstruction of bacteriophage P22 reveals organization of its DNA packaging and infecting machinery. *Structure* **14,** 1073–1082.

Chapman, M. S. (1995). Restrained real-space macromolecular atomic refinement using a new resolution-dependent electron density function. *Acta Crystallogr.* **A51,** 69–80.

Chen, J. Z., Settembre, E. C., Aoki, S. T., Zhang, X., Bellamy, A. R., Dormitzer, P. R., Harrison, S. C., and Grigorieff, N. (2009). Molecular interactions in rotavirus assembly and uncoating seen by high-resolution cryo-EM. *Proc. Natl. Acad. Sci. USA* **106,** 10644–10648.

Cheng, Y., and Walz, T. (2009). The advent of near-atomic resolution in single-particle electron microscopy. *Annu. Rev. Biochem.* **78,** 723–742.

Cherrier, M. V., Kostyuchenko, V. A., Xiao, C., Bowman, V. D., Battisti, A. J., Yan, X., Chipman, P. R., Baker, T. S., Van Etten, J. L., and Rossmann, M. G. (2009). An icosahedral algal virus has a complex unique vertex decorated by a spike. *Proc. Natl. Acad. Sci. USA* **106,** 11085–11089.

Choi, K. H., McPartland, J., Kaganman, I., Bowman, V. D., Rothman-Denes, L. B., and Rossmann, M. G. (2008). Insight into DNA and protein transport in double-stranded DNA viruses: the structure of bacteriophage N4. *J. Mol. Biol.* **378,** 726–736.

Crowther, R. A., Amos, L. A., Finch, J. T., De Rosier, D. J., and Klug, A. (1970). Three dimensional reconstructions of spherical viruses by Fourier synthesis from electron micrographs. *Nature* **226,** 421–425.

DeRosier, D., and Klug, A. (1968). Reconstruction of three dimensional structures from electron micrographs. *Nature* **217,** 130–134.

Emsley, P., and Cowtan, K. (2004). Coot: Model-building tools for molecular graphics. *Acta Crystallogr. D Biol. Crystallogr.* **60,** 2126–2132.

Fernandez, J. J., Luque, D., Caston, J. R., and Carrascosa, J. L. (2008). Sharpening high resolution information in single particle electron cryomicroscopy. *J. Struct. Biol.* **164,** 170–175.

Finch, J. T., and Klug, A. (1965). The structure of viruses of the papilloma-polyoma type 3. Structure of rabbit papilloma virus, with an appendix on the topography of contrast in negative-staining for electron-microscopy. *J. Mol. Biol.* **13,** 1–12.

Fokine, A., Chipman, P. R., Leiman, P. G., Mesyanzhinov, V. V., Rao, V. B., and Rossmann, M. G. (2004). Molecular architecture of the prolate head of bacteriophage T4. *Proc. Natl. Acad. Sci. USA* **101,** 6003–6008.

Frank, J. (2009). Single-particle reconstruction of biological macromolecules in electron microscopy–30 years. *Q. Rev. Biophys.* **42,** 139–158.

Frank, J., Radermacher, M., Penczek, P., Zhu, J., Li, Y., Ladjadj, M., and Leith, A. (1996). SPIDER and WEB: Processing and visualization of images in 3D electron microscopy and related fields. *J. Struct. Biol.* **116,** 190–199.

Grigorieff, N. (2007). FREALIGN: High-resolution refinement of single particle structures. *J. Struct. Biol.* **157,** 117–125.

Guasch, A., Pous, J., Ibarra, B., Gomis-Ruth, F. X., Valpuesta, J. M., Sousa, N., Carrascosa, J. L., and Coll, M. (2002). Detailed architecture of a DNA translocating machine: the high-resolution structure of the bacteriophage $\phi$29 connector particle. *J. Mol. Biol.* **315,** 663–676.

Gurda, B. L., Parent, K. N., Bladek, H., Sinkovits, R. S., Dimattia, M. A., Rence, C., Castro, A., McKenna, R., Olson, N., Brown, K., Baker, T. S., and Agbandje-McKenna, M. (2010). Human bocavirus capsid structure: Insights into the structural repertoire of the parvoviridae. *J. Virol.* **84,** 5880–5889.

Havelka, W. A., Henderson, R., and Oesterhelt, D. (1995). Three-dimensional structure of halorhodopsin at 7 Å resolution. *J. Mol. Biol.* **247,** 726–738.

Heymann, J. B. (2001). Bsoft: Image and molecular processing in electron microscopy. *J. Struct. Biol.* **133,** 156–169.

Ji, Y., Marinescu, D. C., Zhang, W., Zhang, X., Yan, X., and Baker, T. S. (2006). A model-based parallel origin and orientation refinement algorithm for cryoTEM and its application to the study of virus structures. *J. Struct. Biol.* **154,** 1–19.

Jiang, W., Chang, J., Jakana, J., Weigele, P., King, J., and Chiu, W. (2006). Structure of epsilon15 bacteriophage reveals genome organization and DNA packaging/injection apparatus. *Nature* **439,** 612–616.

Johnson, J. E., and Chiu, W. (2007). DNA packaging and delivery machines in tailed bacteriophages. *Curr. Opin. Struct. Biol.* **17,** 237–243.

Jones, T. A., Zou, J. Y., Cowan, S. W., and Kjeldgaard, M. (1991). Improved methods for building protein models in electron density maps and the location of errors in these models. *Acta Crystallogr. A* **47**(Pt 2), 110–119.

Jonic, S., Sorzano, C. O., and Boisset, N. (2008). Comparison of single-particle analysis and electron tomography approaches: An overview. *J. Microsc.* **232,** 562–579.

Kostyuchenko, V. A., Chipman, P. R., Leiman, P. G., Arisaka, F., Mesyanzhinov, V. V., and Rossmann, M. G. (2005). The tail structure of bacteriophage T4 and its mechanism of contraction. *Nat. Struct. Mol. Biol.* **12,** 810–813.

Lander, G. C., Stagg, S. M., Voss, N. R., Cheng, A., Fellmann, D., Pulokas, J., Yoshioka, C., Irving, C., Mulder, A., Lau, P. W., Lyumkis, D., Potter, C. S., and Carragher, B. (2009). Appion: An integrated, database-driven pipeline to facilitate EM image processing. *J. Struct. Biol.* **166,** 95–102.

Lander, G. C., Tang, L., Casjens, S. R., Gilcrease, E. B., Prevelige, P., Poliakov, A., Potter, C. S., Carragher, B., and Johnson, J. E. (2006). The structure of an infectious P22 virion shows the signal for headful DNA packaging. *Science* **312,** 1791–1795.

Ludtke, S. J., Baldwin, P. R., and Chiu, W. (1999). EMAN: Semiautomated software for high-resolution single-particle reconstructions. *J. Struct. Biol.* **128,** 82–97.

Marinescu, D. C., and Ji, Y. (2003). A computational framework for the 3D structure determination of viruses with unknown symmetry. *J. Parallel Distrib. Comput.* **63,** 738–758.

Morais, M. C., Choi, K. H., Koti, J. S., Chipman, P. R., Anderson, D. L., and Rossmann, M. G. (2005). Conservation of the capsid structure in tailed dsDNA bacteriophages: The pseudoatomic structure of $\phi$29. *Mol. Cell* **18,** 149–159.

Morais, M. C., Tao, Y., Olson, N. H., Grimes, S., Jardine, P. J., Anderson, D. L., Baker, T. S., and Rossmann, M. G. (2001). Cryoelectron-microscopy image reconstruction of symmetry mismatches in bacteriophage $\phi$29. *J. Struct. Biol.* **135,** 38–46.

Olson, N. H., Kolatkar, P. R., Oliveira, M. A., Cheng, R. H., Greve, J. M., McClelland, A., Baker, T. S., and Rossmann, M. G. (1993). Structure of a human rhinovirus complexed with its receptor molecule. *Proc. Natl. Acad. Sci. USA* **90,** 507–511.

Peterson, C., Simon, M., Hodges, J., Mertens, P., Higgins, L., Egelman, E., and Anderson, D. (2001). Composition and mass of the bacteriophage $\phi$29 prohead and virion. *J. Struct. Biol.* **135,** 18–25.

Pettersen, E. F., Goddard, T. D., Huang, C. C., Couch, G. S., Greenblatt, D. M., Meng, E. C., and Ferrin, T. E. (2004). UCSF Chimera–a visualization system for exploratory research and analysis. *J. Comput. Chem.* **25,** 1605–1612.

Rayment, I., Holden, H. M., Whittaker, M., Yohn, C. B., Lorenz, M., Holmes, K. C., and Milligan, R. A. (1993). Structure of the actin-myosin complex and its implications for muscle contraction. *Science* **261,** 58–65.

Rosenthal, P. B., and Henderson, R. (2003). Optimal determination of particle orientation, absolute hand, and contrast loss in single-particle electron cryomicroscopy. *J. Mol. Biol.* **333,** 721–745.

Rossmann, M. G., Bernal, R., and Pletnev, S. V. (2001). Combining electron microscopic with X-ray crystallographic structures. *J. Struct. Biol.* **136,** 190–200.

Schroder, G. F., Brunger, A. T., and Levitt, M. (2007). Combining efficient conformational sampling with a deformable elastic network model facilitates structure refinement at low resolution. *Structure* **15,** 1630–1641.

Smith, T. J. (2003). Structural studies on antibody–virus complexes. *Adv. Protein Chem.* **64,** 409–453.

Smith, T. J., Olson, N. H., Cheng, R. H., Chase, E. S., and Baker, T. S. (1993). Structure of a human rhinovirus-bivalently bound antibody complex: Implications for viral neutralization and antibody flexibility. *Proc. Natl. Acad. Sci. USA* **90,** 7015–7018.

Sorzano, C. O., Marabini, R., Velazquez-Muriel, J., Bilbao-Castro, J. R., Scheres, S. H., Carazo, J. M., and Pascual-Montano, A. (2004). XMIPP: A new generation of an open-source image processing package for electron microscopy. *J. Struct. Biol.* **148,** 194–204.

Steven, A. C., Heymann, J. B., Cheng, N., Trus, B. L., and Conway, J. F. (2005). Virus maturation: Dynamics and mechanism of a stabilizing structural transition that leads to infectivity. *Curr. Opin. Struct. Biol.* **15,** 227–236.

Stewart, A., and Grigorieff, N. (2004). Noise bias in the refinement of structures derived from single particles. *Ultramicroscopy* **102,** 67–84.

Stewart, P. L., Dermody, T. S., and Nemerow, G. R. (2003). Structural basis of nonenveloped virus cell entry. *Adv. Protein Chem.* **64,** 455–491.

Stewart, P. L., Fuller, S. D., and Burnett, R. M. (1993). Difference imaging of adenovirus: Bridging the resolution gap between X-ray crystallography and electron microscopy. *EMBO J.* **12,** 2589–2599.

Suloway, C., Pulokas, J., Fellmann, D., Cheng, A., Guerra, F., Quispe, J., Stagg, S., Potter, C. S., and Carragher, B. (2005). Automated molecular microscopy: The new Leginon system. *J. Struct. Biol.* **151,** 41–60.

Tama, F., Miyashita, O., and Brooks, C. L., 3rd (2004). Normal mode based flexible fitting of high-resolution structure into low-resolution experimental data from cryo-EM. *J. Struct. Biol.* **147,** 315–326.

Tang, J., Ochoa, W. F., Sinkovits, R. S., Poulos, B. T., Ghabrial, S. A., Lightner, D. V., Baker, T. S., and Nibert, M. L. (2008a). Infectious myonecrosis virus has a totivirus-like, 120-subunit capsid, but with fiber complexes at the fivefold axes. *Proc. Natl. Acad. Sci. USA* **105,** 17526–17531.

Tang, J., Olson, N., Jardine, P. J., Grimes, S., Anderson, D. L., and Baker, T. S. (2008b). DNA poised for release in bacteriophage $\phi$29. *Structure* **16,** 935–943.

Tao, Y., Olson, N. H., Xu, W., Anderson, D. L., Rossmann, M. G., and Baker, T. S. (1998). Assembly of a tailed bacterial virus and its genome release studied in three dimensions. *Cell* **95,** 431–437.

Topf, M., Lasker, K., Webb, B., Wolfson, H., Chiu, W., and Sali, A. (2008). Protein structure fitting and refinement guided by cryo-EM density. *Structure* **16,** 295–307.

Trabuco, L. G., Villa, E., Mitra, K., Frank, J., and Schulten, K. (2008). Flexible fitting of atomic structures into electron microscopy maps using molecular dynamics. *Structure* **16,** 673–683.

van Heel, M., and Schatz, M. (2005). Fourier shell correlation threshold criteria. *J. Struct. Biol.* **151,** 250–262.

Volkmann, N., and Hanein, D. (2003). Docking of atomic models into reconstructions from electron microscopy. *Methods Enzymol.* **374,** 204–225.

Wang, B. C. (1985). Resolution of phase ambiguity in macromolecular crystallography. *Methods Enzymol.* **115,** 90–112.

Wang, G. J., Porta, C., Chen, Z. G., Baker, T. S., and Johnson, J. E. (1992). Identification of a Fab interaction footprint site on an icosahedral virus by cryoelectron microscopy and X-ray crystallography. *Nature* **355,** 275–278.

Wikoff, W. R., Liljas, L., Duda, R. L., Tsuruta, H., Hendrix, R. W., and Johnson, J. E. (2000). Topologically linked protein rings in the bacteriophage HK97 capsid. *Science* **289,** 2129–2133.

Wolf, M., Garcea, R. L., Grigorieff, N., and Harrison, S. C. (2010). Subunit interactions in bovine papillomavirus. *Proc. Natl. Acad. Sci. USA* **107,** 6298–6303.

Wriggers, W., and Birmanns, S. (2001). Using situs for flexible and rigid-body fitting of multiresolution single-molecule data. *J. Struct. Biol.* **133,** 193–202.

Xiang, Y., Morais, M. C., Battisti, A. J., Grimes, S., Jardine, P. J., Anderson, D. L., and Rossmann, M. G. (2006). Structural changes of bacteriophage $\phi$29 upon DNA packaging and release. *EMBO J.* **25,** 5229–5239.

Yan, X., Dryden, K. A., Tang, J., and Baker, T. S. (2007a). Ab initio random model method facilitates 3D reconstruction of icosahedral particles. *J. Struct. Biol.* **157,** 211–225.

Yan, X., Sinkovits, R. S., and Baker, T. S. (2007b). AUTO3DEM–an automated and high throughput program for image reconstruction of icosahedral particles. *J. Struct. Biol.* **157,** 73–82.

Zhang, X., Jin, L., Fang, Q., Hui, W. H., and Zhou, Z. H. (2010). 3.3 Å cryo-EM structure of a nonenveloped virus reveals a priming mechanism for cell entry. *Cell* **141,** 472–482.

Zhang, X., Settembre, E., Xu, C., Dormitzer, P. R., Bellamy, R., Harrison, S. C., and Grigorieff, N. (2008). Near-atomic resolution using electron cryomicroscopy and single-particle reconstruction. *Proc. Natl. Acad. Sci. USA* **105,** 1867–1872.

Zhou, Z. H. (2008). Towards atomic resolution structural determination by single-particle cryo-electron microscopy. *Curr. Opin. Struct. Biol.* **18,** 218–228.

CHAPTER EIGHT

# Single Particle Analysis at High Resolution

Yao Cong *and* Steven J. Ludtke

## Contents



## Abstract

Electron cryomicroscopy (cryo-EM) and single particle analysis is emerging as a powerful technique for determining the 3D structure of large biomolecules and biomolecular assemblies in close to their native solution environment. Over the last decade, this technology has improved, first to sub-nanometer resolution, and more recently beyond 0.5 nm resolution. Achieving sub-nanometer resolution is now readily approachable on mid-range microscopes with straightforward data processing, so long as the target specimen meets some basic requirements. Achieving resolutions beyond 0.5 nm currently requires a

National Center for Macromolecular Imaging, The Verna and Marrs McLean Department of Biochemistry and Molecular Biology, Baylor College of Medicine, Houston, Texas, USA

*Methods in Enzymology*, Volume 482
ISSN 0076-6879, DOI: 10.1016/S0076-6879(10)82009-9

high-end microscope and careful data acquisition and processing, with much more stringent specimen requirements. This chapter will review and discuss the methodologies for determining high-resolution cryo-EM structures of nonvirus particles to sub-nanometer resolution and beyond, with a particular focus on the reconstruction strategy implemented in the EMAN software suite.

# 1. Introduction

The basic concept of single particle reconstruction dates back to the early 1970s where the first 3D reconstructions of human wart virus and bushy stunt virus were performed using images of fewer than 10 particles (Crowther *et al.*, 1970; DeRosier and Moore, 1970). Since those early beginnings, the technique has steadily improved, and in recent years has demonstrated the ability to produce structures of even asymmetric objects at better than 0.5 nm resolution (Cong *et al.*, 2010).

In this method, purified biomolecules (protein, RNA, and DNA) or assemblies in aqueous buffer are deposited on the surface of a electron cryo-microscope (cryo-EM) grid and then flash frozen to produce a thin layer of vitreous (glassy) ice (Chapter 3, Vol. 481). This ice mimics a solution environment, producing particles effectively in a snapshot of their solution conformation. Instrumentation improvements such as field emission guns (FEG), improved lens systems, and improved cryoholders/stages have lead to increased high-resolution contrast in the raw image data. However, such improvements would have been useless without the simultaneous developments in computer technology and the software improvements that they enabled. The fact that a $1000 PC now exceeds the capacity of a $1 million supercomputer 10 years ago has dramatically altered both the algorithms which can be applied in performing reconstructions and the amount of data that can be processed. The combination of instrumentation improvements, larger computers, better software, and better specimen preservation methods have combined to make the current near-atomic resolution structures possible.

As the popularity of the technique expanded, available software for performing single particle reconstructions also became more widely available. The most widely used image processing packages capable of single particle reconstruction are SPIDER (Frank *et al.*, 1996; Shaikh *et al.*, 2008), IMAGIC (van Heel *et al.*, 1996), XMIPP (Sorzano *et al.*, 2004), B-soft (Heymann and Belnap, 2007), SPARX (Hohn *et al.*, 2007), and EMAN (Ludtke *et al.*, 1999; Tang *et al.*, 2007). In addition, FREALIGN (Grigorieff, 2007) and PFT3DR (Baker and Cheng, 1996) are dedicated packages for refinements, and IMIRS (Liang *et al.*, 2002) is an integrated package targeted at icosahedral structures. These packages each possess their own strengths and weaknesses. With the exception of IMAGIC, all are freely available. The majority of the published structures beyond 0.5 nm resolution are large icosahedral virus particles (Jiang *et al.*, 2008; Wolf *et al.*, 2010; Yu *et al.*, 2008; Zhang *et al.*, 2008, 2010b), which have been discussed in an earlier chapter (Chapter 14). For smaller, less symmetric particles, the three published to date have used EMAN for processing (Cong *et al.*, 2010; Ludtke *et al.*, 2008; Zhang *et al.*, 2010a). While the majority of this chapter will be broadly applicable and software independent, we focus on EMAN when specific details are required.

## 2. Specimen Requirements

Regardless of which software package is used for the reconstruction, there are a number of basic requirements for any specimen being used for high-resolution single particle reconstruction. Clearly the specimen must be solubilized in an aqueous buffer and monodisperse. In addition, the components of the buffer itself should be carefully chosen. Simple salts are generally well tolerated in cryo-EM images. While contrast may decrease somewhat, concentrations up to ~0.5 *M* are generally workable. However, certain common buffer additives, such as cryoprotectants like glycerol, are disastrous for cryo-EM imaging, due to the prodigious contrast reduction they cause, even at low concentrations (even 2% can be detrimental).

When imaging integral membrane proteins, special care must be taken for detergent solubilized preparations. It is typical to keep such specimens in a detergent containing buffer above the critical micelle concentration (CMC), due to improved preservation of protein assemblies under these conditions. However, as the micelles themselves will appear in the images, this is not permissible for cryo-EM. There have been many instances where researchers have begun single particle reconstructions of images of their "protein" only later to discover that they were actually looking at micelles. Detergent concentration must thus be somewhat below CMC. In addition, the detergent surrounding hydrophobic regions of the protein will also appear in the images, and may be difficult to distinguish from the protein it protects. For all of these reasons, membrane protein resolutions tend to lag behind those for soluble proteins using this technique (Ludtke *et al.*, 2005; Samsó *et al.*, 2005).

## 3. Data Collection Requirements

### 3.1. Orientation

The image of each individual particle represents a 2D projection of the molecule/assembly. As the particles were tumbling in solution prior to vitrification, ostensibly each particle is randomly oriented and positioned in the micrograph. Each particle must be located, its orientation determined, and any microscope artifacts corrected to make a 3D reconstruction possible. Resolution will be limited by the accuracy to which each particle can be aligned and oriented.

That said, it is critical to understand that resolution in structural biology is not a measure of the traditional Gaussian blur implied by resolution/resolvability in optical microscopy. In structural biology, resolution refers to the level of noise (Chapter 3) present at a specified level of detail (spatial frequency) in the reconstruction. To achieve isotropic resolution in a structure, an equal amount of information must be available in all particle orientations. If the image data included particles in orientations covering a single 180° rotation around a single tilt axis, it would yield complete information about the target's internal structure. This is the basic principle involved in CT scans. However, this geometry yields an anisotropic resolution, with higher resolution along the tilt axis, and lower resolution in the plane orthogonal to the axis. In situations where the target adopts a preferred orientation due to interactions with the air–water interface or substrate, there may be a problem similar to the missing wedge problem encountered in electron tomography. In some cases, controlling the hydrophobicity of the grid or adding a very small amount of detergent, sufficient only to coat the air–water interface, can help with particle orientation. A complete discussion of this topic is beyond the scope of this chapter, but it is important to be aware that the preferred orientation can cause reconstruction artifacts (Chapter 9).

## 3.2. High-resolution image acquisition

While high quality data can be poorly processed to produce a low quality structure, low quality data can never produce a high-resolution structure. Optimizing imaging conditions and specimen preparation is at least as critical as optimizing image processing. Toward this end, images must be carefully assessed as part of the preprocessing process, and imaging conditions (primarily defocus) should generally be planned out in advance of an experiment.

For high-resolution imaging, the microscope should be as well aligned as possible, ideally with coma-free alignment, negligible astigmatism, and a stable holder/stage with minimal drift. In addition, images with significant amounts of drift or astigmatism (Fig. 8.1), which may be due to charging or holder instability, should generally be excluded from the reconstruction. If overall contrast appears low, the most probable cause is thick ice. Optimizing ice thickness and particle distribution during the freezing process, is, in fact, one of the most time-consuming portions of the experimental process, and requires trial and error for each specimen. While reproducibility has been improved by development of computer controlled vitrification robots, parameters such as grid hydrophobicity and buffer constituents also play a strong role.

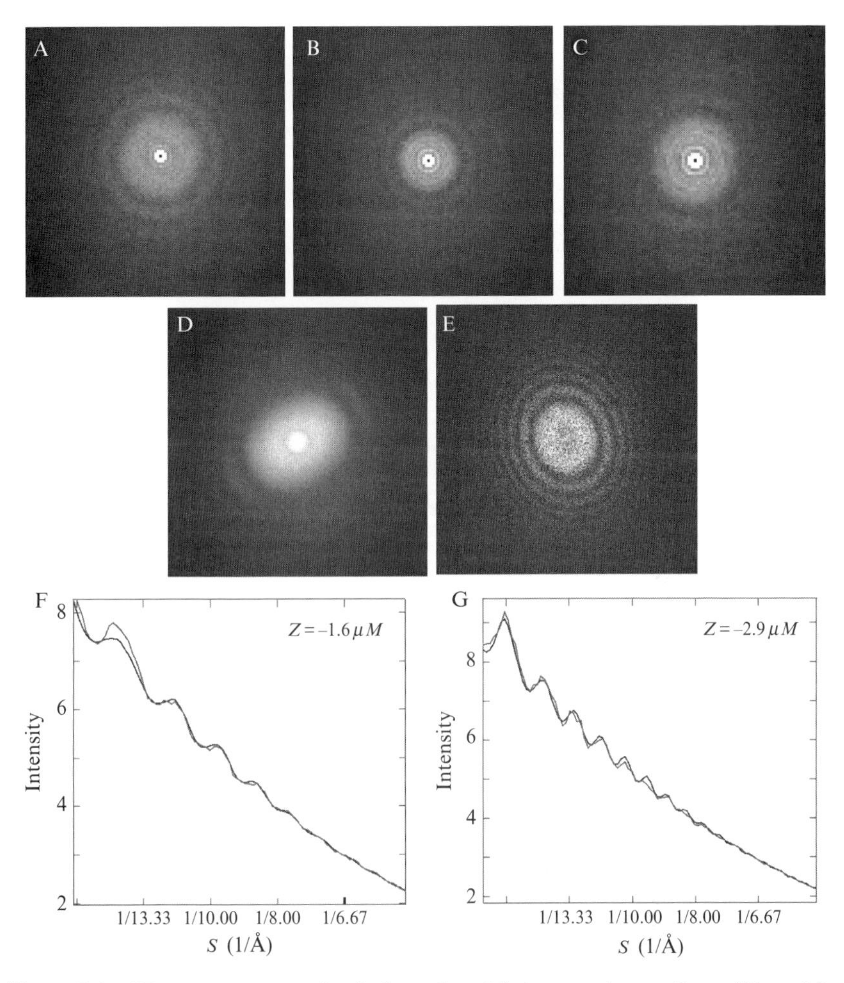


**Figure 8.1** 2D power spectra of typical good particle images close to focus (A), and far from focus (B), and images with drift (C), vibration (D), and astigmatism (E). (F) and (G) are 1D averaged power spectra corresponding to (A) and (B), respectively, with overlaid theoretical fits to the data.

There are several factors to consider when planning defocus ranges for single particle reconstruction. In general, there is an inverse relationship between image resolution and defocus even on modern FEG microscopes. This suggests that smaller defocus settings should be used. However, moving further from focus will enhance low resolution and overall image contrast. This makes particles easier to locate and increases the reliability of orientation determination. In general, images should be collected over a $\sim 1$ $\mu$m range starting as close to focus as possible, while still being able to clearly identify particles in the images.

To provide a concrete example of data collection for high-resolution reconstruction, consider the case of TRiC at ~4 Å resolution (Figs. 8.2A and 8.3; Cong *et al.*, 2010). Data were collected on a JEM3200FSC electron cryomicroscope operated at 300 kV and liquid nitrogen temperature, and equipped with an in-column omega energy filter. The energy filter serves to improve low-resolution contrast, improving accuracy in particle orientation determination. Data was collected at 50,000× magnification and <20 electrons/Å$^2$ total dose. Due to the relatively low particle concentration, images were recorded on film rather than CCD to increase the particle numbers per micrograph. For most specimens, CCDs are quite acceptable even for high-resolution reconstructions (Booth *et al.*, 2004; Yu *et al.*, 2008; Zhang *et al.*, 2010a). The micrographs were digitized on a Nikon 9000ED scanner producing 1.2 Å/pixel sampling. Following the traditional 3× oversampling rule for single particle analysis, this yields a maximum achievable resolution of ~3.6 Å. Most defocuses in this set were in the 1.2–2.7 μm range. Defocuses lower than this provided insufficient contrast to clearly identify particles and would have produced inaccurate orientations (Cong *et al.*, 2003; Joyeux and Penczek, 2002).

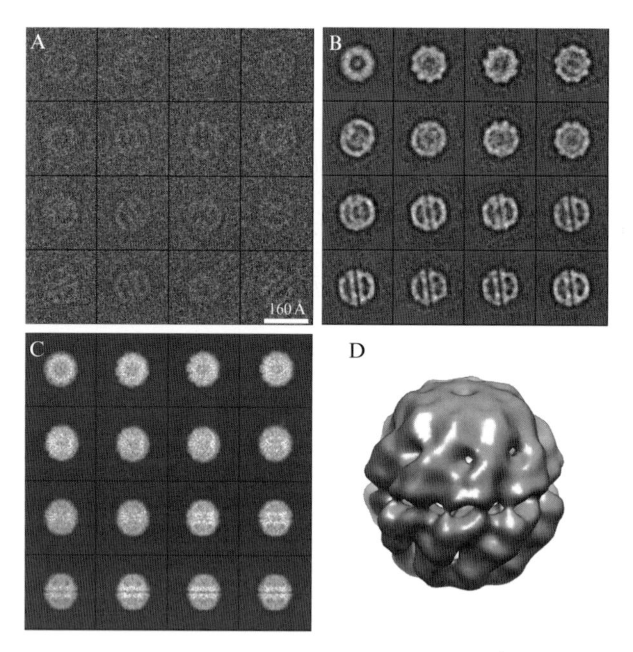


**Figure 8.2** (A) Representative raw particles used in the 4.7 Å resolution asymmetric cryo-EM reconstruction of closed bovine TRiC. Reference-free 2D class-averages (B), and 2D projections of the reconstructed map (C) of the same dataset. (D) Initial model used for this TRiC asymmetric reconstruction, which is a previously determined 15-Å resolution eightfold symmetrized map of closed TRiC.

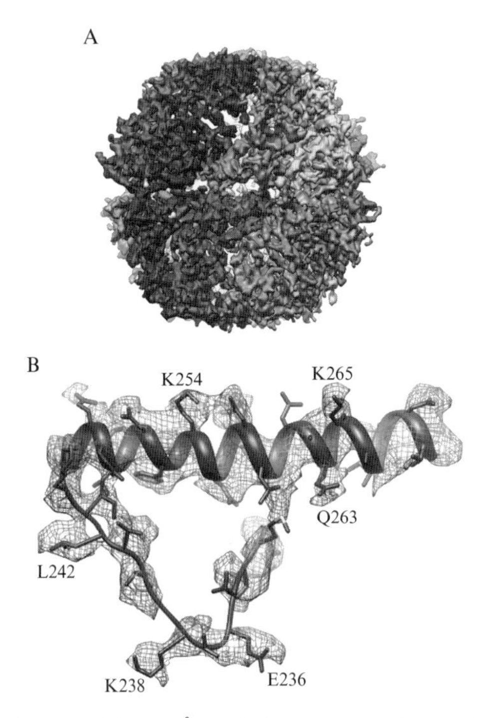


**Figure 8.3** (A) Side view of the 4.7 Å resolution asymmetric cryo-EM map of TRiC in the closed conformation, with the eight distinct subunits in each of the two rings highlighted in different colors. This map reveals the location of an unenforced twofold symmetry between its two rings. By enforcing this twofold symmetry, the map was further extended to 4.0 Å resolution (Cong *et al.*, 2010). (B) A portion of the 4.0 Å resolution reconstruction illustrating the visibility of side chain densities. Side chain densities were sufficient to unambiguously distinguish among the eight highly homologous subunits forming TRiC. (See Color Insert.)

## 4. Introduction to EMAN

EMAN was first developed in 1998 to deal with a number of issues not addressed by other available software at that time, and to provide a more modern framework for developing new algorithms to take advantage of rapidly expanding computational capacity (Ludtke *et al.*, 1999, 2001). It rapidly gained popularity for its accurate CTF correction, relative ease of use, and GUI capabilities. For the last several years, development has focused on EMAN2 (Tang *et al.*, 2007), a complete refactoring of EMAN1, designed with a truly modular design and an OpenGL based GUI interface, providing much needed 3D capabilities, and greatly expending potential for easily incorporating new algorithms and modules. While

EMAN2 is now producing high-resolution structures, the highest resolution structures already published have been completed using EMAN1, including GroEL (Ludtke *et al.*, 2008), Mm-cpn (Zhang *et al.*, 2010a), TRiC (Cong *et al.*, 2010), and several virus particles: ε15 (Jiang *et al.*, 2008) and P-SSP7 (Liu *et al.*, 2010). EMAN2 is now beginning to supplant EMAN1, and is recommended for new users, due to its integrated workflow interface and other advanced capabilities.

A typical reconstruction strategy is outlined in Fig. 8.4 including the names of the corresponding EMAN1/2 programs invoked at each step.

## 5. Particle Selection/Boxing

The process of locating individual particles of interest within the overall micrograph is generally referred to as boxing. While it seems like a fairly straightforward process for a human, despite years of effort, there are still no computer programs which can reliably perform this task completely automatically on arbitrary cryo-EM data (Zhu *et al.*, 2004). This statement

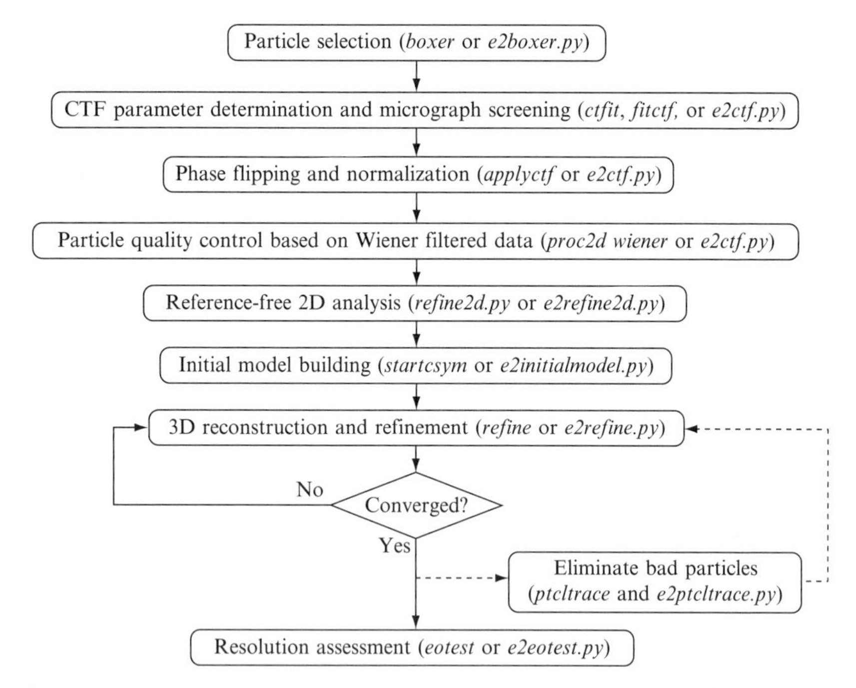


**Figure 8.4** A diagram outlining the general process for a typical high-resolution refinement. The names of EMAN1/2 programs for performing each step are shown.

has many caveats, of course. For specific types of specimens, such as icosahedral viruses, fully automated methods are quite reliable. However, when particles have a less regular shape and/or a smaller size, producing lower contrast, automated boxing methods are usually imperfect. In most cases, a human is required to manually verify the results, and eliminate false positives. For data sets now reaching hundreds of thousands of particles, this process can be quite tedious, but is generally regarded as necessary for achieving optimal results.

The EMAN1 programs "boxer" and "batchboxer" provide reference-based automatic particle selection. The EMAN2 program "e2boxer.py" incorporates an algorithm called "swarm" (Woolford *et al.*, 2007), which uses a trainable heuristic based approach. Many other programs are available for this task as well (Zhu *et al.*, 2004). In the TRiC example, approximately 160,000 ostensibly good particles were selected from over 1500 micrographs for the 3D reconstruction (Fig. 8.2A). In general, the number of particles required to achieve a given resolution is proportional to the number of asymmetric units present in each particle. A particle with C4 symmetry (a single fourfold axis) will require 1/4 as many particles as an asymmetric object, and a particle with icosahedral symmetry will require only 1/60th as many particles.

The relationship between number of particles and resolution, however, is highly nonlinear (Rosenthal and Henderson, 2003; Saad *et al.*, 2001). This fact also leads to a somewhat controversial statement: It is generally better to be conservative when picking particles for single particle reconstruction. It is important to eliminate particle-like objects which are not real particles (degraded particles, overlapping particles, ice contamination, etc.), even if doing so costs 30–40% of the total particle count. Halving the size of a data set, particularly if the excluded half contains the least clear particles, is not likely to damage resolution very significantly. The one caveat to this statement is that care must be taken not to accidentally exclude all particles in any specific orientation due to an unanticipated shape.

## 6. CTF and Envelope Correction

The contrast transfer function (CTF) is a mathematical description of the imaging process in the TEM (Erickson and Klug, 1970), expressed in Fourier space. Ideally, TEM images would represent true projections of the electron density of the specimen. Instead, images are distorted by the microscope optics, representing primarily phase contrast rather than amplitude contrast, and contain high levels of noise. In this section, we consider CTF correction to include corrections for all of these effects, even those not technically part of the CTF. While a thorough understanding of this section will require familiarity with Fourier transforms, it is possible to perform the necessary corrections in existing software without a detailed understanding of the underlying theory.

Without any CTF correction, the model produced by a reconstruction may contain significant local density displacements. Even with basic phase-flipping corrections, a lack of low-resolution amplitude correction can result in effects such as solid objects appearing hollow and exaggeration of extended domains. Each of the various available software packages handle CTF correction in slightly different ways, and these different methodologies can have an impact on the final 3D reconstruction, though the differences may be subtle.

Even before considering fitting CTF parameters for individual images, images should be assessed qualitatively for drift, astigmatism, and overall resolution. This is generally performed by examining the 1D and 2D power spectrum of each image (Fig. 8.1). This power spectrum may be generated from the entire micrograph, boxed out regions, or from sets of selected particles, depending on the convention of the individual software. Regardless, drift will be characterized by a directional fall-off in the pattern of Thon rings (Fig. 8.1C and D), and astigmatism will be characterized by elliptical rather than circular rings (Fig. 8.1E). Estimated potential resolution of the data can be assessed by examining the 1D power spectrum and determining the maximum resolution at which believable CTF oscillations are visible (Fig. 8.1F and G). While it may be possible through averaging to achieve resolutions slightly beyond this point, this value will serve as a good estimate of the limiting resolution for any given data set. While in principle, including some data with poorer resolution should not be harmful to a reconstruction, neither will it do much good, as the higher resolution images will already provide adequate contrast at lower resolutions.

In the EMAN CTF correction approach, the goal is to characterize the SSNR (spectral signal to noise ratio) of the images of the actual particles being used in the reconstruction, then use this information at various points in the reconstruction. Toward this aim, EMAN1 and 2 both use the averaged power spectrum of the boxed out particles from a single image to characterize the CTF. This approach has a number of advantages over working with the averaged power spectrum of random areas in the image, except in cases where each frame contains a very small number of particles.

A discussion of EMAN1's CTF correction can be found in (Ludtke *et al.*, 1999, 2001; Chapters 1–3), but briefly, the measured image power spectrum can be described, for our purposes (Fig. 8.5A), as:

$$M^2(s,\theta) = F^2(s,\theta)C^2(s)E^2(s) + N^2(s)$$

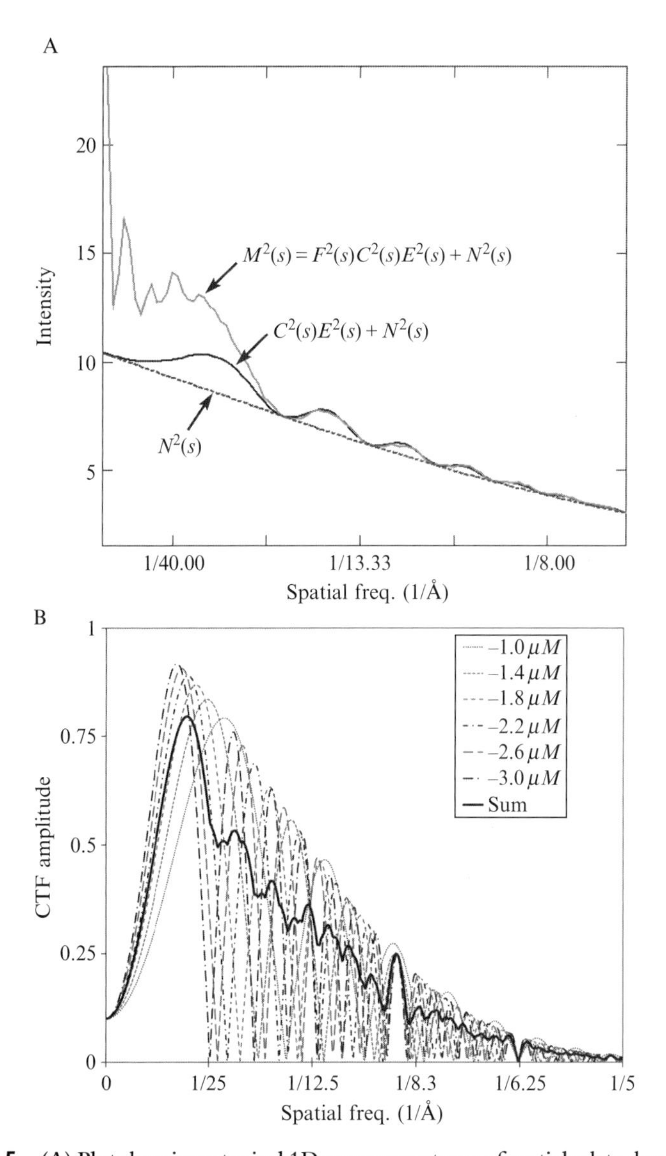


**Figure 8.5** (A) Plot showing a typical 1D power spectrum of particle data decomposed in terms of the model described in the text. (B) Plots of $|C(s)|$ for a range of defocuses, and a darker line representing the average of the individual curves. This darker line represents the shape of the filter that is effectively applied to a single particle reconstruction when CTF amplitude correction is not applied. The most important effect in terms of structure interpretation is the high-pass filter effect at low spatial frequencies, as this can result in substantial misinterpretations of connectivity and solidity of structures.

where $M^2(s)$ represents the TEM image power spectrum, $F^2(s,\theta)$ is the power spectrum of the true, undistorted particle, $C^2(s)$ is the CTF, $E^2(s)$ is the envelope function and is the main resolution limiting term, and $N^2(s)$ represents random additive noise. "$s$" represents spatial frequency, and $\theta$ is direction in polar coordinates. In the case of well stigmated drift-free images (Fig. 8.1A and B), we can represent the power spectra as 1D curves (Fig. 8.1F and G), since $C$, $E$, and $N$ are circularly symmetric. $C(s)$ is a defocus-dependent sinusoidal oscillation with increasing frequency at higher $s$. $E^2(s)$ decays with $s$ and is often approximated as a single aggregate term as a Gaussian (Rosenthal and Henderson, 2003; Saad *et al.*, 2001). The form of $N^2(s)$ is more complicated to characterize. On older microscopes with film data, often $N^2(s)$ is fairly well approximated as an exponential decay, but a more general form is required for arbitrary TEM data.

Assuming the images are well stigmated and have minimal drift, at least to the targeted resolution, the absolute minimal CTF correction needed is phase flipping. The fact that $C(s)$ is oscillatory and becomes negative at specific frequency ranges implies a "phase flipping" at those frequencies. This will result in inversion of features, but only over particular length scales. A decade ago, rather than trying to correct these phase flips, it was common practice to simply low-pass filter the image data at the first zero-crossing, and thus eliminate the inaccurate high-resolution data. This process limits resolution to ~20–30 Å at typical defocus ranges, but as this concept primarily originated with negative stain data. It was considered that staining artifacts would limit resolutions beyond this range, so further corrections were unnecessary. Given the relative ease of use of current CTF correction software, this method is no longer common. All high-resolution reconstructions will have performed phase-flipping corrections at a minimum.

Phase-flipping corrections simply consist of multiplying the Fourier transform of each particle image by $-1$ over appropriate frequency ranges. Even so, this leaves the issue of having $C(s) \sim 0$ at specific spatial frequencies. Fortunately, this issue can be handled by incorporating images over a range of different defocuses which will compensate for missing information in any one image (Fig. 8.5B).

In some reconstruction strategies only phase-flipping corrections are performed. As long as particle images with a variety of defocuses are used, the zeroes will be filled in during the reconstruction process, and a map very similar to the correct structure will be achieved. The remaining inaccuracy is that the structure will appear to be filtered by a function consisting of the sum of the various $|C(s)|$ curves (Fig. 8.5B). SPIDER, EMAN, and a few other software packages correct for this effect by dividing the images by $C(s)$, combined with a Wiener filter to compensate for predicted noise levels in the final structure and prevent infinities. EMAN1 and 2 additionally take another step, and measure the SSNR of the individual micrographs, using this information to perform a weighted average of the data with SSNR weights. Images will contribute to the average in direct proportion to their contrast at each resolution. Near $C(s) = 0$, a specific image will not contribute at all. In addition, the SSNR provides a reasonably accurate estimate of the final SSNR of the combined average permitting a more accurate Wiener filter to be performed.

How necessary is all of this complicated correction? The few comparative reconstructions from SPIDER and EMAN using the same data have been, reassuringly, nearly identical. However, tests have also shown that when the SSNR is used in computing particle alignments in EMAN, alignment accuracy can be improved, leading to modestly higher resolutions.

Each software package (even EMAN1 vs. EMAN2) uses a slightly different model to characterize the CTF of each image. Other than defocus, these parameterizations of the CTF are not generally compatible, presenting one of the largest impediments to moving partially analyzed data among the various software packages. CTF correction needs to be handled entirely within a single package to ensure optimal results. EMAN1 utilizes an eight parameter model for characterizing the CTF, with four of these parameters for characterizing noise. Two programs in EMAN1, "ctfit" and "fitctf," are available to determine CTF parameters manually and semiautomatically, respectively. EMAN2 on the other hand replaces the four parameter noise model from EMAN1 with a curve automatically measured directly from the data, and eliminates one additional redundant parameter in the eight parameter model of EMAN1. This leaves only three free parameters for the CTF determination in EMAN2 (defocus, B-factor and % amplitude contrast), which can be determined fully automatically in most cases. In addition, EMAN1's model of SSNR cannot deal properly with cases where either a continuous carbon substrate was used, or the solute (buffer) contained strongly scattering elements. In such cases, EMAN1 would inaccurately characterize a portion of the noise in the image as signal, and thus overestimate the SSNR, producing a textured appearance in class-averages. In EMAN2, the CTF model rectifies this problem.

The general process for CTF correction can be broken down into the following steps: (1) compute 1D and 2D averaged power spectra, (2) parameterize observed CTF for each micrograph/CCD frame (Fig. 8.1F and G), (3) flip phases of individual particles and (in some software) store the CTF parameters with the particles for later use, and (4) reconstruct the phase-flipped particles, performing automatic CTF amplitude correction at the appropriate stage. More detailed descriptions of how to perform these tasks are left to the manuals for the individual software packages.

## 7. Reference-Free 2D Analysis

Reference-free 2D analysis is a strategy to sort raw particles in different orientations or conformations into different groups based solely on the features present in the 2D images (Fig. 8.2B). In this process, there is no requirement that a self-consistent 3D structure result, and no prior knowledge is invoked. These averages are thus less biased than any produced via the 3D reconstruction methodology, and can eventually be used as a cross-check of the final determined 3D structure. In EMAN1/EMAN2, the class-averages thus produced are normally used to assess the structural variability of the specimen and to create an initial model for refinement. These reference-free class-averages are not used in any way during the iterative refinement process described below. This is, nonetheless, an important step, as it is critical to be aware of any structural variability before attempting a high-resolution refinement. A variety of mathematical techniques are used in performing this analysis (Chen *et al.*, 2006; Frank, 2006; Scheres *et al.*, 2007; van Heel and Frank, 1981). EMAN uses an iterative algorithm incorporating alignment and multivariate statistical analysis (MSA) in conjunction with *k*-means classification. The programs for performing this task are "refine2d.py" in EMAN1 (Chen *et al.*, 2006) and "e2refine2d.py" in EMAN2. Most of the other available software suites offer equivalent functionality.

## 8. Initial Model Building

The model refinement process used in single particle reconstruction to achieve the final high-resolution structures is iterative, beginning with a set of raw particle images and an initial "guess" at the 3D structure of the molecule. The initial 3D model is then repeatedly refined and updated against the raw data until convergence. It has been demonstrated that for the vast majority of structures especially those with a uniform distribution of particle orientations, even an extremely poor starting model will converge to the correct final structure eventually. Unfortunately, there will still often be a few "local minima" characteristic of each structure. These minima are incorrect models which the refinement process will converge to if certain incorrect starting models are used. For this reason, an initial 3D model similar in overall shape to the true structure will facilitate achieving the correct high-resolution structure, but may not be strictly necessary.

There are four basic approaches for determining initial models in cryo-EM. Cross common-lines and angular reconstitution (Van Heel, 1987) is an approach particularly well suited toward objects with high symmetry, but has also been applied to asymmetric objects, and involves searching for common information among groups of three or more images in Fourier space. The EMAN1 program "startAny" can perform cross common-lines to build an initial model. While this approach will frequently produce accurate initial models, it is not recommended for most particles without high symmetry, due to its susceptibility to finding incorrect structures which lie close to local minima. That is, if it does produce the incorrect structure, it will still be sufficiently self-consistent that iterative refinement may be unable to achieve the correct structure.

Random conical tilt (Radermacher *et al.*, 1987) and the related orthogonal tilt method (Leschziner and Nogales, 2006, Chapter 9) measure each particle in two different orientations with a known rotation between them. Then by finding particles with similar appearance in one image, the particles in the other image can be combined to produce a rough 3D structure. With the orthogonal tilt method, these structures can actually have full sampling if there is no preferred orientation. With the random conical tilt method, each 3D model thus generated will have a missing cone (Chapter 13). This method can work quite well for initial model generation, but is labor intensive and requires collecting data specifically for this process, which is useless for later high-resolution refinement.

Single particle tomography (Walz *et al.*, 1997) provides a robust approach where a tomogram is collected from the same grid normally used for traditional single particle reconstruction. 3D particles are then extracted from the reconstructed tomogram, and aligned and averaged in 3D. This technique is quite reliable, but is currently considered to be limited to ~20–50 Å resolution, and again, is fairly labor intensive. EMAN1/2 offer sets of programs for single particle tomography (Schmid and Booth, 2008), and support for random conical tilt is planned for EMAN2, but details of these methods are beyond the scope of this chapter.

The recommended approach in EMAN1 and the primary method in EMAN2 are based on a process similar to Monte Carlo methods. The approach is to generate a set (typically 10–20) of completely random low-resolution 3D density patterns, and use each as the initial model for a rapid 3D refinement. While it is unlikely that all of these refinements will find the global minimum (correct structure), generally at least one of them will. To use this method, the raw particle data is heavily downsampled to increase the particle contrast and refinement speed. Unless the particle is unusually large, many complete refinements can typically be done in under an hour on even a single CPU desktop computer. Then each of these obtained putative 3D structures is evaluated for how well its 2D projection (Fig. 8.2C) agrees with the class-averages from the reference-free 2D analysis (Fig. 8.2B), and solutions can thus be ranked. Ideally the top scoring results will agree with each other and provide a sufficiently correct starting 3D model. For the majority of structures, regardless of symmetry, this method is completely adequate, and requires no additional data collection. For instance, it has been demonstrated to produce the correct structure for the ribosome (unpublished), $Ca^{2+}$ release channel (Ludtke *et al.*, 2005), hemocyanin (Cong *et al.*, 2009), a variety of chaperonins (Cong *et al.*, 2010; Ludtke *et al.*, 2008; Zhang *et al.*, 2010a), and a range of other specimens. In EMAN2, the program implementing this method is called "e2initialmodel.py."

There are two possible issues which can produce unexpected results from this method. First, if the particle has a strongly preferred orientation, specifically, if there are large ranges of Euler angles which are barely represented in the data, there is a risk that these missing regions will be filled in with particles which did not fit perfectly in other orientations, producing a deformed, but convergent structure. The second risk, exacerbating the first, is that of structural heterogeneity/flexibility. If the specimen is sufficiently flexible in solution to cause gross changes in quaternary structure, such as the mammalian fatty acid synthase, which undergoes motions as large as 30–40 Å in solution (Brignole *et al.*, 2009; Brink *et al.*, 2004), or carboxysomes which naturally exist in a range of distorted and variable size icosahedra (Schmid *et al.*, 2006), then the algorithm can confuse changes in conformation with changes in orientation, producing distorted, but apparently reasonably self-consistent structures. In such instances, both the Monte Carlo method and the common-lines based approach are very likely to produce incorrect structures. Either single particle tomography, orthogonal tilt, or random conical tilt experiments must be performed in such cases. The underlying structural variability will normally be detectable either directly in the 2D reference-free averages or in the lack of consistency of the final 3D structure.

## 9. Refinement

Refinement is the most time consuming and critical stage in single particle analysis. While there are a number of different strategies for refinement, such as cross common-lines (Van Heel, 1987), maximum likelihood methods (Scheres *et al.*, 2008; Chapter 10), and global minimization (Yang *et al.*, 2005), to date, all high-resolution reconstructions have used a projection matching approach, with the exception of FREALIGN-based reconstructions which use a related method in Fourier space (Grigorieff, 2007). In EMAN1 and EMAN2, projection matching is the standard method. This method is iterative, with four steps in each cycle: (1) Projections of the current 3D model are generated over all possible orientations within an asymmetric unit; (2) 2D image classification, in which each of the raw 2D particles is computationally compared with the set of projections and assigned to the most similar projection; (3) Class-averaging. For each projection, the particles found to be most similar to it are iteratively aligned to each other and averaged together with CTF amplitude correction; and (4) 3D reconstruction, in which the class-averages in known orientations are combined to produce a new 3D reconstruction. These four steps are then iterated until some convergence criterion has been met.

## 9.1. 2D image classification

Key to the entire refinement procedure is how the very noisy 2D particle images (Fig. 8.2A) are aligned to and compared with the set of projection images (Fig. 8.2C). Inaccurate alignment leads to unreliable classification, and will produce an inaccurate reconstruction (Cong *et al.*, 2003; Joyeux and Penczek, 2002). In EMAN1, there are several specific choices to make. For 2D alignment, either the standard aligner, or the recently developed FRM2D aligner (Cong *et al.*, 2003, 2005, 2009) may be used, either of which can be combined with a final iterative refinement of the alignment parameters. Both methods have been successfully used in achieving high-resolution structures. In EMAN2, the user is permitted to select among all available alignment algorithms and similarity metrics both for alignment and classification of particles. A list of and documentation for the available alignment routines can be provided by the "e2help.py aligner -v 1" command.

Once the particles have been aligned, their similarity to each corresponding projection must be assessed. Again, several methods are available for measuring similarity. One common metric is a normalized dot-product between the particle and reference, however, in cases where the projections and particle have not been filtered in the same way, this metric is susceptible to a deterministic bias in orientation parameters (Ludtke *et al.*, 2001), and thus this metric is NOT normally used in EMAN1/2 without additional adjustments. The other three criteria in common use are mean phase error, integrated Fourier ring correlation, and filtered/rescaled real-space RMSD (Ludtke *et al.*, 2001). This final method has been used for all published high-resolution structures in EMAN1. In EMAN2, these algorithms are available, and in addition, each has a number of available options. The "e2help.py cmp -v 1" command will provide a list and documentation for each.

## 9.2. 3D reconstruction

The refinement process outlined above is implemented as a single program in both EMAN1 ("refine") and EMAN2 ("e2refine.py"). In EMAN2, this program can also be executed using the built-in workflow GUI ("e2workflow.py"), which provides help in selecting values for the large number of available options. A full discussion of all options available in each program would be too extensive for this chapter, but a representative EMAN1 refinement command, specifically the one used for the final refinement of the 4.7 Å TRiC asymmetric structure (Cong *et al.*, 2010), will be dissected as an example. In this case, the box size of the particle image is 288 × 288 pixels, with a sampling of 1.2 Å/pixel. A previously determined 15Å resolution eightfold symmetrized map of closed TRiC was used as the initial model (Fig. 8.2D).

*refine 6 mask=128 frm2d=90,160 maxshift=10 hard=23 sym=c1 pad=360 ang=2.5 classkeep=0.7 classiter=3 ctfcw=<structure factor file> dfilt xfiles=1.2,1000,99 amask=24,0.9,20 refine proc=<number of CPUs>*

- The "6" specifies six iterations of the overall refinement loop.
- "mask=128" is the radius in pixels of a circular mask to apply in various stages and is required, though it is supplanted by the tighter mask produced by the "amask" option.
- "frm2d=90,160" specifies that the FRM2D alignment algorithm should be used. Here the first parameter, "90" specifies the approximate maximum radius of the particle, which can be estimated from reference-free 2D analysis for unknown structures. The second value is the number of angular sampling points at each radius when resampling the 2D particle image into polar coordinates; here, "160" points corresponds to an angular step of 2.25° (360°/160 = 2.25°). Usually 128 is a reasonable choice for sub-nanometer resolution reconstruction where box size < 200 pixels. Too large a value will increase the computational time. This value is closely related to the "ang=" option.
- "maxshift=10" specifies the 1D translation search up to 10 pixels, which actually closely related to the center of mass determination uncertainty during boxing process. "maxshift" is usually suggested to use when FRM2D method is specified, so as to limit the translational search range in FRM2D.
- "hard=23" is a criterion for excluding class-averages which are not sufficiently consistent with the overall structure, 20–25 are typical values, though for high-resolution structures in the presence of somewhat preferred orientation, values as small as 6–10 may be used.
- "sym=c1" may be omitted in this case (no symmetry), but is shown since it must be specified if symmetry is to be enforced.
- "pad=360" specifies an expanded box size in pixels to be used temporarily during 3D reconstruction to reduce Fourier artifacts. It is typically 25–50% larger than the normal box size.
- "ang=2.5" specifies the angular sampling in degrees to use in generating projections of the model. Smaller values will produce more projections and thus more class-averages, but will increase the time required for the refinement, and decrease the number of particles in each class-average. This value must be sufficiently small to provide adequate angular

sampling at the desired resolution, and is dependent on both resolution and particle size. For a large virus this may be less than 1, and for low-resolution work on smaller particles may be as large as 7.5 or 9.

- "classkeep=0.7" is a coefficient for excluding "bad" particles from class-averages. Each particle is compared to the final average, and particles worse than 0.7•sigma of this distribution are excluded from the final average. Typical values are (0.5–3).
- "classiter=3" indicates the number of times particles within a single class should be iteratively aligned to each other. For historical reasons, the actual number of iterations is $N - 2$, actually "3" performs only a single iteration. Large values (6–8) will eliminate problems with initial model bias, but will degrade resolution, so it is typical to begin with a few rounds of refinement with classiter = 8, then reduce it to 3 for final high-resolution refinement.
- "ctfcw=<sf file>" tells the refinement to perform full amplitude CTF correction with a Wiener filter. The structure factor file is generally determined as part of the CTF fitting process, and may be generated automatically in EMAN2.
- "dfilt" specifies that the optimized real-space RMSD similarity metric should be used for classification. This option has been used for all published structures from EMAN at beyond 0.5 nm resolution.
- "xfiles=<apix>,<mass>,<ali>" is a convenience function to permit easier visualization of final results, and is a prerequisite for "amask=." <apix> is the Å/pixel sampling of the data, <mass> is the particle mass in kDa, and <ali> is normally set to 99.
- "amask=<r>,<thr>,<shells>" causes a mask to be produced for the structure tighter than the default spherical mask. In addition, this mask is a "soft" mask with a Gaussian decay at the edge, preventing resolution exaggeration due to mask correlations. Selecting the parameters for this mask is a slightly complex issue described in the EMAN1/EMAN2 documentation and wiki (http://blake.grid.bcm.edu).
- "refine" invokes local refinement of the 2D alignments of particles to projections, and produces more accurate classification of particles, but at a significant computational cost.
- "proc=" specifies how many processors to use in EMAN1's parallel processing methodology.

## 9.3. Eliminating particles

In many cases, single particle reconstruction is a straightforward process. However, some fraction of projects will have specific issues which need to be deal with to achieve optimal results. The asymmetric structure of TRiC represents one such case (Fig. 8.2).

Unlike most chaperonins, which have multiple copies of one to three different subunits, TRiC has eight distinct, but highly homologous subunits. The subunits are sufficiently structurally similar that breaking the eightfold rotational symmetry is possible only at very high resolutions. This means that each individual macromolecule must have sufficient contrast to unambiguously determine which of the eight possible orientations is correct, otherwise the final reconstruction would be effectively symmetrized and can not provide enough structural features to distinguish the differences among them.

The solution to this problem was to perform a standard single particle refinement with no symmetry enforced for several iterations, then examine each particle to see if it was assigned to a consistent asymmetric orientation over several iterations after convergence had been achieved. This process invoked specialized programs "ptcltrace" and "e2ptcltrace.py" to make the necessary assessments. In the end, ~37% of the particles in this data set (~60,000 particles) were excluded due to this ambiguity, and the final high-resolution refinement included only the 100,000 best particles under this criterion.

While this specific process for removing "bad" particles is not broadly applicable, often some method for using a refined reconstruction to revisit the issue of particle quality can be beneficial. This issue is addressed differently in the various software packages. While clearly desirable to use only the best particles, actually identifying "bad" particles in a useful way can be challenging. In many cases sound-seeming approaches for eliminating particles prove to do little to improve the final structure.

## 10. Resolution and Assessing Results

Once a structure has been refined, its veracity and resolution must be assessed. While many standard methods are available for this process, none yet developed is immune to issues of noise/model bias (Stewart and Grigorieff, 2004), and there are numerous "tricks" which can result in either intentional or unintentional resolution exaggeration.

The first question to ask when running a refinement is how to decide when the iterative process has converged and can be stopped. For this purpose, a common method is to compare each 3D model to the 3D model from the previous iteration using the same Fourier shell correlation (FSC) algorithm used to assess resolution. This sequence of curves (Fig. 8.6) will generally gradually converge to higher and higher values until no further improvement is possible and the same curve is produced over several sequential iterations. While these curves may have a similar appearance to the final FSC curve used to assess resolution, they cannot be used for this purpose.

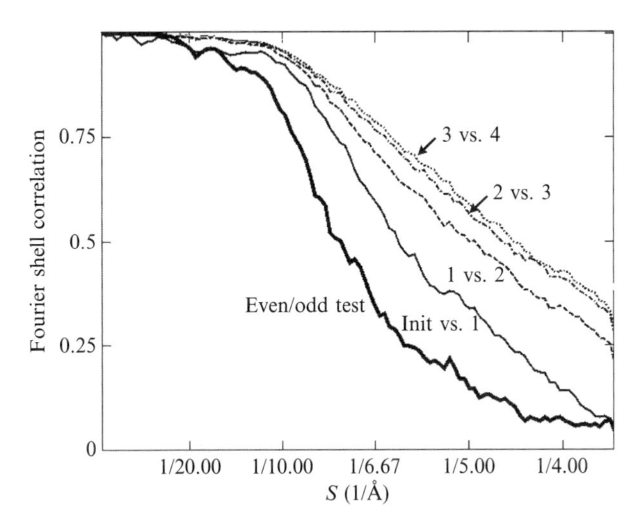


**Figure 8.6** Monitoring convergence and final resolution assessment using Fourier shell correlation for a representative refinement of hemocyanin. Thin lines represent comparisons between successive refinement iterations, and ideally would become 1.0 if true convergence were achieved. In this case, pseudoconvergence had almost been achieved after the final shown iteration. The darker solid line represents a resolution evaluation FSC curve between models generated from even and odd halves of the data. Note that this resolution curve is slightly suboptimal as it falls almost, but not quite, to zero. This is likely due to use of a final mask which is slightly too tight around the structure.

The next important assessment to perform is self-consistency of the data and refined model. Necessary, but not sufficient, evidence for a correct reconstruction is that class-averages must match projections of the final structure in all orientations where sufficient data exists. A single class-average with good contrast, which does not match the corresponding projection, is an indication of a problem with the reconstruction. The other test is more difficult in many cases due to the low contrast of the particles, but the class-averages should contain representations of all observed raw particle conformations. Some level of misclassification of particles is inevitable due to high noise levels, so this is not a requirement that all particles in a specific class have the correct shape, but rather a statement that for each particle, it should be possible to find a class-average with a similar shape.

Resolution assessment is described in detail in another chapter of this volume (Chapters 1–3), but briefly, the standard method is to split the particle set into even and odd halves, and compute a FSC between 3D models generated "independently" for each set. While ideally these determinations would be completely independent, this is rarely performed in practice, and generally the already determined orientations are used for this process. EMAN at least takes the step of recomputing 2D alignments as part of this process, to help reduce the influence of model bias, but does not generally perform completely independent refinements.

A "healthy" FSC curve will start at ~1.0 at low resolution, then at some resolution will begin decaying toward zero, which it will asymptotically approach (Fig. 8.6). In this case, the resolution is typically assessed as the frequency at which the FSC curve falls below some specific threshold value. There has been a great deal of controversy in the community the "correct" threshold value (or curve), with no real consensus. However, as these various thresholds all produce resolution values over a fairly narrow range, common practice has become to select some criterion for describing the resolution, and to include the FSC curve itself in the supplementary data when publishing, to permit the reader to interpret as they choose.

Publishing the FSC curve has the additional benefit of permitting an expert reader to assess the quality of the curve itself. Many common artifacts can appear in the FSC curve, and are indicative of problems with the refinement. One common issue is evidenced by a FSC curve which begins falling toward zero, but then at some spatial frequency above zero it actually begins increasing (toward 1). This effect can be the result of several issues, but the most common causes are insufficient angular sampling ("ang=" too large in EMAN1), or overaggressive masking resulting in mask correlation.

## 11. Required Computational Capacity and Emerging Methods

Computational requirements for single particle reconstruction can span several orders of magnitude. A low-resolution structure of a moderately sized particle (~1 MDa) could be performed in <100 CPU-hr on a typical workstation. At the other extreme, the asymmetric TRiC refinement (Cong *et al.*, 2010) required an aggregate 1,000,000+ CPU-hr and involved over a year of processing on several large Linux clusters. Common projects of moderate size with some symmetry will require ~1000–10,000 CPU-h and are generally performed on small Linux clusters, or in some cases sets of multicore workstations already available in a lab. However, an alternative strategy has been emerging over the last 2–3 years. The graphics processing unit (GPU) in most modern PC's are actually extremely capable high performance processors in their own right. In fact, for many algorithms common in image processing, the GPU can outperform the CPU by as much as a factor of 100. This means that use of a GPU could potentially make a single quad-core desktop PC the equivalent of a 25-node cluster. Several of the available single particle refinement programs have already announced GPU capabilities for at least a subset of the necessary algorithms in single particle analysis, including SPIDER, FREALIGN, and EMAN2. It is likely that these capabilities will expand in coming years. While reconstructions at the highest possible resolution are likely to require the use of clusters for some years to come, this emerging technology may make subnanometer resolution accessible to those with only desktop workstations, and potentially facilitate the cryo-EM study of biomolecules with low symmetry, or heterogeneity.

## 12. Conclusions

From a historical perspective, it is clear that single particle analysis is finally coming of age, producing structures which are beginning to directly compete with X-ray crystallography. As seen in other chapters, this methodology also presents capabilities for assessing structural flexibility, ligand binding, and numerous other problems difficult to address using traditional structural biology methods. The availability of biological TEMs has also been expanding in recent years, increasing the size of the single particle processing community dramatically. With the rapid development of instrumental and computational techniques, it is now widely expected that single particle analysis at 3–5 Å resolution will become routine in coming years.

## *ACKNOWLEDGMENT*

We would like to acknowledge Joanita Jakana for providing image data. The work and software described in this chapter were supported by the National Institutes of Health (PN1EY016525, P41RR02250, and R01GM080139).

## *REFERENCES*

Baker, T. S., and Cheng, R. H. (1996). A model-based approach for determining orientations of biological macromolecules imaged by cryoelectron microscopy. *J. Struct. Biol.* **116,** 120–130.

Booth, C. R., *et al.* (2004). A 9 angstroms single particle reconstruction from CCD captured images on a 200 kV electron cryomicroscope. *J. Struct. Biol.* **147,** 116–127.

Brignole, E. J., *et al.* (2009). Conformational flexibility of metazoan fatty acid synthase enables catalysis. *Nat. Struct. Mol. Biol.* **16,** 190–197.

Brink, J., *et al.* (2004). Experimental verification of conformational variation of human fatty acid synthase as predicted by normal mode analysis. *Structure* **12,** 185–191.

Chen, D. H., *et al.* (2006). An expanded conformation of single-ring GroEL–GroES complex encapsulates an 86 kDa substrate. *Structure* **14,** 1711–1722.

Cong, Y., *et al.* (2003). 2D fast rotational matching for image processing of biophysical data. *J. Struct. Biol.* **144,** 51–60.

Cong, Y., *et al.* (2005). Fast rotational matching of single-particle images. *J. Struct. Biol.* **152,** 104–112.
Cong, Y., *et al.* (2009). Structural mechanism of SDS-induced enzyme activity of scorpion hemocyanin revealed by electron cryomicroscopy. *Structure* **17,** 749–758.
Cong, Y., *et al.* (2010). 4.0-Å resolution cryo-EM structure of the mammalian chaperonin TRiC/CCT reveals its unique subunit arrangement. *Proc. Natl. Acad. Sci. USA* **107,** 4967–4972.
Crowther, R. A., *et al.* (1970). Three dimensional reconstructions of spherical viruses by Fourier synthesis from electron micrographs. *Nature* **226,** 421–425.
DeRosier, D. J., and Moore, P. B. (1970). Reconstruction of three-dimensional images from electron micrographs of structures with helical symmetry. *J. Mol. Biol.* **52,** 355–369.
Erickson, H. P., and Klug, A. (1970). The Fourier transform of an electron micrograph: Effects of defocusing and aberrations, and implications for the use of underfocus contrast enhancement. *Philos. Trans. R. Soc. Lond. B* **261,** 221–230.
Frank, J. (2006). Multivariate data analysis and classification of images. *In* Three-Dimensional Electron Microscopy of Macromolecular Assemblies: Visualization of Biological Molecules in their Native State. pp. 145–192. Oxford University Press, New York.
Frank, J., *et al.* (1996). SPIDER and WEB: Processing and visualization of images in 3D electron microscopy and related fields. *J. Struct. Biol.* **116,** 190–199.
Grigorieff, N. (2007). FREALIGN: High-resolution refinement of single particle structures. *J. Struct. Biol.* **157,** 117–125.
Heymann, J. B., and Belnap, D. M. (2007). Bsoft: Image processing and molecular modeling for electron microscopy. *J. Struct. Biol.* **157,** 3–18.
Hohn, M., *et al.* (2007). SPARX, a new environment for Cryo-EM image processing. *J. Struct. Biol.* **157,** 47–55.
Jiang, W., *et al.* (2008). Backbone structure of the infectious epsilon15 virus capsid revealed by electron cryomicroscopy. *Nature* **451,** 1130–1134.
Joyeux, L., and Penczek, P. A. (2002). Efficiency of 2D alignment methods. *Ultramicroscopy* **92,** 33–46.
Leschziner, A. E., and Nogales, E. (2006). The orthogonal tilt reconstruction method: An approach to generating single-class volumes with no missing cone for ab initio reconstruction of asymmetric particles. *J. Struct. Biol.* **153,** 284–299.
Liang, Y., *et al.* (2002). IMIRS: A high-resolution 3D reconstruction package integrated with a relational image database. *J. Struct. Biol.* **137,** 292–304.
Liu, X., *et al.* (2010). Structural changes in a marine podovirus associated with release of its genome into Prochlorococcus. *Nat. Struct. Mol. Biol.* **17,** 830–836.
Ludtke, S. J., *et al.* (1999). EMAN: Semiautomated software for high-resolution single-particle reconstructions. *J. Struct. Biol.* **128,** 82–97.
Ludtke, S. J., *et al.* (2001). A 11.5 A single particle reconstruction of GroEL using EMAN. *J. Mol. Biol.* **314,** 253–262.
Ludtke, S. J., *et al.* (2005). The pore structure of the closed RyR1 channel. *Structure* **13,** 1203–1211.
Ludtke, S. J., *et al.* (2008). De novo backbone trace of GroEL from single particle electron cryomicroscopy. *Structure* **16,** 441–448.
Radermacher, M., *et al.* (1987). Three-dimensional reconstruction from a single-exposure, random conical tilt series applied to the 50S ribosomal subunit of *Escherichia coli*. *J. Microsc.* **146,** 113–136.
Rosenthal, P. B., and Henderson, R. (2003). Optimal determination of particle orientation, absolute hand, and contrast loss in single-particle electron cryomicroscopy. *J. Mol. Biol.* **333,** 721–745.
Saad, A., *et al.* (2001). Fourier amplitude decay of electron cryomicroscopic images of single particles and effects on structure determination. *J. Struct. Biol.* **133,** 32–42.

Samsó, M., *et al.* (2005). Internal structure and visualization of transmembrane domains of the RyR1 calcium release channel by cryo-EM. *Nat. Struct. Mol. Biol.* **12,** 539–544.
Scheres, S. H., *et al.* (2007). Disentangling conformational states of macromolecules in 3D-EM through likelihood optimization. *Nat. Methods* **4,** 27–29.
Scheres, S. H. W., *et al.* (2008). Image processing for electron microscopy single-particle analysis using XMIPP. *Nat. Protoc.* **3,** 977–990.
Schmid, M. F., and Booth, C. R. (2008). Methods for aligning and for averaging 3D volumes with missing data. *J. Struct. Biol.* **161,** 243–248.
Schmid, M. F., *et al.* (2006). Structure of *Halothiobacillus neapolitanus* carboxysomes by cryo-electron tomography. *J. Mol. Biol.* **364,** 526–535.
Shaikh, T. R., *et al.* (2008). SPIDER image processing for single-particle reconstruction of biological macromolecules from electron micrographs. *Nat. Protoc.* **3,** 1941–1974.
Sorzano, C. O., *et al.* (2004). XMIPP: A new generation of an open-source image processing package for electron microscopy. *J. Struct. Biol.* **148,** 194–204.
Stewart, A., and Grigorieff, N. (2004). Noise bias in the refinement of structures derived from single particles. *Ultramicroscopy* **102,** 67–84.
Tang, G., *et al.* (2007). EMAN2: An extensible image processing suite for electron microscopy. *J. Struct. Biol.* **157,** 38–46.
Van Heel, M. (1987). Angular reconstitution: A posteriori assignment of projection directions for 3D reconstruction. *Ultramicroscopy* **21,** 111–123.
van Heel, M., and Frank, J. (1981). Use of multivariate statistics in analysing the images of biological macromolecules. *Ultramicroscopy* **6,** 187–194.
van Heel, M., *et al.* (1996). A new generation of the IMAGIC image processing system. *J. Struct. Biol.* **116,** 17–24.
Walz, J., *et al.* (1997). Electron tomography of single ice-embedded macromolecules: Three-dimensional alignment and classification. *J. Struct. Biol.* **120,** 387–395.
Wolf, M., *et al.* (2010). Subunit interactions in bovine papillomavirus. *Proc. Natl. Acad. Sci. USA* **107,** 6298–6303.
Woolford, D., *et al.* (2007). SwarmPS: Rapid, semi-automated single particle selection software. *J. Struct. Biol.* **157,** 174–188.
Yang, C., *et al.* (2005). Unified 3-D structure and projection orientation refinement using quasi-Newton algorithm. *J. Struct. Biol.* **149,** 53–64.
Yu, X., *et al.* (2008). 3.88 A structure of cytoplasmic polyhedrosis virus by cryo-electron microscopy. *Nature* **453,** 415–419.
Zhang, X., *et al.* (2008). Near-atomic resolution using electron cryomicroscopy and single-particle reconstruction. *Proc. Natl. Acad. Sci. USA* **105,** 1867–1872.
Zhang, J., *et al.* (2010a). Mechanism of folding chamber closure in a group II chaperonin. *Nature* **463,** 379–383.
Zhang, X., *et al.* (2010b). 3.3 Å Cryo-EM structure of a nonenveloped virus reveals a priming mechanism for cell entry. *Cell* **141,** 472–482.
Zhu, Y., *et al.* (2004). Automatic particle selection: Results of a comparative study. *J. Struct. Biol.* **145,** 3–14.

CHAPTER NINE

# The Orthogonal Tilt Reconstruction Method

Andres Leschziner

## Contents



## Abstract


Generating reliable initial models for novel asymmetric molecules, particularly heterogeneous ones, remains a major challenge in cryo-electron microscopy. Geometric reconstruction methods, relying on the ability to tilt the microscope stage to obtain two or more views of each molecule, are arguably the most robust for these types of samples as they generate independent reconstructions for each characteristic view obtained.

Random Conical Tilt (RCT) is the classic geometric reconstruction method. Pairs of images are collected at high tilt (around 50°) and 0°. The latter are used to sort the data into characteristic views of the molecule and the former are used for their reconstruction. RCT's greatest strength is its ability to generate structures regardless of the number of orientations adopted by the sample on


Department of Molecular and Cellular Biology, Harvard University, Cambridge, Massachusetts, USA

*Methods in Enzymology*, Volume 482
ISSN 0076-6879, DOI: 10.1016/S0076-6879(10)82010-5

the support. Its major drawback stems from the limited tilt of the microscope stage; this results in an incomplete sampling of the structure in Fourier space and artifacts in its real space representation. Orthogonal Tilt Reconstruction (OTR), a modification of this data collection strategy, results in fully sampled structures. It relies on collecting data at $-45°$ and $+45°$ and treating the tilt pairs as equivalent to the ideal $0°/90°$ that cannot be collected directly in the microscope. OTR requires a sample that adopts a large number of orientations on the support.

Here, the RCT and OTR methods are reviewed and their performances with a biological test sample are compared. The steps required to apply OTR are also discussed.

# 1. Introduction

Cryo-electron microscopy (cryo-EM) of single particles has made spectacular progress in recent years, with several projects reaching secondary structure and near-atomic resolution (Ludtke *et al.*, 2008; Schuette *et al.*, 2009; Seidelt *et al.*, 2009; Yu *et al.*, 2008; Zhang *et al.*, 2008). Alongside this progress, invariably achieved with well-behaved or benchmark samples, we have seen a growing interest in the analysis of heterogeneity (biochemical and conformational) as a source of invaluable biological information that cryo-EM is uniquely well suited to address. The fact that several chapters in this volume are devoted to the subject is a testimony to its rising prominence.

A number of computational tools have been developed to sort out multiple species coexisting in a sample (see Spahn and Penczek, 2009; for a recent review). A common aspect of most of these approaches is the requirement for at least one initial model of the macromolecule in question. As a result, the challenge of obtaining initial models, particularly of potentially heterogeneous asymmetric molecules, has seen a revived interest.

There are currently four approaches to generating initial models from experimental data: Angular Reconstitution (Van Heel, 1987), Random Conical Tilt (RCT) (Radermacher *et al.*, 1987), Orthogonal Tilt Reconstruction (OTR) (Leschziner and Nogales, 2006), and tomography. The first of these methods, Angular Reconstitution, is an analytical approach that can directly determine the relative orientations of different molecular views present in a sample without additional geometric information. The method is based directly on the Central Section theorem (described in Section 2.1), a consequence of which is that any two projections (images) of a given structure share a common central line in Fourier space (as well as a common one-dimensional projection in real space). While elegant, and extremely successful when applied to structures with high symmetry, Angular Reconstitution is less robust with asymmetric and/or heterogeneous samples.

The difficulty resides in the need for the user to distinguish between different views of the same species (as found in a homogeneous sample) and different views of different species (as found in a heterogeneous one). Without *a priori* structural information, this distinction becomes a serious challenge and mistakes may lead to incorrect reconstructions. Provided these problems can be overcome, and that a large number of views of the molecule are available in the sample, a major advantage of Angular Reconstitution is that it generates reconstructions that are fully sampled (the issue of sampling will be discussed further below).

Geometric methods are intrinsically more robust for two reasons. First, two (or more, in the case of tomography) views of each molecule are obtained experimentally and their spatial relationship is therefore known. Second, a reconstruction is generated for each characteristic view of the molecule(s) present in the sample. This removes the need to decide whether different views correspond to the same or different molecular species. Comparisons are postponed until after reconstructions have been generated.

Tomography, the one true "single particle" method, can generate three-dimensional (3D) reconstructions for each molecule present in the sample without the need for averaging. This is both its strength and its weakness: avoiding averaging means that each reconstruction is truly homogeneous but the price to be paid for this is the low signal-to-noise ratio of the reconstructions. The solution to this involves bringing averaging back in, this time at the level of the reconstructed volumes. Unlike Angular Reconstitution, tomography generates incompletely sampled reconstructions. This is a consequence of the limit in the extent to which samples can be tilted in the microscope during data collection (this is discussed further in Section 2.1).

The RCT method (Radermacher *et al.*, 1987) was originally developed to reconstruct samples adopting preferred orientations on the support. While it remains the method of choice with samples taking one or a few orientations on the grid, our ability to collect and process ever larger data sets has made the method equally useful, and robust, with samples adopting a larger number of orientations. RCT's main limitation is the same one faced by tomography: our inability to collect data with the sample tilted to 90°. As a result, the structure is incompletely sampled and suffers from artifacts (discussed in detail in Section 2.1). There are solutions to this problem but these are often not trivial and require the user to make judgments about the molecular identity of a number of different (noisy) reconstructions, a considerable challenge in the presence of heterogeneity.

The OTR method was developed recently (Leschziner and Nogales, 2006) to circumvent the limitations found with RCT. The method, which requires a sample that adopts a large number of orientations on the support, combines strengths of both RCT and Angular Reconstitution: like the former, it generates reconstructions for every characteristic view present in the sample regardless of how they relate to each other molecularly and, like the latter, the reconstructions are fully sampled and thus free of artifacts. These two properties make OTR particularly amenable to automation, as user intervention is potentially unnecessary until after initial reconstructions have been refined to higher resolution. Techniques capable of this type of automation will become increasingly necessary as we focus our attention on complex samples containing multiple molecular species and conformations.

In this article, I review the principles and applications of OTR. I begin with a more in-depth discussion of the geometry behind the method and its similarities and differences with RCT. I follow this with a description of the steps required to generate reconstructions using OTR. Throughout this description I focus mainly on those aspects unique to the application of OTR, mentioning (but not describing) those that are more general to all reconstruction approaches. I illustrate the different steps, whenever possible, using our current data. I conclude by comparing reconstructions of the same macromolecule obtained with RCT and OTR to highlight the properties of the two methods.

# 2. Orthogonal Tilt Reconstruction: Principles and Application

## 2.1. Random Conical Tilt and the "missing cone" problem

The RCT and OTR methods rely on the same geometric principle. If two views related by a known angle are available for each molecule in the sample, one set of images can be used to sort out the entire data set into characteristic views through alignment and classification and their "tilt mates," having "fanned out" as a result of the alignment rotations, sample the 3D structure of the molecule (see Fig. 9.1). This sampling is a consequence of the Central Section Theorem, which states that the Fourier transform of a 2D projection of a 3D structure is equivalent to a central section through the 3D Fourier transform of that structure orthogonal to the projection direction (Frank, 1996). It follows that the closer the angle between the views used for alignment and reconstruction is to 90°, the more complete the sampling of the structure will be. A useful visual analogy is that of a coin being spun on a tabletop, where the coin corresponds to a projection and therefore a central section through the 3D Fourier transform of a molecule. Initially, as the coin starts to spin, it sweeps space fully. As it loses momentum and begins to tilt, the volume not being sampled by the coin grows. In this example, the position of the coin at a given point in time (its rotation around an axis perpendicular to the table) would be the result of aligning and classifying its "tilt mate" (some imaginary coin lying flat on the table) and the angle of the coin relative to the table at any point during the spin is the angle between the two views recorded from the sample.

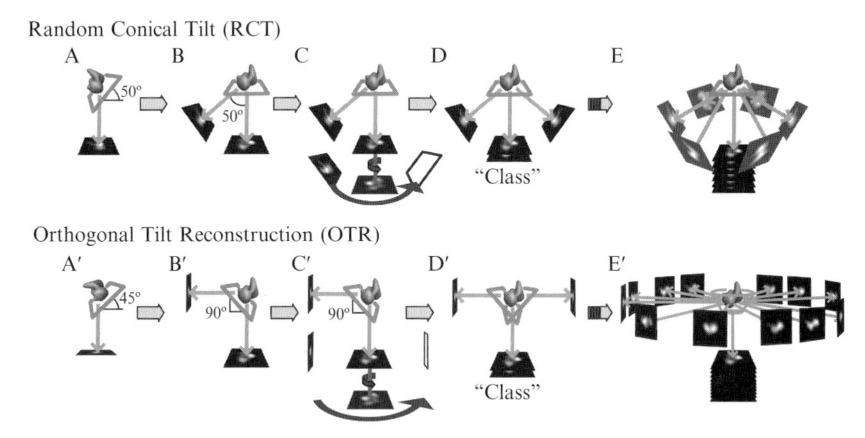


**Figure 9.1** Geometry of the Random Conical Tilt (RCT) and Orthogonal Tilt Reconstruction (OTR) methods. The basic steps in the RCT (top) and OTR (bottom) methods are illustrated in this figure. Throughout the figure, the green object represents a molecule; the orange rhomboid the support; blue arrows the direction of projection (imaging); and red arrows the in-plane rotations applied during alignment and classification. (A and A′) An image is collected with the sampled tilted, either to 50–60° (RCT) or 45° (OTR). (B and B′) A second image is recorded from the same area after the sample has been either returned to 0° (RCT) or tilted to −45° (i.e., 45° in the opposite direction) (OTR). The images collected in A and B (or A′ and B′) constitute a "tilt pair." (C and C′) The 0° (RCT) or −45° (OTR) image from a second pair of images can be aligned to the first one if the images represent the same (but rotated) view of the molecule. The in-plane rotation applied to the image being aligned determines the new spatial location of its tilt mate (represented by the empty black frames). (D and D′) The two images are now members of the same "class" (i.e., they represent the same view and are aligned to each other). As indicated in the previous step, the alignment results in their tilt mates "fanning" out in a cone (RCT) or equator (OTR) around the molecule that gave rise to them. This step also illustrates one of the main differences between RCT and OTR: the two molecules giving rise to the two images in the "class" have the same orientation in RCT (although different in-plane rotations on the support) (D) while their orientations are entirely different in the case of OTR (D′). (E and E′) Once there are enough images in a class to fully sample the desired structure, it can be reconstructed. The orange truncated cone shown in E′ emphasizes the fact that every image in this arrangement has originated from a molecule adopting a different orientation on the support. (See Color Insert.)

The ideal situation is one where this angle is 90° (the beginning of the spin) as this is the only geometry that leads to a full sampling of the 3D Fourier Transform of the structure.

The RCT method, originally designed for samples adopting a preferred orientation on the support, involves the following basic steps (Radermacher *et al.*, 1987):

1. Images are collected at 0° and at high tilt, typically in the 50–60° range (Fig. 9.1A and B). The tilted images, to be used for reconstruction, are

collected first to limit the dose to which they are exposed (Fig. 9.1A). The 0° images, used for alignment and classification, will have been exposed twice (Fig. 9.1B).

2. The 0° images are sorted into groups representing characteristic views of the molecule through cycles of alignment and classification (Fig. 9.1C and D). The in-plane rotation angles required to align images that show the same view determine the relative positions of the corresponding high-tilt images. The alignment and classification also generates the "class" files, indicating which images were grouped together.
3. The tilt geometry parameters must be determined. These are the tilt angle (e.g., 50° in Fig. 9.1A) and the tilt axis angle (the actual position of the tilt axis relative to the coordinate system). These are typically calculated by the software package used for extracting the tilt pairs from the micrographs.
4. The in-plane rotation angles (2) and tilt geometry parameters (3) are combined into a set of Euler angles describing the relative positions of the tilted images (Fig. 9.1E).
5. 3D Reconstructions are calculated for each "class" obtained in (2) (Fig. 9.1E). The "class" files are used to select the appropriate tilted images and their corresponding Euler angles for each reconstruction.

The fact that the tilted images in RCT are not collected at 90° (impossible to do in the microscope) means that they sample Fourier space incompletely (a spinning coin that is tilting does not sweep space fully). The volume that fails to be sampled is known as the "Missing Cone" due to its shape in Fourier space. In the coin analogy, the region of space not swept by the coin once it begins to tilt has a conical shape as well (remember that the coin represents a central section through the 3D Fourier transform of the molecule). The missing data are an important challenge in RCT as they lead to artifacts in the final reconstruction (Fig. 9.2). Typically, a loss of internal detail and an elongation in a direction parallel to the incident electron beam are observed (Fig. 9.2B). The standard method for addressing this problem is to combine independent reconstructions corresponding to molecules having adopted different orientations on the support. Because the missing data are a function of the orientation of the molecules relative to the electron beam, two reconstructions representing different orientations will be missing information in different regions of Fourier space. In order to fully fill each other's Missing Cones, the two reconstructions must be related by a "tilt" angle (around an axis perpendicular to the electron beam) equal to or larger than $[(90 - \Theta)^\circ \times 2)]$, where $\Theta$ is the tilt angle used during data collection (e.g., if the tilted data were collected at 50° the two reconstructions must be related by 80°). This solution, while easy in principle, can be far from trivial in practice. The deformations resulting from the Missing Cone, combined with the noisy nature of the individual reconstructions can

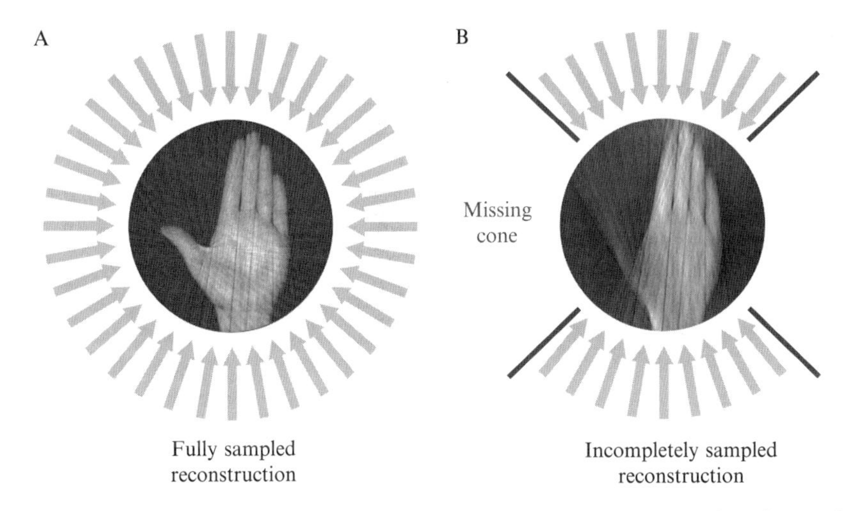


**Figure 9.2** The "Missing Cone": Incomplete sampling of a structure leads to distorted reconstructions. This figure shows a two-dimensional representation of the "Missing Cone" problem. As can be seen in Fig. 9.1E, views beyond a certain angle (past which useful data cannot be collected in the microscope) are not present in a RCT reconstruction. As a consequence, the structure is incompletely sampled in Fourier space, leading to artifacts in the reconstruction. (A) A hand was reconstructed using evenly spaced projections of itself (denoted by the arrows). (B) A subset of projections was removed beyond a certain angle (indicated by the blue lines) and the hand was reconstructed again. The area (or volume in a 3D reconstruction) lacking information is known as the "Missing Cone" due to its shape in Fourier space (see text). It can be seen that the incompletely sampled hand shows an elongation in the vertical direction (corresponding to the beam axis), a loss of internal detail and the partial disappearance of the thumb. (For interpretation of the references to color in this figure legend, the reader is referred to the Web version of this chapter.)

make it difficult to determine whether any two of them truly represent the same molecular species and can therefore be merged. This can be further complicated by the presence of heterogeneity (biochemical and/or conformational) in the sample (Fig. 9.3). An additional challenge arises from the fact that cross-correlation coefficients are typically used to score the goodness of the match between any two reconstructions during the rotational search used for their alignment. Because one is looking for volumes that will fill each other's missing data, these will by definition have the least amount of overlap in Fourier space, resulting in a lower cross-correlation coefficient. The desired answers can be, and often are, found but the process requires a significant level of expertise and interaction with the data.

The OTR method bypasses the Missing Cone problem entirely by taking advantage of the realization that obtaining the ideal 90° (orthogonal) relationship between the two views of the sample does not require that one of them be collected at 0°. Since it is only the angle between the views we

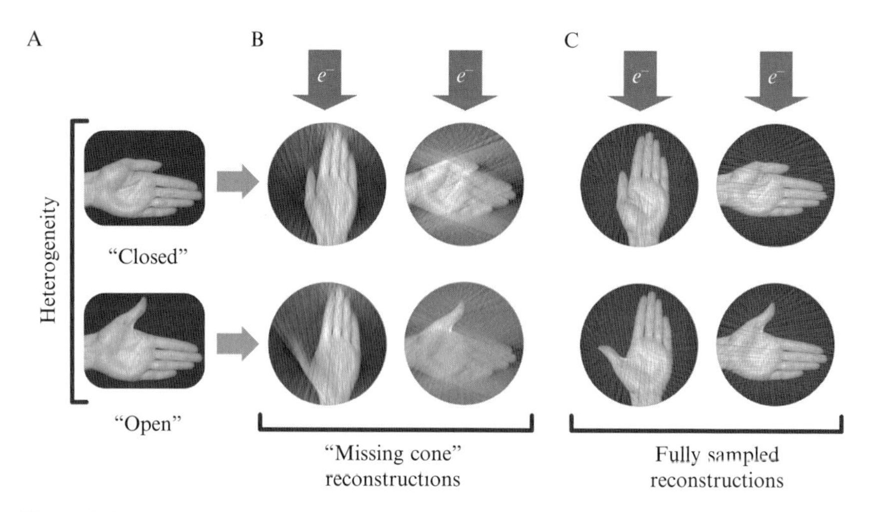


**Figure 9.3** Incomplete sampling is particularly challenging in the presence of heterogeneity. This figure illustrates the difficulties in solving the "Missing Cone" problem by merging reconstructions when heterogeneity is present in the sample. (A) The hypothetical sample contains two conformations of the hand, "closed" and "open." (B) Two reconstructions, for hands adopting two different orientations relative to the electron beam (arrow on top), were generated for each conformation. It can be seen that the "Missing Cone" results in structures with the same orientation being more similar to each other than those representing the same conformation. (C) The fully sampled reconstructions of the same orientations for the two conformations are shown.

are interested in, a 90° angle can easily be achieved by collecting data at −45° and +45° (Fig. 9.1A′–E′). Any combination resulting in 90° will work but the ±45° option is often the most practical (discussed in Section 2.4).

A fundamental difference between RCT and OTR lies in the images used for alignment and classification (see Fig. 9.1). The alignments applied to the particles during this step consist of $X$, $Y$ shifts and an in-plane rotation (about an axis perpendicular to the image plane). In RCT, where 0° images are used at this stage, these rotations also correspond to rotations of the molecules about an axis perpendicular to the support (Fig. 9.1B and C). Therefore, all particles within a given class come from molecules that adopted the same orientation on the support, differing only on their trivial in-plane rotation (Fig. 9.1E). Consequently, a 3D reconstruction (albeit with a Missing Cone) can be obtained from a sample showing even a single orientation, a powerful feature of RCT. In OTR, where the images being aligned and classified come from a tilted sample (e.g., −45° images), the in-plane rotations applied during alignment do not correspond to rotations of the molecules about an axis perpendicular to the support. This is due to the fact that the axes for the in-plane rotation of the molecule on the support and for the rotational alignment of images are no longer parallel (Fig. 9.1B′ and C′). Therefore, every particle in a class with a unique in-plane rotation angle has a different orientation on the support (Fig. 9.1E′). Hence one of the most important requirements for OTR: the sample must adopt a large number of orientations. In principle, a molecule adopting orientations that correspond to a precession with a 45° angle would be sufficient to obtain a fully sampled (i.e., no Missing Cone) reconstruction. I discuss below an approach to determining whether a given sample appears to satisfy this requirement and is therefore amenable to OTR. While preferred orientations may often be a limitation with negatively stained samples, we expect OTR to be far more generally applicable with cryo-EM samples.

Another important difference between OTR and RCT stems from the fact that OTR deals exclusively with tilted images and, as outlined above, particles assigned to a given class represent molecules with different orientations on the support. A consequence of this is that the −45° and +45° particles are interchangeable. If we use the −45° set for alignment and classification, the +45° particles are effectively their +90° tilt mates. Conversely, if the +45° particles are used to generate the classes, the −45° particles become their −90° tilt mates. It follows that there is no need to use only one half of the data for alignment and classification, saving the other half for reconstruction. Although this is the situation we simulated when we first introduced the method (Leschziner and Nogales, 2006), we have since begun pooling the +45° and −45° particles for alignment and classification, treating them as a single data set (Leschziner *et al.*, 2007). The data presented here to illustrate different aspects of OTR have been treated in this way as well. The only potential disadvantage to this approach is that the images coming from the second tilt have been subjected to two exposures, the reason behind the fact that tilted images are collected first in RCT (Radermacher *et al.*, 1987). While this will result in some loss of resolution, this is likely beyond what is expected from initial models and is far outweighed by the advantage of doubling the size of the data set.

## 2.2. The test sample: The yeast exosome

When we first introduced the OTR method, we illustrated its features using synthetic data (Leschziner and Nogales, 2006). While this approach allowed us to assess the performance of the technique quantitatively by comparing our results with the known structure that gave rise to the synthetic data (Leschziner and Nogales, 2006), it bypassed many of the difficulties associated with real samples. Subsequently, we applied OTR to the reconstruction of the ATP-dependent chromatin remodeling complex RSC from the yeast *S. cerevisiae* in negative stain (Leschziner *et al.*, 2007). However, a few single-particle electron microscopy reconstructions of this complex have been published to date and they disagree with each other (Asturias *et al.*, 2002; Chaban *et al.*, 2008; Leschziner *et al.*, 2007; Skiniotis *et al.*, 2007). The moderate resolution of all these reconstructions and the absence of high-resolution structures of any significant components of the RSC complex make it difficult, if not impossible, to validate any of them at this point. We were therefore still interested in testing the performance of OTR with a biological sample of known structure.

Our current test sample is the yeast Rrp44-exosome complex, an RNA-processing assembly from *S. cerevisiae*. We chose this 398 kDa macromolecule for several reasons: (1) The structure of both the core exosome and its complex with the Rrp44 subunit have been solved by the RCT method (Wang *et al.*, 2007); (2) A high-resolution structure of the yeast core exosome is available (Liu *et al.*, 2006); and (3) We had access to the same grid used for the determination of the RCT structure of the exosome (Wang *et al.*, 2007) (courtesy of Hongwei Wang, Yale University), thus eliminating sample preparation as a variable when comparing reconstructions obtained with the OTR and RCT methods.

I use data from our current work on the exosome to illustrate the different steps along a reconstruction using OTR. I focus on those aspects that are unique to OTR, bypassing those that are general to any reconstruction process.

## 2.3. Sample preparation

Although we are currently working on implementing OTR with cryo-EM samples, we have so far tested it only on negatively stained samples. I therefore focus exclusively on the latter.

The main goals when preparing negatively stained samples to be reconstructed by OTR are (1) to minimize flattening and (2) to maximize stain thickness. While these may be desirable properties for any sample, they are particularly important in OTR because their absence (i.e., flattened, thinly stained samples) manifests itself most severely in data collected from a tilted sample.

As is discussed below, it seems possible to detect and eliminate particles affected by flattening during alignment and classification of tilted data. However, if a sample is severely affected by flattening, it becomes equivalent to one with preferred orientations and OTR is no longer applicable (discussed in Section 2.9). Similarly, samples embedded in a shallow layer of stain will exhibit stain pooling on one side when imaged at a tilt (Fig. 9.4C and D). Because the location of the stain pool is a function of the tilt geometry, particles with different orientations on the support that would otherwise give identical projections become different due to the staining. This will negatively affect alignment and classification.

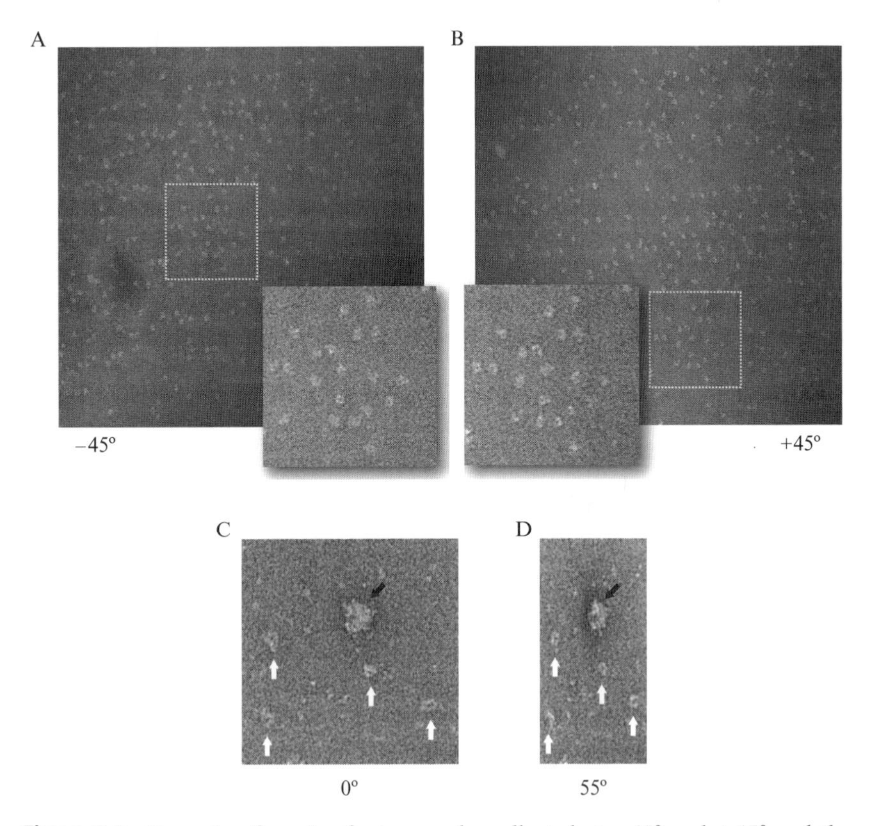


**Figure 9.4** Example of a pair of micrographs collected at −45° and +45° and the effect of stain thickness. (A and B) A representative pair of micrographs of the yeast Rrp44-exosome complex collected at −45° (A) and +45° (B) for OTR processing. The images were recorded using automated OTR data collection as implemented in Leginon (Yoshioka *et al.*, 2007) at NRAMM, The Scripps Research Institute. The insets show enlarged versions of corresponding small areas in the two micrographs (indicated by the stippled squares) to highlight the absence of stain pooling around the particles, one of the criteria we use to select micrographs for further processing. The fact that particles are not elongated along the tilt axis (running vertically on the page) is also an indication that the sample is not severely flattened. Due to the fact that the two micrographs were collected at the same absolute angle (but different sign), the positions of the tilt mates are virtually identical. (C and D) This tilt pair illustrates the effect of stain thickness on the appearance of "pooling" around particles. The same region was cut out from micrographs collected at 0° (C) and 55° (D). The sample imaged contains both yeast dynein motor heads (white arrows) and a much larger yeast ribosome (black arrow). Due to their size difference and the fact that the dynein motor head (ring-shaped) tends to lie flat on the grid, dynein molecules are fully embedded in the stain while a stain meniscus surrounds the ribosome. This incomplete embedding results in the ribosome showing asymmetric "pooling" of stain in the tilted micrograph (see darker rim on the left side in D). This artifact is not seen for either the dynein molecules in D or the exosome molecules in A and B.

We are able to routinely satisfy these requirements using a staining protocol based on the deep staining described by Ohi *et al.* (2004). Briefly, our protocol consists of the following steps:

1. Approximately, 50–70 $\mu$L of freshly prepared 2% uranyl formate is drawn into a pipette tip, followed by a small air gap and then about 5 $\mu$L of sample.
2. A freshly glow-discharged grid (home made holey carbon or Quantifoil® with a layer of thin continuous carbon) held by tweezers is set up in a stand so it is positioned at a tilt.
3. The sample is applied to the grid in a continuous motion; the grid's tilt helps in draining the stain as it is applied immediately after the sample.
4. The grid is rinsed for 10 s (without blotting) in four consecutive drops of about 70 $\mu$L of 2% uranyl formate.
5. The grid is allowed to sit in stain for another minute, either on a drop or with a droplet of stain on it.
6. If the "sandwich" method is to be used, a small square of thin carbon is floated on a pool of 2% uranyl formate in a well and picked up from underneath with the grid (again, without blotting it before going into the stain pool).
7. Whether sandwiching has been used or not, the grid is now carefully blotted from the side, stopping while a thin layer of liquid is still clearly visible on the grid.
8. The grid is allowed to air-dry.

We have found that the "rapid" stain (i.e., applying the sample and the stain, separated by an air gap, in one continuous motion) results in particles that are both more homogeneous and display a larger number of orientations. Homogeneity also seems to be improved by blotting only once at the end of the staining. We often include a small percentage of trehalose (3–6%) in our samples, as this seems to increase the depth of the stain as well. For a detailed discussion on staining protocols, see Ohi *et al.* (2004).

We use two criteria to determine, visually, whether a sample appears to satisfy the flattening and stain depth requirements. We collect a few tilted images at $\pm 45°$ and look for shadows around the particles (indicating thin staining) or a general elongated appearance in the direction of the tilt axis (indicating flattening). Only samples that show neither are used for data collection. We use the same criteria to select micrographs after data collection. Figure 9.4 shows an example of a $\pm 45°$ tilt pair with the desired characteristics: deep staining and no evidence of severe flattening.

## 2.4. Data collection

In RCT and OTR, two images must be collected from each field of molecules at two different angles in order to provide the necessary information for 3D reconstruction. In OTR, the goal is to obtain pairs of images related by a 90° angle. Although many combinations for the first and second tilt angles can satisfy this requirement, there are certain advantages to using +45° and −45°. First, the overlap in the number of particles that are present in both micrographs is maximized, as the image compression due to the tilt is the same in both. Second, this geometry keeps the tilt angle for both micrographs in a pair to a minimum, helping to lessen the effect that stain-related artifacts may have during alignment and classification when negatively stained samples are used (discussed below). Third, because all particles come from micrographs having the same tilt angle, the effects of stain and tilt on their appearance will also be similar; the particles coming from the −45° and +45° micrographs can be aligned and classified together without the risk of tilt angle-related artifacts driving the process.

We have collected OTR data both manually and automatically. The data used for our reconstruction of the ATP-dependent chromatin remodeling complex RSC were collected manually (Leschziner *et al.*, 2007). The data for the Rrp44-exosome reconstruction shown here were collected at NRAMM (The Scripps Research Institute) in an automated manner using the OTR option in the Leginon software package (Yoshioka *et al.*, 2007).

For manual data collection, we have found it easier to collect all the images from a square at a given tilt before moving to the second tilt and collecting their tilt mates. This minimizes drift due to constant tilting of the goniometer and reduces data collection time. Our strategy typically involves the following steps: (1) Select good squares at very low magnification and store their coordinates, (2) Go to the first square and tilt (e.g., to −45°), (3) Take an image at a magnification that includes the entire square, (4) Print the image and mark the target holes, (5) Collect the full magnification images from all the holes, (6) Tilt to the second angle (+45°), and (7) Collect the tilt mates. Because one must be able to identify the target holes during data collection, we have found it convenient to use homemade holey carbon grids (with a continuous carbon support) as they provide unique and easily identifiable patterns.

For a thorough description of automated data collection with RCT and OTR geometries using the Leginon software package, see Yoshioka *et al.* (2007). The Rrp44-exosome data shown in this article were collected from samples that had been prepared on Quantifoil® grids.

## 2.5. Selection of particles

Of the few programs available for the selection of particles from micrograph tilt pairs (Frank *et al.*, 1996; Scheres *et al.*, 2008; Voss *et al.*, 2009) only TiltPicker (Voss *et al.*, 2009) is capable of handling both RCT and OTR geometries. Due to the underlying assumption of RCT geometry, the others cannot properly process OTR data. Whereas small deviations from the 0° assumed for the untilted micrograph in RCT result in negligible compression of the image in the direction perpendicular to the tilt axis, this effect is far more severe with 45° images. (For example, an image collected at an actual 5° shows a ~0.4% compression relative to a 0° image, whereas one collected at 50° is compressed by ~9.1% relative to the 45° image.)

We have semiautomated the picking of particle pairs from OTR micrographs: the particles from one micrograph in each pair are selected manually and their corresponding tilt mates are obtained automatically from the second micrograph. TiltPicker (Voss *et al.*, 2009) is now capable of extracting OTR tilt pairs in a fully automated way but we have not tested it with our data yet. We typically do the initial selection from one micrograph (either the +45° or −45°) using EMAN's Boxer (Tang *et al.*, 2007), either in interactive or semiautomated (Autoboxer) mode, but we have used SPIDER's WEB (Frank *et al.*, 1996) as well. The coordinates from the initial selection are then used to automatically find their tilt mates using a series of SPIDER scripts that perform the following steps:

1. An initial search (performed with the operation AP SH) over in-plane rotations (tilt axis angles) and "tilts"—manifested as cosine stretches or compressions of the micrographs—finds the transformation that results in the best match between the two micrographs (+45° or −45°) in a tilt pair. These transformations are useful only in the context of these scripts, as they do not represent a real geometric relationship between the two micrographs. The parameters describing the tilt geometry must be obtained separately (see Section 2.7).
2. The transformations are used to determine the overlap between the two micrographs in a tilt pair (i.e., what particles are present in both micrographs); particles that were originally selected but would not have mates are ignored in subsequent processing.
3. The coordinates of the originally selected particles are used to window out relatively large areas—several times the size of the box that will be used for processing—centered on each particle.
4. The transformation calculated in (1) is used to generate a "guess" for where the tilt mate of the particle windowed out in (3) will be located.
5. An area of the same size as the one in (3) is windowed out centered on the "guess" coordinates calculated in (4).
6. A cross-correlation is calculated between the two windows and the position of the peak is used to refine the initial guess and provide the final tilt mate coordinates. These coordinates are used to window out the tilt mates.

We have tested these scripts with a few different data sets and the success rate (measured as the fraction of tilt mates that are both correct and centered in their windows) is around 90%. (Scripts are available upon request.)

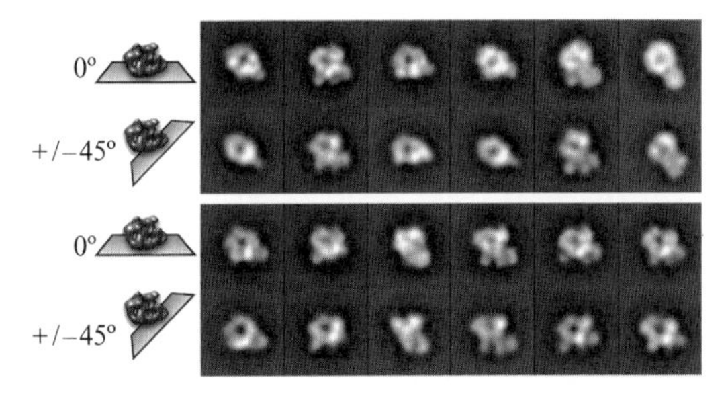


**Figure 9.5** A qualitative assessment of whether a sample adopts enough orientations to be amenable to OTR. Experimental class averages obtained from either 0° data for RCT (first and third rows) or ±45° data for OTR (second and fourth rows) were aligned to each other and the best matching pairs determined from the cross-correlation coefficients. Aligned pairs are shown in this figure. The RCT data are from Wang *et al.* (2007) and were a courtesy of Hongwei Wang (Yale University).

## 2.6. Does the sample adopt enough orientations to satisfy OTR requirements?

As discussed in Section 2.1, one of the main requirements for the application of the OTR method is the adoption by the sample of a large number of orientations on the support. Although it is simple to determine with alignment and classification whether a sample adopts one or a few preferred orientations, it is not possible to unambiguously establish that enough orientations are present without *a priori* knowledge about the structure.

We have addressed this limitation using the approach illustrated in Fig. 9.5. The basic idea behind this approach is that a sample adopting random orientations on the support will present the same overall set of views regardless of whether or not it is tilted during data collection. Therefore, the set of class averages obtained from untilted data should be matched by that obtained from tilted data. This ideal scenario is unlikely to ever be true: random orientations are seldom, if ever, observed and the lower quality of tilted data (due to staining artifacts or increased thickness in the stain) will affect the class averages. However, one can still expect to observe a good match between class averages obtained from tilted and untilted data, provided the sample adopts a large number of orientations. Conversely, failure to observe such a match can be taken as an indication that the sample is not amenable to OTR.

The untilted data required for this test are almost always available at the beginning of a project when 0° images are collected for an initial characterization of the sample. Once the 0° and ±45° particles have been separately aligned and classified an alignment is performed between the two sets of class averages to find the best-matching pairs. Although we usually judge the quality of the matches visually (Fig. 9.5), more quantitative comparisons between class averages (such as Fourier Ring Correlation (FRC); Harauz and van Heel, 1986) could be used.

## 2.7. Determination of tilt geometry parameters

As I discussed in Section 2.5, it is not possible to obtain the tilt geometry parameters directly from particle-picking software. The parameters we obtain for the transformation between the two micrographs in a tilt pair as part of our automated particle picking are simply a tool to make this automation possible but they do not bear any relationship with the real tilt geometry. The same is true for TiltPicker, which automates the extraction of the molecular images but does not calculate the tilt geometry parameters for OTR data (Voss *et al.*, 2009).

We obtain the tilt angle, tilt axis angle (the position of the tilt axis relative to the coordinate system), and defocus (at the center) for each micrograph using the program CTFTILT (Mindell and Grigorieff, 2003). An additional advantage of determining the parameters using CTFTILT is that the program outputs both degree and sign of the tilt, preventing bookkeeping errors if the user has switched the order in which the two tilts are collected. Of course, this is not an issue when automated data collection is used (as was the case for the exosome data shown here, obtained using Leginon) as the database stores this information.

We store the output from CTFTILT in a SPIDER-format file to be used in subsequent processing.

Although we have not implemented it, one could use the output from CTFTILT to constrain the search over tilt angles and tilt axis angles used to find the tilt mates in our semiautomated particle picking (see Section 2.5)

## 2.8. CTF correction

Images collected from a tilted sample have, by definition, a defocus gradient. This is quite significant at 45°; an image collected at a magnification of 50,000 on a 4 × 4 k CCD camera will have a defocus gradient of 1.23 $\mu$m. This is clearly more severe on film or larger CCD cameras. Since OTR relies on the alignment and classification of tilted data, it is to some extent affected by the wide range of defoci present in the data. In order to minimize this effect, we always correct the CTF of our particles by multiplying by the calculated CTF.

Our CTF correction is done particle-by-particle using the defocus value and tilt geometry parameters obtained from CTFTILT (Mindell and Grigorieff, 2003). The coordinates of each particle are used to calculate the defocus at that particular location and that defocus is then used to generate the estimated CTF for correction.

## 2.9. Alignment and classification

Alignment and classification of OTR data are similar to those used with other reconstruction methods. The data used in this article were aligned and classified using multivariate statistical analysis and hierarchical ascendant classification as implemented in the IMAGIC software suite (van Heel *et al.*, 1996). There are two aspects specific to OTR: (1) the utilization of alignment information to detect and eliminate classes that are likely to be enriched in flattened particles and/or particles with preferred orientations and (2) the fact that the $-45°$ and $+45°$ data can be combined rather than be treated as two separate sets used for alignment/classification and reconstruction (discussed in Section 2.1).

As I discussed in Section 2.1, OTR requires samples that adopt a large number of orientations on the support. This was illustrated in Fig. 9.1E′, where each image shown in the equator surrounding the structure arises from a molecule adopting a different orientation on the grid. If the sample adopted one or a limited number of orientations, all the equatorial images used for reconstruction would be located at a single point on this equator, on a line perpendicular to the tilt axis (the left image shown in Fig. 9.1B′). This situation can result from a sample adopting preferred orientations (Fig. 9.6C) and/or from flattening of the sample (Fig. 9.6B). Even with samples where flattening has been minimized and the number of orientations maximized (see Section 2.3 above), one is likely to see certain orientation bias and a distribution in the severity of flattening in different parts of the sample. Whenever these factors result in unique species, a situation like that in Fig. 9.1B′ will arise. The noisy nature of the data, however, will lead to the equatorial images spreading out around the line perpendicular to the tilt axis (i.e., not all images will coincide exactly with that shown in Fig. 9.1B′ but will rather lie in its vicinity). This unusual distribution is immediately apparent in a plot of the in-plane rotation angles from alignment and classification (Fig. 9.6D) and we have observed these biased angular distributions with both real (Leschziner *et al.*, 2007) and synthetic data (Leschziner and Nogales, unpublished). We take advantage of such plots to detect and remove classes that are likely to contain preferred orientations and/or flattened particles. Regardless of their origin, these classes would not lead to useful reconstructions given their incomplete sampling of Fourier space. Whether this potential selection against flattened particles is responsible for the fact that OTR reconstructions appear relatively unaffected by flattening even when generated from the same grid that gave rise to apparently flattened RCT reconstructions (see Section 2.11 below and Fig. 9.7) is something we have not established yet.

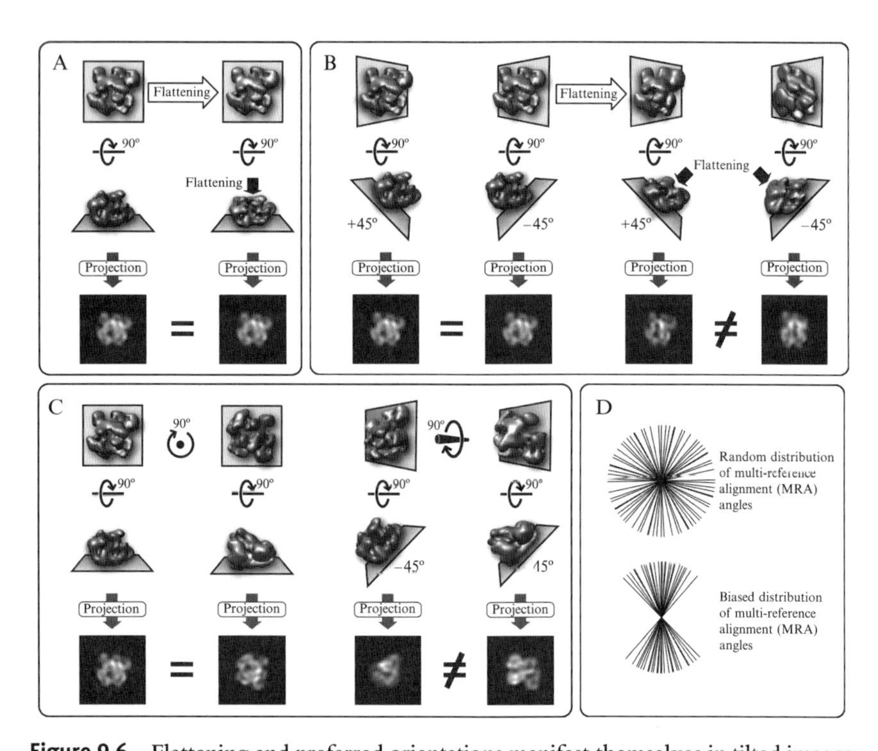


**Figure 9.6** Flattening and preferred orientations manifest themselves in tilted images. In panels A–C, a molecule (the exosome) is first shown looking down towards the support ("top view," top row) and then tilted 90° ("side view," middle row). The bottom row shows the resulting projection in an orientation equivalent to that of the "top view." (A) Flattening is mostly not apparent when images are collected at 0°. The molecule on the right has been flattened by 1/3 along the $Z$-axis (parallel to the projection direction), an effect that is not seen in the projection. (B) This situation changes when images are collected from tilted samples. In the absence of flattening, the same projection can be obtained from molecules adopting different orientations on the support (first two columns); this is what makes OTR possible. If flattening occurs, however, the two projections are now different because the direction of flattening and projection are no longer parallel (last two columns). Therefore, flattening that was mostly unnoticed in an untilted sample (A) becomes obvious in a tilted one (B). (C) Samples with preferred orientations are not amenable to reconstruction by OTR. In an untilted sample, the random in-plane rotations of the molecules on the support result in trivial in-plane rotations between projections that are otherwise identical (first two columns). Upon tilting, however, an in-plane rotation of the molecule results in projections that are no longer the same (second two columns). It can be seen in the last two columns in C that the projections shown can be obtained with one and only one orientation of the molecule on the support. It therefore becomes impossible to obtain a class containing a full set of equatorial views of the molecule as shown in Fig. 9.1E′. A comparison of panels B and C shows that flattening is, in a sense, a particular case of preferred orientation in that flattening turns each orientation into a unique species. (D) Representative plots of in-plane rotation angles from alignment and classification showing classes with random (top) and biased (bottom) distributions of angles. Each line in the plots represents the in-plane rotation angle (from multireference alignment) for a given particle in a class. Panels B and C show why flattening and/or preferred orientations will give rise to biased distributions of in-plane rotation angles.

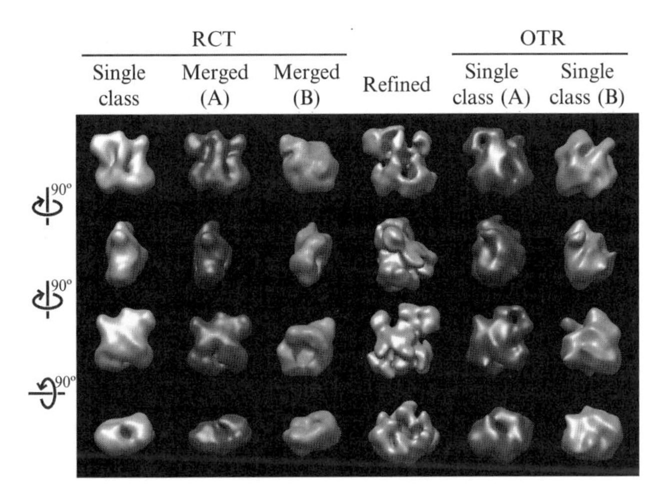


**Figure 9.7** Comparison between RCT and OTR initial reconstructions: surface rendering. This figure shows different views of the two initial RCT reconstructions reported by Wang *et al.* (2007) (dark (A) and medium (B) green) as well as one of the single-class reconstructions (light green) merged into volume A (H. Wang, personal communication); two of our OTR initial models (red (A) and orange (B)) and the final, refined exosome structure from Wang *et al.* (2007) (gray). The "Merged" RCT models are the result of combining 6 (Merged A) and 4 (Merged B) single-class volumes and contain a total of 701 and 633 particles, respectively. The OTR models are both single-class volumes and were generated from 222 (Single Class A) and 240 (Single Class B) particles. All four initial volumes were filtered to 20 Å. The final exosome structure (gray) (EMDB entry EMD-1438) was obtained by refining the Merged A volume against approximately 4000 0° images of the complex in negative stain and is shown at 19 Å as published by Wang *et al.* (2007). The arrows on the left indicate the relationships among the four views shown for each of the volumes. The volumes were displayed using Chimera (Pettersen *et al.*, 2004). (See Color Insert.)

## 2.10. Reconstruction

Once alignment and classification have been completed, the final in-plane rotations obtained from the alignment must be combined with the tilt geometry parameters (see Section 2.7) to generate the Euler angular file for reconstruction. Particular attention must be paid to any conversions required to account for the different conventions adopted by the programs used during data processing. In our case, we must combine rotation angles from alignment with IMAGIC (van Heel *et al.*, 1996) with tilt geometry parameters obtained with CTFTILT (Mindell and Grigorieff, 2003) to generate an Euler angular file that can be used for reconstruction in SPIDER (Frank *et al.*, 1996). The following formulas are used to calculate the final Euler angles ($\Phi$, $\Theta$, $\Psi$), which should be written out in that order in the angular file to be used for reconstruction with the SPIDER commands BP RP or BP 32F:

$$\Phi = \gamma_{\text{“untilted”}} - 90^{\circ} - \text{MRA}$$
$$\Theta = \Theta_{\text{“tilted”}} - \Theta_{\text{“untilted”}}$$
$$\Psi = 90^{\circ} - \gamma_{\text{“tilted”}}$$

where $\gamma_{\text{“untilted”}}$ is the tilt axis angle (i.e., the in-plane rotation of the actual tilt axis relative to the coordinate system) obtained from CTFTILT for the “untilted” micrograph; “MRA” is the in-plane rotation obtained from alignment and classification; $\Theta_{\text{“tilted”}}$ and $\Theta_{\text{“untilted”}}$ are the tilt angles from the “tilted” and “untilted” micrographs, respectively, from CTFTILT and $\gamma_{\text{“tilted”}}$ is the tilt axis angle from CTFTILT for the “tilted” micrograph. The “90°” terms correct for the different conventions used by SPIDER and CTFTILT. The “MRA” angle is subtracted in this case to calculate $\Phi$ to account for the different rotation conventions used by SPIDER and IMAGIC. If the alignment and classification were performed in SPIDER (or any other package with the same convention—a positive angle being a counterclockwise in-plane rotation), the “MRA” angle would be added instead.

It is important to keep in mind that in OTR, unlike RCT, "tilted" and “untilted” are relative terms. We use “untilted” to refer to the particles in a class and “tilted” to refer to their tilt mates, which are used for reconstruction. Because the $+45°$ and $-45°$ particles can all be used for alignment and classification (see Section 2.1), every particle will get to play both roles. For example, if a $-45°$ particle A is assigned to class X and its 45° tilt mate B is used for reconstruction, we would say that A is “untilted” and B is “tilted”. B is the tilt mate of A with an effective tilt angle ($\Theta$) of 90°. At the same time, particle B will have been assigned to a different class (Y) with A now being its tilt mate. In this context, B is the “untilted” particle and A is its “tilted” mate with a tilt angle ($\Theta$) of $-90°$ (note the reversal in sign when the two particles switch roles).

Once the Euler angular file has been created, reconstruction of OTR single-class volumes proceeds in the same manner used for RCT volumes. The main difference, introduced in Section 2.1, lies in the fact that in OTR the same set of particles used for alignment and classification can be used for reconstruction, while in RCT one switches from the untilted to the tilted set of particles. As a result of this, a typical OTR class would consist of particles coming from both $+45°$ and $-45°$ micrographs and the tilt ($\Theta$) angles used for reconstruction of that class volume would contain values around both $+90°$ and $-90°$.

We perform all data processing following alignment and classification using SPIDER (Frank *et al.*, 1996). Class volumes are generated by back-projection (using the commands BP RP or BP 32F). The translational parameters of the particles used for reconstruction are iteratively refined using projection matching.

## 2.11. Comparison between OTR and RCT reconstructions

In order to illustrate the different properties of OTR and RCT reconstructions, I will end with a direct comparison of volumes generated from Rrp44-exosome data with both methods. As I indicated in Section 2.2, the OTR data presented here were obtained from the same EM grid Wang *et al.* (2007) used for their RCT reconstruction of the yeast exosome, thus removing sample preparation as a potential source of variability. The main differences between the two data sets are the following: (1) While both were collected at 120 kV, the OTR data were collected on a Tecnai F20 microscope (NRMM, The Scripps Research Institute) and the RCT data set on a Tecnai G2 ($LaB_6$) instrument (Wang *et al.*, 2007), (2) The RCT data were collected on film (Wang *et al.*, 2007) and the OTR data on a $4 \times 4$ k CCD camera, (3) The OTR data set is significantly larger (12,692 tilt pairs vs. 3,872 tilt pairs for the RCT reconstruction), as would usually be the case due to the requirement for a large number of orientations. Data processing was otherwise comparable, with IMAGIC (van Heel *et al.*, 1996) used for alignment and classification and SPIDER (Frank *et al.*, 1996) for reconstruction and refinement in both cases.

The comparison presented here is between the two RCT reconstructions reported by Wang *et al.* (2007) and two of our best initial reconstructions. We selected these with a combination of visual inspection and by looking at the match between the 0° projection of the volume and the class average giving rise to it as well as the match between a set of pseudo-evenly spaced projections of the volume and the experimental class averages. It should be noted that the two RCT volumes are the result of merging either six or four different class volumes, while the OTR reconstructions correspond to single classes (see Figs. 9.7 and 9.8). This is a more relevant comparison than looking at single-class RCT reconstructions because the merging of volumes (to try to solve the Missing Cone problem) is an integral part of RCT. Conversely, we have not attempted to merge OTR reconstructions because the ability to use single-class volumes as initial models for refinement is one of the strengths of the method. Partly as a consequence of the difference in the type of volume used for the comparison—merged versus single-class—the RCT reconstructions contain a larger number of particles: they were generated from 701 (volume A in Figs. 9.7 and 9.8) and 633 (volume B in Figs. 9.7 and 9.8) particles, while the OTR ones are the result of 222 (volume A in Figs. 9.7 and 9.8) and 240 (volume B in Figs. 9.7 and 9.8) particles.

Figure 9.7 shows a few views of surface renderings of the two RCT and two OTR reconstructions along with the corresponding views of the final refined reconstruction from Wang *et al.* (2007). This figure illustrates the fact that single-class OTR reconstructions display more of the features observed in the refined structure than even the RCT merged reconstructions. Most

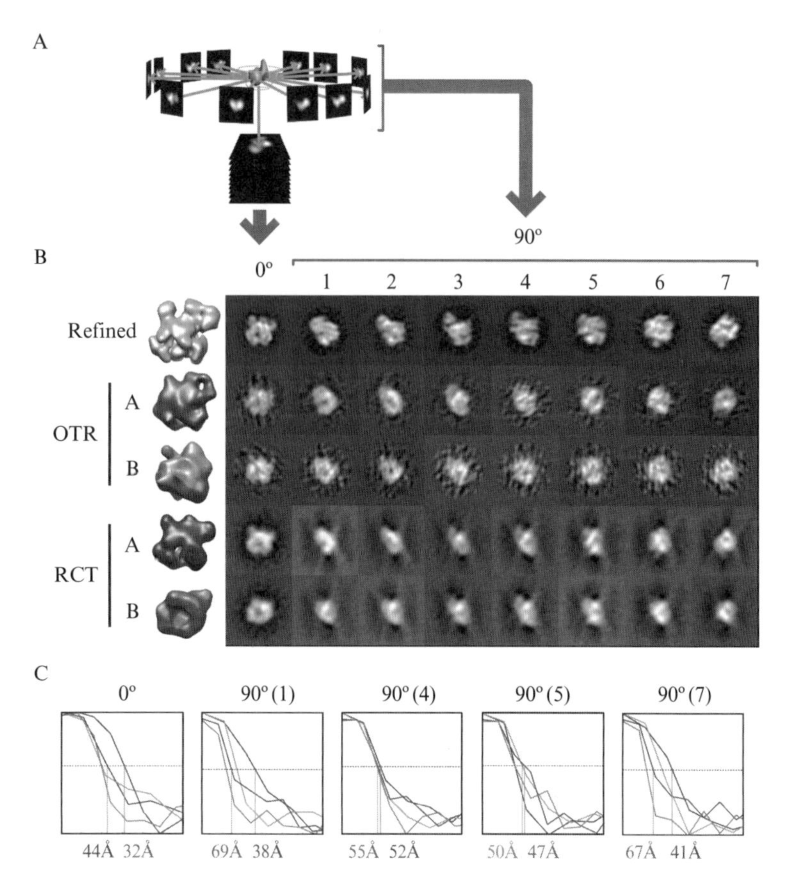


**Figure 9.8** Comparison between RCT and OTR initial reconstructions: projections. In this figure, we have used FRCs to compare projections from the RCT and OTR reconstructions with the corresponding projection from the refined exosome structure (EMDB entry EMD-1438) (Wang *et al.*, 2007). This measure can be taken as an indication of the amount of information about the final structure already present in the initial reconstructions. All four initial models (the same ones introduced in Fig. 9.7) were aligned to the refined structure (shown in grey). (A and B) projections were calculated at 0° and 90° (0° being the direction of the class average that gave rise to the volume and 90° being directions orthogonal to it). The volumes in (B) are shown in the orientation corresponding to the 0° projection. FRCs were calculated between the 0° and 90° projections and the corresponding ones from the refined structure; a few representative plots are shown in (C). The FRC plots are color-coded following the coloring of the volumes in (B) (and Fig. 9.7). I have indicated, below each plot, the resolution (in Å) corresponding to 0.5 FRC for the better-performing RCT and OTR reconstruction in each case. The RCT and OTR volumes were low-pass filtered to 20 Å and displayed using Chimera (Pettersen *et al.*, 2004). (For interpretation of the references to color in this figure legend, the reader is referred to the Web version of this chapter.)

strikingly, the OTR reconstructions do not show any sign of flattening despite the fact that all the data originated from the same grid. As I mentioned in Section 2.9, we expect that alignment and classification of tilted data and the removal of classes that show a biased distribution of rotation angles (see Fig. 9.6) could lead to reconstructions that are enriched in particles unaffected by flattening. Additionally, since an OTR class is comprised of projections coming from particles adopting different orientations on the support, it is possible that flattening is "averaged out" to some extent in the reconstructions. Whatever its origin, this relatively more faithful representation of the structure's dimensions appears to be a feature of OTR volumes.

Although surface renderings provide some information in terms of comparing reconstructions, it is the projections of the initial volumes that are most important, as these will be used for refinement to higher resolution. The best initial models will contain the highest amount of information relative to the final structure and will yield good representations of the structure (i.e., references) in all possible projection directions. As a result of the geometry of the Missing Cone, projections generated from RCT reconstructions in a direction perpendicular to the beam axis (orthogonal to the class average that gave rise to the volume) would be most affected by artifacts. On the other hand, the 0° projection, corresponding to a section in Fourier space that is fully sampled, should be of a high quality. OTR volumes, because of their fully sampled nature, should give rise to projections of equal quality regardless of their direction. To determine whether these predictions held for the exosome reconstructions, we measured FRC (Harauz and van Heel, 1986) between projections generated from the four initial reconstructions (both OTR and RCT) against the corresponding projection from the refined exosome structure (Wang *et al.*, 2007) (EMDB entry EMD-1438). We generated projections in the 0° direction (corresponding to the class average) as well as a number of 90° directions (Fig. 9.8A and B). Figure 9.8C also shows a few representative plots of the FRCs along with an indication of the "resolution" of the projections (defined as the frequency at which the correlation between a projection from an RCT or OTR volume and the corresponding one from the refined exosome structure reached 0.5) (see Chapter 3).

As can be seen in Fig. 9.8C (first plot) and even visually in Fig. 9.8B (0° projection), RCT outperforms OTR in the direction of the class average giving rise to the reconstruction. We had observed the same phenomenon with synthetic data and had attributed it to the absence of a defocus gradient in the 0° data used for alignment and classification of RCT data. Additional factors with experimental data that would further favor the alignment and classification of RCT data are the staining and data collection artifacts associated with tilted images.

All other directions, however, show either very minor differences among the different projections (plots 4 and 5 in Fig. 9.8C) or significantly higher resolution in the OTR projections (plots 1 and 7 in Fig. 9.8C).

We are yet to find a 90° projection of an RCT volume that significantly outperforms its OTR equivalent. It should be emphasized that the RCT volumes contain approximately three times more particles than the OTR ones. Furthermore, the FRCs were calculated using projections from the refined exosome structure (Wang *et al.*, 2007), which should, if anything, favor the merged RCT volume A, which was used as its initial reference. Our OTR single-class volumes can be refined, by projection matching, to structures very similar to that published by Wang *et al.* (2007) (not shown).

## 3. Conclusion

I have presented here an overview of the main steps required to generate reconstructions using the OTR method. All the steps can be performed using tools available in the field and even data collection, arguably the most challenging aspect of the method, has been automated (as shown by the data presented here and discussed in further detail in Chapter 15 of vol. 483).

A comparison of reconstructions obtained from the same sample using the RCT and OTR methods shows the ability of the latter to generate fully sampled, artifact-free initial models. This makes OTR particularly useful for novel asymmetric samples with potential heterogeneity where geometric methods are required but the merging of volumes to solve the Missing Cone problem poses a serious challenge. OTR volumes, due to their full sampling, can be directly refined to higher resolution without the need for user intervention.

## Acknowledgments

I thank Hongwei Wang for generously providing us with the sample grid used for his exosome reconstructions, for sharing data necessary for our comparisons, and for discussions on the differences among the reconstructions. I thank NRAMM, where the data shown here were collected and, in particular, Craig Yoshioka and Neil Voss for their help during data collection. I also thank Preethi Chandramouli for generating most of the data presented in this review. AL is supported by a fellowship from the Alfred P. Sloan Foundation.

## References

Asturias, F. J., Chung, W. H., Kornberg, R. D., and Lorch, Y. (2002). Structural analysis of the RSC chromatin-remodeling complex. *Proc. Natl. Acad. Sci. USA* **99,** 13477–13480.

Chaban, Y., Ezeokonkwo, C., Chung, W. H., Zhang, F., Kornberg, R. D., Maier-Davis, B., Lorch, Y., and Asturias, F. J. (2008). Structure of a RSC-nucleosome complex and insights into chromatin remodeling. *Nat. Struct. Mol. Biol.* **15,** 1272–1277.

Frank, J. (1996). Three-Dimensional Electron Microscopy of Macromolecular Assemblies. Academic Press, Inc., San Diego, CA.

Frank, J., Radermacher, M., Penczek, P., Zhu, J., Li, Y., Ladjadj, M., and Leith, A. (1996). SPIDER and WEB: processing and visualization of images in 3D electron microscopy and related fields. *J. Struct. Biol.* **116,** 190–199.

Harauz, G., and van Heel, M. (1986). Direct 3D reconstruction from projections with initially unknown angles. Elsevier, Amsterdam.

Leschziner, A. E., and Nogales, E. (2006). The orthogonal tilt reconstruction method: an approach to generating single-class volumes with no missing cone for ab initio reconstruction of asymmetric particles. *J. Struct. Biol.* **153,** 284–299.

Leschziner, A. E., Saha, A., Wittmeyer, J., Zhang, Y., Bustamante, C., Cairns, B. R., and Nogales, E. (2007). Conformational flexibility in the chromatin remodeler RSC observed by electron microscopy and the orthogonal tilt reconstruction method. *Proc. Natl. Acad. Sci. USA* **104,** 4913–4918.

Liu, Q., Greimann, J. C., and Lima, C. D. (2006). Reconstitution, activities, and structure of the eukaryotic RNA exosome. *Cell* **127,** 1223–1237.

Ludtke, S. J., Baker, M. L., Chen, D. H., Song, J. L., Chuang, D. T., and Chiu, W. (2008). De novo backbone trace of GroEL from single particle electron cryomicroscopy. *Structure* **16,** 441–448.

Mindell, J. A., and Grigorieff, N. (2003). Accurate determination of local defocus and specimen tilt in electron microscopy. *J. Struct. Biol.* **142,** 334–347.

Ohi, M., Li, Y., Cheng, Y., and Walz, T. (2004). Negative staining and image classification—Powerful tools in modern electron microscopy. *Biol. Proced. Online* **6,** 23–34.

Pettersen, E. F., Goddard, T. D., Huang, C. C., Couch, G. S., Greenblatt, D. M., Meng, E. C., and Ferrin, T. E. (2004). UCSF Chimera—A visualization system for exploratory research and analysis. *J. Comput. Chem.* **25,** 1605–1612.

Radermacher, M., Wagenknecht, T., Verschoor, A., and Frank, J. (1987). Three-dimensional reconstruction from a single-exposure, random conical tilt series applied to the 50 S ribosomal subunit of Escherichia coli. *J. Microsc.* **146**(Pt. 2), 113–136.

Scheres, S. H., Nunez-Ramirez, R., Sorzano, C. O., Carazo, J. M., and Marabini, R. (2008). Image processing for electron microscopy single-particle analysis using XMIPP. *Nat. Protoc.* **3,** 977–990.

Schuette, J. C., Murphy, F. V.t., Kelley, A. C., Weir, J. R., Giesebrecht, J., Connell, S. R., Loerke, J., Mielke, T., Zhang, W., Penczek, P. A., *et al.* (2009). GTPase activation of elongation factor EF-Tu by the ribosome during decoding. *Embo. J.* **28,** 755–765.

Seidelt, B., Innis, C. A., Wilson, D. N., Gartmann, M., Armache, J. P., Villa, E., Trabuco, L. G., Becker, T., Mielke, T., Schulten, K., *et al.* (2009). Structural insight into nascent polypeptide chain-mediated translational stalling. *Science* **326,** 1412–1415.

Skiniotis, G., Moazed, D., and Walz, T. (2007). Acetylated histone tail peptides induce structural rearrangements in the RSC chromatin remodeling complex. *J. Biol. Chem.* **282,** 20804–20808.

Spahn, C. M., and Penczek, P. A. (2009). Exploring conformational modes of macromolecular assemblies by multiparticle cryo-EM. *Curr. Opin. Struct. Biol.* **19,** 623–631.

Tang, G., Peng, L., Baldwin, P. R., Mann, D. S., Jiang, W., Rees, I., and Ludtke, S. J. (2007). EMAN2: an extensible image processing suite for electron microscopy. *J. Struct. Biol.* **157,** 38–46.

Van Heel, M. (1987). Angular reconstitution: a posteriori assignment of projection directions for 3D reconstruction. *Ultramicroscopy* **21,** 111–123.

Van Heel, M., Harauz, G., Orlova, E. V., Schmidt, R., and Schatz, M. (1996). A new generation of the IMAGIC image processing system. *J. Struct. Biol.* **116,** 17–24.

Voss, N. R., Yoshioka, C. K., Radermacher, M., Potter, C. S., and Carragher, B. (2009). DoG Picker and TiltPicker: Software tools to facilitate particle selection in single particle electron microscopy. *J. Struct. Biol.* **166,** 205–213.

Wang, H. W., Wang, J., Ding, F., Callahan, K., Bratkowski, M. A., Butler, J. S., Nogales, E., and Ke, A. (2007). Architecture of the yeast Rrp44 exosome complex suggests routes of RNA recruitment for 3' end processing. *Proc. Natl. Acad. Sci. USA* **104,** 16844–16849.

Yoshioka, C., Pulokas, J., Fellmann, D., Potter, C. S., Milligan, R. A., and Carragher, B. (2007). Automation of random conical tilt and orthogonal tilt data collection using feature-based correlation. *J. Struct. Biol.* **159,** 335–346.

Yu, X., Jin, L., and Zhou, Z. H. (2008). 3.88 A structure of cytoplasmic polyhedrosis virus by cryo-electron microscopy. *Nature* **453,** 415–419.

Zhang, X., Settembre, E., Xu, C., Dormitzer, P. R., Bellamy, R., Harrison, S. C., and Grigorieff, N. (2008). Near-atomic resolution using electron cryomicroscopy and single-particle reconstruction. *Proc. Natl. Acad. Sci. USA* **105,** 1867–1872.

CHAPTER TEN

# An Introduction to Maximum-Likelihood Methods in Cryo-EM

Fred J. Sigworth,* Peter C. Doerschuk,† Jose-Maria Carazo,‡ *and* Sjors H. W. Scheres‡,[1]

## Contents



* Department of Cellular and Molecular Physiology, Yale University, New Haven, Connecticut, USA
† Department of Biomedical Engineering, Cornell University, Weill Hall, Ithaca, New York, USA
‡ Biocomputing Unit, Centro Nacional de Biotecnología – CSIC, Cantoblanco, Madrid, Spain
[1] Current address: MRC Laboratory of Molecular Biology, Hills Road, Cambridge, UK.

*Methods in Enzymology*, Volume 482

ISSN 0076-6879, DOI: 10.1016/S0076-6879(10)82011-7

**Abstract**

The maximum-likelihood method provides a powerful approach to many problems in cryo-electron microscopy (cryo-EM) image processing. This contribution aims to provide an accessible introduction to the underlying theory and reviews existing applications in the field. In addition, current developments to reduce computational costs and to improve the statistical description of cryo-EM images are discussed. Combined with the increasing power of modern computers and yet unexplored possibilities provided by theory, these developments are expected to turn the statistical approach into an essential image-processing tool for the electron microscopist.

# 1. Introduction

The cryo-electron microscopy (cryo-EM) single-particle reconstruction (SPR) problem is a very difficult one. Given a large number of very noisy electron-microscope images, each showing a macromolecular "particle" in a random position and orientation, the problem is to deduce the three-dimensional (3D) structure of the particles that were imaged. It is amazing that standard software packages now allow even casual users to perform such reconstructions, in a numerical process that seems little short of magical. The goal of this chapter is to describe the basis of a particularly powerful approach to the SPR problem and related tasks in two-dimensional (2D) crystallography and electron tomography.

## 1.1. Maximum-likelihood estimates

When we use an SPR algorithm to obtain a 3D model from a set of images $\mathcal{X}$, how do we know that we have found the best model? Formally, we would want to know what quantity is optimized to yield the model. Traditional SPR refinement algorithms use an iterative process. The images are aligned and sorted to give the best match to a set of projections of the $n$th model. Given the angles of each projection, a synthesis of the 3D map from the images is performed by standard methods, to yield the $(n + 1)$st model. This process is repeated until the model does not change. Arguably this process yields a least-squares estimate, as the traditional cross-correlation-based projection-matching step is equivalent to a least-squares optimization. In the optimization process, a numerical density value is obtained for each voxel of the model, and as a by-product the values of orientation angles for each particle are also found. There is however no theory that says that the model obtained in this way is the most reliable one.

The statistical approaches discussed in this chapter seek to maximize a probability function. Suppose we group the desired 3D map along with any other quantities that we wish to estimate into a more generalized model $\Theta$. In a sense, what we would like to maximize is the probability $P(\Theta|\mathcal{X})$ that this model is the correct one, given the data. Unfortunately, there are both philosophical and practical problems surrounding this quantity. Some people would say that $\Theta$ is not a random variable in the first place, so how can one define a probability function. But if the philosophical problems are bypassed, perhaps by imagining an ensemble of possible true structures $\Theta$ that could have given us our data set $\mathcal{X}$, there is still the problem of computing this quantity. An easy way around these problems is to compute the probability of observing $\mathcal{X}$ given $\Theta$. This is a valid and computable probability, and it is given a special name, the likelihood $\mathcal{L}(\Theta) = P(\mathcal{X}|\Theta)$. What is unusual here is that the likelihood $\mathcal{L}$ is expressed as a function of the model $\Theta$ rather than of the data $\mathcal{X}$.

The choice of the likelihood as the quantity to be maximized can be understood by applying Bayes' rule,

$$P(\Theta|\mathcal{X}) = P(\mathcal{X}|\Theta)\frac{P(\Theta)}{P(\mathcal{X})}. \tag{10.1}$$

We know what $P(\mathcal{X}|\Theta)$ is, that is just the likelihood. $P(\mathcal{X})$ is an imponderable: what is the probability of obtaining this data set instead of some other? However, since it does not depend on $\Theta$, we would not have to worry about it, and for our present purposes we can replace it by a constant. The resulting quantity to be maximized is $P(\mathcal{X}|\Theta)P(\Theta)$ and is called the posterior probability. The model $\Theta$ that maximizes it is called the maximum a posteriori (MAP) estimate, and we discuss briefly the use of this estimation approach at the end of this chapter. The term $P(\Theta)$ is called the prior probability, as it reflects any knowledge we might have about the model in the absence of any data. For now, suppose we believe that all possible models are equally likely, so that $P(\Theta)$ is also a constant. Then, maximizing $P(\Theta|\mathcal{X})$ becomes the same as maximizing the likelihood.

Finding the model $\Theta$ that gives the maximum value of the likelihood—this is called the maximum-likelihood estimate or MLE—is a very good way to find the best model. The MLE is asymptotically unbiased and efficient; that is, in the limit of very large data sets, the MLE is as good or better than any other estimate of the true model.

## 1.2. Introduction to the EM–ML algorithm

Finding the model $\Theta$ that maximizes the likelihood is a daunting task. In principle, the density value of each voxel in the 3D map must be varied until the best set of densities is found—an optimization problem with on the order of $10^6$ unknowns! Fortunately, there is a straightforward way to find the ML estimate, called the *expectation-maximization algorithm* (Dempster *et al.*, 1977). We abbreviate this as the expectation-maximization maximum-likelihood (EM–ML) algorithm to avoid confusion with the terminology of EM for "electron microscopy." The idea behind the EM–ML algorithm is illustrated here with a simple example based on the Gaussian mixture model (Redner and Walker, 1984), while a formal presentation of the theory is considered in the next section of this chapter. Additional examples designed to mimic cryo-EM are worked out in Yin *et al.* (2003) and continued to consider resolution in Prust *et al.* (2009). More sophisticated miniature examples are considered in Yin *et al.* (2004).

Imagine a series of position measurements $x_i$ with $i = 1, \ldots, N$ coming from a single-molecule optical tracking experiment. Typically, the observed position shows Gaussian-distributed errors from measurement noise. One can record a large number of position measurements and compute, for example, the MLE of the mean and variance directly from the measurements.

Suppose that the single molecule under observation undergoes a conformational change between two states, which we call state 0 and state 1. With this change the reporter group changes its true $x$-coordinate between two discrete values. We would like to estimate these two values. One approach would be to make a histogram of the measured values, as in Fig. 10.1, and do some sort of fit to the histogram. A more powerful way however is to find the MLE of the two mean values directly from the measured values.

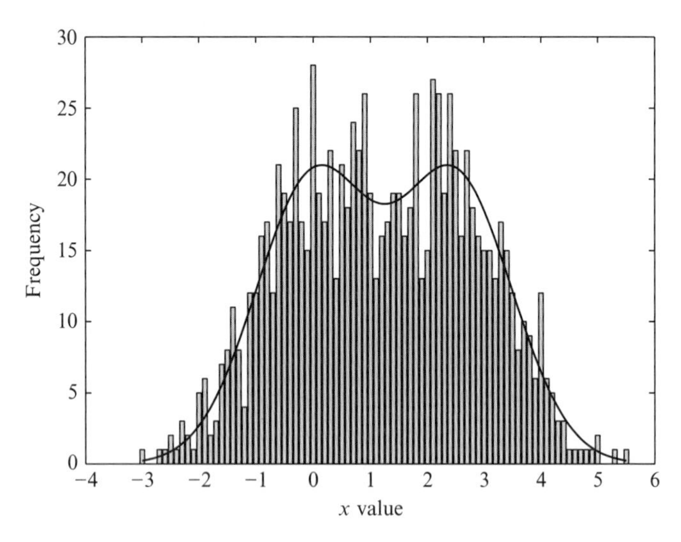


**Figure 10.1** Histogram of values of a random variable $x$. The values were drawn from a distribution consisting of a mixture of two Gaussians with means of 0 and 2.5.

Computing the likelihood starts with the probability density function (PDF) of the observed values, a mixture of two Gaussians,

$$f_\Theta(x) = \frac{1}{\sqrt{2\pi}}\left(a_0\,\mathrm{e}^{(x-\mu_0)^2/2\sigma^2} + a_1\,\mathrm{e}^{(x-\mu_1)^2/2\sigma^2}\right). \tag{10.2}$$

In this example, we are interested in determining the model parameters $\Theta = \{\mu_0, \mu_1\}$, assuming that the other parameters $a_0$, $a_1$, and $\sigma$ are already known. The probability of a particular measured value would be vanishingly small were it not for the fact that any physical measurement has a finite resolution. Letting $\varepsilon$ be the resolution of measurement, the probability of measuring the first value is

$$P_m(x_i|\Theta) = \varepsilon f_\Theta(x_1) \tag{10.3}$$

and the probability of the measurement of the entire data set is

$$P_m(\mathcal{X}|\Theta) = \varepsilon^N f_\Theta(x_1) f_\Theta(x_1) f_\Theta(x_2) \cdots f_\Theta(x_N). \tag{10.4}$$

Since $\varepsilon$ does not vary when $\Theta$ is changed, it is irrelevant to the maximization process. Thus in the literature this factor is traditionally ignored; the likelihood is instead written simply as the PDF, which in this case, is a product of simpler PDFs:

$$P(\mathcal{X}|\Theta) = f_\Theta(x_1) f_\Theta(x_2) \cdots f_\Theta(x_N). \tag{10.5}$$

Nevertheless maximizing this product of PDFs appears daunting.

In the case of a single Gaussian distribution, the maximization of the likelihood turns out to have a very simple form. The ML estimates of the mean and standard deviation (SD) are simply equal to the mean and SD of all the measured values. In this special case, ML and least-squares estimations give the same answer. Is there a way we can exploit this simple ML estimation approach to the present problem involving a mixture of two Gaussians?

We could just set a threshold and divide the data set into two halves. We would then compute the average of all the $x_i$ values that fell below the threshold, and do the same for all those above the threshold, and call these our estimates of the means. However, taking the averages of these two "classes" of measurements will produce estimates that are biased, being spread more widely than the two correct means.

Estimating the two means would be very simple if we had independent information about the underlying state of the molecule. Let $y_i$ be a set of "switch" variables such that when $y_i = 0$ we know that the corresponding $x_i$

is obtained from the molecule in state 0, and when $y_i = 1$ the molecule is in its state 1. Then, it is very easy to obtain the means $\mu_0$ and $\mu_1$ which are the MLEs of the positions:

$$\mu_0 = \frac{\sum_i (1 - y_i) x_i}{\sum_i 1 - y_i}, \qquad \mu_1 = \frac{\sum_i y_i x_i}{\sum_i y_i}. \tag{10.6}$$

Unfortunately, we do not have access to the $y_i$ values; they are so-called "hidden variables." The EM–ML algorithm however provides a simple way to iteratively converge on the ML mean values. It starts with an initial guess of the parameters of the two underlying Gaussian distributions. For example, the initial guess could have means that are too far apart, as in the top panel of Fig. 10.2A. The first step of the EM algorithm is to provide an estimate of the hidden $y_i$ variable corresponding to each measurement $x_i$. The estimate is computed as the expectation value $\hat{y}_i$, which in this case is simply equal to the probability of the molecule being in state 1. It is computed from each measurement $x_i$ based on the current estimate of the Gaussian distributions.

The second step is the maximization step, where we compute the MLEs of the means based on the inferred $\hat{y}$ values. In this case, the MLEs are simply computed as weighted averages of the $x_i$ values, weighted by either $1 - \hat{y}_i$ or $\hat{y}_i$ to yield the two estimates $\mu_0$ and $\mu_1$. The calculation is exactly the same as in Eq. (10.6) except that $\hat{y}_i$ replaces $y_i$. This gives us two new mean values, which we can then use for another round of the EM–ML algorithm.

The remarkable property of the expectation-maximization iteration is that the new estimated values of the means are guaranteed to result in an increased likelihood of the model. Figure 10.2 provides an illustration of the EM–ML algorithm. In the figure, histograms are used as a visualization device, but the underlying computations do not involve the binning or sorting of events at all. The figure shows one EM–ML iteration, and also the result after 24 iterations, when the parameters have reached a fixed value.

This simple example demonstrates three important features of the EM–ML algorithm. First, the algorithm "fills in" missing information through the computation of the expectation values of hidden variables. Second, it exploits the hidden variable values to make the process of computing the MLE much easier. Third, perhaps the most interesting aspect of the EM–ML algorithm is that it makes use of "fuzzy" estimates of the hidden variables. Even though the underlying $y_i$ values, could we measure them, take only the values 0 and 1, the expectation values $\hat{y}_i$ vary continuously over the range of 0–1. The maximization of the likelihood nevertheless converges correctly when these expectation values are used.

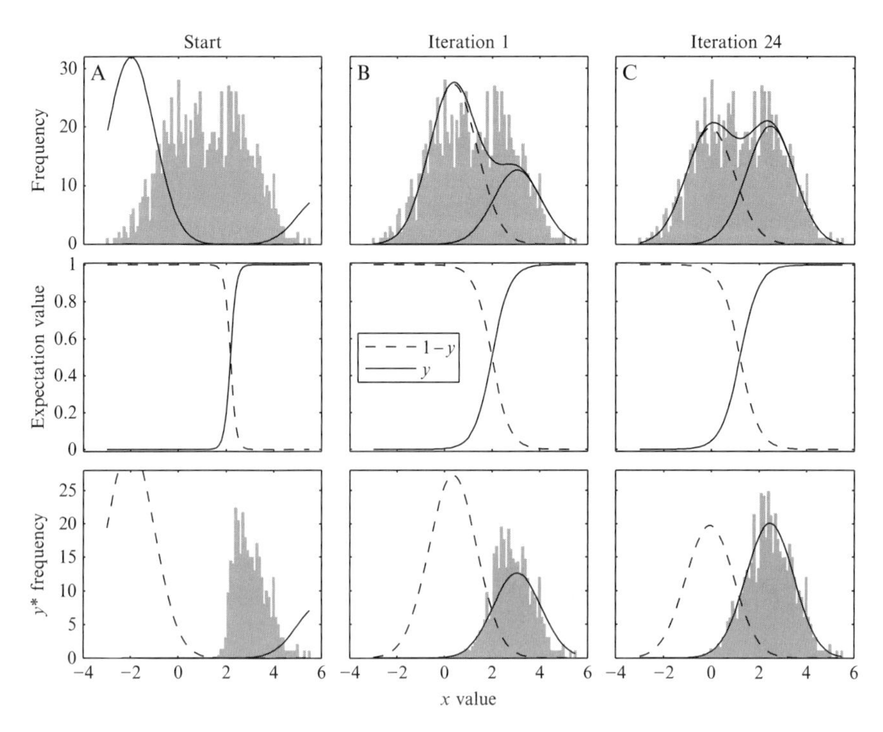


**Figure 10.2** EM–ML estimation of the means and amplitudes in a mixture of two Gaussians as in Fig. 10.1. (A, top panel) A histogram of 1000 simulated measurements is shown, along with two Gaussian components computed on the basis of initial guesses for the mean and amplitude of the components. The middle panel shows how the expectation $\hat{y}$ varies with the measured value $x$. Instead of there being a strict classification of $x$ values into one or the other component, these $\hat{y}$ values take on intermediate values in the range of $x$ values where the components overlap. The bottom panel illustrates how a weighted average can give rise to the mean value for the right-hand component. The bins of the histogram have been scaled by the values of $\hat{y}$, while the initial guess of the PDF is plotted as a smooth curve. The center of mass of the weighted histogram yields the new estimate for $\mu_1$. (B) The same plots are shown after the two Gaussian components have been updated with new means and amplitudes by one EM–ML iteration. (C) The result after convergence with 24 iterations. The simulation was based on means of 0 and 2.5. The EM–ML estimated means were $-0.06$ and 2.46.

## 1.3. Relevance to SPR problems

The features of this simple example are mirrored in the much more complex computations involved in SPR. In EM–ML processing of particle images, "hard" values for the orientation angles of each particle are not assigned. Instead, the orientation of each particle is described in a fuzzy way as a probability density function, giving the probability that the particle assumes each possible orientation. Similarly, when EM–ML algorithm is used to simultaneously reconstruct several different conformations (i.e., distinct 3D maps) from images of mixed populations of particles, the assignment of a given particle image to a given conformation is treated as a hidden variable, and is made in a "fuzzy" way.

This chapter reviews the applications of the EM–ML algorithm to a range of image processing tasks across various cryo-EM modalities. First, a formal description of the theory is presented along with a typical example from cryo-EM image processing. This theoretical framework is then used as a common basis to describe existing EM–ML approaches for the analysis of single particles (with and without symmetry), 2D crystals, and subtomograms. In addition, the validity of the most common statistical data model and possible alternatives is discussed and an overview of approaches to accelerating the intensive EM–ML computations is presented. The chapter concludes with a discussion of the perspectives of the statistical approach to cryo-EM image processing.

## 2. Theoretical Basis

### 2.1. The maximum-likelihood estimator

At the heart of the ML method lies a parameterized, statistical model that is used to describe the underlying physics of the data formation process. Again, we denote the set of model parameters by $\Theta$, and our data set of $N$ independent measurements by $\mathcal{X} = (X_1, X_2, \ldots, X_N)$. Then, the statistical model is expressed in the form of the probability density function $P(\mathcal{X}|\Theta)$, the probability of observing the data given $\Theta$. For a given set of model parameters, the PDF will show that some data are more probable than others. In the practical situation, however, we are interested in the opposite. We have already observed the data and are looking for those model parameters that best fit the data. In particular, we want to find those model parameters that make the data "more likely" than any other parameter set would make them.

To that purpose, and as explained in more detail in Section 1, we define the *likelihood function* $\mathcal{L}(\Theta) = P(\mathcal{X}|\Theta)$ as a function of $\Theta$. Whereas the PDF is defined as a function of the data given a particular set of model parameters, the likelihood function is a function of the model parameters for a given data set. The method of ML aims at finding the set of parameters that maximizes $\mathcal{L}(\Theta)$. This is the maximum-likelihood estimator (MLE) of $\Theta$:

$$\Theta^{MLE} = \arg\max_{\Theta} \mathcal{L}(\Theta). \tag{10.7}$$

Assuming that all observations are statistically independent, the likelihood can be written as a product of the PDFs of the individual observations:

$$\mathcal{L}(\Theta) = \prod_{i=1}^{N} P(X_i|\Theta) \tag{10.8}$$

and since maxima are unaffected by monotone transformations, for computational convenience, one often takes the logarithm of this expression:

$$L(\Theta) = \log(\mathcal{L}(\Theta)) = \sum_{i=1}^{N} \log P(X_i|\Theta). \tag{10.9}$$

Depending on the form of $P(X_i|\Theta)$, finding the maximum of $L(\Theta)$ may be straightforward or extremely difficult. As we describe in the example below, analytical expressions to obtain the MLE may be obtained directly for simple problems. For many other problems, direct optimization of the likelihood function is analytically intractable and more elaborate techniques, like the EM–ML algorithm, must be employed.

## 2.2. An example of direct MLE calculation in cryo-EM

One example of straightforward likelihood optimization in cryo-EM is the estimation of the underlying signal from a series of noisy, structurally homogeneous, and aligned 2D images. Let us assume the following data model:

$$X_i = A + \sigma G_i, \quad \text{with } i = 1, \ldots, N, \tag{10.10}$$

where $X_i$ are the observed images of $J$ pixels each and with pixel values $X_{ij}$; $A$ is the underlying 2D image (with pixel values $A_j$) that is common to all images; and $G_i$ are the images of independent noise with pixel values $G_{ij}$ taken from a Gaussian distribution with zero mean and unity SD.

It is to be noted that in this case our set of model parameters $\Theta$ consists of image $A$ and parameter $\sigma$. The PDF of observing pixel value $X_{ij}$ is then given by a Gaussian distribution centered at $A_j$ and with SD $\sigma$ (Fig. 10.3). Furthermore, because we assume independence between all pixels, the PDF of observing the entire image $X_i$ may be expressed as the multiplication over the PDFs of all individual pixels:

$$P(X_i|\Theta) = \prod_{j=1}^{J} \frac{1}{\sqrt{2\pi}\sigma} \exp\left\{\frac{\left(X_{ij} - A_j\right)^2}{-2\sigma^2}\right\} = \left(\frac{1}{\sqrt{2\pi}\sigma}\right)^J \exp\left\{\frac{\| X_i - A\|^2}{-2\sigma^2}\right\}, \tag{10.11}$$

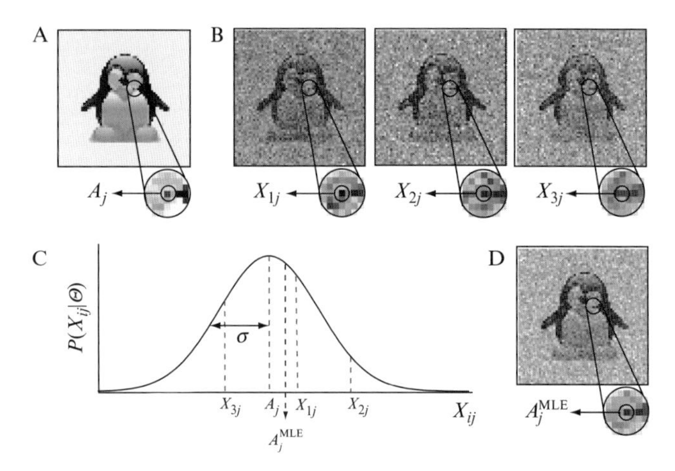


**Figure 10.3** An example of direct MLE calculation. (A) An image $A$ with a zoom window centered on one individual pixel value with value $A_j$ is shown. (B) Three copies of image $A$ with different instances of white Gaussian noise are shown together with zoom windows on the same pixel $j$ with pixel values $X_{1j}$, $X_{2j}$, and $X_{3j}$. (C) the PDF of the $j$th pixel in the noisy images is shown as a Gaussian curve centered at $A_j$ and with SD $\sigma$. Direct calculation of $A_j^{\mathrm{MLE}}$ is performed by averaging over $X_{1j}$, $X_{2j}$, and $X_{3j}$. (D) The MLE of the entire image $A$ is shown together with a zoom window, centered on $A_j^{\mathrm{MLE}}$. Note that the MLE of the image will approach the image in (A) if larger numbers of noisy images are available.

where $\|X_i - A\|^2$ denotes the sum of the squared residuals over all pixels $\sum_{j+1}^{J}(X_{ij} - A_j)^2$.

Thereby, the log-likelihood function, as defined in Eq. (10.9), reduces to:

$$L(\Theta) = -JN\log\left(\sqrt{2\pi}\sigma\right) - \frac{1}{2\sigma^2}\sum_{i=1}^{N} \| X_i - A\|^2 \qquad (10.12)$$

The MLE for the underlying signal may then be obtained directly by setting the partial derivatives of Eq. (10.12) with respect to $A_j$ equal to zero and solving for $A_j$. This yields the well-known equation to calculate the average image:

$$A^{\mathrm{MLE}} = \frac{1}{N}\sum_{i=1}^{N} X_i. \qquad (10.13)$$

Similarly, the MLE for $\sigma$ would yield the formula for calculating the root mean square deviation between the average and the observed images.

Equation (10.12) illustrates that under the assumption of independent, Gaussian noise with equal SD, the MLE is equal to the *least-squares estimator*, which aims at minimizing $\sum_{i+1}^{N} \| X_i - A \|^2$. This equality has been used erroneously to argue against the application of ML methods in cryo-EM. However, as we see below, for more complicated problems the method of ML will actually yield very different results from conventional approaches based on least-squares estimation.

## 2.3. Incomplete data problems

For many problems, the likelihood function cannot be optimized directly, but it can be simplified by assuming the existence of additional, "hidden" variables. Without the hidden variables, the data are considered to be *incomplete*. The complete data would comprise both the observed and the hidden variables, and finding the MLE for the complete data problem would be a trivial task. In some cases, the hidden variables actually correspond to incompleteness in the data vectors themselves, due to problems in the data collection process. More often however, the hidden variables correspond to aspects of physical reality that could in principle be measured but are not observed for practical reasons.

Let $\mathcal{X}$ be the incomplete, observed data and assume that a complete data set $(\mathcal{X}, \mathcal{Y})$ exists. Then, the MLE is determined by the so-called *marginal likelihood* of the observed data:

$$\mathcal{L}(\Theta) = P(\mathcal{X}|\Theta) = \int_{\mathcal{Y}} P(\mathcal{X}|\mathcal{Y}, \Theta)P(\mathcal{Y}|\Theta)\mathrm{d}\mathcal{Y}, \qquad (10.14)$$

where $P(\mathcal{X}|\Theta)$ is the unconditional probability of $\mathcal{X}$, regardless of the values of $\mathcal{Y}$, and this probability is obtained by integrating the joint probabilities over all possible values of $\mathcal{Y}$. This is called *marginalization*. $P(\mathcal{X}|\mathcal{Y}, \Theta)$ is the probability of $\mathcal{X}$ given that $\mathcal{Y}$ has happened, and $P(\mathcal{Y}|\Theta)$ is the *prior probability* of $\mathcal{Y}$ happening.

It is important to note that because of the marginalization Eq. (10.14) is only a function of $\Theta$ and not of the hidden variables $\mathcal{Y}$. Thereby, the problem at hand is only to find those parameter values $\Theta^{\mathrm{MLE}}$ that maximize the marginal-likelihood function.

## 2.4. An example of an incomplete data problem in cryo-EM

The first description of a ML approach to an incomplete data problem in cryo-EM was described by Sigworth (1998). He considered the problem of finding $A^{\mathrm{MLE}}$, cf. Eq. (10.13), for a set of noisy 2D images with unknown in-plane rotations and translations. In other words, he presented a ML approach to the 2D alignment of a structurally homogeneous set of images. In this case, the data are modeled as:

$$X_i = R_{\phi_i} + A + \sigma G_i, \quad \text{with}\, i = 1, \ldots, N, \tag{10.15}$$

where $R_{\phi i}$ denotes the in-plane transformation that brings the common underlying signal in register with the *i*th image. This transformation comprises a single rotation $\alpha_i$ and two translations in perpendicular directions $x_i$ and $y_i$.

The incompleteness of the observed data lies in the fact that the relative orientations of all images have remained unobserved in the experiment. The complete data set would be $(\mathcal{X}, \mathcal{Y})$, with $\mathcal{Y} = (\phi_1, \phi_2, \ldots, \phi_N)$, and finding the MLE for the complete data set would be as trivial as described in the simple example above.

For the incomplete case, the marginal log-likelihood function, cf. Eq. (10.14), is given by:

$$L(\Theta) = \sum_{i=1}^{N} \log \int_{\phi} P(X_i|\phi, \Theta) P(\phi|\Theta) \mathrm{d}\phi. \tag{10.16}$$

For any given transformation $\phi$ and parameter set $\Theta$, the conditional probability of observing image $X_i$ is again expressed as a multiplication over $J$ Gaussian distributions, this time centered at the correspondingly oriented reference image $R_\phi A$:

$$P(X_i|\phi, \Theta) = \left(\frac{1}{\sqrt{2\pi}\sigma}\right)^J \exp\left\{\frac{\| X_i - R_\phi A \|}{-2\sigma^2}\right\}. \tag{10.17}$$

Note that, apart from the model parameters $\Theta$, this probability is also a function of the hidden variable $\phi$. This concept is further illustrated in Fig. 10.4.

An additional advantage of the ML approach is the natural way in which prior knowledge about the hidden variables may be handled. The term $P(\phi|\Theta)$ provides a statistical description of the distribution of the hidden variables. If we assume that the in-plane rotations are evenly distributed and that particle picking has left residual origin offsets with Gaussian distributions in both directions, the probability density of $\phi$ may be calculated as:

$$P(\phi|\Theta)\mathrm{d}\phi = \frac{1}{2\pi\sigma_{xy}^2} \exp\left\{\frac{(x - \xi_x)^2 + (y - \xi_y)^2}{-2\sigma_{xy}^2}\right\} \frac{\mathrm{d}\alpha}{2\pi} \mathrm{d}x\mathrm{d}y, \tag{10.18}$$

where $\xi_x$ and $\xi_y$ are the expected values for the in-plane translations in directions $x$ and $y$, and $\sigma_{xy}$ is the SD in the translations in either direction.

Thereby, prior knowledge that large origin offsets are less likely than small offsets is translated in an effective downweighting of larger offsets in the probability calculations of all possible orientations. Although a similar

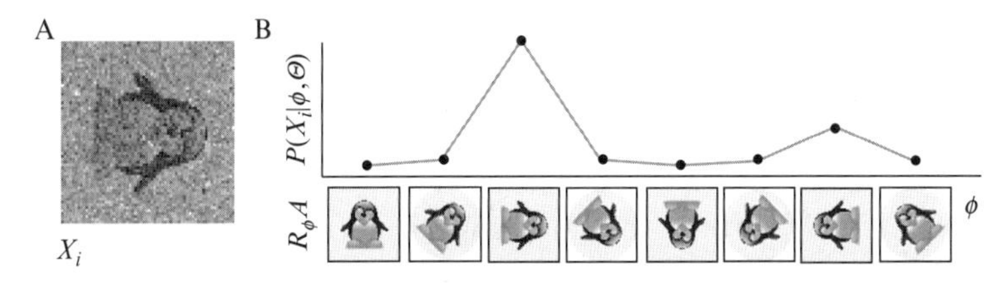


**Figure 10.4** The PDF of a noisy image as a function of its relative orientation with respect to a reference image. (A) $X_i$, a rotated and noisy version of reference image $A$ from Fig. 10.3A is shown. (B, top panel) The probability $P(X_i|\phi, \Theta)$ of observing $X_i$ given a model $\Theta$ that comprises image $A$ is shown (on an arbitrary scale) as a function of the relative orientation $\phi$. The bottom panel shows $R_\phi A$, that is, image $A$ rotated according to $\phi$. Note that $P(X_i|\phi, \Theta)$ is highest when $\phi$ corresponds to the correct orientation of $X_i$.

term may also be incorporated into maximum cross-correlation approaches (Sigworth, 1998), this is not common practice in conventional approaches to the alignment of cryo-EM images. Instead, one often expresses this prior knowledge in a less powerful way: by limiting the searches for the optimal offsets to a user defined maximum value in all directions.

From Eqs. (10.17) and (10.18), we can see that $\Theta = (A, \sigma, \xi_x, \xi_y, \sigma_{xy})$ and the task at hand is to find those parameters $\Theta^{\text{MLE}}$ that maximize the marginal-likelihood function as defined in Eq. (10.16).

## 2.5. The EM–ML algorithm

The EM–ML algorithm is a general tool to find MLEs for incomplete data problems (Dempster *et al.*, 1977). The intuitive idea behind this algorithm is an old one. Because one does not know parameter estimates $\Theta$ nor the hidden variables $\mathcal{Y}$, one iteratively alternates between estimating both. For a given set of model parameters $\Theta$ one estimates the hidden variables, for the resulting hidden variables one finds the best model parameters, and one repeats this process until the model parameters no longer change.

However, rather than finding the best $\mathcal{Y}$ given an estimate $\Theta^{(n)}$ at the $n$th iteration, the EM–ML algorithm computes a *distribution* over all possible values of $\mathcal{Y}$. To this purpose, in the so-called expectation ($E$) step one first calculates the expected value of the complete-data log-likelihood function with respect to the missing data $\mathcal{Y}$, given the observed data $\mathcal{X}$ and the current parameter estimates $\Theta^{(n)}$:

$$\begin{aligned} Q(\Theta, \Theta^{(n)}) &= E_{\mathcal{Y}|\mathcal{X},\Theta}[\log P(\mathcal{X}, \mathcal{Y}|\Theta)] \\ &= \int_{\mathcal{Y}} P(\mathcal{Y}|\mathcal{X}|\Theta^{(n)}) \log P(\mathcal{X}|\mathcal{Y}, \Theta) P(\mathcal{Y}|\Theta) \mathrm{d}\mathcal{Y}. \end{aligned} \tag{10.19}$$

Here, $P(\mathcal{Y}|\mathcal{X}, \Theta^{(n)})$ is the conditional probability of the missing variables in terms of the observed measurements and the current model parameters, which is calculated using Bayes' rule:

$$P(\mathcal{Y}|\mathcal{X}, \Theta^{(n)}) = \frac{P(\mathcal{X}|\mathcal{Y}, \Theta^{(n)}) P(\mathcal{Y}|\Theta^{(n)})}{\int_{\mathcal{Y}} P(\mathcal{X}|\mathcal{Y}, \Theta^{(n)}) P(\mathcal{Y}|\Theta^{(n)}) \mathrm{d}\mathcal{Y}}. \tag{10.20}$$

The integration of $P(\mathcal{Y}|\mathcal{X}, \Theta^{(n)})$ over all possible values of $\mathcal{Y}$ represents the above-mentioned distribution of the hidden variables. Given this distribution, in the subsequent maximization (*M*) step one computes new estimates for the model parameters by maximizing the corresponding expectation:

$$\Theta^{(n+1)} = \arg\max_{\Theta} Q(\Theta, \Theta^{(n)}). \tag{10.21}$$

The new model parameters are then used for the next *E*-step and the process is repeated as necessary. It can be shown that each iteration is guaranteed to increase the log-likelihood and the algorithm is guaranteed to converge to a local maximum of the likelihood function (Dempster *et al.*, 1977).

## 2.6. An example of EM–ML in cryo-EM

For the example of aligning a set of structurally homogeneous images described above, the *E*-step of the EM–ML algorithm yields:

$$Q(\Theta, \Theta^{(n)}) = \sum_{i=1}^{N} \int_{\phi} P(\phi|X_i, \Theta^{(n)}) \log\{P(X_i|\phi, \Theta) P(\phi|\Theta)\} \mathrm{d}\phi, \tag{10.22}$$

which can be rewritten by substitution of Eqs. (10.17) and (10.18) and by separating terms $C$ that are not related to $A$:

$$Q(\Theta, \Theta^{(n)}) = \sum_{i=1}^{N} \int_{\phi} P(\phi|X_i, \Theta^{(n)}) \left\{ \frac{1}{2\sigma^2} \| X_i - R_{\phi} A \|^2 + C \right\} \mathrm{d}\phi. \tag{10.23}$$

In the subsequent $M$-step one maximizes $Q(\Theta, \Theta^{(n)})$ with respect to the model parameters. From Eq. (10.23), it can be readily seen that obtaining new estimates for $A$ corresponds to solving a weighted least-squares problem with weights $P(\phi|X_i, \Theta^{(n)})$. The result is a weighted average comprising contributions from all possible values of $\phi$ for every image $X_i$:

$$A^{(n+1)} = \frac{1}{N}\sum_{i=1}^{N}\int_{\phi} P(\phi|X_i, \Theta^{(n)})R_{\phi}^{-1}X_i\,\mathrm{d}\phi. \tag{10.24}$$

The concept of calculating $A^{(n+1)}$ as a probability weighted average is further illustrated in Fig. 10.5. All other model parameters are updated using similar probability-weighted average calculations.

## 2.7. Comparison with conventional methods

It is interesting to compare the EM–ML approach with conventionally more popular methods in cryo-EM. In particular, it has been common practice to include the hidden variables as part of the unknown model parameters. For the example of single-reference 2D-alignment described above, the model thereby comprises the estimate for the underlying signal ($A$) and the optimal orientations for each of the individual observations ($\phi_i^{\star}$ for $i = 1, \ldots, N$). One typically minimizes the following least-squares target:

$$\sum_{i=1}^{N} \| X_i - R_{\phi_i}A\|^2, \tag{10.25}$$

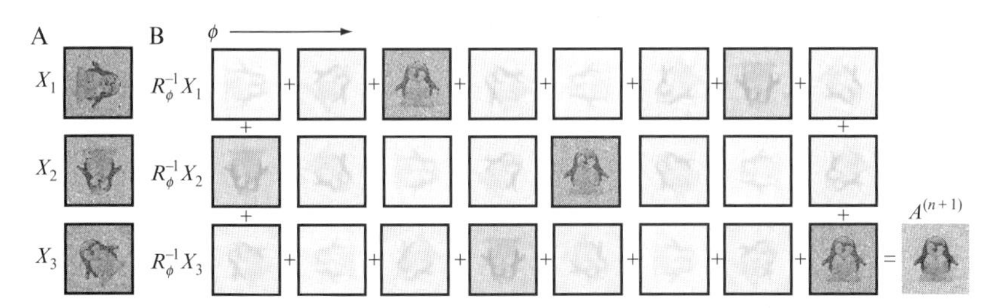


**Figure 10.5** Reference image calculation by probability-weighted averaging. (A) Three noisy versions of reference image $A$ from Fig. 10.3A in different orientations ($X_i$, with $i = 1, 2, 3$). (B) Images $R_{\phi}^{-1}X_i$ for all sampled $\phi$. The opacity of the images is used to illustrate the weight $P(\phi|X_i, \Theta^{(n)})$ in the weighted average calculation that produces the updated estimate of the reference $A^{(n+1)}$, see Eq. (10.24). Note that $X_1$ corresponds to the same image that was shown in Fig. 10.4A and that the columns correspond to the same orientations as in Fig. 10.4B.

by iteratively alternating between estimating the underlying signal and estimating the orientations. For a given estimate $A^{(n)}$, one calculates the optimal orientation $\phi_i^{\star}$ for each image $X_i$ as the one that minimizes the squared difference $\|X_i - R_{\phi i} A\|^2$. Then, one calculates a new estimate for the underlying signal by:

$$A^{(n+1)} = \frac{1}{N}\sum_{i=1}^{N} R_{\phi_i^*}^{-1} X_i. \tag{10.26}$$

The orientation that minimizes the squared difference term mentioned above in turn maximizes the inner product between images $X_i$ and $R_{\phi i} A$. This inner product is also called the cross-correlation and hence the term "maximum cross-correlation" approach.

Comparison of Eqs. (10.25) and (10.16) illustrates that the maximum cross-correlation and ML approaches pursue different objectives. The principal difference lies in the marginalization over the hidden variables in the latter. Consequently, in the ML approach model parameters are calculated as probability-weighted averages over all possible orientations, while only the "best" orientation is considered for each image in the maximum cross-correlation approach, cf. Eqs. (10.26) and (10.24).

The advantages of the ML approach appear at low signal-to-noise ratios. The higher the level of noise in the data, the higher the number of false peaks that occur in the cross-correlation function. Thereby, the orientations that maximize the cross-correlation function are less likely to reflect the optimal ones. The statistical description of the noise in the ML approach allows to deal with these ambiguities in the cross-correlation function. Correspondingly, the ML approach will yield better results than the maximum cross-correlation approach for data with high levels of noise. Interestingly, the two approaches become equivalent in the absence of noise: when $\sigma$ approaches zero, probability distributions $P(\phi|X_i,\Theta^{(n)})$ become delta functions centered at $\phi_i^{\star}$. In other words, the maximum cross-correlation approach may be considered as a special case of the ML method that ignores the presence of noise in the data.

## 3. EM–ML Approaches in Cryo-EM

The EM–ML algorithm has been applied to a variety of tasks in cryo-EM image processing. Although the actual applications differ widely, the theoretical framework described in the previous section may be employed to put all these approaches on a common basis. Most approaches share the assumption of independent Gaussian noise in the data. Thereby, the optimization strategy typically remains the same and the differences between the existing approaches lie in the unknown model parameters ($\Theta$) and the hidden variables ($\mathcal{Y}$).

## 3.1. Single-particle analysis

As described in the example above, it was Sigworth (1998) who introduced the concept of optimizing a marginal-likelihood function to the field. His approach to aligning a structurally homogeneous set of 2D images was tested on simulated data that were in accordance with the assumption of white and Gaussian noise. These simulations showed that the ML approach has a reduced sensitivity toward the choice of the starting model and allows recovering the underlying signal from much noisier data than the maximum cross-correlation method. This chapter was followed by a series of contributions applying similar ideas to single-particle analysis (Table 10.1).

First, Pascual-Montano *et al.* (2001) described a neural network for the classification of a set of prealigned 2D images called *kerdenSOM*, for kernel density self-organizing map. In this work, the data model comprises multiple 2D models ($A_k$ with $k = 1, \ldots, K$) and the assignments $k_i$ of the experimental images to these models are treated as hidden variables. This results in a ML variant of the conventional $k$-means classifier (see Frank, 2006), where experimental images are not assigned to a single model but contribute to all models with varying weights. In addition, this contribution considered the $K$ models to be arranged in a 2D map and defined a regularization term to the log-likelihood in order to impose similarity between neighboring models in the map. Thereby, the behavior of a neural

**Table 10.1** An overview of EM–ML approaches in single-particle analysis

| Approach | Data model | Model parameters $\Theta$ | Hidden variables $\mathcal{Y}$ |
|---|---|---|---|
| Sigworth (1998) | $X_i = R_{\phi i} A + \sigma G_i$ | $A, \sigma, \xi_x, \xi_y, \sigma_{xy}$ | $\alpha_i, x_i, y_i$ |
| kerdenSOM | $X_i = A_{ki} + \sigma G_i$ | $A_k, \sigma, \pi_k$ | $k_i$ |
| ML2D | $X_i = R_{\phi i} A_{ki} + \sigma G_i$ | $A_k, \sigma, \pi_k, \sigma_{xy}$ | $k_i, \alpha_i, \beta_i, \gamma_i, x_i, y_i$ |
| ML3D | $X_i = P_{\phi i} V_{ki} + \sigma G_i$ | $V_k, \sigma, \pi_k, \sigma_{xy}$ | $k_i, \alpha_i, \beta_i, \gamma_i, x_i, y_i$ |
| MLn3D | $X_i = s_i P_{\phi i} V_{ki} + \sigma G_i + n_i$ | $V_k, \sigma, \pi_k, \sigma_{xy}, s_i, n_i$ | $k_i, \alpha_i, \beta_i, \gamma_i, x_i, y_i$ |

Symbols $X_i$, $R_{\phi i}$, $A$, $\sigma$, $\alpha_i$, $x_i$, $y_i$, $\xi_x$, $\xi_y$, and $\sigma_{xy}$ were defined in the section that provided the theoretical basis. Subscripts $k$, with $k = 1, \ldots, K$ are used to indicate the $k$th model in multireference refinement schemes. Then $\pi_k$ are prior probabilities of that model and $k_i$ indicates that model $k$ is the correct one for the $i$th observed image. $V_k$ indicates a 3D model and $P_{\phi i}$ is a projection operation, with $\phi_i = (\alpha_i, \beta_i, \gamma_i, x_i, y_i)$ comprising three Euler angles and two in-plane translations. $s_i$ and $n_i$ are multiplicative and additive parameters for each $i$th image.

network, or self-organizing map, was achieved. This has the additional advantage that $K$ does not necessarily need to reflect the number of different classes in the data. The efficiency of this approach was demonstrated for cryo-EM images of large T-antigen and for rotational spectra of negatively stained particles of hexameric helicase G40P.

Second, Scheres *et al.* (2005a) proposed a 2D multireference alignment scheme, called *ML2D*. As for the kerdenSOM algorithm, in this algorithm $\Theta$ comprises multiple 2D images $A_k$ with $k = 1, \ldots, K$. In this case, however, the problems of alignment and classification were tackled simultaneously by treating both the in-plane transformations $\phi_i = (\alpha_i, x_i, y_i)$ and the class assignments $k_i$ as hidden variables. The ML method was again shown to yield much better model estimates than the maximum cross-correlation approach for simulated data with white Gaussian noise. It was also shown that these advantages were strongly reduced for simulated data with dependent noise. Nevertheless, for experimental cryo-EM images the ML2D algorithm was shown to be robust to the choice of the initial starting models and reference-free alignments and classifications could be obtained by starting the multireference alignments from averages of random subsets of the unaligned images. Application of this approach to cryo-EM images of large-T antigen particles led to the first-time visualization of an overhanging dsDNA probe in this complex.

The next EM–ML approach to single-particle analysis concerned a 3D multireference refinement scheme (Scheres *et al.*, 2007a). In this case, $K$ 3D reference maps $V_k$ are refined simultaneously against a structurally heterogeneous set of projections. The hidden variables of this problem comprise six parameters for every image: its class assignment $k_i$ and its 3D orientation as described by three Euler angles and two in-plane translations $\phi_i = (\alpha_i, \beta_i, \gamma_i, x_i, y_i)$.

For 3D refinements, the maximization step of the EM–ML algorithm is more complicated than in the 2D case. By expressing the 3D electron densities of the $K$ models as weighted sums of smooth radial basis functions called blobs (Marabini *et al.*, 1998), the reconstruction problem was expressed as a system of linear equations. Optimization of the log-likelihood function was shown to correspond to finding a weighted least-squares solution to the reconstruction problem, for which purpose a modified version of the algebraic reconstruction technique (Eggermont *et al.*, 1981) was developed.

Again, the ML approach was shown to be robust to high levels of noise and relatively insensitive to the starting model. Most significantly, it allowed separation of projections from distinct 3D structures by starting multireference refinements from random variations of a single, strongly low-pass filtered initial model. Thereby, the classification protocol, termed *ML3D*, is unsupervised as it does not depend on any prior knowledge about the structural variability in the data. Its efficiency and potentially wide applicability were demonstrated for two challenging cryo-EM data sets. ML3D classification separated projections of 70S *Escherichia coli* ribosomes in a ratcheted conformation and in complex with elongation factor G (EF-G) from unratcheted ribosomes without EF-G. Projections of large-T antigen were classified according to various degrees of bending along the central axis of the dodecameric complex.

More recently, ML3D classification was observed to yield suboptimal results in certain cases. For structurally heterogeneous cryo-EM data sets on *E. coli* 70S ribosomes and on human RNA polymerase II, rather than separating different conformations reconstructions at distinct intensities were obtained. The origin of the problem was found in a typical cryo-EM preprocessing step: image normalization.

In the normalization process, one aims to minimize additive and multiplicative variations in the signal among all images. Since the abundant noise makes it impossible to normalize the signal itself, it is common practice to normalize the noise instead. However, variations in signal-to-noise ratios or artifacts in the images often lead to small, remaining variations in the signal. These variations can often be ignored in conventional refinement schemes because normalized cross-correlation coefficients are invariant to additive or multiplicative factors. In ML refinements, however, the squared distance metric inside the PDF calculation, see Eq. (10.15), is highly sensitive to these variations.

To reduce the corresponding sensitivity of the ML approach to normalization errors, the model parameter set $\Theta$ was extended with a multiplicative and an additive factor for the signal in each experimental image ($s_i$, respectively, $n_i$ in Table 10.1). For both the 70S ribosome and the RNA polymerase II data sets, the corresponding approach, which was termed *MLn3D*, successfully classified distinct structural states (Scheres *et al.*, 2009b). For the 70S ribosome, this resulted in a previously unobserved conformation with spontaneous ratcheting and two tRNAs in hybrid states (Julian *et al.*, 2008).

## 3.2. Icosahedral viruses

The computation of structures having icosahedral symmetry from cryo-EM images is a particular case of single-particle analysis. However, because of the high-order rotational symmetry of the icosahedral group, several special approaches have been developed, which are described in this section.

Incorporation of the icosahedral symmetry is often done by representing the electron scattering intensity of the particle as a weighted sum of basis functions. Because all the operations in the icosahedral group are rotational operations, it is most natural to use the spherical coordinate system ($r$, $\theta$, $\phi$) in which case the symmetry constrains the angular behavior of the basis functions (i.e., $\theta$ and $\phi$) but not the radial behavior of the basis functions (i.e., $r$). The usual choice is to use basis functions that are a product of angular and radial functions where the angular part is a linear combination of spherical harmonics as introduced by Laporte (1948) and developed in subsequent years by many authors including Zheng and Doerschuk (1996). The experimental images are roughly linear transformations of the unknown weights in the weighted sum of basis functions, which is important for practical computation of the 3D reconstruction. However, as in the general single-particle case described above, this 3D reconstruction is dependent on unknown parameters, for example, the projection orientations $\phi_i$ and class assignments $k_i$. Different types of MLEs result depending on the treatment of these parameters.

In the original work of Vogel *et al.* (1986), Provencher and Vogel (1988), Vogel and Provencher (1988), which resulted in the "ROSE" algorithm, a Gaussian MLE is used which estimates both the weights in the weighted sum of basis functions and the parameters that determine the linear transformation. That is, rather than treating the unknown projection orientations $\phi_i$ and class assignments $k_i$ as hidden variables as described for the general single-particle case, this approach includes these parameters in the model $\Theta$. The advantage of this approach is that no probabilistic information on the behavior of the parameters is required with the disadvantage of having to solve a difficult optimization problem, especially as the number of images grows.

This type of approach was further developed by Doerschuk and Johnson (2000) and Yin *et al.* (2001, 2003), who proposed to treat the projection orientations and class assignments as hidden variables ($\mathcal{Y}$) and use an EM–ML algorithm to optimize the corresponding marginal likelihood function of Eq. (10.14). In addition, these authors use alternative radial basis functions that are linear combinations of spherical Bessel functions. These functions have two advantages over the Laguerre polynomials used in the ROSE algorithm. The first advantage is that they are nonzero only for a range of $r$, which allows straightforward masking of the 2D images and 3D reconstruction. The second advantage is that the 3D Fourier transform of the product of the angular and radial basis functions can be computed symbolically, so the projection slice theorem can be used to compute the projection of the electron scattering intensity in an arbitrary direction. Kam and Gafni (1985) describe an alternative approach using much of the same mathematics for the description of the electron scattering intensity. However, the optimization problem that is solved does not appear to be a ML problem.

The approach of locking symmetry into the basis functions is general although, depending on the number of operators in the symmetry group, it is more or less valuable in terms of reducing computation. Zheng and Doerschuk (1996) describe the necessary basis functions for all the platonic symmetries. Prust *et al.* (2009) describe the application of the approach to the rotational symmetry of the tail of infectious bacteriophage P22. Chen (2008) describes the application of the approach to objects with helical symmetry using standard basis functions (Moody, 1990), Lee *et al.* (2009) describe new basis functions focused on helical symmetry, and Lee (2009) uses these basis functions to solve 3D reconstruction problems for objects with helical symmetry. Even in the case where there is no symmetry, describing the electron scattering intensity of the object by a weighted sum of basis functions is possible. Approaches based on weighted sums of basis functions for this case appear to be essentially the same as the single-particle methods described above.

Besides providing numerical values for each of the unknown parameters, ML theory also provides some information on the size of the errors. The key result (Efron and Hinkley, 1978) is that the error between the MLE and the true values of the model parameters is approximately Gaussian distributed with mean vector 0 and a covariance matrix that is the negative inverse of the matrix of mixed second-order partial derivatives, that is, the Hessian, of the log-likelihood function with respect to the model parameter vector. However, there are at least two important challenges in using this result in cryo-EM. First, it may be more or less difficult to connect these error estimates with the most common method for measuring the performance of a cryo-EM 3D reconstruction, which is the Fourier shell correlation (FSC) (Harauz and van Heel, 1986). Second, in the case where the connection can be made, it is quite likely that there are computational complexity issues since the number of model parameters is typically so large that computation of the Hessian matrix is impractical. Still, these ideas have been used to create a ML variant of the FSC for a virus problem with symmetry, specifically, in the *ab initio* 3D reconstruction of the tail of the infectious bacteriophage P22 (Prust *et al.*, 2009). In this situation, the first challenge was solved by a Monte Carlo procedure and the second challenge was not present because the resolution was low, so the number of parameters was relatively small. In a higher resolution problem, it is probably necessary to make a diagonal approximation to the Hessian matrix.

The theory in the previous paragraph describes performance once the data are recorded. An analogous theory may be used to "predict" performance before any data are recorded. Given a PDF $P(\mathcal{X}|\Theta)$, the Cramer-Rao bound (e.g., Marzetta, 1993) will give the minimal achievable variance for any unbiased estimator, including the unbiased MLE. Thereby, one could design the optimal experiment by computing this bound for different experimental strategies and choosing the experiment that makes the bound as small as possible. The required calculations are similar to the calculations described in the previous paragraph and have been demonstrated in a tentative way in Doerschuk and Johnson (2000).

## 3.3. 2D crystallography

Electron crystallography is a cryo-EM method that has provided the highest resolution structures of protein assemblies. Because the penetration distance of electrons is limited to approximately 100 nm, the acquisition of data from

3D protein crystals—as are used for X-ray crystallography—is not practical. However, planar 2D crystals, composed of a single layer or double layer of proteins, can be imaged in the electron microscope. Of particular interest are 2D crystals formed of membrane proteins embedded in a lipid bilayer membrane, as analysis of these crystals provides the most reliable structural information for these proteins. For further details on 2D crystallization and the subsequent structure determination process, the reader is referred to Chapter 4 of this volume and Chapter 5 of Vol. 483.

In a 2D crystal, there are fewer lattice contacts than in 3D crystals and the crystal lattice often shows substantial disorder. Fortunately, in the electron microscope an image can be formed of a 2D crystal and the lattice disorder can be removed by computational "unbending" of the crystal image. In this process, the micrograph is analyzed to locate the center of each unit cell. To the extent that its center deviates from an ideal lattice, each unit cell is then shifted to bring it into the proper position.

In the end, the goal of analysis of a crystal image is to estimate the 2D density of a unit cell of the lattice. The "unbending" method is equivalent to a conventional alignment and averaging of 2D images, and a superior alternative is ML estimation of the unit-cell image as in Eq. (10.24). Zeng *et al.* (2007) implemented the EM–ML algorithm, where each data image $X_i$ is a unit cell in the micrograph, and the transformations $\phi$ are constrained to small translations from lattice points and to a small angular deviations from the lattice directions. That is, the hidden variables $\phi_i$ are modeled as Gaussian-distributed with small SDs which are estimated as part of the model. The result is improved resolution in structures obtained from 2D crystals having substantial disorder. In the ML estimation, it was assumed that the disorder in the crystals was confined to in-plane translations and rotations; however, a further extension can be imagined in which a 3D algorithm similar to ML3D could be employed to account for small out-of-plane deviations of the unit cells as well. In addition, one might envision improvements by relating the disorder parameters of neighboring unit cells to each other, as crystal disorder is often, to some extent, a continuous phenomenon.

## 3.4. Tomography

In electron tomography a reconstruction of a unique 3D object, for example, an entire cell, is obtained by combining a series of projections that are recorded at different tilt angles. The process of tomographic reconstruction is described in more detail in Chapter 14, and its application to HIV-1 is described in Chapter 14 of Vol. 483. Electron tomograms are typically extremely noisy because the electron dose over the whole tilt series needs to be limited in order to prevent radiation damage. Still, averaging over multiple copies of the same macromolecular complex may improve the signal-to-noise ratios provided that the individual subtomograms may be aligned (and classified in the case of structural heterogeneity). Apart from the increased dimensionality of the data vectors, subtomogram averaging is conceptually very similar to 2D averaging approaches in single-particle analysis. Again, the unknown orientations $\phi_i = (\alpha_i, \beta_i, \gamma_i, x_i, y_i, z_i)$ and/or the class assignments $k_i$ of the individual subtomograms may be treated as hidden variables and the EM–ML algorithm may be employed to obtain MLEs of the underlying signals.

The first reported ML approach to subtomogram averaging was the application of the kerdenSOM algorithm to aligned subtomograms of insect flight muscles (Pascual-Montano *et al.*, 2002). In this case, as for the classification of 2D images described above, the class assignments $k_i$ are treated as hidden variables and an additional regularization term to the marginal log-likelihood function results in a neural network-like behavior of the $K$ output classes. However, compared to 2D averaging an additional complication arises in subtomogram averaging that was not taken into account in the kerdenSOM approach. Due to experimental limitations on the tilt angle, electron tomography data cannot be measured in its totality. In the case of single-axis tilting, a wedge-shaped region in Fourier space remains experimentally inaccessible. This region is commonly referred to as the *missing wedge* and subtomogram averaging procedures that do not take the missing wedges into account have been observed to yield suboptimal results (Walz *et al.*, 1997). A variant of the kerdenSOM algorithm that takes missing wedges into account was mentioned in a structural study of cadherins (Al-Amoudi *et al.*, 2007), but details concerning this algorithm were never described.

Conventional alignment and classification approaches for subtomograms have typically restricted the cross-correlation measure to the observed regions in Fourier space. Subtomogram averaging may then be performed by *weighted averaging*, that is, by dividing the sum of all subtomograms by the times that each point in Fourier space has been measured. However, the EM–ML algorithm itself may provide a more elegant solution to the missing wedge problem, since this algorithm was ultimately designed to deal with incomplete data problems. Apart from the unknown orientations $\phi_i$ and the class assignments $k_i$, one may also treat the data points inside the missing wedges as hidden variables. An EM–ML algorithm that optimizes the corresponding marginal log-likelihood function was recently introduced (Scheres *et al.*, 2009a). This algorithm "estimates" the missing data points by calculating probability-weighted averages over all possible values, just as it does for the orientation and class assignments. However, the possible values of each of the missing data points range from minus infinity to infinity, which obviously prohibits the use of numerical integrations. Fortunately, an analytical solution exists where the missing data points for each individual subtomogram are replaced by the corresponding values in the (oriented) reference to which the subtomogram is compared. In this way, the incomplete experimental data are complemented with the currently available information from the model. Therefore, this algorithm does not need to divide the sum of all subtomograms by the number of times each point in Fourier space was observed, which prevents numerical instabilities for data points that were (almost) never observed.

Similar to the ML2D approach, the EM–ML algorithm for subtomogram averaging tackles the problems of alignment and classification simultaneously through a multireference refinement scheme (with hidden variables $\phi_i$ and $k_i$). Moreover, the algorithm may be run in a completely unsupervised manner by starting the multireference refinements from averages of random subsets of subtomograms in random orientations. The advantages of this approach were illustrated using simulated data and reference-free class averages were obtained for experimental subtomograms of groEL and groEL/groES complexes. In addition, the same approach was shown to be effective for the reference-free alignment of random conical tilt (RCT) reconstructions from single-particle experiments on p53. Note that just like subtomograms, RCT reconstructions are 3D data vectors with missing regions in Fourier space, although RCT reconstructions have missing cones rather than missing wedges. An alternative ML approach to the alignment of RCT reconstructions was mentioned by Sander *et al.* (2006), although also for this algorithm the details have not yet been presented.

## 4. The Statistical Data Model

The data model that underlies the PDF calculations plays a crucial role in the EM–ML approach. The PDF defines the log-likelihood target function and is used to calculate the probability distributions over the hidden variables in the model updates. If the model does not describe the data accurately, the benefits of the statistical approach may be lost altogether. All approaches described above share similar assumptions about the signal and they all assume that the noise is independent, additive, and Gaussian. As outlined above, this model results in a computationally attractive algorithm, but how accurately does this model describe cryo-EM images?

Thin samples of biological molecules fulfill the weak phase approximation, the theory of image formation which is used to describe the phase-contrast images of weakly scattering specimens. Under this approximation, the contrast in cryo-EM images is linearly related to the projected object potential. The latter justifies the use of standard X-ray integrals in the projection operators of the 3D approaches described above. The same theory is also used to derive imaging effects of the electron microscope in the form of the contrast transfer function (CTF). The accuracy of this model, at least for thin specimens, is reflected in the (near-)atomic resolutions that have been obtained with it for single particles and 2D crystals (Cheng and Walz, 2009). However, of all the ML approaches described above only those for icosahedral viruses (e.g., Yin *et al.*, 2003) include an explicit description of the CTF. All the other approaches simply ignore the CTF or employ suboptimal CTF correction strategies (e.g., Zeng *et al.*, 2007). Consequently, these approaches may fail to provide good model estimates in cases where the CTF plays an important role, as is often the case in medium-high-resolution refinements. For more details on image restoration and the importance of CTF correction, the reader is referred to Chapter 2.

The noise term in the data model is used to represent all features in the observed images that are not explained by the description of the signal. In cryo-EM images the major source of noise is *shot noise*. This type of noise follows a Poisson distribution and arises from statistical fluctuations in the small number of imaging electrons (typically 10–20 e/Å$^2$). EM–ML algorithms for Poisson distributions are computationally more complicated than those for Gaussians. But as the Poisson distribution approaches the Gaussian for larger numbers of imaging electrons, the latter provides a good approximation for pixels that span many squared Angstroms (Sigworth, 2004). Apart from the abundant shot noise, several other sources of noise exist. *Structural noise* arises from the irreproducible density of the ice that surrounds the particle or the carbon layer that is used to support it. Also, structural variability in the particles that is not described by the data model may be considered as a source of structural noise. *Detector noise* arises from the stochastic nature of the interactions of electrons with the detector—the digital camera or photographic film which is used to acquire the image. The result is a random variation in both the point-spread function and amplitude of the single-electron response of the detector. Baxter *et al.* (2009) experimentally measured the relative contributions of the different types of noise. For typical cryo-EM images on a well-behaved ribosome sample, they found the combination of shot noise and detector noise to be one order of magnitude larger than the structural noise. The latter was found to be approximately of the same power as the underlying signal in the images. Taken together, one may assume that, at least for pixels that span multiple squared Angstroms, Gaussian distributions describe the combination of the various independent sources of noise in cryo-EM images reasonably well.

The assumption of independent noise may be more problematic. The independence between all pixels was used in Eq. (10.11) to relate the PDF of an entire image with the PDFs of the individual pixels, and ultimately allows for fast calculations of the probabilities. Independent noise is uncorrelated from pixel to pixel and has a flat power spectrum. Therefore, it is also called *white noise*. Although shot noise is intrinsically white, the background noise in a cryo-EM micrographs is typically not white for several reasons.

Most importantly, there is a fall-off with resolution of both the signal and the noise due to the point spread functions of the microscope and the electron detector. In addition, various sources of structural noise, such as density for the ice or carbon support or structural variations in the particles are expected to have strong correlations among nearby pixels. Ignoring these correlations in the PDF calculations will lead to a suboptimal behavior of the EM–ML approach.

Two approaches to dealing with nonwhite noise in cryo-EM images have been proposed. In the most straightforward approach, the data are adapted to the white-noise model by so-called *prewhitening* of the observed images (Sigworth, 2004; Zeng *et al.*, 2007). In this preprocessing step, the higher frequencies in the images are boosted based on an estimate for the power spectrum of the noise. The resulting images have a flat power spectrum and are processed using the conventional ML approach. Alternatively, the model itself may be adapted to describe nonwhite data. The latter was achieved by reformulation of the PDF in the Fourier domain. Assuming independent, Gaussian noise on all Fourier components, the PDF may be calculated in a way that is highly similar to Eq. (10.11) but involves Fourier components rather than image pixels. The uniqueness of the expression in the Fourier domain lies in the possibility to estimate the SD of the noise ($\sigma$) as a function of spatial frequency, thereby explicitly modeling nonwhite noise. In addition, the formulation in the Fourier domain allows for a convenient incorporation of the CTF into the data model. This approach was called MLF, for maximum likelihood in the Fourier domain, and it was implemented for the problems of 2D and 3D multireference refinements in single-particle analysis (Scheres *et al.*, 2007b). For simulated CTF-filtered images, the MLF approach was shown to be superior to the ML approach for white noise and its usefulness in the experimental situation was illustrated for structurally heterogeneous cryo-EM data sets on 70S *E. coli* ribosomes and large-T antigen complexes.

Perhaps, the most basic of all assumptions in the data models discussed above is the one that is violated most often. By including an image in the data set one assumes that it contains signal. However, especially for smaller particles (with molecular weights well under 1 MDa) it is often extremely difficult to distinguish genuine particles of interest from artifacts in the micrographs. Consequently, many cryo-EM data sets contain large amounts of particles that do not contain a common underlying signal. This problem typically becomes more severe with the use of (semi-) automated particle selection procedures, which are still outperformed by expert human beings (Zhu *et al.*, 2004). Currently, none of the available ML approaches can distinguish between genuine and artifactual particles, but one approach does aim to provide robustness against outliers in the data (Scheres and Carazo, 2009). This approach uses *t*-distributions, which have wider tails than Gaussians, to accommodate atypical observations. The resulting algorithm downweights images with relatively large residuals. However, application to cryo-EM images did not show clear advantages over the Gaussian algorithm, probably because typical artifacts in the data have relatively small residuals compared to the high levels of noise.

## 5. Reducing Computational Requirements

Perhaps, the major obstacle to a more wide-spread use of the EM–ML approach in cryo-EM image processing is currently its computational load. For many applications, the exhaustive integrations over the hidden variables in the expectation step require large amounts of computing time. For example, in single-particle analysis the integrations over the 3D rotations and in-plane translations span a five-dimensional space that is extended to even six dimensions in the multireference case. That is, for each 3D rotation ($\alpha_i$, $\beta_i$, $\gamma_i$) and each class $k_i$, all $J$ possible in-plane translations ($x_i$, $y_i$) are to be considered. Conventional approaches typically divide this problem into several lower-dimensional ones, for example, by searching in-plane translations only for a single 3D rotation. Although this strategy is not guaranteed to converge, it does provide major speed-ups, for example, of two orders of magnitude, compared to the exhaustive integrals in the ML approach. Consequently, the latter has become known as a computationally expensive technique.

Several contributions have been made to reduce the computational requirements of ML approaches in cryo-EM. Two approaches for single-particle analysis are based on the observation that many of the alignment parameters give rise to near-zero probabilities, especially when the algorithm nears convergence (Scheres *et al.*, 2005b; Tagare *et al.*, 2008). Consequently, the ML approach can be speeded up considerably by *domain reduction* strategies, where the integrals over the hidden variables are restricted to those alignments that contribute significantly to the weighted averages.

The first proposal to domain reduction involved skipping part of the integrations over all in-plane translations (Scheres *et al.*, 2005b). In this approach, optimal translations from the previous iteration are used to calculate probabilities for precentered images as a function of the in-plane rotation for each 2D reference (or projection of the 3D references). Then, for those in-plane rotations where this probability is smaller than a cut-off value times its maximum, the integration over all translations is skipped. Although in a strict sense the resulting algorithm is not guaranteed to converge, in practice a sevenfold acceleration could be obtained for a small 2D problem without notably changing the convergence behavior. For larger 3D problems (Scheres *et al.*, 2007a), the same approach was estimated to yield 10–20-fold accelerations (unpublished results).

Even higher speed-ups of 30–60 times are reported by combining a related domain reduction strategy with an additional *grid interpolation* strategy (Tagare *et al.*, 2008). The latter is based on the observation that the squared difference terms in Eq. (10.17) vary much more smoothly with $\phi$ than the probability terms themselves. Consequently, one may calculate the squared difference terms on a relatively coarse grid and use B-spline interpolation to evaluate the corresponding values on a much finer grid. A similar approach had previously been proposed for the maximum-seeking problem in conventional cross-correlation-based strategies (Sander *et al.*, 2003).

While domain reduction and grid interpolation provide *approximations* of the original EM–ML algorithm, the ML approach for icosahedral viruses in (Doerschuk and Johnson, 2000) may be reformulated in an *exact* manner that is much faster (Lee *et al.*, 2007). In the original approach, the spherical harmonics themselves already provide a highly efficient way to sample 3D rotations and to restrict the resolution of the model. Still, using an idea from Navaza (2003), the same problem may be expressed more efficiently by application of a linear transformation to the observed data, which allows expressing the expectation step in terms of lower-dimensional integrations over the hidden variables. Although the resulting approach thus involves an additional precomputation step, each expectation step itself is accelerated up to 25 times and has reduced storage complexity.

## 6. Outlook

With the rapid increase of available computing power and continuing efforts to accelerate EM–ML approaches, the applicability of ML methods is expected to increase significantly in the near future. Already with relatively modest computer clusters, which will be replaced soon by multicore desktop computers, medium-resolution single-particle multireference refinements are feasible within the time span of only a few days (also see Chapter 11).

Preliminary investigations further indicate that the EM–ML approach may be particularly suitable for acceleration by general purpose computing on the graphical processing unit, which provides high-computing power at relatively low cost (Tagare, Sigworth *et al.*, in preparation).

In addition, apart from continuing developments in domain reduction and grid interpolation strategies, one may employ algorithms to optimize log-likelihood targets with faster convergence rates than the classical EM–ML algorithm. For example, similar to block-Kaczmarz methods for iterative reconstruction (Eggermont *et al.*, 1981), one may perform partial expectation steps by processing the experimental images in subsets, or *blocks*. Thereby, convergence is speeded by incorporating new information faster into the model (Neal and Hinton, 1998). In single-photon emission computed tomography and positron emission tomography, this approach has resulted in an order of magnitude speed-up (Hudson and Larkin, 1994), and investigations into the efficiency of a similar approach for cryo-EM single-particle analysis are currently underway (Scheres *et al.*, unpublished results).

However, the most important future contributions are not expected from making ML methods faster but from efforts to make them better. As explained above, the statistical data model and in particular the assumptions about the noise in the data remain apt for improvement and also the robustness to artifactual particles in the data still needs to be improved. Moreover, there are various aspects of ML theory for which applicability to cryo-EM image processing remains open for more in-depth investigations. While structure validation and resolution assessments remain critical areas of research in the field, ML theory promises reliable error estimates on the model parameters through the use of the Hessian to the log-likelihood. The first indication that these estimates may indeed be useful is the successful definition of a ML equivalent of the FSC as recently explored by Prust *et al.* (2009). In addition, Cramer-Rao lower bounds to the log-likelihood function may be employed to design optimal experimental setups, as was tentatively explored in (Doerschuk and Johnson, 2000).

Finally, there may be improvements to be gained from using MAP estimation instead of ML estimation. As was mentioned in conjunction with Eq. (10.1), the inclusion of prior information in the form of a prior probability $P(\Theta)$ should improve the quality of final models. The advantage of MAP estimation appears when the amount of experimental data is limited, so that the prior probability term makes a substantial contribution. An important example of a prior probability function, well known in X-ray crystal structure determination, is one that penalizes all 3D models in which the density outside the particle boundary (the "solvent" or "ice" density of the map) is nonzero. This choice of a prior probability function is the formal basis of the well-known "solvent flattening" process.

In summary, we have seen that the theory of maximum-likelihood estimation provides improved methods for the processing of cryo-EM data. Of particular interest is its flexible theoretical framework, which allows the limitations of experimental data—heterogeneity, noise, missing wedges, and so on—to be readily and rigorously modeled. With these data models, the process of determining macromolecular structures is both improved and placed on a firmer statistical foundation.

## ACKNOWLEDGMENTS

F. J. S. was supported by NIH Grant P01-GM062580, P. C. D. by the US National Science Foundation (Grant 0735297), and J. M. C. and S. H. W. S. by the Spanish Ministry of Science and Technology (Grants CDS2006-0023, BIO2007-67150-C01, and ACI2009-1022) and the National Heart, Lung and Blood Institute (Grant R01HL070472).

## REFERENCES


Al-Amoudi, A., Diez, D. C., Betts, M. J., and Frangakis, A. S. (2007). The molecular architecture of cadherins in native epidermal desmosomes. *Nature* **450,** 832–837.

Baxter, W. T., Grassucci, R. A., Gao, H., and Frank, J. (2009). Determination of signal-to-noise ratios and spectral SNRs in cryo-EM low-dose imaging of molecules. *J. Struct. Biol.* **166,** 126–132.

Chen, Q. (2008). Nonlinear stochastic tomography reconstruction algorithms for objects with helical symmetry and applications to virus structures, Master's thesis, School of Electrical and Computer Engineering, Cornell University.

Cheng, Y., and Walz, T. (2009). The advent of near-atomic resolution in single-particle electron microscopy. *Annu. Rev. Biochem.* **78,** 723–742.

Dempster, A. P., Laird, N. M., and Rubin, D. B. (1977). Maximum-likelihood from incomplete data via the EM algorithm. *J. R. Stat. Soc. Ser. B* **39,** 1–38.

Doerschuk, P. C., and Johnson, J. E. (2000). Ab initio reconstruction and experimental design for cryo electron microscopy. *IEEE Trans. Inf. Theory* **46,** 1714–1729.

Efron, B., and Hinkley, D. V. (1978). Assessing the accuracy of the maximum likelihood estimator: Observed versus expected fisher information. *Biometrika* **65,** 457–487.

Eggermont, P. P., Herman, G. T., and Lent, A. (1981). Iterative algorithms for large partitioned linear systems, with applications to image reconstruction. *Linear Algebra Appl.* **40,** 37–67.

Frank, J. (2006). Three-dimensional Electron Microscopy of Macromolecular Assemblies. Oxford University Press, New York.

Harauz, G., and van Heel, M. (1986). Exact filters for general geometry three dimensional reconstruction. *Optik* **73,** 146–156.

Hudson, M., and Larkin, R. (1994). Accelerated image reconstruction using ordered subsets of projection data. *IEEE Trans. Med. Imag.* **13,** 601–609.

Julian, P., Konevega, A. L., Scheres, S. H. W., Lazaro, M., Gil, D., Wintermeyer, W., Rodnina, M. V., and Valle, M. (2008). Structure of ratcheted ribosomes with tRNAs in hybrid states. *Proc. Natl. Acad. Sci. USA* **105,** 16924–16927.

Kam, Z., and Gafni, I. (1985). Three-dimensional reconstruction of the shape of human wart virus using spatial correlations. *Ultramicroscopy* **17,** 251–262.

Laporte, O. (1948). Polyhedral harmonics. *Naturforschung* **3a,** 447–456.

Lee, S. (2009). Maximum likelihood reconstruction of 3D objects with helical symmetry from 2D projections of unknown orientation and application to electron microscope images of viruses. PhD thesis, School of Electrical and Computer Engineering, Purdue University West Lafayette, Indiana, USA.

Lee, J., Doerschuk, P. C., and Johnson, J. E. (2007). Exact reduced-complexity maximum likelihood reconstruction of multiple 3-D objects from unlabeled unoriented 2-D projections and electron microscopy of viruses. *IEEE Trans. Image Process.* **16,** 2865–2878.

Lee, S., Doerschuk, P. C., and Johnson, J. E. (2009). Reciprocal space representations of helical objects and their projection images for helices constructed from motifs without spherical symmetry. *Ultramicroscopy* **109,** 253–263.

Marabini, R., Herman, G. T., and Carazo, J. M. (1998). 3D reconstruction in electron microscopy using ART with smooth spherically symmetric volume elements (blobs). *Ultramicroscopy* **72,** 53–65.

Marzetta, T. (1993). A simple derivation of the constrained multiple parameter Cramer-Rao bound. *IEEE Trans. Signal Process.* **41,** 2247–2249.

Moody, M. (1990). Image analysis of electron micrographs. *In* "Biophysical Electron Microscopy: Basic Concepts and Modern Techniques," (P. W. Hawkes and U. Valdre, eds.), pp. 145–287. Academic Press, New York.

Navaza, J. (2003). On the three-dimensional reconstruction of icosahedral particles. *J. Struct. Biol.* **144,** 13–23.
Neal, R., and Hinton, G. E. (1998). A view of the EM algorithm that justifies incremental, sparse, and other variants. *Learning in Graphical Models*, p. 355–368, Kluwer Academic Publishers, Dordrecht The Netherlands.
Pascual-Montano, A., Donate, L. E., Valle, M., Barcena, M., Pascual-Marqui, R. D., and Carazo, J. M. (2001). A novel neural network technique for analysis and classification of EM single-particle images. *J. Struct. Biol.* **133,** 233–245.
Pascual-Montano, A., Taylor, K. A., Winkler, H., Pascual-Marqui, R. D., and Carazo, J. (2002). Quantitative self-organizing maps for clustering electron tomograms. *J. Struct. Biol.* **138,** 114–122.
Provencher, S. W., and Vogel, R. H. (1988). Three-dimensional reconstruction from electron micrographs of disordered specimens. I. method. *Ultramicroscopy* **25,** 209–221.
Prust, C. J., Doerschuk, P. C., Lander, G. C., and Johnson, J. E. (2009). Ab initio maximum likelihood reconstruction from cryo electron microscopy images of an infectious virion of the tailed bacteriophage p22 and maximum likelihood versions of Fourier shell correlation appropriate for measuring resolution of spherical or cylindrical objects. *J. Struct. Biol.* **167,** 185–199.
Redner, R., and Walker, H. (1984). mixture densities, maximum likelihood and the EM algorithm. *SIAM Rev.* **26**(239), 195.
Sander, B., Golas, M. M., and Stark, H. (2003). Corrim-based alignment for improved speed in single-particle image processing. *J. Struct. Biol.* **143,** 219–228.
Sander, B., Golas, M. M., Makarov, E. M., Brahms, H., Kastner, B., Lhrmann, R., and Stark, H. (2006). Organization of core spliceosomal components u5 snRNA loop i and U4/U6 Di-snRNP within U4/U6.U5 Tri-snRNP as revealed by electron cryomicroscopy. *Mol. Cell* **24,** 267–278.
Scheres, S. H. W., and Carazo, J. M. (2009). Introducing robustness to maximum-likelihood refinement of electron-microscopy data. *Acta Crystallogr. D* **65,** 672–678.
Scheres, S. H. W., Valle, M., and Carazo, J. M. (2005a). Fast maximum-likelihood refinement of electron microscopy images. *Bioinformatics* **21**(Suppl. 2), ii243–ii244.
Scheres, S. H. W., Valle, M., Nunez, R., Sorzano, C. O. S., Marabini, R., Herman, G. T., and Carazo, J. M. (2005b). Maximum-likelihood multi-reference refinement for electron microscopy images. *J. Mol. Biol.* **348,** 139–149.
Scheres, S. H. W., Gao, H., Valle, M., Herman, G. T., Eggermont, P. P. B., Frank, J., and Carazo, J. M. (2007a). Disentangling conformational states of macromolecules in 3D-EM through likelihood optimization. *Nat. Methods* **4,** 27–29.
Scheres, S. H. W., Nunez-Ramirez, R., Gomez-Llorente, Y., San Martin, S., Eggermont, P. P. B., and Carazo, J. M. (2007b). Modeling experimental image formation for likelihood-based classification of electron microscopy data. *Structure* **15,** 1167–1177.
Scheres, S. H. W., Melero, R., Valle, M., and Carazo, J. (2009a). Averaging of electron subtomograms and random conical tilt reconstructions through likelihood optimization. *Structure* **17,** 1563–1572.
Scheres, S. H. W., Valle, M., Grob, P., Nogales, E., and Carazo, J. (2009b). Maximum likelihood refinement of electron microscopy data with normalization errors. *J. Struct. Biol.* **166,** 234–240.
Sigworth, F. J. (1998). A maximum-likelihood approach to single-particle image refinement. *J. Struct. Biol.* **122,** 328–339.
Sigworth, F. J. (2004). Classical detection theory and the cryo-EM particle selection problem. *J. Struct. Biol.* **145,** 111–122.
Tagare, H. D., Sigworth, F., and Barthel, A. (2008). Fast, adaptive expectation-maximization alignment for cryo-EM. *Med. Image Comput. Comput. Assist. Interv.* **11,** 855–862.

Vogel, R. H., and Provencher, S. W. (1988). Three-dimensional reconstruction from electron micrographs of disordered specimens. II. implementation and results. *Ultramicroscopy* **25,** 223–239.

Vogel, R. H., Provencher, S. W., von Bonsdorff, C., Adrian, M., and Dubochet, J. (1986). Envelope structure of semliki forest virus reconstructed from cryo-electron micrographs. *Nature* **320,** 533–535.

Walz, J., Typke, D., Nitsch, M., Koster, A. J., Hegerl, R., and Baumeister, W. (1997). Electron tomography of single ice-embedded macromolecules: Three-dimensional alignment and classification. *J. Struct. Biol.* **120,** 387–395.

Yin, Z., Zheng, Y., and Doerschuk, P. C. (2001). An ab initio algorithm for low-resolution 3-D reconstructions from cryoelectron microscopy images. *J. Struct. Biol.* **133,** 132–142.

Yin, Z., Zheng, Y., Doerschuk, P. C., Natarajan, P., and Johnson, J. E. (2003). A statistical approach to computer processing of cryo-electron microscope images: Virion classification and 3-D reconstruction. *J. Struct. Biol.* **144,** 24–50.

Yin, Z., Doerschuk, P. C., and Gelfand, S. B. (2004). Model calculations for joint pattern recognition and signal reconstruction in cryo electron microscopy. *Commun. Inf. Syst.* **4,** 73–88.

Zeng, X., Stahlberg, H., and Grigorieff, N. (2007). A maximum likelihood approach to two-dimensional crystals. *J. Struct. Biol.* **160,** 362–374.

Zheng, Y., and Doerschuk, P. C. (1996). Explicit orthonormal fixed bases for spaces of functions that are totally symmetric under the rotational symmetries of a platonic solid. *Acta Crystallogr. A* **52,** 221–235.

Zhu, Y., Carragher, B., Glaeser, R. M., Fellmann, D., Bajaj, C., Bern, M., Mouche, F., de Haas, F., Hall, R. J., Kriegman, D. J., Ludtke, S. J., Mallick, S. P., *et al.* (2004). Automatic particle selection: Results of a comparative study. *J. Struct. Biol.* **145,** 3–14.

CHAPTER ELEVEN

# Classification of Structural Heterogeneity by Maximum-Likelihood Methods

Sjors H. W. Scheres[1]

## Contents



## Abstract

With the advent of computationally feasible approaches to maximum-likelihood (ML) image processing for cryo-electron microscopy, these methods have proven particularly useful in the classification of structurally heterogeneous single-particle data. A growing number of experimental studies have applied these algorithms to study macromolecular complexes with a wide range of structural variability, including nonstoichiometric complex formation, large

Biocomputing Unit, Centro Nacional de Biotecnología—CSIC, Cantoblanco, Madrid, Spain
[1] Current address: MRC Laboratory of Molecular Biology, Hills Road, Cambridge, United Kingdom

*Methods in Enzymology*, Volume 482
ISSN 0076-6879, DOI: 10.1016/S0076-6879(10)82012-9

conformational changes, and combinations of both. This chapter aims to share the practical experience that has been gained from the application of these novel approaches. Current insights on how to prepare the data and how to perform two- or three-dimensional classifications are discussed together with the aspects related to high-performance computing. Thereby, this chapter will hopefully be of practical use for those microscopists wishing to apply ML methods in their own investigations.

## 1. Introduction

An increasing number of maximum-likelihood (ML) methods for image processing have recently become available to the electron microscopist. These methods hold great potential for the data analysis in a variety of different cryo-electron microscopy (cryo-EM) modalities. However, a literature search for the applications of ML approaches in experimental studies shows that these almost exclusively concern classification tasks in (asymmetric) single-particle analysis.

If one excludes the methodological papers themselves, no reports on the application of ML approaches to two-dimensional (2D) alignment (Sigworth, 1998) or icosahedral virus reconstruction (Doerschuk and Johnson, 2000; Yin *et al.*, 2001, 2003) are available. ML processing of 2D crystals (Zeng *et al.*, 2007) has so far only been applied to cyclic nucleotide-modulated potassium channel, MloK1 (Chiu *et al.*, 2007), and ML classification of sub-tomograms by the kerdenSOM algorithm (Pascual-Montano *et al.*, 2002) has only been reported for cadherins (Al-Amoudi *et al.*, 2007). In contrast, multiple reports are available that describe the applications of ML classification approaches to single-particle analysis. Both the kerdenSOM (Pascual-Montano *et al.*, 2001) and the ML2D algorithms (Scheres *et al.*, 2005) have been applied to multiple 2D studies (Table 11.1), and various reports describing three-dimensional (3D) analysis of macromolecular complexes with different types of structural heterogeneity by ML3D classification (Scheres *et al.*, 2007a) are available (Table 11.2).

Perhaps, an important reason for the relatively widespread use of the ML algorithms for single-particle classification lies in their implementation in the Xmipp package (Sorzano *et al.*, 2004b). This open-source software package has a graphical interface that guides the user through the image processing workflow and eases the task of parallel execution on a variety of hardware architectures (Scheres *et al.*, 2008). This facilitates the testing of new computer programs by the inexperienced experimentalist and enhances the visibility of novel algorithms.

Still, it is often hard to deduce how to use new methods from the technical papers that describe them. Moreover, it is typically not until a

**Table 11.1** ML applications in single-particle 2D analysis

| Reference | Sample | ML approach |
|---|---|---|
| Gomez-Lorenzo *et al.* (2003)[a] | SV40 large T antigen | kerdenSOM |
| Dang *et al.* (2005)[b] | Cytolysin | kerdenSOM |
| Gomez-Llorente *et al.* (2005)[c] | MCM | kerdenSOM |
| Gubellini *et al.* (2006)[d] | Photosynthetic core complex | ML2D + kerdenSOM |
| Nunez-Ramirez *et al.* (2006)[e] | G40P | kerdenSOM |
| Stirling *et al.* (2006)[f] | CCT:PhLP3:tubulin | ML2D |
| Valle *et al.* (2006)[g] | SV40 large T antigen | kerdenSOM |
| Martin-Benito *et al.* (2007a)[h] | CCT:Fab | ML2D |
| Martin-Benito *et al.* (2007b)[i] | Thermosome:prefoldin | ML2D |
| Arechaga *et al.* (2008)[j] | TrwK | ML2D |
| Cuellar *et al.* (2008) | CCT:Hsc70NBD | ML2D |
| Radjainia *et al.* (2008)[k] | Adiponectin | ML2D |
| Rehmann *et al.* (2008)[l] | EPAC2:cAMP:RAP1B | ML2D |
| Tato *et al.* (2007)[m] | TrwA:trwB | ML2D + kerdenSOM |
| Boer *et al.* (2009)[n] | repB | ML2D |
| Greig *et al.* (2009)[o] | Colicin Ia | ML2D |
| Klinge *et al.* (2009)[p] | DNA polymerase α | ML2D |
| Landsberg *et al.* (2009)[q] | AAA ATPase Vps4 | ML2D |
| Recuero-Checa *et al.* (2009)[r] | DNA ligase IV-Xrcc4 | ML2D |
| Reiriz *et al.* (2009)[s] | α,γ-Peptide nanotubes | ML2D |
| Albert *et al.* (2010)[t] | groEL:AGXT | ML2D + kerdenSOM |

[a] *EMBO J.* **22**, 6205–6213.
[b] *J. Struct. Biol.* **150**, 100–108.
[c] *J. Biol. Chem.* **280**, 40909–40915.
[d] *Biochemistry* **45**, 10512–10520.
[e] *J. Mol. Biol.* **357**, 1063–1076.
[f] *J. Biol. Chem.* **281**, 7012–7021.
[g] *J. Mol. Biol.* **357**, 1295–1305.
[h] *Structure* **15**, 101–110.
[i] *EMBO Rep.* **8**, 252–257.
[j] *J. Bacteriol.* **190**, 4572–5479.
[k] *J. Mol. Biol.* **381**, 419–430.
[l] *Nature* **455**, 124–127.
[m] *J. Biol. Chem.* **281**, 7012–7021.
[n] *EMBO J.* **28**, 1666–1678.
[o] *J. Biol. Chem.* **284**, 16126–16134.
[p] *EMBO J.* **28**, 1978–1987.
[q] *Structure* **17**, 427–437.
[r] *DNA Repair* **8**, 1380–1389.
[s] *J. Am. Chem. Soc.* **131**, 11335–11337.
[t] *J. Biol. Chem.* **285**, 6371–6376.

**Table 11.2** ML applications in single-particle 3D analysis

| Reference | Sample | Heterogeneity type |
|---|---|---|
| Recuero-Checa *et al.* (2009)[a] | DNA ligase IV-Xrcc4 | Data cleaning |
| Wang *et al.* (2009)[b] | RISC loading complex | Substoichiometric complex |
| Nickell *et al.* (2009)[c] | 26S proteasome | Substoichiometric complex |
| Klinge *et al.* (2009)[d] | DNA polymerase α | Flexible arm |
| Greig *et al.* (2009)[e] | Colicin Ia | Unclear |
| Julian *et al.* (2008) | Hybrid state 70S Ribosome | Ratcheting and ligand occupation |
| Rehmann *et al.* (2008)[f] | EPAC2:cAMP: RAP1B | Data cleaning |
| Cuellar *et al.* (2008) | CCT:Hsc70NBD | Substoichiometric complex |
| Cheng *et al.* (2010) | Stalled 70S Ribosome | Ligand occupation |

[a] *DNA Repair* **8**, 1380–1389.
[b] *Nat. Struct. Mol. Biol.* **16**, 1148–1153.
[c] *Proc. Natl. Acad. Sci. USA* **106**, 11943–11947.
[d] *EMBO J.* **28**, 1978–1987.
[e] *J. Biol. Chem.* **284**, 16126–16134.
[f] *Nature* **455**, 124–127.

new method has been applied to a variety of experimental data sets that a more profound understanding of its optimal processing strategies is obtained. This chapter aims to facilitate the use of ML methods by sharing the experience obtained so far with single-particle classification. It first describes how one typically prepares the data for ML refinement, and then discusses how to perform 2D and 3D classification. In addition, as ML approaches may require large amounts of computing time, relevant aspects of high-performance computing are highlighted.

This chapter focuses on a range of issues that have arisen in different experimental studies. Some of these issues are not only restricted to ML classification but also play a role in other refinement approaches. Many of these issues were never published, either because they concerned negative results or because they were not deemed relevant for the biologically oriented publications. Taken together, I hope that this contribution will be of practical help to others who want to apply ML image processing approaches in their investigations.

## 2. Data Preprocessing

In general terms, data preprocessing aims to provide a set of images that is as closely as possible in accordance with the statistical model that underlies the ML approach. All currently available ML approaches in single-particle analysis assume that each experimental image is a noisy projection of one or more 3D objects in unknown orientations. In most of the approaches, the noise is assumed to be independent, additive, and Gaussian. Only in two of the currently available ML approaches (Doerschuk and Johnson, 2000; Scheres *et al.*, 2007b), the effects of the contrast transfer function (CTF) are included in the data model. In these approaches, the data model is expressed in Fourier space. All other approaches use a real-space data model.

This section gives a stepwise description of the operations that are typically performed to convert a collection of experimental micrographs into a data set of single-particle images that is suitable for ML refinement. A schematic overview of these steps is given in Fig. 11.1.

## 2.1. Micrograph preprocessing

The first preprocessing step is to estimate the CTF effects by fitting a theoretical model to the power spectra of the experimental micrographs. This is a common step in many cryo-EM image processing strategies, and a variety

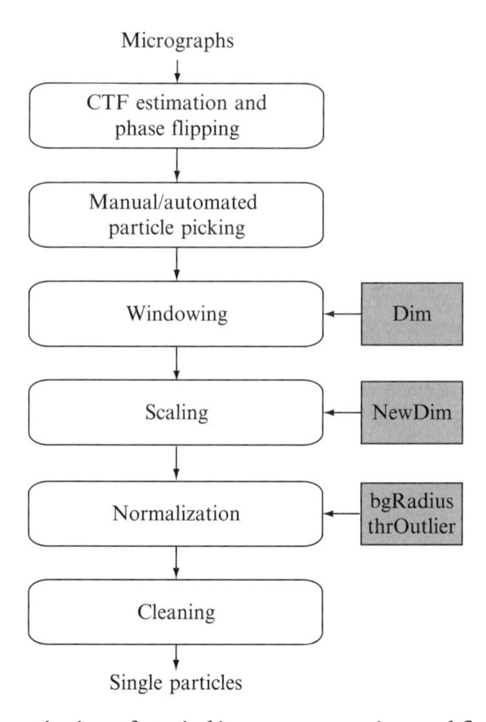


**Figure 11.1** Schematic view of a typical image preprocessing workflow. Important parameters that are discussed in the main text are highlighted in gray. These are the dimension of the boxed particles (Dim), the dimension of the downscaled particles (newDim), the radius of the circle that determines the particle noise area (bgRadius), and the threshold in standard deviations of the image that is used to discard outlier pixel values (thrOutlier).

of programs are available for this task (Frank, 2006). The reciprocal-space variant of single-particle ML refinement (Scheres *et al.*, 2007b) is capable of handling theoretical CTF models, provided they are rotationally symmetric. Therefore, micrographs with strong drift or astigmatism should be discarded at this stage, and nonastigmatic CTF models should be calculated for the remaining micrographs. Currently available real-space ML approaches do not model CTF effects. Still, it is also useful for these approaches to discard micrographs with strong astigmatism or drift as they lead to inconsistencies in the data. In addition, to partially make up for the absence of a CTF model, real-space ML approaches often benefit from a phase-flipping correction in the micrographs.

## 2.2. Particle selection

One then needs to identify individual particles of interest in the micrographs. One can perform this task manually or use (semi)automated procedures. This choice will often depend on the amount and quality of the data, and on the size and shape of the particles under study. Automated procedures are typically a much faster and less tedious option, but one should be aware that human experts still outperform automated approaches in most of the cases (Zhu *et al.*, 2004). Data sets obtained from (semi)automated particle selection procedures often contain more false positives than those selected manually. False positives may comprise a wide range of artifacts, such as irreproducibly damaged or aggregated particles, contaminations or spurious background features. The number of false positives should be minimized because they violate the most basic assumption of the data model: that every image contains a reproducible signal.

## 2.3. Particle extraction

Once the particles of interest have been selected, they are to be extracted (boxed) from the micrographs as individual images. The only parameter to be adjusted at this step is the size of the squared images. Several considerations play a role in deciding on the image size. On one hand, the extracted images should obviously be large enough to accommodate the particles in all directions and should include sufficient space to account for residual origin offsets. On the other hand, smaller images will reduce the computational load and have the advantage that fewer neighboring particles are present in the images. Neighboring particles are artifacts that are not accounted for in the data model and should thereby be avoided. In conventional refinement approaches, one often deals with neighboring particles by masking the experimental images. However, masks also fall outside the scope of the data model of currently available ML approaches. Masking the experimental images would lead to a systematic underestimation of the standard deviation

(SD) in the experimental noise because this estimation is performed using all pixels of the image. By extracting the particles in smaller images, one reduces the amount of neighboring particles without violating the data model, which is the main reason why one often performs ML refinements with relatively tightly boxed particles.

## 2.4. Scaling

Often cryo-EM micrographs are sampled at higher frequencies than needed. According to information theory, the highest resolution that can be obtained from the images is two times the pixel size, while limits of at least three times are more common in practice. That means that a pixel size of say 2 Å allows estimating the underlying signal up to a resolution of approximately 6 Å. However, while 2 Å pixels are relatively common in cryo-EM, very few reconstructions up to 6 Å resolution have been reported. In many cases, other factors like CTF envelope functions or structural heterogeneity put stronger limitations on the resolution than the sampling frequency. In those cases, the images may be resampled onto larger pixels, that is, one may use smaller images, without compromising the attainable resolution.

Other problems may be separated into subtasks that may be solved at distinct resolutions. A typical example is 3D refinement of structurally heterogeneous projection data. Often, the structural variability can be described at low-intermediate resolutions so that ML3D classification (see below) may be performed with downscaled images. Then, after the structural heterogeneity has been dealt with, the resulting classes may be refined separately to higher resolutions using the original images.

Downscaling has the obvious advantage that subsequent image processing will be computationally cheaper, which is especially relevant for ML approaches that require large amounts of computing time and memory. However, using small images has yet another, perhaps less obvious, advantage. Since signal-to-noise ratios in cryo-EM data tend to fall off with resolution, downscaling typically results in an increase of the overall signal-to-noise ratio. Consequently, refinements with smaller images tend to be more robust to overfitting and model bias. Similar effects could be achieved by low-pass filtering of the data. But low-pass filtering introduces correlations among the pixels, which conflicts with the data model and is therefore not recommended in combination with ML approaches.

## 2.5. Normalization

All images in the data set are assumed to have equal powers in the noise and signal. Since this may not be necessarily true for a cryo-EM data set, one typically normalizes the images. Image normalization aims to bring the entire data set to a common grayscale by applying additive and multiplicative factors to each of the individual images. Perhaps the most straightforward approach to image normalization would be to subtract the image means and to divide by the SDs, resulting in zero mean and unity SD for all images in the data set. Although this is a reasonable approximation for more or less spherically shaped particles, it will yield suboptimal results for strongly elongated particles (Sorzano *et al.*, 2004a). For the latter, projections along the long axis of the particle will be significantly more intense than projections perpendicular to that axis, which will give rise to systematic differences in the image means and SDs. Therefore, it is better to calculate the mean and SD values for a defined area of the images that only contains noise. One often uses the area outside a circle with a diameter that is typically several pixels smaller than the image size (Fig. 11.2). Additional advantages of this procedure are that the signal will be positive, the expected value for the solvent will be 0 and the SD in the noise will be 1. Based on the latter, one may start the likelihood optimization with an initial estimate for the SD in the noise of unity.

Several modifications of the normalization procedure may yield better results in specific cases (Fig. 11.2). Often, locally ramp-shaped gradients in background intensity are visible in the micrographs. In those cases, rather than subtracting a constant value one may fit a least squares plane through the pixels in the noise area of each image and subtract the resulting plane. Alternatively, one could apply high-pass filters to deal with these low-resolution effects. However, high-pass filters fail to describe the physics of the underlying problem and introduce correlations among the pixels in the filtered image, thereby again violating the assumption of independency in real-space ML approaches.

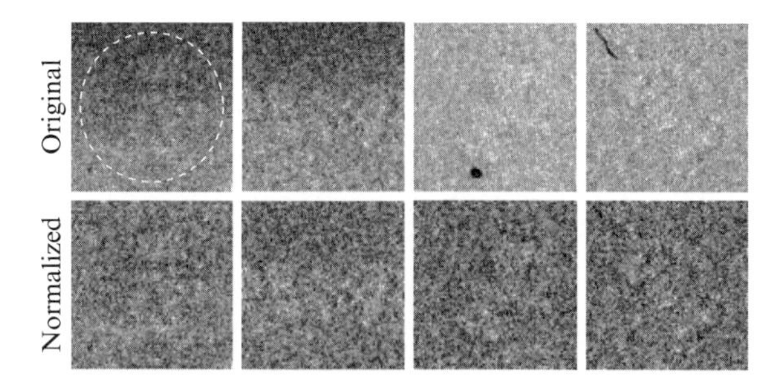


**Figure 11.2** An illustration of image normalization. All pixels outside the white circle in the upper right image are assumed to contain only noise. These pixels are used to fit a least-squares background plane and to calculate the standard deviation of the noise. The upper row shows original images as they were extracted from the micrographs. The two images on the left show ramping backgrounds, the two images on the right show pixels with extremely low (i.e., black) intensity values. The lower row shows the same four images after normalization.

Other micrographs contain exceptionally white or black pixel values, which may result from broken pixels in the digital camera, from X-rays that hit the film or camera, or from dust particles or other artifacts in the process of scanning photographic film. These extremely high- or low-pixel values correspond to outliers in the assumedly Gaussian distributions of the experimental noise and should be removed from the data. One possibility to do so is to replace pixels with values larger than a given number (e.g., four or five) times the SD in the image by a random instance from a Gaussian with zero mean and unity SD. Typically, one does not perform this correction if there are no visual indications of these artifacts in the data, since statistically a small fraction of the pixels is always expected to have such large or small values.

## 2.6. Data cleaning

Finally, it is often useful to check (once more) for remaining outliers in the data. In particular, automated particle selection procedures tend to yield data sets with large amounts of artifacts but also manually selected data may still contain features that are not described by the statistical data model. One could again propose completely automated procedures for the detection of atypical features, but one should keep in mind that human experts are typically much better at this task than computer algorithms. Therefore, visual inspection of the selected particles is often the best way to remove remaining artifacts from the data.

Still, computer algorithms may alleviate the tedious task of looking at all particles. One example is an *ad hoc* sorting procedure that was implemented in the *xmipp_sort_by_statistics* program and that has resulted useful on multiple occasions. This simple algorithm sorts all particles on a continuous scale from typical to atypical particles. To this purpose, the program calculates a large number of features for every individual particle. These features include mean, SD, minimum, and maximum values but also features like the number of pixels with values above or below one SD and features related to differences between image quadrants. The current implementation contains a total of 14 features and this list could in principle be expanded or reduced to better reflect specific cases. For every feature $f$, the program calculates a mean ($\mu_f$) and a SD ($\sigma_f$) values over all particles in the data set. (Alternatively, in some cases, improved results may be obtained by calculating $\mu_f$ and $\sigma_f$ over a subset of the particles that one is confident about.) For each $i$th image, one then calculates the average $Z$-score over all $F$ features:

$$\bar{Z}_i = \frac{1}{F}\sum_{f=1}^{F}\frac{|f_i - \mu_f|}{\sigma_f}, \tag{11.1}$$

where $f_i$ is the feature value for the $i$th image. Subsequently, one sorts the images on $\bar{Z}_i$ and visualizes them in typical matrix views where images with

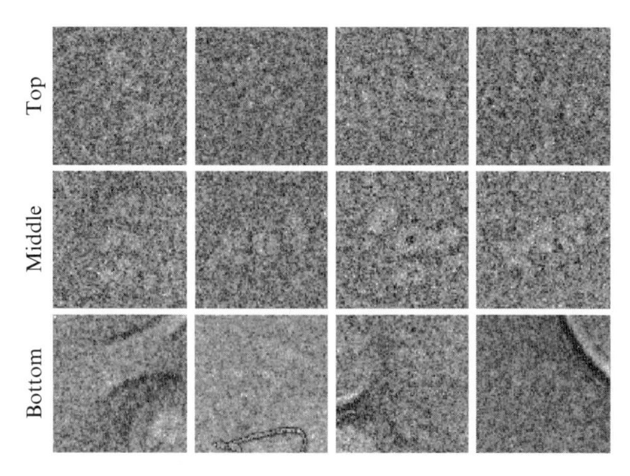


**Figure 11.3** The results of automated image sorting based on average $Z$-scores. The top row shows images with relatively low $Z$-scores, the middle row images with average $Z$-scores, and the bottom row shows images with relatively high $Z$-scores. The number of bad particles typically increases with higher $Z$-scores, which may facilitate the interactive removal of outliers by the user.

low-average $Z$-scores are at the top and images with high values are at the bottom. It is still up to the human expert to decide which particles to discard, but often this process is much easier as nice particles tend to be at the top and many more artifacts are present near the bottom of the sorted particle list (Fig. 11.3).

## 3. 2D Classification

2D analysis of cryo-EM data may be a useful tool to answer a wide range of biological questions. Most often, however, it is used as a means of data quality assessment or as a preprocessing step prior to 3D reconstruction. In general, 2D averaging procedures are much faster and more robust than their 3D counterparts. The lower complexity of the problem often allows retrieving 2D signals (class averages) from the noisy data in a reference-free manner, that is, without a prior estimate of the signal. This greatly reduces the pitfalls of model bias as encountered in 3D procedures. Still, images cannot be aligned well without separating distinct classes and it is hard to separate classes when the images are not aligned well. Conventionally, this *chicken and egg* problem has been addressed by iterative schemes of alignment, classification, and realignment of the resulting classes (see Frank, 2006 for a comprehensive overview). Two ML approaches have recently emerged as powerful add-ons to existing approaches: a multi-reference alignment scheme called ML2D (Scheres *et al.*, 2005) and a neural network

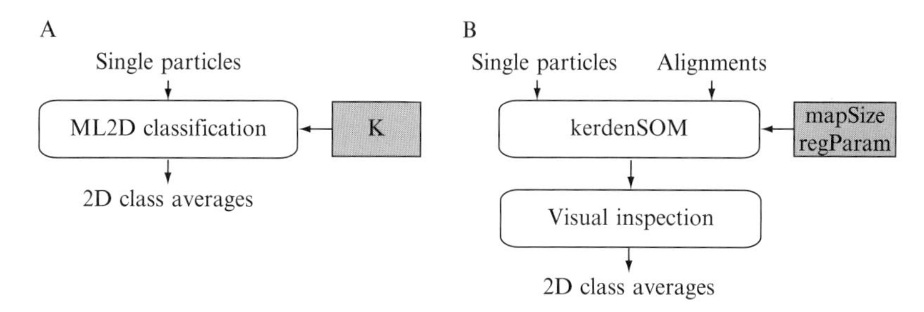


**Figure 11.4** Schematic view of the use of the ML2D (A) and kerdenSOM (B) algorithms. Important parameters that are discussed in the main text are highlighted in gray. These are the number of classes to be used in ML2D classification (*K*) and the size of the map (mapSize) and its regularization parameters (regParam) for the kerdenSOM algorithm.

called kerdenSOM (Pascual-Montano *et al.*, 2001). A schematic overview of how these approaches are used is shown in Fig. 11.4 and both approaches are described in more detail below.

## 3.1. ML2D

The ML2D algorithm may be used to simultaneously align and classify single-particle images (Scheres *et al.*, 2005). Reference-free class averages are obtained in a completely unsupervised manner by starting multi-reference alignments from average images of random subsets of the unaligned data. The only parameter that is adjusted by the user is the number of references (*K*). This number should ideally reflect the number of different 2D structures that are present in the data, but that number is typically not known. In practice, one affronts this problem by running the algorithm multiple times with different values for *K*. Higher values for *K* are accepted if interpretation of the resulting class averages leads to new information. However, a maximum number of classes does exist. The more classes, the fewer particles participate in each class (or more strictly, the lower the weighted sums of particle contributions to each class). Averaging over a low number of particles leads to noisy averages, which result in suboptimal alignments and classifications. Therefore, in practice, one often limits *K* so that on average there are at least 200–300 cryo-EM particles per class, while these numbers are typically much lower for negative stain data.

An intrinsic characteristic of the ML approach is that it does not assign images to one particular class or orientation. Instead, images are compared with all references in all possible orientations and probability weights are calculated for each possibility. Class averages are then calculated as weighted averages over all possible assignments and used for the next iteration. Still, in practice the probability distributions often converge to approximate delta functions. In this situation, each image effectively contributes to only a single class and orientation, and division of the data into separate classes or the assignment of optimal orientations is justified. The sharpness of the probability distributions may be monitored by their maximum value. Since the distributions integrate to unity, maximum values close to 1 indicate near-delta functions and values close to 0 indicate broad distributions. This value may also serve to identify outliers as those particles that still have relatively broad distributions upon convergence.

Two theoretical drawbacks of the real-space ML2D approach are that it does not take the CTF into account and that it assumes white noise in the pixels. Although this does not withhold ML2D from obtaining useful results in many cases, more precise descriptions of the CTF-affected signal and nonwhite noise are employed by the MLF2D algorithm (Scheres *et al.*, 2007b). An additional advantage of this algorithm is its intrinsic multiresolution approach where higher frequencies are only included in the optimization process if the class averages extend to such resolutions. For some cases, the MLF approach has been observed to yield much improved results compared to its real-space counterpart. Often, these cases concern data with a large spread in defocus values or with relatively low-resolution signals. For other cases, however, the MLF algorithm has been observed to converge to suboptimal solutions with a few highly populated and many empty classes. Sometimes, these problems are alleviated by running the MLF algorithm without CTF correction. In general, it is difficult to predict whether the real-space or the reciprocal-space approach will be better for a given data set and one typically performs tests with both algorithms. The approach that works best in 2D is often also the optimal choice in 3D. Thereby, the relatively cheap tests in 2D may save computing time in subsequent 3D refinements.

An additional advantage of the unsupervised character of ML(F)2D is that it is easily incorporated into high-throughput data processing pipelines. One example is the implementation of the ML2D approach inside the Appion pipeline (Lander *et al.*, 2009). This interface aims to streamline cryo-EM data processing by facilitating the use of flexible image processing workflows that use multiple programs from various software packages (for more information, see Chapter 16, this volume). Typical applications of ML2D inside this pipeline include the generation of templates for automated particle picking, data cleaning (by discarding images that give rise to bad class averages or with relatively flat probability distributions), and the generation of class averages for subsequent random conical tilt or common lines reconstructions (Voss *et al.*, 2010).

## 3.2. kerdenSOM

The kerdenSOM algorithm, which stands for kernel density self-organizing map, is a neural network based on ML principles (Pascual-Montano *et al.*, 2001). The kerdenSOM algorithm may be used to classify images that have been prealigned. However, rather than classifying the particles in a predefined number of classes, the algorithm outputs a 2D map of average images called code vectors. A regularized likelihood target function results in an output map that is *organized*. This means that the differences between neighboring code vectors are relatively small and that structural differences vary smoothly across the map. Then, it is up to the user to decide how many structural classes are present in the map.

Although the user does not have to choose a fixed number of classes, he does have to decide on the size of the output map and on the parameters that determine its regularization. Larger output maps can accommodate more different structures, but too large maps may have very few particles contributing to each code vector. Too strong regularization results in very smooth output maps that cannot describe the structural variability in the data, while too weak regularization leads to unorganized maps where it is difficult to identify continuous structural variations among the data. Therefore, the kerdenSOM algorithm is often run multiple times with different sizes of the output map and different regularization parameters. This is facilitated by the reduced computational costs compared to ML2D classification, but interpretation of the corresponding results does make the kerdenSOM algorithm relatively user-intensive.

An important advantage of using prealigned particles is that the classification may be focused on a particular region of interest in the images. In this scenario, only pixels inside a binary mask are included in the likelihood optimization. This speeds up the calculations and prevents structural variations in other regions of the images to interfere with the classification. This strategy has proven particularly useful in combination with the ML2D algorithm (e.g., see ML2D + kerdenSOM entries in Table 11.1). Here, one first uses ML2D to align the images and separate classes corresponding to relatively large structural differences. Then, one focuses the kerdenSOM on a particular area of a given class. For example, ML2D may be used to separate top and side views of a complex and the kerdenSOM may be focused on the nonstoichiometrical binding of a factor that is only visible in the side view. Apart from its implementation in the Xmipp package, this workflow is also accessible through the Appion pipeline.

## 4. 3D Classification

Because molecules are 3D objects, 3D reconstructions from cryo-EM single-particle projections typically provide much more information than 2D class averages. However, 3D reconstruction is a much more complex mathematical problem than 2D averaging. One consequence of this increased complexity is that 3D reconstructions often depend on the availability of an initial estimate of the underlying signal. Suitable initial models may be derived from known structures of similar complexes or they may be determined *de novo* by angular reconstitution or random conical or orthogonal tilt experiments (see Frank, 2006 and Chapter 10, this volume for more details). Still, even for structurally homogeneous data sets obtaining a reliable model is often far from straightforward, and bias toward incorrect models may introduce important artifacts in the results.

The situation is even more complex if multiple 3D structures are present in the data. The combination of distinct conformations in a single reconstruction leads at best to a general loss of resolution or to the loss of electron density in specific areas for partially flexible or nonstoichiometric complexes. Larger conformational changes may cause much more prominent artifacts, even up to the point where the refined structure no longer reflects any of the conformations in the sample. Nevertheless, structural heterogeneity also offers a unique opportunity to obtain information about multiple functional states, provided that projections from different structures can be classified. ML methods have recently emerged as powerful tools for this complicated task. 3D multi-reference refinement by ML (ML3D; Scheres *et al.*, 2007a) has provided the first unsupervised classification tool that is applicable to a wide range of structurally heterogeneous data sets (Table 11.2). An overview of the image processing workflow for ML3D classification is shown in Fig. 11.5 and the separate steps are discussed in detail below.

## 4.1. Preparing the starting model

ML3D refinement depends on an initial 3D reference structure and the selection of a suitable model has been found to be a pivotal step for successful classification. The expectation-maximization algorithm that underlies the ML3D approach is a local optimizer, that is, it converges to the nearest local minimum. Despite the marginalization over the orientations and class assignments, model bias has still been observed to play an important role in ML3D classification. Therefore, it is recommended to start ML3D classifications from a consensus model that ideally reflects to some extent all the different structures in the heterogeneous data set. This is almost a contradiction in terms. Perhaps, a hypothetical example illustrates the role of the consensus model: if the data were already separated into $K$ structurally homogeneous subsets, separate refinements of the consensus model against each of the subsets should be able to converge to the $K$ different structures that are present in the data.

In many cases, a suitable starting model may be obtained by refinement of the complete data set against a single 3D reference, followed by a strong low-pass filter. The effective resolution of the low-pass filter plays a crucial role here, as high frequencies in the starting model are prone to induce local

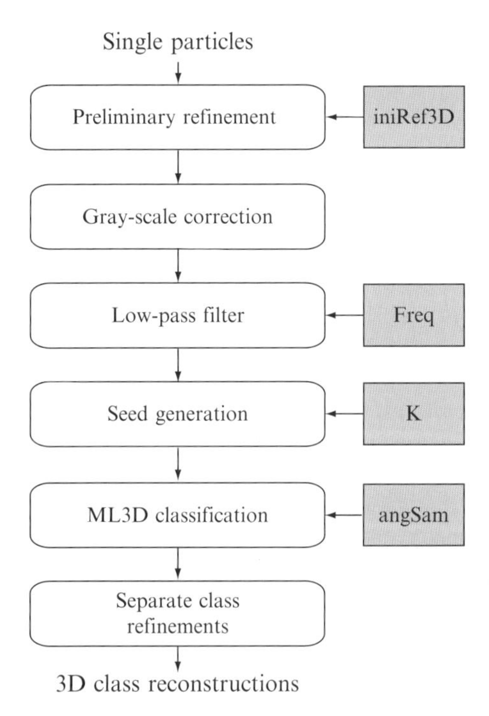


**Figure 11.5** Schematic view of the ML3D classification workflow. Important parameters that are discussed in the main text are highlighted in gray. These are the initial 3D model (iniRef3D), the frequency of the low-pass filter (Freq), and the number of classes ($K$) and the angular sampling (angSam) to be used in the ML3D refinement.

minima and too restrictive filters result in featureless blobs that cannot be refined. One typically filters the consensus model "as much as possible." That is, one filters the consensus model to the lowest possible resolution for which refinement against the heterogeneous data still converges to a solution that is similar to the unfiltered model. For many cases, useful low-pass filters have been observed to lie in the range of 0.05–0.07 pixel$^{-1}$ (using downscaled images as described above).

The direct use of low-pass filtered models from the Protein Data Bank (PDB), negative stain reconstructions or geometrical phantoms as starting models in ML3D classification has been observed to yield suboptimal results. It is often better to first refine such a model against the complete, structurally heterogeneous data set. In principle, any program could be used for this task. Rather than aiming at high-resolution information (the resulting model will be low-pass filtered anyway), this refinement should aim at removing false low-resolution features from the model. For example, negative stain models may be flattened, the dimensions of geometrical phantoms may be off, atomic models may be incomplete or have an unrealistically high contrast, and so on. Another common pitfall is a small difference in pixel size between the starting model and the actual data set, which may arise from suboptimal calibrations of the microscope magnification. The latter may also be checked by comparing projections of the starting model with reference-free class averages of the structurally heterogeneous data, as for example obtained using ML2D.

Although any refinement program could be used to generate the starting model, one should keep in mind that some software packages provide reconstructions for which the absolute intensity (or grayscale) is not consistent with the intensity of the experimental images. This may be because the reconstruction algorithm itself is not implemented in a consistent way, or because the reconstruction is normalized internally. Reprojections of such maps are on a different grayscale than the signal in the experimental images. This is typically harmless in conventional refinements where maximum cross-correlation searches are insensitive to additive or multiplicative factors. However, the squared difference terms inside the Gaussian distributions are highly sensitive to these factors and ML refinements of models with inconsistent grayscales may give rise to extreme artifacts. A typical observation in such cases is that all experimental images contribute to only a single projection direction and the corresponding reconstruction is severely artifactual.

All algorithms from the Xmipp package yield reconstructions with consistent grayscales. Consequently, in case of doubt one may correct the grayscale by performing a reconstruction inside the Xmipp package. To this purpose, one could transfer the orientations from the other software package to Xmipp, but this may involve cumbersome conversion issues. Often it is easier to subject the refined model from the other package to a single, additional iteration of conventional projection matching refinement inside the Xmipp package. As the resulting model will be low-pass filtered anyway, this step may be performed in a quick manner using a coarse angular sampling and a fast reconstruction algorithm.

## 4.2. Multi-reference refinement

ML3D classification is a multi-reference refinement scheme and thus requires multiple starting models. A key achievement of the ML3D approach is that distinct structures can be separated in an unsupervised manner, that is, without prior knowledge about the structural variability in the data. In particular, multi-reference ML refinements were observed to converge to useful solutions when starting from initial models (seeds) that are random variations of a single consensus model. To generate randomly different seeds, one typically divides the structurally heterogeneous data set into random subsets and performs a single iteration of ML3D refinement of the consensus model against each of the subsets separately. Perhaps, alternative ways like adding different instances of random noise to the consensus model would also work. However, the division of the data in random subsets is more closely related to traditional *k*-means and does not introduce any additional parameters (e.g., how much noise to add).

Apart from the starting model itself, the most important parameter in ML3D classification is $K$, the number of 3D models that are refined simultaneously. As in ML2D classification, $K$ should ideally reflect the number of different structures in the data, but because that number is unknown $K$ is typically varied over multiple runs. Again, the maximum number of references is ultimately determined by the amount of experimental data available. Often, (asymmetric) reconstructions from less than 5000 to 10,000 cryo-EM particles become too noisy to allow reliable alignment and classification. In practice, $K$ may also be limited by available computing resources. In general, 3D ML refinement is computationally demanding and the current implementation was optimized for speed by storing the reference projections of all models in memory. Thereby, memory requirements scale linearly with $K$ so that high numbers may require more computer memory than physically available.

Hardware limitations also put stringent limitations on the angular sampling rate. Because memory requirements scale quadratically with the angular sampling rate, and computing time scale even cubically, high-angular sampling rates quickly become computationally prohibitive. Therefore, ML3D classification is typically performed with relatively coarse angular sampling rates. In most of the applications reported thus far an angular sampling rate of approximately 10° was used. This intrinsically limits the resolution that can be obtained by ML3D, but fortunately in many cases the structural variability can be resolved at medium-low resolutions. An additional effect of the coarse sampling is that otherwise similar reference structures may rotate several degrees with respect to each other during the optimization process. These rotated references may better accommodate particles with orientations that fall in between the coarse angular sampling used in the refinement (Scheres *et al.*, 2007a). The latter should also be taken into account when choosing $K$, because the presence of different rotated versions of the same conformation reduces the number of different conformations that can be separated for a given number of $K$.

ML3D multi-reference refinement is typically performed for a user-defined number of iterations, often around 25. Convergence may be monitored by analysis of the log-likelihood value during the iterations and the number of particles that change their optimal orientations or class from one iteration to the next. Typically, one stops the calculations when the log-likelihood increase has leveled off and when the number of the particles that still change their optimal orientation or class has stabilized to a small fraction of the data set. As in the ML2D case, the width of the probability distributions may be monitored by their maximum values, and these distributions tend to converge to approximate delta functions. The latter again allows one to divide the data set into separate classes, which ideally should be structurally homogeneous. These classes may then be refined separately to higher resolution using conventional refinement algorithms and the original images without downscaling.

### 4.3. Interpretation of the classification results

As mentioned above, one of the main advantages of ML3D classification is its unsupervised character. This circumvents the main pitfall of biasing the classification toward a false assumption about the classes in supervised approaches. Still, the starting model may lead to bias in the alignment of the particles, which will in turn affect classification (Fig. 11.6). The problem of model bias is not unique to ML3D classification. It plays an important role in many 3D refinement programs for cryo-EM single-particle data. A typical sign that the refinement process is affected by model bias is that the references do not gain new structural details. Often, the refined structures remain similar to the initial seeds at intermediate-low resolution and only seem to accumulate noise at higher frequencies. In other cases, the references may even disintegrate during the refinement process. A good indication that the refinement is not affected by model bias is that various, different starting models all converge to a similar solution. Still, especially for relatively small particles with no or low symmetry, the absence of a good starting model may be an important obstacle for successful 3D alignment and classification.

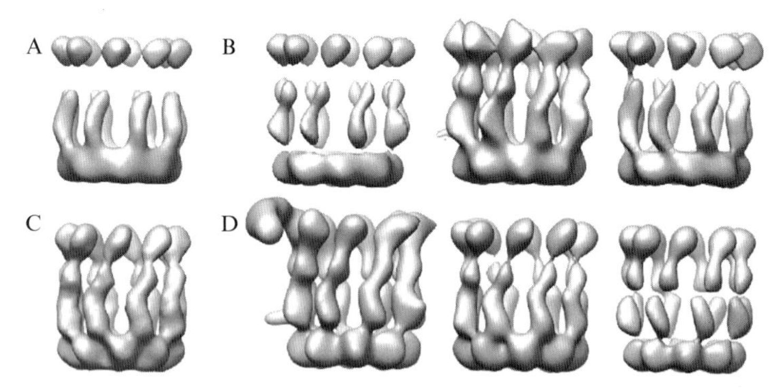


**Figure 11.6** An example of model bias affecting ML3D classification. Using a suboptimal initial reference (A), ML3D classification yielded suboptimal results (B). Using an improved initial model (C), ML3D classification separated uncomplexed CCT from CCT:Hsc70NBD complexes (see Cuellar *et al.*, 2008 for more details).

Even if a good starting model is available, the results of 3D classifications of cryo-EM data should be interpreted with care. Many hypothetical divisions of the data may give rise to 3D reconstructions with plausible structural differences. However, due to the high-noise levels in the data, these differences are not necessarily related to actual structural variability in the data. Although interpretation of the classes in the light of prior biochemical and structural knowledge may be an extremely powerful criterion to decide on the plausibility of a classification, it is often also a highly subjective endeavor with a considerable risk of overinterpretation. Fortunately, there are several, more objective tests that one can (and perhaps should) perform.

Firstly, the differences between the refined structures could have arisen from the random variations among the initial starting seeds. Although the initial variations are typically small, they may be amplified during the refinement of the noisy data. However, in that case a second classification starting from different random seeds will not likely result in the same classes. Therefore, it is often useful to check the reproducibility of the classification by comparing classes from multiple classification runs that were started from different random seeds. Significant class overlaps (e.g., of more than 75%) are usually an indication of reliable classification.

Secondly, structurally homogeneous data sets should behave better in refinement than structurally heterogeneous data sets. Therefore, to test whether classes obtained by ML3D are more homogeneous than the original structurally heterogeneous data set, one could compare conventional refinement statistics. However, this comparison is complicated by the fact that the classes obtained by ML3D classification are per definition smaller than the original data set. A solution to this problem is to randomly divide the original data set into subsets of identical size as those obtained by ML3D classification. The random division is not expected to resolve any of the structural heterogeneity. Therefore, refinements of the subsets obtained by ML3D classification should then yield better statistics, for example, Fourier shell correlations, than refinements of the random subsets of identical size. Similarly, the refined ML3D classes should give rise to less intense 3D variance maps (Penczek *et al.*, 2006) than the refined random subsets.

Indications for remaining structural heterogeneity in the classes are the same ones as those that were used to identify the heterogeneity in the first place. Low-density values for factors that bind nonstoichiometrically, fuzzy density or relatively low resolution for flexible domains, or overall too low resolutions may indicate that the classes are still heterogeneous. In that case, one could opt to repeat the original classification with a higher number of classes, but this may not be feasible because of limited computing resources. Alternatively, one may also divide one or more of the classes separately into multiple subclasses. In this way, the structural heterogeneity may be removed in a hierarchical manner, focusing on ever smaller details in subsequent steps (Fig. 11.7).

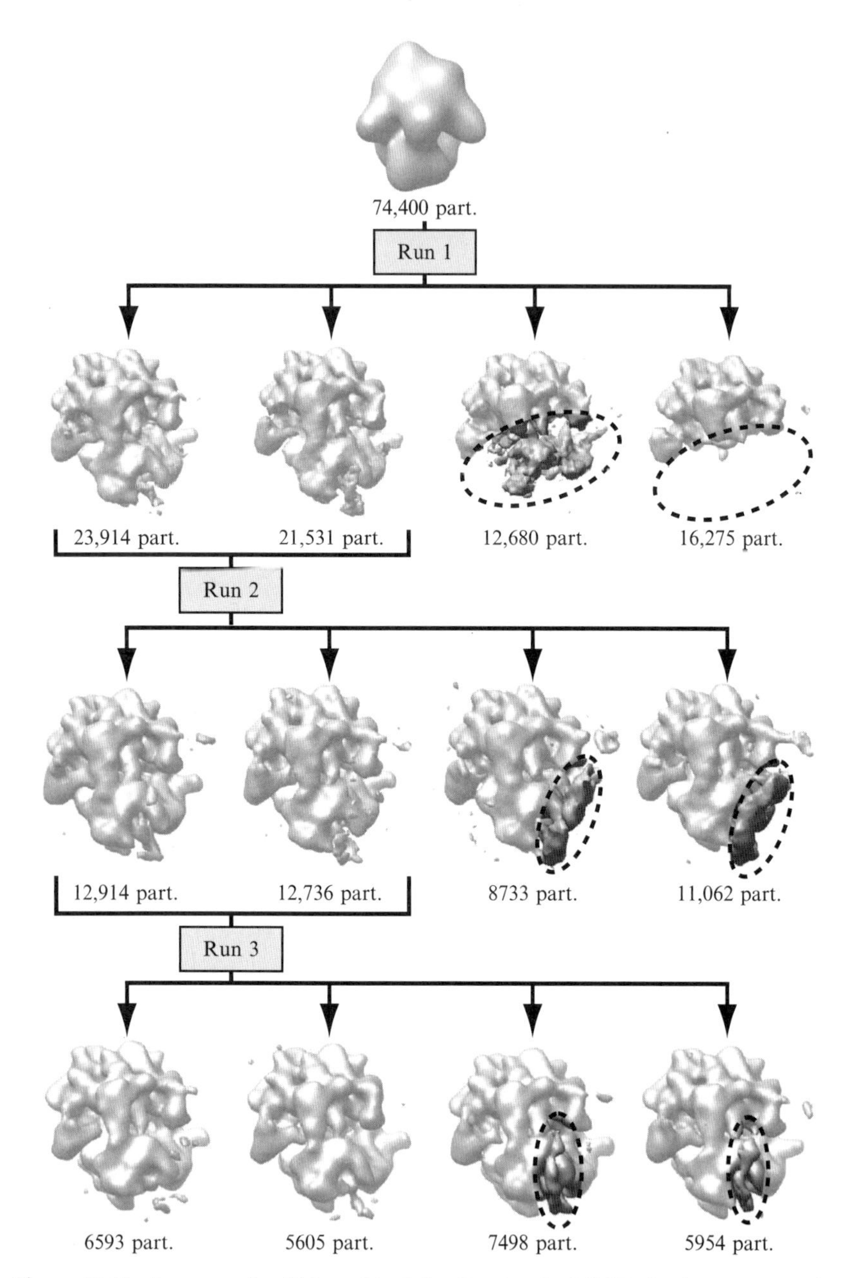


**Figure 11.7** An example of hierarchical classification by ML3D. In three consecutive runs (run1–3), ML3D classification is used to separate a dataset of 74,400 ribosome particles into multiple classes. Run1 separates intact ribosomes from disintegrated, 50S particles; run2 separates particles with strong density for tmRNA; and run3 classifies the remaining intact ribosomes into particles without apparent tmRNA density and particles with weak tmRNA density in an alternative conformation (see Cheng *et al.*, 2010 for more details).

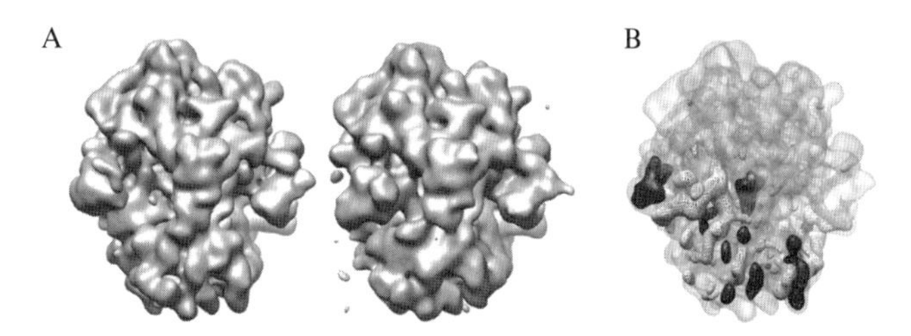


**Figure 11.8** An illustration of the use of difference maps in the analysis of distinct structural classes. Side-by-side visualization of maps rendered at the same threshold does not readily reveal the structural differences between them (A). Positive (black) and negative (white) difference maps rendered at 5 standard deviations better reveal compositional and conformational differences (B).

A final comment on the analysis of structural differences between classes concerns the use of normalized difference maps. Rather than comparing two maps that are rendered at a given threshold side by side in a 3D visualization program, it is often much more informative to visualize the positive and negative differences between both. If one assumes that the differences between two independent reconstructions of identical structures are zero-mean and Gaussian distributed, one may interpret the difference map in a statistical manner. In such an interpretation, areas of positive or negative difference density above a certain threshold (e.g., three times the SD) may be considered as significant. In the difference maps, presence or absence of factors typically shows up as isolated peaks, domain movements are characterized by positive density on one side of the domain and negative density on the other side of the domain, and the solvent region should not contain strong difference density (Fig. 11.8).

However, in order to subtract one map from the other several issues need to be taken into consideration. Firstly, it is important that both maps are aligned. As maps may rotate with respect to each other during ML3D refinement, realignment of the output maps is often necessary. Secondly, both maps should be on the same resolution. Typically, subsets of different sizes yield reconstructions at different resolutions, and both maps should be filtered to the lowest common resolution. Finally, both maps should be on the same intensity scale and have similar power spectra, or *B*-factor decays. This will generally be true for maps that were reconstructed from subsets of a single data set, but special care should be taken when subtracting maps obtained from other microscopy experiments or atomic structures.

## 5. High-Performance Computing

Because of their elevated computational costs, 3D ML approaches rely heavily on high-performance computing approaches. For example, classification of 91,000 ribosome particles into four classes was reported to take more than 6 months of computing time (Scheres *et al.*, 2007a), while these calculations were actually performed within a few days using 64 processors in parallel. In order to make efficient use of available computing resources, it may be necessary to have some understanding of the available hardware and the parallelization strategies employed.

The most expensive part of the expectation-maximization algorithm is the *E*-step, where for every particle an integration over all possible orientations and classes is evaluated. This step is similar to the alignment step in conventional refinement approaches. Fortunately, each of the typically thousands of individual particles can be processed independently within each *E*-step. In parallel computing such problems are called *embarrassingly parallel*, as very little effort is typically required to split them into a number of parallel tasks. Only when reaching the *M*-step (i.e., the reconstruction step) the information from all particles is to be combined, but this step is typically much less expensive.

Parallelization strategies may be divided into two categories depending on the computer hardware that is used. *Shared-memory parallelization* employs multicore processors that share a single centralized memory,[1] while *distributed parallelization* is used for computing nodes with their own local memory that are connected via a network. As parallel processes in a multicore processor have access to the same memory, the implementation of shared-memory parallelization is often relatively easy. Single process may be programmed to launch multiple *threads* to perform separate tasks simultaneously. Apart from the relative ease of software development (which may be of little concern to the experimentalist) multithreaded programs are also relatively easy to use. Executing these programs in parallel does not involve additional complications and they can be run on commonly available multicore desktop computers.

In distributed parallelization, parallel tasks are executed on different *nodes* that cannot see each other's memory, and information exchange between nodes is typically performed by *message passing*. Thereby, the efficiency of distributed parallelization schemes is often a trade-off between the costs of computation and communication. The main advantage of distributed parallelization is scalability. Many thousands of nodes may be used simultaneously in large computing clusters, whereas shared-memory parallelization is limited by the number of cores on a single processor (currently up to eight). However, even when using common standards like the Message Passing Interface, or MPI (Gropp *et al.*, 2009), the installation of message passing programs is often more complicated than for sequential programs.

[1] Parallel computing on the graphics processing unit (GPU) may also be considered as a form of shared-memory parallelization. This recent trend in computer science is promising much increased computing capacities with relatively cheap hardware. However, as GPU implementations of ML2D or ML3D are not yet available, this development is not discussed here. For a recent review on this topic, the reader is referred to Schmeisser *et al.* (2009).

Moreover, not all electron microscopy laboratories have access to a computing cluster, and jobs are often executed through specific queuing systems that may present an additional stumbling block for the experimentalist.

The ML2D and ML3D programs in the XMIPP package have been implemented using a *hybrid parallelization* approach using both threads and MPI.[2] Thereby, one can take full advantage of modern, multicore computer clusters. In this scenario, MPI takes care of passing messages between a possibly high number of multicore nodes, each of which runs multiple threads in parallel. The shared-memory parallelization is particularly useful in 3D classification where memory resources pose important limitations on the angular sampling and the number of classes. By sharing the memory inside a multicore node one prevents the duplication of part of the memory as would be the case in distributed parallelization. Thereby, within a single node much more memory is available so that a higher number of classes may be refined and with finer angular samplings.

In most of the cases, parallelization of the $E$-step can be done with excellent efficiency, but this is not the case for the $M$-step. The $M$-step of ML2D is very quick, basically the calculation of $K$ 2D average images, and thus does not play an important role. The $M$-step in ML3D comprises $K$ 3D reconstructions that are performed using a modified weighted least-squares (wls) ART algorithm (Scheres *et al.*, 2007a). These reconstructions are relatively fast compared to the $E$-step, but a distributed parallelization scheme is not straightforward for this type of algorithm (Bilbao-Castro *et al.*, 2009). Therefore, the current wlsART implementation only uses threads, although the $K$ independent reconstructions may each be performed by a different MPI process. The $M$-step may thereby become a bottle neck if the number of MPI processes is much larger than $K$. Therefore, it is important to realize that if the $M$-step starts taking similar amounts of time as the $E$-step, using more MPI processes will no longer lead to important gains in speed.

## 6. Outlook

Image classification plays an increasingly important role in single-particle cryo-EM as it offers the unique opportunity to characterize multiple structural states from a single sample. While these potentials are being illustrated for a growing number of macromolecular complexes, it is becoming clear that these complexes are even more flexible than anticipated. For example, using biophysical single-molecule techniques and single-particle cryo-EM it was shown that even at room temperature thermal energy alone appears to be sufficient to induce major conformational

[2] Note that current implementations of the MLF2D and MLF3D programs only use MPI.

changes in the 70S ribosome (Agirrezabala *et al.*, 2008; Cornish *et al.*, 2008; Julian *et al.*, 2008). These observations are changing our view of molecular machines in a profound way and will eventually lead to even bigger challenges in image classification. In response to these insights, many different 3D classification tools have recently emerged (see Spahn and Penczek, 2009 for a recent review). Compared to existing alternatives, ML classification may be an attractive choice in many experimental studies because of its unsupervised character, its robustness to high-noise levels, and its simultaneous assignment of orientations and classes.

As mentioned in Section 1, their implementation in the Xmipp package (Sorzano *et al.*, 2004b) may have played an important role in the relatively widespread use of ML single-particle classification approaches. Therefore, it is promising that similar programming efforts are being made for ML approaches in other cryo-EM modalities as well. Recently, a dedicated program has been written for MLF of icosahedral viruses (Prust *et al.*, 2009), while the ML approach for 2D crystals has been implemented in the 2dx software package (Gipson *et al.*, 2007) that provides a user friendly interface to the MRC package (Crowther *et al.*, 1996).

New results with ML methods in other cryo-EM modalities are other promising indicators that ML approaches may eventually become of general use in many aspects of cryo-EM data processing. Recently, the ML approach for icosahedral virus refinement by Doerschuk and Johnson (2000) and Yin *et al.* (2001, 2003) was successfully applied to classify an assembly mutant of CCMV (Lee, 2009), and ML sub-tomogram averaging (Scheres *et al.*, 2009) provided reference-free alignment and classification of 100S ribosome particles (Ortiz *et al.*, 2010). Together with the continuing increase in available computing power, ML methods and related statistical approaches are thereby expected to play progressively central roles in a wide range of experimental cryo-EM studies.

## ACKNOWLEDGMENTS

I thank Dr. Carmen San Martin for critically reading the manuscript and Dr. Jose-Maria Carazo for support. Part of the work described was funded by the Spanish Ministry of Science and Technology (Grants CDS2006-0023, BIO2007-67150-C01, and ACI2009-1022) and the National Heart, Lung and Blood Institute (Grant R01HL070472).

## REFERENCES

Agirrezabala, X., Lei, J., Brunelle, J. L., Ortiz-Meoz, R. F., Green, R., and Frank, J. (2008). Visualization of the hybrid state of tRNA binding promoted by spontaneous ratcheting of the ribosome. *Mol. Cell* **32,** 190–197, PMID: 18951087.

Al-Amoudi, A., Diez, D. C., Betts, M. J., and Frangakis, A. S. (2007). The molecular architecture of cadherins in native epidermal desmosomes. *Nature* **450,** 832–837.

Bilbao-Castro, J., Marabini, R., Sorzano, C., Garcia, I., Carazo, J., and Fernandez, J. (2009). Exploiting desktop supercomputing for three-dimensional electron microscopy reconstructions using ART with blobs. *J. Struct. Biol.* **165,** 19–26.

Cheng, K., Ivanova, N., Scheres, S. H., Pavlov, M. Y., Carazo, J. M., Hebert, H., Ehrenberg, M., and Lindahl, M. (2010). tmRNASmpB complex mimics native aminoacyl-tRNAs in the A site of stalled ribosomes. *J. Struct. Biol.* **169,** 342–348.

Chiu, P., Pagel, M. D., Evans, J., Chou, H., Zeng, X., Gipson, B., Stahlberg, H., and Nimigean, C. M. (2007). The structure of the prokaryotic cyclic nucleotide-modulated potassium channel MloK1 at 16 Å resolution. *Structure* **15,** 1053–1064.

Cornish, P. V., Ermolenko, D. N., Noller, H. F., and Ha, T. (2008). Spontaneous intersubunit rotation in single ribosomes. *Mol. Cell* **30,** 578–588.

Crowther, R. A., Henderson, R., and Smith, J. M. (1996). MRC image processing programs. *J. Struct. Biol.* **116,** 9–16.

Cuellar, J., Martin-Benito, J., Scheres, S. H. W., Sousa, R., Moro, F., Lopez-Vinas, E., Gomez-Puertas, P., Muga, A., Carrascosa, J. L., and Valpuesta, J. M. (2008). The structure of CCT-Hsc70NBD suggests a mechanism for hsp70 delivery of substrates to the chaperonin. *Nat. Struct. Mol. Biol.* **15,** 858–864.

Doerschuk, P. C., and Johnson, J. E. (2000). Ab initio reconstruction and experimental design for cryo electron microscopy. *IEEE Trans. Inf. Theory* **46,** 1714–1729.

Frank, J. (2006). Three-Dimensional Electron Microscopy of Macromolecular Assemblies. Oxford University Press, New York.

Gipson, B., Zeng, X., Zhang, Z. Y., and Stahlberg, H. (2007). 2dx—User-friendly image processing for 2D crystals. *J. Struct. Biol.* **157,** 64–72.

Gropp, W., Lusk, E., and Skjellum, A. (2009). Using MPI, Portable Parallel Programming with the Message Passing Interface. MIT Press, Cambridge, MA.

Julian, P., Konevega, A. L., Scheres, S. H. W., Lazaro, M., Gil, D., Wintermeyer, W., Rodnina, M. V., and Valle, M. (2008). Structure of ratcheted ribosomes with tRNAs in hybrid states. *Proc. Natl. Acad. Sci. USA* **105,** 16924–16927.

Lander, G. C., Stagg, S. M., Voss, N. R., Cheng, A., Fellmann, D., Pulokas, J., Yoshioka, C., Irving, C., Mulder, A., Lau, P., Lyumkis, D., Potter, C. S., and Carragher, B. (2009). Appion: An integrated, database-driven pipeline to facilitate EM image processing. *J. Struct. Biol.* **166,** 95–102.

Lee, S. (2009). Maximum likelihood reconstruction of 3D objects with helical symmetry from 2D projections of unknown orientation and application to electron microscope images of viruses. PhD thesis School of Electrical and Computer Engineering, Purdue University West Lafayette, Indiana.

Ortiz, J. O., Brandt, F., Matias, V. R. F., Sennels, L., Rappsilber, J., Scheres, S. H. W., Eibauer, M., Hartl, F. U., Baumeister W. (2010). "Structure of hibernating ribosomes studied by cryo-electron tomography in vitro and in situ." *J. Cell Biol.* (in press).

Pascual-Montano, A., Donate, L. E., Valle, M., Barcena, M., Pascual-Marqui, R. D., and Carazo, J. M. (2001). A novel neural network technique for analysis and classification of EM single-particle images. *J. Struct. Biol.* **133,** 233–245.

Pascual-Montano, A., Taylor, K. A., Winkler, H., Pascual-Marqui, R. D., and Carazo, J. (2002). Quantitative self-organizing maps for clustering electron tomograms. *J. Struct. Biol.* **138,** 114–122.

Penczek, P. A., Yang, C., Frank, J., and Spahn, C. M. (2006). Estimation of variance in single-particle reconstruction using the bootstrap technique. *J. Struct. Biol.* **154,** 168–183.

Prust, C., Wang, K., Zheng, Y., and Doerschuk, P. (2009). Special purpose 3-D reconstruction and restoration algorithms for electron microscopy of nanoscale objects and an enabling software toolkit. Proceedings—2009 IEEE International Symposium on Biomedical Imaging: From Nano to Macro, ISBI 2009, pp. 302–305.

Scheres, S. H. W., Valle, M., Nunez, R., Sorzano, C. O. S., Marabini, R., Herman, G. T., and Carazo, J. M. (2005). Maximum-likelihood multi-reference refinement for electron microscopy images. *J. Mol. Biol.* **348,** 139–149.

Scheres, S. H. W., Gao, H., Valle, M., Herman, G. T., Eggermont, P. P. B., Frank, J., and Carazo, J. M. (2007a). Disentangling conformational states of macromolecules in 3D-EM through likelihood optimization. *Nat. Methods* **4,** 27–29.

Scheres, S. H. W., Nunez-Ramirez, R., Gomez-Llorente, Y., Martin, C. S., Eggermont, P. P. B., and Carazo, J. M. (2007b). Modeling experimental image formation for likelihood-based classification of electron microscopy data. *Structure* **15,** 1167–1177.

Scheres, S. H. W., Nunez-Ramirez, R., Sorzano, C. O. S., Carazo, J. M., and Marabini, R. (2008). Image processing for electron microscopy single-particle analysis using XMIPP. *Nat. Protoc.* **3,** 977–990.

Scheres, S. H. W., Melero, R., Valle, M., and Carazo, J. (2009). Averaging of electron subtomograms and random conical tilt reconstructions through likelihood optimization. *Structure* **17,** 1563–1572.

Schmeisser, M., Heisen, B. C., Luettich, M., Busche, B., Hauer, F., Koske, T., Knauber, K., and Stark, H. (2009). Parallel, distributed and GPU computing technologies in single-particle electron microscopy. *Acta Cryst. D* **65,** 659–671.

Sigworth, F. J. (1998). A maximum-likelihood approach to single-particle image refinement. *J. Struct. Biol.* **122,** 328–339.

Sorzano, C. O. S., de la Fraga, L. G., Clackdoyle, R., and Carazo, J. M. (2004a). Normalizing projection images: A study of image normalizing procedures for single particle three-dimensional electron microscopy. *Ultramicroscopy* **101,** 129–138.

Sorzano, C. O. S., Marabini, R., Velazquez-Muriel, J., Bilbao-Castro, J. R., Scheres, S. H. W., Carazo, J. M., and Pascual-Montano, A. (2004b). XMIPP: A new generation of an open-source image processing package for electron microscopy. *J. Struct. Biol.* **148,** 194–204.

Spahn, C. M., and Penczek, P. A. (2009). Exploring conformational modes of macromolecular assemblies by multiparticle cryo-EM. *Curr. Opin. Struct. Biol.* **19,** 623–631.

Voss, N. R., Lyumkis, D., Cheng, A., Lau, P., Mulder, A., Lander, G. C., Brignole, E. J., Fellmann, D., Irving, C., Jacovetty, E. L., Leung, A., Pulokas, J., *et al.* (2010). A toolbox for ab initio 3-D reconstructions in single-particle electron microscopy. *J. Struct. Biol.* **169,** 389–398.

Yin, Z., Zheng, Y., and Doerschuk, P. C. (2001). An ab initio algorithm for low-resolution 3-D reconstructions from cryoelectron microscopy images. *J. Struct. Biol.* **133,** 132–142.

Yin, Z., Zheng, Y., Doerschuk, P. C., Natarajan, P., and Johnson, J. E. (2003). A statistical approach to computer processing of cryo-electron microscope images: Virion classification and 3-D reconstruction. *J. Struct. Biol.* **144,** 24–50.

Zeng, X., Stahlberg, H., and Grigorieff, N. (2007). A maximum likelihood approach to two-dimensional crystals. *J. Struct. Biol.* **160,** 362–374.

Zhu, Y., Carragher, B., Glaeser, R. M., Fellmann, D., Bajaj, C., Bern, M., Mouche, F., de Haas, F., Hall, R. J., Kriegman, D. J., Ludtke, S. J., Mallick, S. P., *et al.* (2004). Automatic particle selection: Results of a comparative study. *J. Struct. Biol.* **145,** 3–14.

CHAPTER TWELVE

# Methods for Three-Dimensional Reconstruction of Heterogeneous Assemblies

Elena V. Orlova *and* Helen R. Saibil

## Contents



## Abstract

Electron microscopy (EM) has developed into an important method for determining the three-dimensional (3D) structures of biological complexes, in particular of isolated macromolecular complexes in vitrified solution (cryo-EM of "single particles"). One of the consequences of studying complexes in solution rather than in a crystal lattice is that they are less constrained to adopt a single conformation. It is a common problem in single-particle analysis that samples of purified macromolecules can be structurally heterogeneous, with molecules adopting different conformations, corresponding to different functional states. In the case of multisubunit assemblies, there may also be heterogeneity of assembly or ligand binding. Heterogeneity limits the accuracy and resolution of 3D structures, since different conformations will contribute to a single 3D map and variable parts of the structure will be smeared out. Therefore, a new group of image processing methods has been developed to deal with the problems of detecting and sorting structural heterogeneity. The basic problem is to discriminate the

Crystallography and Institute of Structural Molecular Biology, Birkbeck College, London, United Kingdom

*Methods in Enzymology*, Volume 482
ISSN 0076-6879, DOI: 10.1016/S0076-6879(10)82013-0

source of image variations, and then to separate the images into homogeneous subsets for separate reconstruction. Variations in image features can arise from different particle orientations, variations in conformation and/or ligand binding, and noise fluctuations in the low signal-to-noise ratio images typical of cryo-EM. Here, we present a review of approaches developed to deal with these problems, along with examples of the application of a method based on multivariate statistical analysis to both model and real data. The methods have been used to discriminate small differences in size, conformation and ligand binding, and to obtain high quality, reliable reconstructions of multiple structures from mixed data sets.

## 1. Introduction

An advantage of cryo-electron microscopy (EM) structural analysis is its ability to capture isolated macromolecular complexes in their native state in solution. However, functioning macromolecular complexes often coexist in multiple states in solution and these are captured in the vitrified sample. Structural heterogeneity can be caused by dynamic structural states, the presence of reaction intermediates, variable ligand binding, multiple oligomeric states, or changes in environmental conditions such as temperature and solutes. Heterogeneity imposes significant limitations on the achievable resolution, since information from different molecular states in a heterogeneous ensemble will be combined into one reconstruction. In some cases, particular conformations can be trapped biochemically before EM imaging, for example, by using antibiotics to stabilize selected states of ribosomes. However, this is not always experimentally possible, so that other means are needed to solve the problem of structural analysis of heterogeneous assemblies. Therefore, separating images of molecules in different conformations is increasingly important for EM structural analysis, not only for improving the resolution of reconstructions but also to reveal different functional states.

There are three main approaches currently used to identify and sort molecular heterogeneity, shown schematically in Fig. 12.1. All of them are based on statistical methods but differ by the steps in which statistical analysis is applied. The next level of divergence in the methods is in the types of statistical methods used, and finally in their implementation. The first approach is based on statistical analysis of raw images (in 2D)—*a priori* analysis to detect the heterogeneity of the sample in its images. Here, the initial sorting is done on 2D images only, prior to any three-dimensional (3D) reconstruction (Fig. 12.1, left panel). In the second approach, an initial 3D map is required in order to separate the images into subsets of images presenting a molecular complex in similar orientations. Analysis of structural

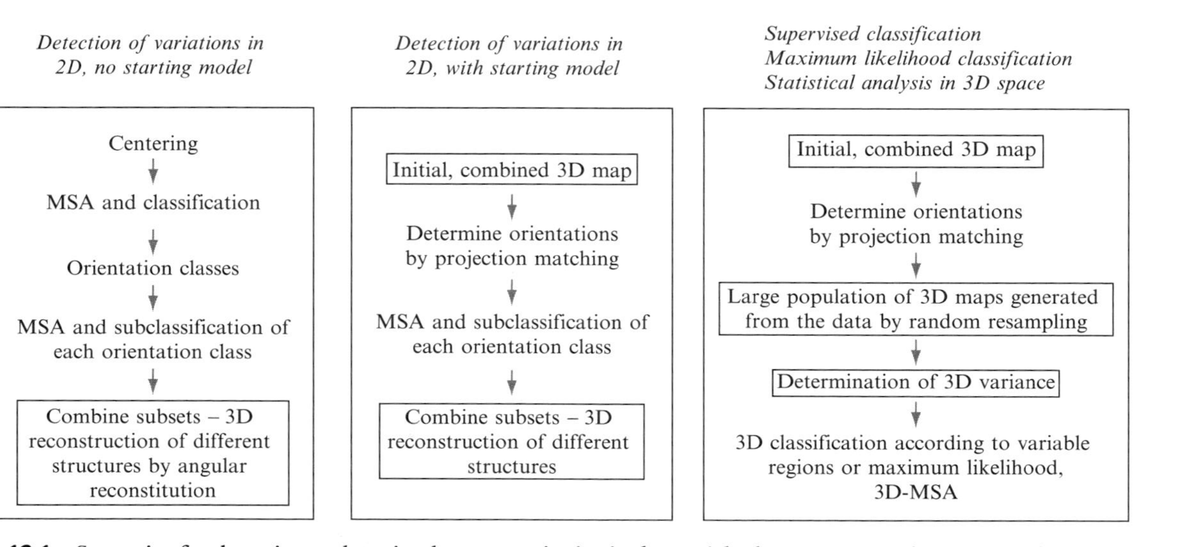


**Figure 12.1** Strategies for detecting and sorting heterogeneity in single-particle data sets. Steps done on 3D data are boxed.

heterogeneity is then done in 2D for each orientation subset. This approach minimizes orientation variation within classes, and as a result facilitates recognition of conformational variations. It has been applied to the reconstruction of heterogeneous ribosome complexes (Fu *et al.*, 2007; Klaholz *et al.*, 2004) and to icosahedral viruses with symmetry mismatches or partial occupancy (Briggs *et al.*, 2005) (Fig. 12.1, middle panel). The third approach is based on *a posteriori* analysis of 3D reconstructions by considering a population (many as possible) of 3D reconstructions in order to examine variations in 3D maps (Ohi *et al.*, 2004; Penczek *et al.*, 2006a,b; Scheres *et al.*, 2007) (Fig. 12.1, right panel). It should be noted that all these methods in their present form require involvement of the researcher in the separation procedure, to decide which parameters must be taken into account as significant during the analysis.

Here, we will focus on the *a priori* group of methods based on double multivariate statistical analysis (MSA) of features in the 2D images to detect image variations that reflect changes both in orientation and conformation without an initial model (Fig. 12.1, left panel). This 2D statistical analysis has been used to separate particles of different size (White *et al.*, 2004) and also with different ligand occupancy (Elad *et al.*, 2007, 2008).

## 2. Theoretical Background

EM images of biological complexes differ from each other not only because of variations in their orientations and possible conformational changes but also due to noise caused by the effects of uneven background (e.g., from different thickness of ice or support film), or variations in detector sensitivity. A low signal-to-noise ratio makes the visual comparison of single particle images and characterization of important features such as size and orientation difficult or impossible for small complexes in ice. Consequently, statistical analysis is needed to analyze differences within large data sets. The essence of any statistical analysis is in distinguishing significant variations related to differences in molecular structure from noise, and a major question is how to decide which parameters should be taken into consideration. Sorting images according to their features can be done by cross-correlation between all possible pairs of images to assess their similarities and differences (covariance matrix of images). However, this apparently simple idea is computationally prohibitive. Therefore, statistical methods are essential for the separation of noisy images according to their structural features.

Statistical approaches make it possible to reduce the number of variables used to compare images by finding relationships between the variables. Differences related to changes in molecular orientation or conformation are characterized by synchronized density variations in particular image locations, whereas randomly distributed differences are usually caused by noise. MSA techniques enable us to identify image variations due to differences in orientation or structural features and distinguish them from insignificant differences produced by noise. The principles of MSA are described in detail in van Heel *et al.* (2009).

The aim of MSA is to find the strongest characteristic features of the entire data set, the combination of which can be used to describe each image in the data set. These characteristic features correspond to the principal variances within the data set or major dissimilarities between the images. The variances will have different levels of significance, or in other words, different weights. The characteristic features (patterns) of image variance are known as eigenvectors (or eigenimages) (Frank, 1990; van Heel and Frank, 1981; van Heel *et al.*, 2009). In MSA, the first eigenvector corresponds to the average of all images in the data set. Each successive eigenimage characterizes the variances in images that are not reflected by the preceding eigenimages. Thus, these eigenimages are independent (do not correlate). Typically, the number of eigenimages needed to describe the data set is much smaller than the number of pixel densities in the images being compared. The eigenvalues indicate the strength of variations in the data set reflected by the corresponding eigenvectors. The eigenimages are normally presented in the order of their significance, which is quantified by their eigenvalues. The first few eigenimages are usually related to the shape of the complex under study, reflecting differences in orientation. For example, variance can arise from the shape of extended objects which are centered but not rotationally aligned. Eigenimages relating to fine details in the images appear after the mayor variations. An advantage of statistical analysis of 2D images is that it does not require an initial 3D model, thus avoiding problems of model bias. Reduction of the number of parameters to be analyzed (compression of the data) makes these approaches fast and efficient, computationally cheap and therefore applicable to large data sets.

To ensure that statistical analysis will reflect features related to the structural organization of the object, it is necessary to center all images within their frames. All images must be normalized to the same standard deviation and the same average density. Otherwise, statistical analysis will reflect variations in image displacement, contrast, and background. This is a common requirement in all types of statistical approaches. The population of images subjected to analysis can be separated according to the major differences identified by the eigenimages. By selectively weighting the eigenimages, it is possible to dissect different sources of variation in the data set and classify images accordingly. Images can be classified initially according to their major features and then the resulting subclasses can be analyzed according to finer variations. Such double MSA reduces the influence of variance caused by different molecular orientations and highlights variations induced by additional small factors such as ligand binding. Here, we will focus on applications of MSA in analysis of heterogeneous samples.

Determination of the angular orientations of the images is essential for 3D reconstruction. The two main computational approaches for orientation determination are angular reconstitution and projection matching (Frank, 2006; van Heel *et al.*, 2000). Angular reconstitution is based on finding similar one dimensional (1D) projections (known as "common lines") between different 2D projections (classes) of the 3D structure. The angle between common lines of any two projections and a third projection corresponds to the angle between the first two projections. Analysis of the cross-correlation of 1D projections from several views helps to refine their angular orientations. The theoretical basis of this approach is described in van Heel *et al.* (2000). The use of class averages improves the signal/noise ratio and consistency of the angular search.

Another, very widely used approach is projection matching. This is based on comparison of raw images with a large set (often several thousands) of projections of the initial model over a range of different directions. Assignment of the projection orientation is made by identifying a particular model projection having the best match with the image (the highest correlation coefficient) (Penczek *et al.*, 1994). This technique requires an initial model.

## 3. Recognition and Separation of Particles with Variable Sizes

### 3.1. Model data

We now address the question of how to separate a mixture of images with some model calculations. Let us demonstrate the technique using a 3D model data set with size variation. Two asymmetric 3D models were created, one model 10% smaller than the other. Both were projected in directions evenly distributed on the Euler sphere with an angular increment of 3°. From each regular set 1500 images were randomly selected and randomly rotated in the plane, and noise was added to the resulting images. The two sets of final images were combined into one data set (Fig. 12.2A), which was normalized and band pass filtered according to standard procedures of single-particle analysis, and then subjected to MSA. The resulting eigenimages are presented in Fig. 12.2B.

The first eigenimage represents the sum of all centered images (Fig. 12.2B, eigenimage 1). Because the images were randomly rotated and no angular alignment was so far performed, this eigenimage represents an approximation of the projection of the rotationally averaged object. The following eigenimages mainly reflect features, such as protrusions or

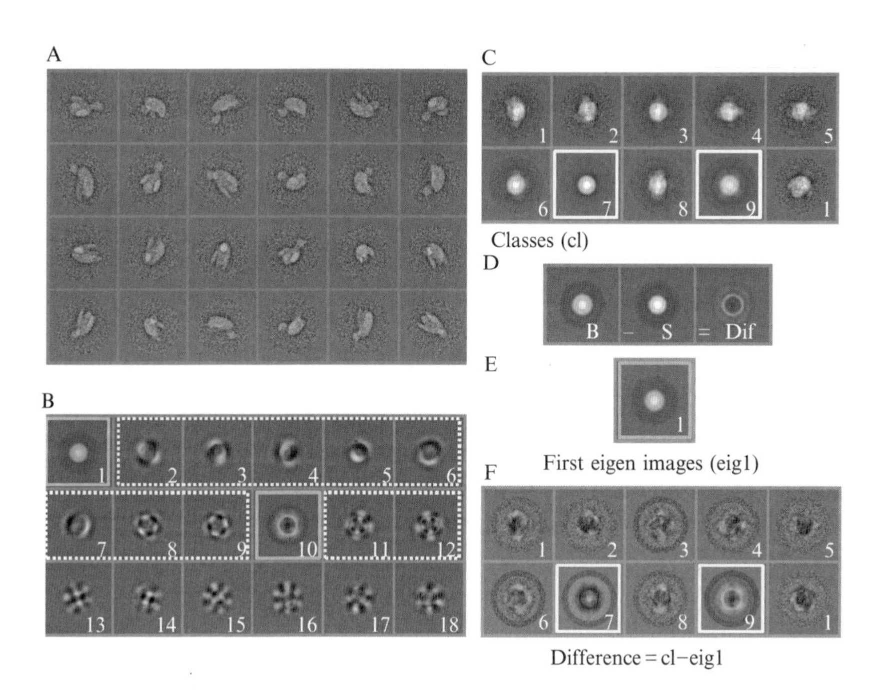


G

| | Big | Small | Superimposed structures | Differences |
|---|---|---|---|---|
| Model | | | | |
| Iteration 1 | | | | |
| Iteration 3 | | | | |

**Figure 12.2** Separation of a mixed population of model particles with size variation. (A) Representative images of particles in different orientations. (B) Eigenimages of the mixed data set. Eigenimage 1 is the sum of all images and eigenimage 10 reports on size heterogeneity in the data set. (C) Result of classification using eigenimages 1 and 10 at full weight and eigenimages outlined with the white dotted line downweighted by a factor of 0.001. Eigenvectors above 12 were not used. (D) Difference between rotationally averaged total sums of the original big and small subsets. (E) The first eigenimage from (B) was subtracted from all classes shown in (C) to generate the differences

elongated elements in different orientations, of the molecular shape in projection. The eigenimage with the ring (Fig. 12.2B, eigenimage 10), reporting on size differences, would have higher significance for an isometric object (White *et al.*, 2004). However, the randomly oriented elongated objects give rise to spherical harmonic features and the ring eigenimage is present with lower significance. Classification of images using the whole set of eigenimages yields class averages containing images of both small and big molecules, resulting in the loss of fine details. In this case, the class averages obtained differed only by rotation in the plane and the size difference was not detected.

MSA of the model data set shows that eigenimage 10 (Fig. 12.2B) has a ring feature, indicative of size differences. The main variations represented in the first 12 eigenimages reflect variations in the object shape and also all possible rotations of the molecule in the image plane. Finer details are reflected in the subsequent eigenimages. Therefore, we performed a classification in which eigenimages 2–9 and 11–12 were suppressed by a factor of 0.01 and classification was based mainly on eigenimages 1 and 10, to crudely sort the data into 10 classes (Fig. 12.2C). To check the validity of this approach for size separation, we calculated averages of all images of the small and big molecules separately, and then determined the sum and difference of their averages (Fig. 12.2D). The sum of projections of the rotationally averaged objects reproduces the first eigenimage and the difference corresponds to eigenimage 10 with the ring, the width of which matches the size difference.

To evaluate the size difference, we subtracted the first eigenimage from all 10 classes (after normalization of all classes and the eigenimage). In this case, the difference corresponding to the class with small particles (7 in Fig. 12.2C and F) has negative density while the class of large particles (9 in Fig. 12.2C and F) produces a positive difference density. The other classes do not reveal any clear differences, representing mixed populations. Images in class averages 7 and 9 were therefore identified as corresponding to the small and large objects respectively, and were extracted and separately analyzed. The reconstructions obtained from images selected after the first separation based on classes 7 (small) and 9 (big) are shown in Fig. 12.2G, middle row. The first models obtained were used to create references for multi-reference alignment (MRA) of the whole mixed data set (van Heel

shown in (F). Images outlined in white in (C) and (F) indicate a class formed by small particles (7), which has a negative difference with the first eigenimage, and class 9, comprised of large particles, which has a positive difference with the first eigenimage. (G) Surface representations of the original models (upper row). Middle row, 3D reconstructions obtained after the first round of separation by MSA using the ring eigenimages. The big reconstruction was obtained from the images in class 9 and the small reconstruction from class 7. Bottom row, 3D reconstructions after competitive alignment and angular refinement. (See Color Insert.)

*et al.*, 2000). Reprojections were calculated in uniformly spaced directions over the Euler sphere. The models were refined by competitive alignment, in which images with better correlation to the first model were extracted into the first subset and those that best matched the second model were extracted into the second subset (see supplementary information in Rye *et al.*, 1999). New classes for each subset representing different views were used to obtain refined reconstructions of the big and small objects. The reconstructions after three iterations of competitive alignment and angle determination using angular reconstitution are shown in Fig. 12.2G, bottom row. Comparison of the resulting models reveals the expected size differences (Fig. 12.2D).

## 3.2. Small heat shock proteins

The small heat shock proteins (sHsps) are important in preventing aggregation of nonnative proteins by binding them in stable complexes (Haslbeck *et al.*, 2005). Hsp26, one of the two sHsps from yeast, forms well-defined 24-mers which are reversibly activated to bind nonnative substrate proteins at elevated temperatures (Haslbeck *et al.*, 2005). Raw cryo-EM images of Hsp26 showed round, oligomeric particles of apparently uniform size. Attempts at 3D reconstruction from a data set of 15,000 images of Hsp26 did not lead to a consistent map, and it was eventually realized that the particles were not all the same size. Typical class averages are shown in Fig. 12.3A. MSA of centered images (translational alignment only) shows concentric rings in the second eigenimage (Fig. 12.3B), indicating variations in particle size. Therefore, the same approach was applied as in the model experiments. Using the first four eigenimages, with the first one suppressed to 0.1 of its original value, the whole data set was grouped into four classes (Fig. 12.3C). This first round of classification allowed us to distinguish the size difference. Images were then extracted into three subsets. Those images in the class with the largest diameter (Fig. 12.3C, panel 1 outlined with a continuous white circle) were assigned to the big subset while those in the class with the smallest diameter (Fig. 12.3C, panel 2 outlined with a broken white circle) became the small subset. The third, mixed, subset was formed from images in the other two classes. MSA was run on each subset to check for the ring pattern in the eigenimages that would indicate mixed sizes within a subset. This analysis did not reveal any heterogeneity within the subsets of big and small particles. The initial models for each subset were obtained using the filtered back-projection approach with angles determined by angular reconstitution (van Heel *et al.*, 2000). The refined structures from the big and small subset were obtained using the strategy outlined above, with alternate competitive alignment and angular reconstitution for orientation determination followed by 3D reconstruction to refine the separation.

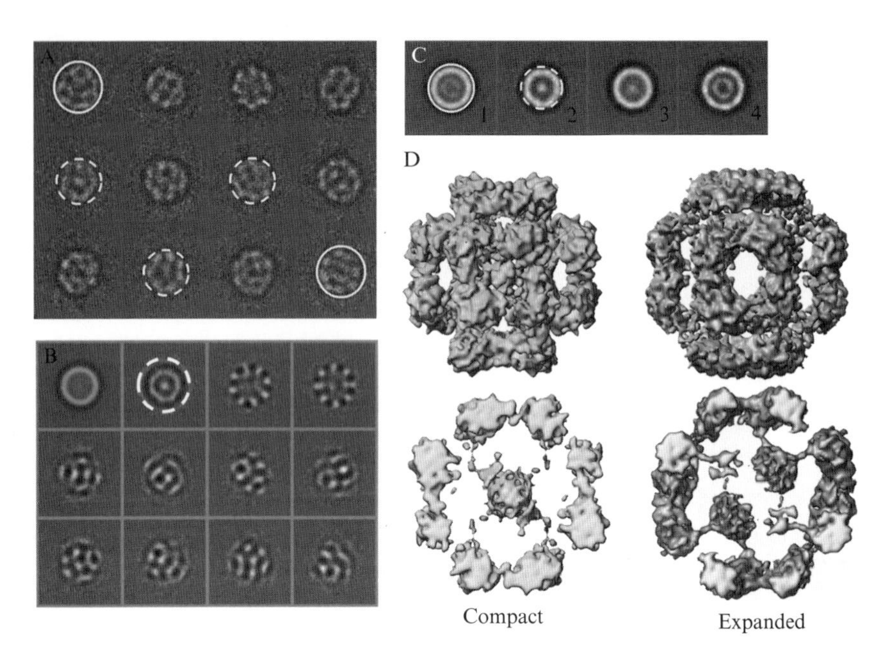


**Figure 12.3** Real data with size variation: Cryo-EM data of Hsp26. (A) Typical class averages of the whole data set. Big classes are outlined with a continuous white line and small classes with a dashed line. (B) Eigenimages of the whole data set. Note the ring in the second eigenimage is outlined with a broken white line. (C) The whole data set was sorted into four classes by MSA using only the first four eigenimages. The big class (1) is circled with a continuous white line, the small class (2) with a broken white line and the other two classes (3 and 4) are mixed. (D) Surface rendered views of the compact and expanded maps of Hsp26. The bottom row shows sliced open views of the compact and expanded maps, showing the inserted densities. Figures reproduced from White *et al.*, 2006.

These two refined structures allowed the separation of the mixed size subset into big and small subpopulations. The overall procedure was iterated until no significant movement of particles between classes was observed. When the separation was completed, the 3D maps were refined by angular reconstitution with 700 classes in each set. The two maps are shown in Fig. 12.3D (White *et al.*, 2006). They reveal interesting differences, mainly in the internal domains, in addition to a 5% difference in diameter.

## 4. Statistical Analysis of Particles with Variable Ligand Occupancy

2D statistical analysis can also be used to reveal heterogeneity and separate mixed populations with conformational changes triggered by ligand binding. In contrast to the previous application, this approach needs preliminary alignment to separate differences in images caused by various orientations from differences produced by variable binding of ligands. Without rotational alignment, the initial eigenimages are dominated by variations produced by differences in molecular orientation. The initial classification according to angular orientation minimizes orientation variations and increases the weight of localized changes induced by partial ligand occupancy. Therefore, MSA comprises two successive classifications, the first based on eigenimages showing global variance of structural features due to different orientations, that we call "orientational" classes. This is followed by a second classification based on eigenimages showing localized variance arising from changes induced by partial ligand occupancy, that are termed "structural" classes.

## 4.1. Model data

We generated a model of variable chaperonin–substrate binding based on the atomic coordinates of the 800 kDa GroEL oligomer with an additional 33 kDa substrate protein in the binding sites of one ring in half of the complexes. In GroEL–substrate complexes, the substrate is unfolded or partially folded; for modeling purposes, the native structure of malate dehydrogenase (MDH) was used. Gaussian noise was added to the images to give a 2:1 signal-to-noise ratio. After alignment, images were classified with ~15 images per class based on 40 eigenimages. MRA is a procedure in which each image is compared to several (sometimes as in projection matching several thousand) projections of a model. The position of the highest correlation peak between a model projection and the image analyzed provides the information on how the image should be shifted. The MRA procedure was subsequently performed with three representative classes as references—top, side, and tilted views.

Rotationally aligned images were subjected to statistical analysis to separate them into orientational classes representing molecules in different orientations (Fig. 12.4A). The first 12 eigenimages of the entire population of aligned images mainly show density variations distributed over the area occupied by the molecule (Fig. 12.4B). However, some eigenimages such as 5 and 9 demonstrate prominent positive or negative peaks (bright or dark spots, Fig. 12.4B). The overall pattern of all these eigenimages indicates at two sources of variation: one is due to different orientations and the other is related to the presence/absence of ligand (most notably seen in eigenimages 5 and 9), which reflects the addition of ligand to only half of the molecules.

To separate the data set into groups of images with the same orientation, most of the eigenimages obtained by MSA must be included (Fig. 12.4C). These steps do not require determination of angular orientations, so the technique is not biased to any 3D model. The images contained in each orientational class are extracted and subjected to further classification into 4–6 structural subclasses each containing at least 30 images. This step of

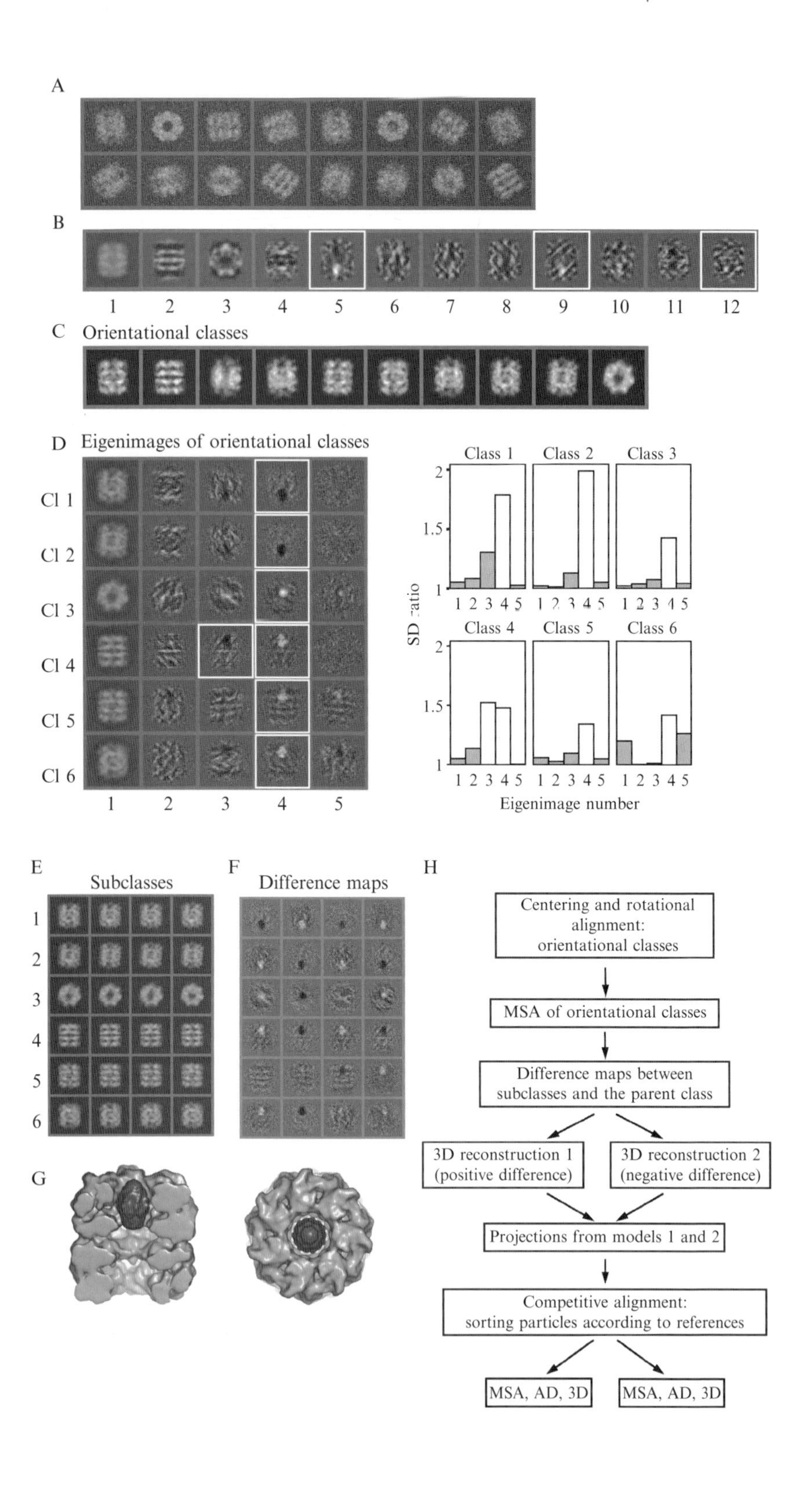
A
B
1
2
3
4
5
6
7
8
9
10
11
12
C Orientational classes
D Eigenimages of orientational classes
Cl 1
Cl 2
Cl 3
Cl 4
Cl 5
Cl 6
Class 1
Class 2
Class 3
Class 4
Class 5
Class 6
SD ratio
Eigenimage number
E
Subclasses
F
Difference maps
G
H
Centering and rotational alignment: orientational classes
MSA of orientational classes
Difference maps between subclasses and the parent class
3D reconstruction 1 (positive difference)
3D reconstruction 2 (negative difference)
Projections from models 1 and 2
Competitive alignment: sorting particles according to references
MSA, AD, 3D
MSA, AD, 3D

classification is based on eigenimages with strong local variations (Fig. 12.4D, rows 1–6). Separation into only two structural subclasses is usually insufficient to resolve clear differences, whereas a large number of subclasses reduces the number of images per class and increases the influence of noise. Variations due to partial ligand occupancy appear as areas of high variance in the eigenimages (local peaks) (Fig. 12.4D, white boxes). The significance of these variations can be quantified by measuring the ratios of standard deviation within and outside local peaks (Fig. 12.4D, graphs). Measurement of these ratios can be used to automate the analysis of local variance in the eigenimages.

Determining whether a subclass should be attributed to the "full" (containing ligand) or "empty" (not containing ligand) group cannot be reliably assessed by eye (Fig. 12.4A). Calculation of difference maps between each structural subclass and the corresponding orientational class average provides a more objective evaluation of the separation and indicates the presence or absence of ligand (Fig. 12.4E and F). With this information, full and empty subclasses from different orientation groups can be combined for separate 3D reconstructions (Fig. 12.4G).

## 4.2. Chaperonin–substrate complexes

The strategy described in the preceding section for model data was applied to an experimental data set of 40,000 particles obtained from a sample of GroEL complexes with nonnative, disordered MDH, compared to a similar analysis of a dataset of GroEL without substrate (Elad *et al.*, 2007). For image processing, side views were selected and first sorted into orientational classes. Then these were sorted into smaller, more homogeneous, subsets on the basis of eigenimages reporting on substrate occupancy and distribution (Fig. 12.5A). Two 20 × 20 pixel subregions were selected from each eigenimage, one surrounding the most extreme density value and the other in the central part of the eigenimage. Subregions with the highest standard

**Figure 12.4** Model data with variable ligand occupancy. Results of double MSA of mixed images of GroEL and GroEL–MDH. (A) Representative images in different orientations. (B) Eigenimages obtained after the first round alignment and MSA. (C) Representative classes obtained using all eigenimages. (D) Eigenimage analysis of images within orientational classes (six representative classes are shown), with corresponding plots of the standard deviation ratio of the subregion with high variance relative to a neighboring area of the eigenimage. Eigenimages with a high standard deviation ratio are highlighted in (D) and the corresponding ratios are shown as white bars in the plots. These eigenimages were used for subclassification. (E) Subclass averages and (F) the corresponding difference images between the orientation classes and each of their subclasses. (G) 3D reconstruction of the ligand-occupied class. Left, cutaway view of the map showing the MDH ligand density in dark gray. Right, top view of MDH in the GroEL cavity. Figures reproduced from Elad *et al.* (2008). (H) Flowchart of the processing. AD, angle determination.

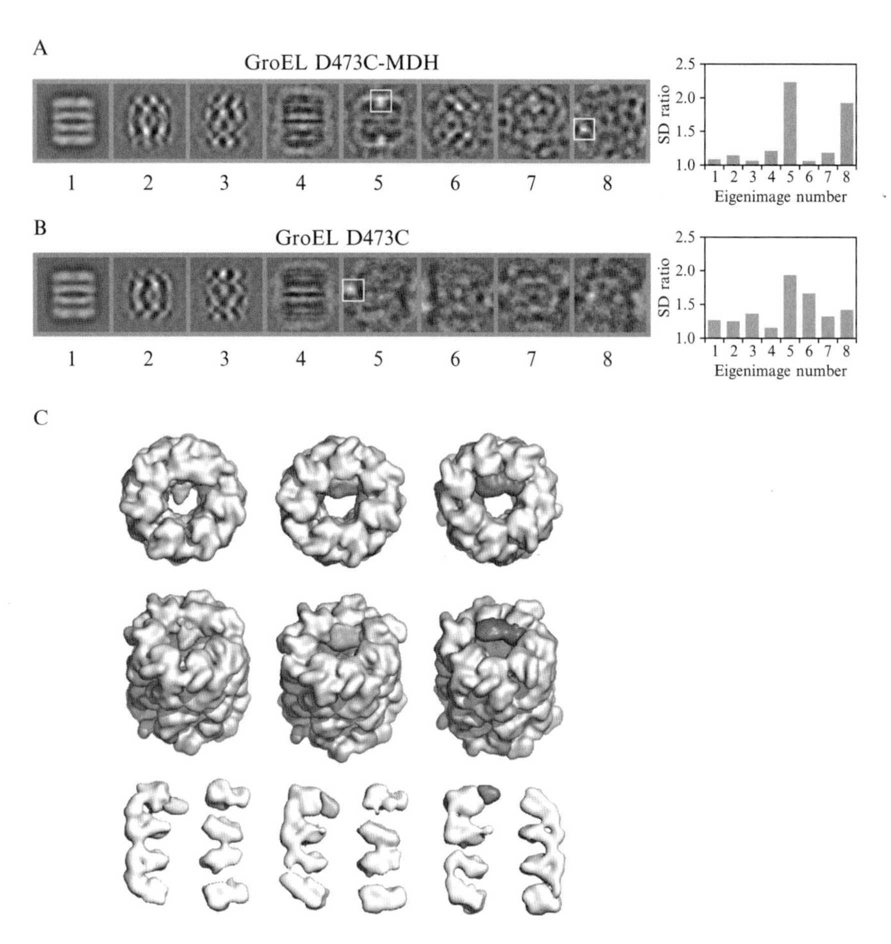


**Figure 12.5** Real data with variations in ligand occupancy: chaperonin–substrate complexes. Eigenimages of (A) the initial 8000 image data set of GroEL–MDH complexes and of (B) 6800 images of empty GroEL. Standard deviation ratios calculated as in Fig. 12.4D are plotted in the corresponding graphs on the right. (C) Surface representation of three different asymmetric cryo-EM maps of GroEL complexed with nonnative MDH obtained from sorted images (Elad *et al.*, 2007). Shown are end views of the maps in (top row), tilted views (middle row) followed by central sections (bottom row). The GroEL subunits are generated from the atomic structures of subunit domains fitted into the EM densities. Substrate densities were extracted from the experimental density maps and are highlighted in different colors. Panels (A) and (B) are reproduced from Elad *et al.* (2007) and (C) is reproduced from Saibil (2008). (See Color Insert.)

deviation are boxed in white (Fig. 12.5A). Eigenimages 1–4 are similar in both substrate-containing and empty GroEL data sets and are typical of GroEL images. The first one corresponds to the average of all images (which were all possible side views), and 2–4 reflect image variance due to differences in particle rotational orientation around the sevenfold axis and small tilts out of the image plane. In this case, the density variations are evenly distributed over the particle area and there are no prominent local maxima. Eigenimage 5 of the GroEL–MDH data set shows an exceptionally high peak at the position of one end cavity of GroEL and a similar, somewhat weaker peak at the opposite end cavity. These peaks reflect variation in substrate occupancy, and they do not appear in the corresponding apo GroEL eigenimages. Relatively high peaks appear also in eigenimages of both data sets at the outer surface of the equatorial domains (Fig. 12.5A, eigenimage 8 and 5B, eigenimage 5). This feature is attributed to the presence of a surface cysteine introduced to generate more side view orientations (Elad *et al.*, 2007). It interacts with other molecules, causing independent variations at that position.

The orientational classes were progressively subdivided into structural subclasses, using substrate-related eigenimages to assess their heterogeneity. Further subdivision was done until no localized maxima were observed in the eigenimages. Competitive alignment was used to refine the separation at each stage. Ultimately, the data set was separated into five major structural groups. These showed a high degree of stability in consecutive rounds of competitive alignment and angle assignment. We concluded that this is the maximum number of distinct structural groups that we can discriminate in this data set. Of the five final groups, three produced structures with well-defined extra densities revealing different positions of nonnative MDH binding (Fig. 12.5C). The other two groups did not exhibit significant extra density in the cavity (not shown). All groups refined to a resolution of 10–11 Å at 0.5 threshold of the Fourier shell correlation (FSC), and sharpened maps showed features of the α-helical substructure, as expected in this resolution range. Attempts at further classification based on any of the eigenimages that were calculated within classes did not produce new stable subgroups. These results are reported in Elad *et al.* (2007).

## 4.3. 70S ribosome with elongation factor G

The double MSA method was also tested on real cryo-EM images of 70S *E. coli* ribosomes complexed with the ligand elongation factor G (EF-G), an asymmetric structure without preferred orientations (Elad *et al.*, 2008). Images were provided by H. Gao and J. Frank (Gao *et al.*, 2003). The population of complexes contained a mixture of molecules in bound and nonbound states. 10,000 preprocessed and centered images were subjected to statistical analysis and classified into 100 general orientational classes. Ten different characteristic views that revealed clear features reflecting different orientations of the ribosome were selected and used for MRA. After rotational and translational alignment, the images were grouped into 30 orientational classes (~350 images/class), from which 23 with the best contrast were subclassified. Representative orientational classes and their eigenimages after subclassification are shown in Fig. 12.6A and B. Analysis of local variations in the eigenimages indicates variations due to heterogeneity in ligand binding. The first three eigenimages are still mainly related to small differences in the angular orientations. Because of the combination of ligand-induced conformational changes in the structure and

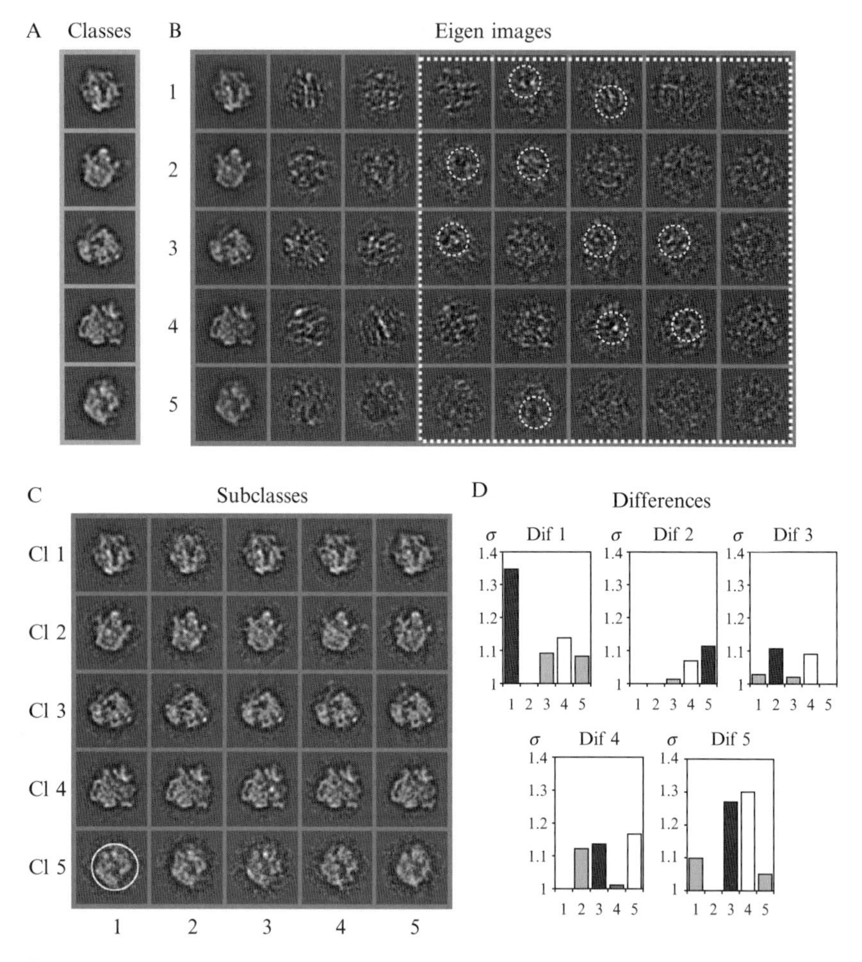


**Figure 12.6** Double MSA of ribosome–EF-G complexes. (A) Representative classes after the first round of MSA and classification. (B) First eight eigenimages from the second round of MSA (within classes). Regions of high variance are circled with white dashes. (C) Subclass averages from classification on the basis of eigenimages 4–8. (D) Average differences between subclasses and parent class averages. The area used for assessment of standard deviations is indicated by the white circle in the bottom left image in (C). Black peaks indicate a negative average difference and white ones a positive difference. Figures reproduced from Elad *et al.* (2008).

heterogeneity in ligand binding, not all groups of images demonstrated localized variation in one specific eigenimage. To automate the separation procedure for each class, the same group of eigenimages, 4–8 inclusive, was used to partition the parent class into five structural subclasses (Fig. 12.6B and C). Next, the parent orientational class averages were subtracted from their respective structural subclasses. Separation of the ribosome images was based on two parameters: the highest standard deviation in the difference map and their average difference density (Fig. 12.6C and D; sigma peaks from classes with negative average differences are shown in black, positive differences in white). 1651 images were extracted from the subclasses that showed positive average difference density (EF-G present) and 1532 images were extracted from the subclasses that showed negative difference density (EF-G absent). The remaining images belonged to subclasses which did not show clear positive or negative differences in their difference map and were therefore not used for this step of the analysis. The sorted images were used to obtain initial 3D maps with and without EF-G, using angular reconstitution without symmetry. Reprojections from these 3D maps were used for competitive alignment of the complete data set followed by orientational classification. According to this alignment, 4161 images were assigned to the ribosome without elongation factor, while 5839 images were assigned to the EF-G bound state. Reconstructions are shown in Fig. 12.7. The upper panel shows several sections of the two structures and their differences (Fig. 12.7A). The white arrow points to the main region of variations in these sections. The difference maps clearly reveal the position of EF-G as additional density in the cleft between subunits. Therefore structure A corresponds to the ribosome with bound EF-G and structure B the ribosome without EF-G (Fig. 12.7B and C). The density of EF-G and other, more minor differences are in excellent agreement with the original papers on this system (Gao *et al.*, 2003; Valle *et al.*, 2002).

## 5. Conclusions

The method described here provides a framework for analysis of EM images in which data sets can be tested for heterogeneity arising from local changes in a macromolecule and also accompanied by global conformational changes. White *et al.* (2004) described a method that allows separation of particle images based on size variation, with subsequent 3D analysis. The technique was shown to work with overall size variations as small as 5% (White *et al.*, 2006). Size variations are more readily recognized in populations of isometric particles. With an elongated structure, variance between different orientations is more dominant, but size variation can still be detected. In this case, more steps of alignment and classification may be needed.

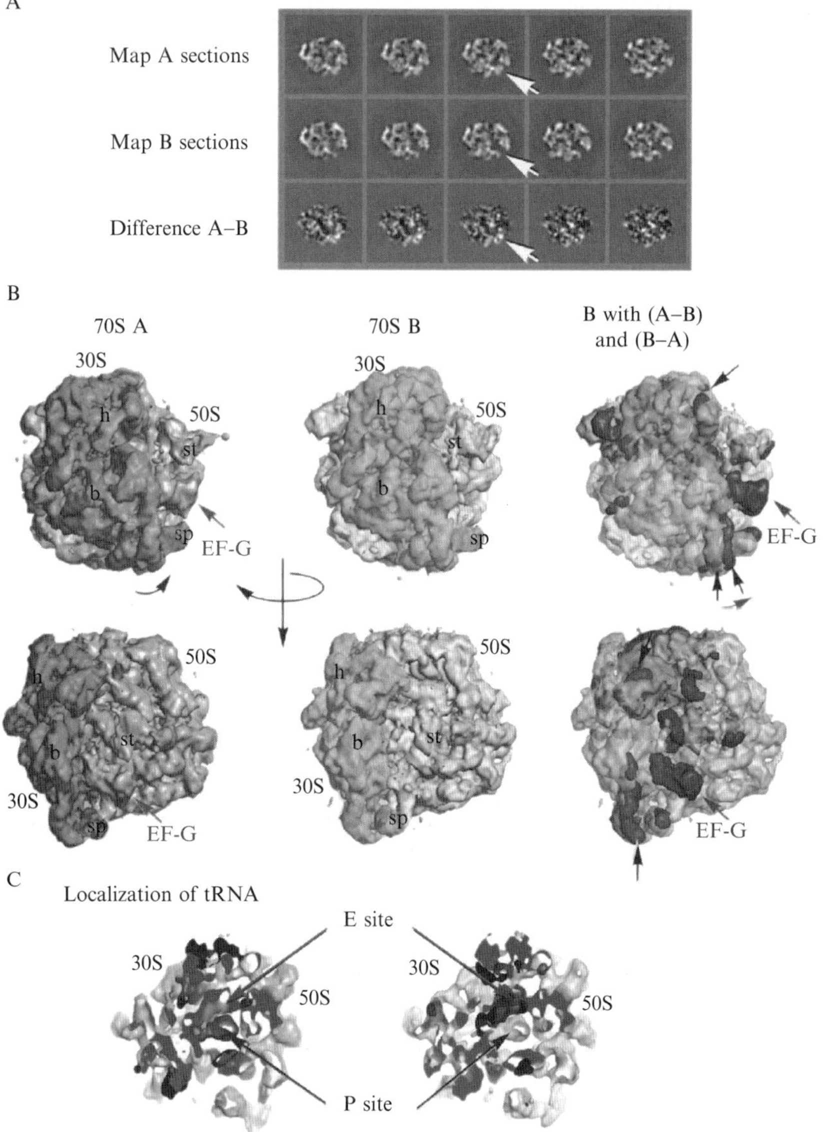


**Figure 12.7** Reconstructions of ribosome–EF-G complexes. (A) Sections of the 3D maps and their difference. The 70S ribosome map A reveals extra density in the cleft between large and small subunits, whereas map B does not show such density. White arrows point to the major differences in the sections. (B) In map A, the small subunit is shown in blue and the large subunit in light blue. In structure B, the subunits are orange (small) and gold (large). The difference map between structures A and B (in red) is superimposed on the B map (right column). Red arrows point to the EF-G location. Black arrows point to differences related to movement of the small subunit. (C) Cutaway views of the 70S ribosome with and without EF-G. Red arrows indicate the E-site tRNA that is only occupied in the presence of elongation factor. Blue arrows indicate the P-site tRNA that is occupied in the absence of elongation factor. Labels: h, head; b, body; sp, spur; st, stalk base. (See Color Insert.)

The approach of *a priori* analysis of images by double MSA proved to be successful on model and real data. It worked well for different mixtures of two states, for which subgroups of images were identified with better than 90% accuracy, leading to reliable reconstructions. These reconstructions provided good quality models for further refinement. The principles developed here have been successfully applied to cryo-EM data on GroEL complexes with MDH, in which five different structural states of the nonnative ligand were resolved from a data set of ~35,000 images (Elad *et al.*, 2007, 2008) and for GroEL complexes with a folding bacteriophage

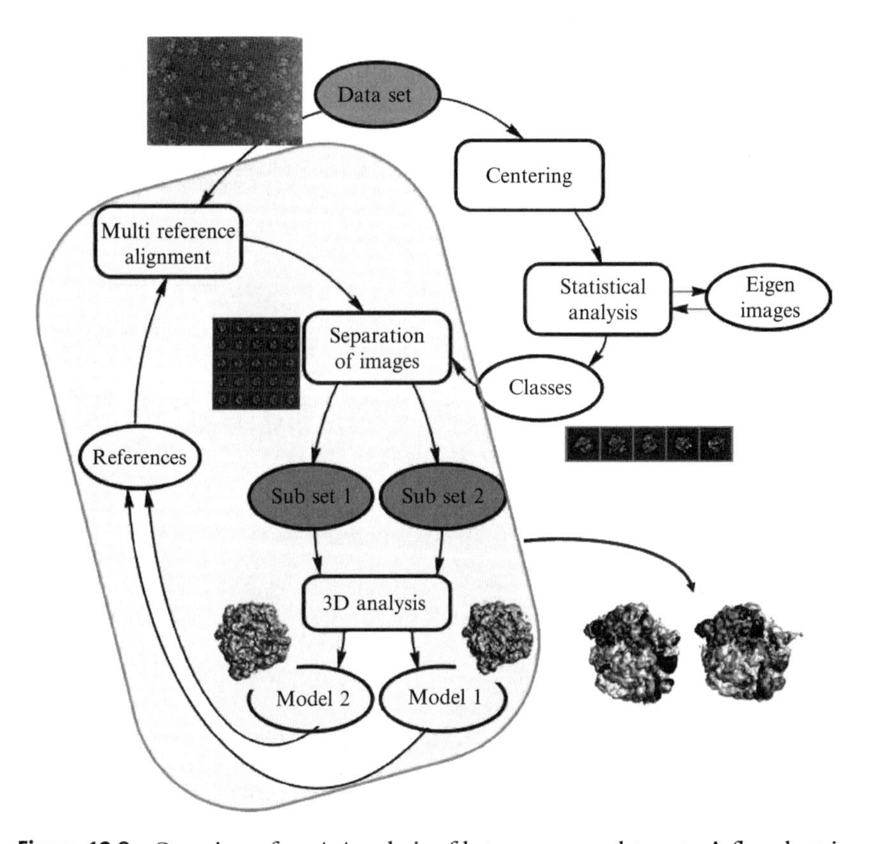


**Figure 12.8** Overview of *a priori* analysis of heterogeneous data sets. A flowchart is shown for the separation of heterogeneous data into subsets and their separate 3D reconstruction. The initial steps include centering and, for the ligand binding case, rotational alignment, followed by the first statistical analysis. Subsequent iterative steps are enclosed in the gray area. They include competitive multi-reference alignment, statistical analysis, reconstruction, and reprojection. It is important to note that competitive alignment can be performed against several models (for simplicity, only two models are shown). Subdivisions of these models can include internal cycles of competitive alignment.

capsid protein (Clare *et al.*, 2009). Heterogeneity due to partial occupancy of a ligand can be revealed, as long as the changes are statistically significant, as in ribosomes (Elad *et al.*, 2008). It can be applied to structures of unknown conformation with any symmetry. Therefore, any set of projections that contains a range of views of a macromolecular complex can be tested in this way for local variance, and if necessary classified into more homogeneous subsets. This approach is summarized in Fig. 12.8, with the steps in refinement of the sorting by separate 3D reconstruction and competitive projection matching highlighted in grey. The images can then be separated into more homogeneous subsets on the basis of their significant structural variations.

## ACKNOWLEDGMENTS

EVO and HS are grateful to the UK BBSRC, the EU 3DEM Network of Excellence and the Wellcome Trust for funding, Luchun Wang for EM support, and David Houldershaw and Richard Westlake for computing support.

## REFERENCES

Briggs, J. A. G., Huiskonen, J. T., Fernando, K. V., Gilbert, R. J. C., Scotti, P., Butcher, S. J., and Fuller, S. D. (2005). Classification and three-dimensional reconstruction of unevenly distributed or symmetry mismatched features of icosahedral particles. *J. Struct. Biol.* **150,** 332–339.

Clare, D. K., Bakkes, P. J., van Heerikhuizen, H., van der Vies, S. M., and Saibil, H. R. (2009). Chaperonin complex with a newly folded protein encapsulated in the folding chamber. *Nature* **457,** 107–110.

Elad, N., Farr, G. W., Clare, D. K., Orlova, E. V., Horwich, A. L., and Saibil, H. R. (2007). Topologies of a substrate protein bound to the chaperonin GroEL. *Mol. Cell* **26,** 415–426.

Elad, N., Clare, D., Saibil, H. R., and Orlova, E. V. (2008). Detection and separation of heterogeneity in molecular complexes by statistical analysis of their two-dimensional projections. *J. Struct. Biol.* **162,** 108–120.

Frank, J. (1990). Classification of macromolecular assemblies studied as 'single particles'. *Q. Rev. Biophys.* **23,** 281–329.

Frank, J. (2006). Three-Dimensional Electron Microscopy of Macromolecular Assemblies. Oxford University Press, New York.

Fu, J., Gao, H., and Frank, J. (2007). Unsupervised classification of single particles by cluster tracking in multi-dimensional space. *J. Struct. Biol.* **157,** 226–239.

Gao, H., Sengupta, J., Valle, M., Korostelev, A., Eswar, N., Stagg, S. M., Van Roey, P., Agrawal, R. K., Harvey, S. C., Sali, A., Chapman, M. S., and Frank, J. (2003). Study of the structural dynamics of the *E. coli* 70S ribosome using real-space refinement. *Cell* **113,** 789–801.

Haslbeck, M., Franzmann, T., Weinfurtner, D., and Buchner, J. (2005). Some like it hot: The structure and function of small heat-shock proteins. *Nat. Struct. Mol. Biol.* **12,** 842–846.

Klaholz, B. P., Myasnikov, A. G., and Van Heel, M. (2004). Visualization of release factor 3 on the ribosome during termination of protein synthesis. *Nature* **427,** 862–865.

Ohi, M., Li, Y., Cheng, Y., and Walz, T. (2004). Negative staining and image classification—Powerful tools in modern electron microscopy. *Biol. Proc. Online* **6,** 23–34.

Penczek, P. A., Grassucci, R. A., and Frank, J. (1994). The ribosome at improved resolution: New techniques for merging and orientation refinement in 3D cryo-electron microscopy of biological particles. *Ultramicroscopy* **53,** 251–270.

Penczek, P. A., Frank, J., and Spahn, C. M. T. (2006a). A method of focused classification, based on the bootstrap 3D variance analysis, and its application to EF-G-dependent translocation. *J. Struct. Biol.* **154,** 184–194.

Penczek, P. A., Yang, C., Frank, J., and Spahn, C. M. T. (2006b). Estimation of variance in single particle reconstruction using the bootstrap technique. *J. Struct. Biol.* **154,** 168–183.

Rye, H. S., Roseman, A. M., Chen, S., Furtak, K., Fenton, W. A., Saibil, H. R., and Horwich, A. L. (1999). GroEL-GroES cycling: ATP and non-native polypeptide direct alternation of folding-active rings. *Cell* **97,** 325–338.

Saibil, H. R. (2008). Chaperone machines in action. *Curr. Opin. Struct. Biol.* **18,** 35–42.

Scheres, S. H., Gao, H., Valle, M., Herman, G. T., Eggermont, P. P., Frank, J., and Carazo, J. M. (2007). Disentangling conformational states of macromolecules in 3D-EM through likelihood optimization. *Nat. Methods* **4,** 27–29.

Valle, M., Sengupta, J., Swami, N. K., Grassucci, R. A., Burkhardt, N., Nierhaus, K. H., Agrawal, R. K., and Frank, J. (2002). Cryo-EM reveals an active role for aminoacyl-tRNA in the accommodation process. *EMBO J.* **21,** 3557–3567.

van Heel, M., and Frank, J. (1981). Use of multivariate statistics in analysing the images of biological macromolecules. *Ultramicroscopy* **6,** 187–194.

van Heel, M., Gowen, B., Matadeen, R., Orlova, E. V., Finn, R., Pape, T., Cohen, D., Stark, H., Schmidt, R., Schatz, M., and Patwardhan, A. (2000). Single-particle electron cryo-microscopy: Towards atomic resolution. *Q. Rev. Biophys.* **33,** 307–369.

van Heel, M., Portugal, R., and Schatz, M. (2009). Multivariate statistical analysis in single particle (cryo) electron microscopy. *In* "Handbook on DVD "3D-EM in Life Sciences," (E. Orlova and A. Verkleij, eds.), NoE 3DEM, London.

White, H. E., Saibil, H. R., Ignatiou, A., and Orlova, E. V. (2004). Recognition and separation of single particles with size variation by statistical analysis of their images. *J. Mol. Biol.* **336,** 453–460.

White, H. E., Orlova, E. V., Chen, S., Wang, L., Ignatiou, A., Gowen, B., Stromer, T., Franzmann, T. M., Haslbeck, M., Buchner, J., and Saibil, H. R. (2006). Multiple distinct assemblies reveal conformational flexibility in the small heat shock protein Hsp26. *Structure* **14,** 1197–1204.

CHAPTER THIRTEEN

# Alignment of Cryo-Electron Tomography Datasets

Fernando Amat,*[,1] Daniel Castaño-Diez,[†] Albert Lawrence,[‡] Farshid Moussavi,* Hanspeter Winkler,[§] *and* Mark Horowitz*

## Contents



## Abstract

Data acquisition of cryo-electron tomography (CET) samples described in previous chapters involves relatively imprecise mechanical motions: the tilt series has shifts, rotations, and several other distortions between projections. Alignment is the procedure of correcting for these effects in each image and requires the estimation of a projection model that describes how points from the sample in three-dimensions are projected to generate two-dimensional images. This estimation is enabled by finding corresponding common features between images. This chapter reviews several software packages that perform alignment and

* Department of Electrical Engineering, Stanford University, Stanford, California, USA
† Center for Cellular Imaging and Nanoanalytics, Department of Structural Biology and Biophysics, University Basel, Basel, Switzerland
‡ National Center for Microscopy and Imaging Research, Center for Research in Biological Structure, University of California at San Diego, La Jolla, California, USA
§ Institute of Molecular Biophysics, Florida State University, Florida, USA
1 Current address: Janelia Farm Research Campus, Howard Hughes Medical Institute, Ashburn, Virginia, USA

*Methods in Enzymology*, Volume 482
ISSN 0076-6879, DOI: 10.1016/S0076-6879(10)82014-2

reconstruction tasks completely automatically (or with minimal user intervention) in two main scenarios: using gold fiducial markers as high contrast features or using relevant biological structures present in the image (marker-free). In particular, we emphasize the key decision points in the process that users should focus on in order to obtain high-resolution reconstructions.

## 1. Introduction

In order to reconstruct a three-dimensional (3D) density map from a set of tilt series images, we need to understand how each point in 3D was projected onto each 2D image that was acquired. The alignment process estimates this set of projections, transforming each image to a common reference frame and enabling 3D reconstruction (Penczek *et al.*, 1995). The typical projection model solves for shifts, rotations, and magnification changes as well as other distortions that occur during the data acquisition phase. The resulting model allows us to obtain the density for each 3D location as a linear combination of the intensities from the corresponding projections in 2D. Thus, the quality of the reconstruction is directly dependent on the alignment accuracy. While reconstructing a 3D scenario or density from multiple 2D projections is a common problem in many fields, such as structure from motion in computer vision (Hartley and Zisserman, 2004; Trucco and Verri, 1998; Ma *et al.*, 2003) and computerized tomography in medical imaging (Herman, 2009; Kak and Slaney, 2001), the low signal-to-noise ratio (SNR) in cryo-electron tomography (CET) images presents a challenge for standard tracking methods. This chapter describes how to overcome these tracking problems to obtain the parameters of the projection model for CET and how these parameters affect the reconstruction process.

Experimental conditions such as presence or absence of fiducial markers, ice thickness, microscope distortions, and acquisition scheme are just some of the variables one needs to consider when choosing an alignment method. In this chapter, we will point out key issues that determine which alignment procedure will work best for a given CET situation, and will cite and describe the methodology of several software packages freely available that align most CET images automatically (or with minimal user intervention). Most of these alignment packages follow a common pipeline (Fig. 13.1):

*Preprocessing*: Basic preprocessing of images usually includes denoising and the deletion of hot pixels caused by the charge-coupled device (CCD). The denoised images are only used during the alignment to facilitate feature detection; to avoid losing information, they are not used for reconstruction.

*Coarse alignment*: Despite reliable automatic data acquisition (as described in previous chapters), raw tilt series can contain large shifts and rotations.

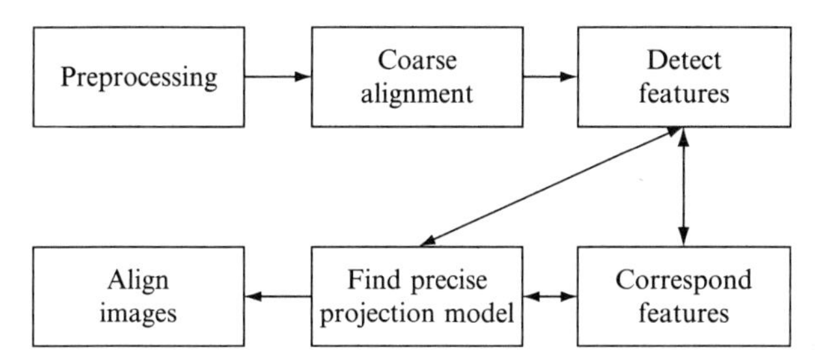


**Figure 13.1** Alignment pipeline. There are two main steps to align CET samples: tracking of features of interest (detection and correspondence) and estimation of projection model.

Using cross-correlation to coarsely align adjacent images is advisable (Guckenberger, 1982). After coarse alignment, we can assume a smooth trajectory of features from image to image, facilitating subsequent alignment steps.

*Feature tracking*: The most common technique for feature detection is to define a patch around the feature as a template, and cross-correlate this template with other images to search for the feature of interest. Assuming we have templates, we seek to detect and correspond those features among multiple images. Due to high noise in CET images, it is not uncommon to have many false positive detections. Here, contextual information is crucial to match the templates more robustly and is encoded differently depending on the software package.

*Find precise projection model*: Once we establish the 2D location of each feature and its correspondence across multiple images, we can use standard optimization techniques to fit a model of how the points in 3D are projected onto the 2D images. It is standard to iterate between projection modeling and feature tracking to increase the overall robustness of the alignment process.

The following sections describe how different packages implement these steps for different types of samples. The main distinction is whether the samples need fiducial markers to align the images (Section 3) or not (Section 4). Section 5 presents how the alignment choice affects the reconstruction process, specifically in samples with nonlinear distortions.

## 2. Notation

We define three Cartesian coordinate systems to describe the tilt series acquisition: $S = \{O_s, \boldsymbol{s}_1, \boldsymbol{s}_2, \boldsymbol{s}_3\}$, which is fixed with respect to the specimen, $M = \{O_m, \boldsymbol{m}_1, \boldsymbol{m}_2, \boldsymbol{m}_3\}$, which is the microscope coordinate system, and finally, $B_i = \{O_i, \boldsymbol{u}_i, \boldsymbol{v}_i\}$ for each projection image, where $i = 1, \ldots, n_P$ and $n_P$ is the number of images in the tilt series. $B_i$ is defined by the pixel raster, and the origin $O_i$ is usually located at the bottom left of the image. A fiducial point $j$ in the coordinates of $S$ is denoted as a coordinate vector $\boldsymbol{r}_j = (x_j, y_j, z_j)^T$, and its projection in the coordinates of $B_i$ is the vector $\boldsymbol{p}_{ij} = (u_{ij}, v_{ij})^T$, where $j = 1, \ldots, n_T$ and $n_T$ is the number of features or fiducials identified in the sample. We assume that the coordinate axes $\boldsymbol{u}_i$, $\boldsymbol{v}_i$ and $\boldsymbol{m}_1$, $\boldsymbol{m}_2$ lie in the same plane, the projection plane, which is perpendicular to the projection direction and the axis $\boldsymbol{m}_3$, the optical axis of the microscope. The origins $O_m$ and $O_s$ are implicitly defined by choosing one of the projection images as a reference image, and by assuming that they both lie on the tilt axis. Unless stated otherwise, the descriptions presented in this chapter refer to data acquired following a single-tilt axis scheme, where the sample is rotated around one axis to generate the tilt series.

# 3. Alignment Using Fiducial Markers

Spherical gold beads are ideal markers to guide the alignment process in CET samples. They create high contrast point-like features in each image (Fig. 13.3) since their projection looks like a disk from any angle and gold is opaque for the electron beam. Therefore, their center always represents the same 3D point. If we assume that fiducial markers are fixed in the ice of the CET sample and we are able to track several gold beads across different projections, we can recover the 3D location of each gold bead and the position of the microscope for each projection (Fig. 13.2). In this section, we explain in detail all the steps required to achieve that.

## 3.1. Marker detection

When the sample contains fiducial markers, we are looking for high contrast disks of a particular diameter (Fig. 13.3). Most software packages generate a bead template from an average of a few fiducials selected from the same sample. Then, normalized cross-correlation (NCC) is used for template matching to avoid errors due to change of illumination from image to image. As a rule of thumb, it is desirable that markers have at least 8 pixels in diameter in each image. Otherwise, detection can be difficult due to high noise.

There are two ways to create an initial template of what a marker looks like in each image. On one hand, programs like Kremer *et al.* (1996), Heymann *et al.* (2008), and Nickell *et al.* (2005) allow users to select all the markers to be tracked. This action is known as "seeding the fiducial model" and the user is generally presented with the image at 0° tilt where contrast is higher, thus it is easier to click on the center of each marker. Since the seeded markers are the only ones that are going to be tracked, it is advisable to pick

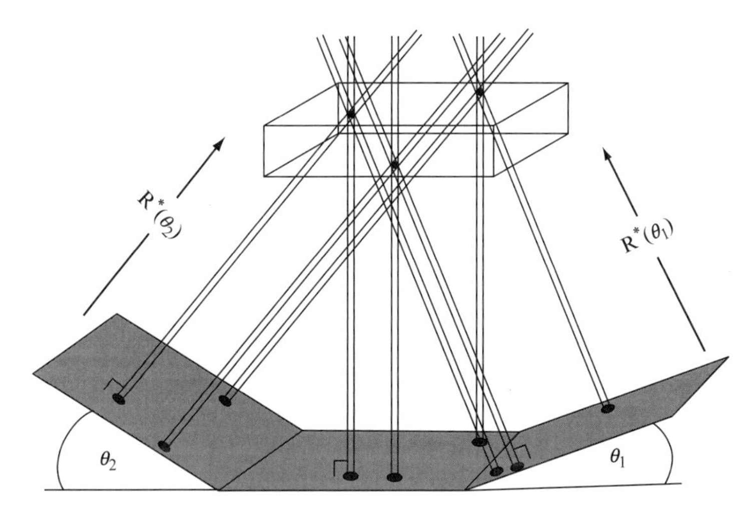


**Figure 13.2** Schematic showing how gold beads are projected in 2D images. Tracking gold beads across images enables the recovery of the 3D location of the gold beads and the motion of the microscope.

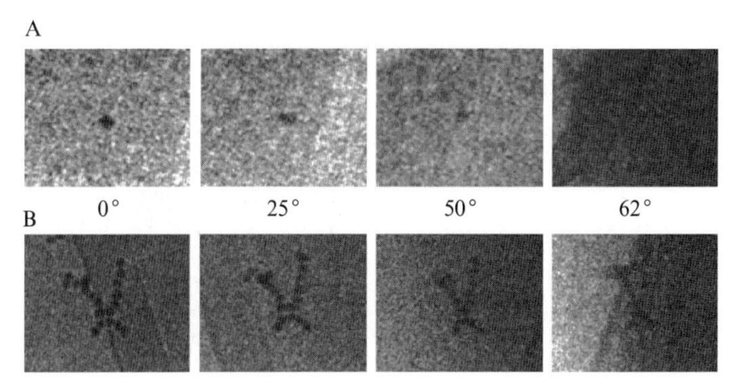


**Figure 13.3** (A) Example of the same gold bead at different tilt projections. The contrast declines in higher tilt angles due to thickness. (B) Example of gold markers clumping together, which makes it difficult to find all the centers and correspond each individual marker from image to image. Gold beads are 10 nm in diameter. Image by Luis R. Comolli at Lawrence Berkeley Laboratory, California.

markers all over the image to obtain a global alignment. For samples with very low SNR, it is also advisable to seed four or more markers across all the projections (not only at zero tilt), in order to give more initial contextual information to the algorithm and improve the tracking results.

On the other hand, programs like Brandt *et al.* (2001b) and Amat *et al.* (2008) try to detect markers automatically without user intervention. Briefly, they create a binary image of a perfect gold bead and select the top NCC scores from few low tilt images. Using these locations, they can generate a more realistic template. In this case, the user does not have control of which beads are going to be tracked and it is not guaranteed that the gold beads are spread across the sample. However, because fully automatic approaches try to track as many markers as possible, they will usually succeed in imposing this condition.

Once we have a template, several problems need to be taken into account during the marker detection process:

*Image contrast reduces with thickness*: Fig. 13.3 shows a clear example of how thickness affects contrast in CET images. For example, in single-tilt axis acquisition, thickness increases as $1/\cos(\beta)$, where $\beta$ is the angle of rotation around the tilt axis. At high tilt angles, marker detection can be difficult even for the human eye.

*Markers tend to cluster*: Fig. 13.3B shows a typical example of a cluster of markers. In that situation, it is difficult to distinguish the center of each individual marker and difficult not to make mistakes when matching markers from image to image. As a rule of thumb, it is not advisable to try to track clusters with more than three markers on them.

*Markers might have low NCC scores*: High NCC scores generally correspond to real markers, while low NCC scores might be false positives or real markers. Therefore, NCC alone does not give a clear indication of good matching. We need extra information, such as geometric context, to discern correct correspondences from false positives.

*Occlusion*: Markers can disappear in some images due to occlusion from another object in the sample or because they are out of the field of view after tilting the sample. Therefore, the system has to account for the fact that some markers might not be found in all the images.

## 3.2. Corresponding features between images

All available packages use some sort of contextual information on top of the template matching to make the detection process robust. The main idea is that the spatial location of markers in contiguous images is highly correlated, since there are few degrees of separation between projections. Moreover, packages also follow a sequential approach where they start with images at zero tilt, where SNR is better, and move their way up to higher tilts, where SNR is lower and more contextual information is needed.

### 3.2.1. IMOD: Iterate projection model and marker detection

Packages like IMOD (Kremer *et al.*, 1996) and Bsoft (Heymann *et al.*, 2008) try to estimate the projection model incrementally after each image to predict the location of the markers in the next image. Imagine we have tracked $n_T$ markers in the first four images and we want to solve for its 3D location using $M$ parameters in the projection model. In order to be able to solve for all the unknowns, we need $2n_{\mathrm{T}} \cdot 4 > 3n_{\mathrm{T}} + 4M$. Simplifying the expression, we have that $n_{\mathrm{T}} \geq \frac{4}{5}M$. Therefore, we need roughly as many markers as parameters to obtain an initial projection model and 3D location of the beads.

In particular, Beadtrack in IMOD starts at zero tilt, where little bead movement is expected between images because of the coarse prealignment, and it predicts positions on the next image first by simple extrapolation, then by fitting to a projection model once positions are available on at least four images. It increases the complexity of the model being fit as more data becomes available over a wider tilt range. Rather than correlating to the same bead template for all the markers, it correlates a small patch of image (roughly the size of 5 bead diameters) with the average of the image around that particular bead over the last few images. A preliminary position from correlation is refined by finding the position that maximizes the integrated intensity of the bead relative to the background just surrounding the bead. This intensity is required to reach a threshold level set as a system parameter before accepting a density as a bead. To find beads on one image, the program works from the center of the image outward, using the disparity between predicted and actual positions of beads already found to refine the prediction. After tracking as many beads as it can this way, it includes all of the new positions in the projection model; beads that do not fit well enough are discarded and searched for again at the positions predicted by this new projection model. This approach takes advantage of the prior knowledge of where beads should be located to help in the detection of beads. This iterative procedure between marker detection and projection model estimation can be conducted as many times as desired.

### 3.2.2. RAPTOR: Corresponding groups of markers

The above procedure works well for many CET samples. However, it is not robust in thick samples with low SNR, since an error in the initial correspondence at low tilts will affect the first projection model estimation and the iterative process will not be able to recover.

Packages like Brandt *et al.* (2001b) and RAPTOR (Amat *et al.*, 2008) try to make the correspondence step more robust to false detections. The main idea is to correspond all the possible detected markers in image $i$ and image $i + 1$ at once, instead of using an iterative process. The input is a set of 2D points detected as markers in image $i$ and a set of 2D points detected in $i + 1$. The output is a correspondence between the two sets of points. The intuition behind both approaches is that since two consecutive projections are very similar, point sets of corresponding markers should have similar spatial conformations (Fig. 13.4).

Brandt *et al.* (2001b) use epipolar constraints plus local graph matching to perform the alignment. Epipolar constraints refer to the geometric relations between two images that represent projections of the same 3D object

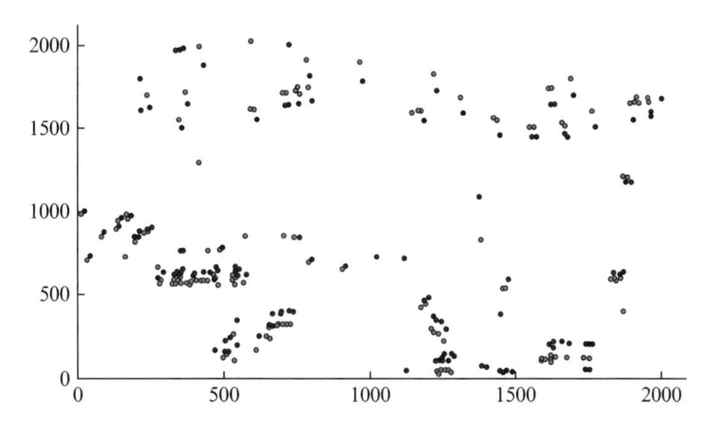


**Figure 13.4** Extracted possible locations of markers using template matching (red circle for image A and blue for image B). The correspondence is obvious for the human eye because of similar local spatial conformations (geometric contextual information) even in the presence of outliers. Picking only the closest blue to red point does not guarantee the correct assignment. (For interpretation of the references to color in this figure legend, the reader is referred to the Web version of this chapter.)

from different angles. These relations lead to constraints of where each region from image $i$ can be found in image $i + 1$, effectively limiting the set of possible candidates for each marker. The local graph matching helps disambiguate between different matching possibilities that the epipolar constraint could not rule out when markers are clumped together.

RAPTOR (Amat *et al.*, 2008) extends the approach using global graph matching.[1] The advantage of global graph matching is that it tolerates local distortions and higher number of false positives. Moreover, it does not assume a particular acquisition geometry. The only assumption is that image $i$ and $i + 1$ are similar projections, so the relative difference in spatial distribution between neighboring markers can be assumed to be minimal and can be exploited. We construct a graph in which each node is a detected point in image $i$, and we connect two nodes in the graph if their Euclidean distance is less than a given threshold. We can do the same operation for image $i + 1$ and obtain a similar graph. The goal is to find which nodes correspond between graphs, allowing for null correspondences in the case of occlusions and false positives.

The exact solution to graph matching is combinatorial with exponential complexity, and is only feasible for small graphs. Alternatively, this problem can be posed as a probabilistic inference over a Markov Random Field (MRF), in which a joint distribution over all possible correspondence

[1] In this chapter the notion of graph matching deviates from the classical notion in graph theory because we impose constraints between pairs of nodes.

assignments is modeled using local dependencies from the constructed graph. The goal is to define a joint probability in which the highest probability corresponds to the most likely matching between the two graphs. RAPTOR defines the joint probability as a product of two terms: image similarity between patches around each marker location and relative position between neighboring markers. These two quantities can be computed from the data directly to build the joint distribution. However, the complexity of estimating high probability assignments is exponential and approximate inference is used instead, which calculates the marginal distributions for each individual assignment from the entire distribution. Efficient algorithms such as Loopy Belief Propagation (Elidan *et al.*, 2006; Yedidia *et al.*, 2005), which takes advantage of conditional independence within the graph, have proved to find good solutions to graph matching problems in practice (Torresani *et al.*, 2008).

This global graph matching approach has proved very robust in low SNR conditions. However, when the number of markers in the sample exceeds 100, the correspondence starts to fail due to several factors. First, it is difficult to locate all the markers in every image, which generates many short partial trajectories. Second, even if all the markers can be found, they are so close to one another that it is easier to make mistakes. Extending the methodology to crowded samples is still an open problem.

### 3.2.3. Number of markers to track

The number of markers that need to be placed and tracked in the sample to achieve an accurate alignment is a crucial point, because the more distortions you want to account for, the more markers you need to track. Using the same calculations as above, when we have $n_{\mathrm{P}}$ images, we need $n_{\mathrm{T}}\left(2 - \frac{3}{n_p}\right) \geq M$. Since $n_{\mathrm{P}} > 100$ in most cases, we need roughly half as many markers as unknown parameters in the projection model per image. For example, if a user believes that the electron microscope produces no distortions and there are no mechanical imperfections in the rotations, two markers are sufficient to calculate translations and the tilt angle.

However, due to noise in the images tracked, gold bead positions can be imprecise, and it is recommended to have many more observations than the theoretical minimum to fit the model. A good rule of thumb is to have five times as many observations as the theoretical constraint. Fig. 13.5 shows a simulation to illustrate this point: we generate projections of $n_{\mathrm{T}}$ markers across $n_{\mathrm{P}} = 122$ images following a single-axis tilt scheme, where for each projection, we modify the rotation angle around the tilt axis and add 2D image shifts. Finally, we add zero mean Gaussian noise[2] with $\sigma = 5/3$ to

[2] Assuming a gold bead diameter of 10 pixels, $\sigma = \frac{5}{3}$ implies that 99.7% of detections are inside the gold bead.

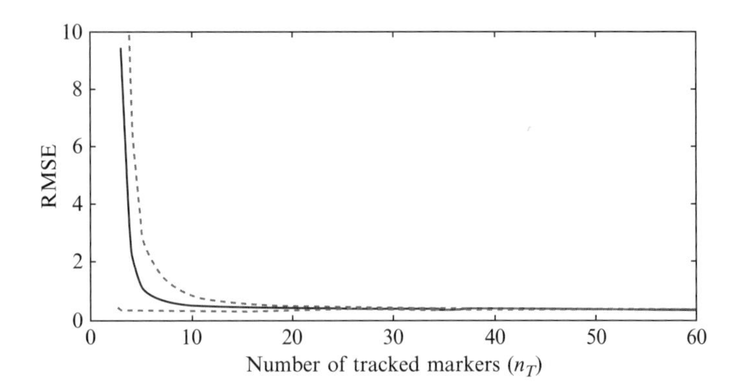


**Figure 13.5** Simulation of reprojection error when fitting a projection model with three unknowns per image: tilt angle and image shifts in 2D. Zero mean Gaussian noise with $\sigma = 5/3$ was added to each $u_i$ and $v_i$ coordinate. 1000 simulations where conducted for each value of $n_T$ to obtain mean (continuous line) and confidence 5th and 95th percentile intervals (dashed lines).

each $p_{ij}$ to simulate detection errors in the center of the gold beads. Theoretically, only two markers are necessary to recover the unknown parameters. However, the simulation clearly shows how the root mean square error (RMSE) of the reprojected points improves dramatically if we track more markers. Roughly, for $n_T > 10$, the improvement reaches a plateau, which illustrates the rule of thumb explained above.

Following this line of reasoning, one could think of placing as many markers as possible in the sample. However, this can lead to several problems: first, the markers can occlude important structural details in the sample. Second, the markers are so close together that it is hard to establish correct correspondences. Thus, a range of tracked markers between 20 and 100 should be sufficient to obtain an accurate alignment for any projection model, since even a complicated projection model with 20 parameters accounting for nonlinear distortions terms can be fitted properly with a 100 markers.

## 3.3. Projection model estimation

Once we have found the correspondence between gold markers in different images we need to find the 3D location of each marker and the projection model for each image. Frank (2006, Chap. 5) contains a very detailed description of the projection models used for alignment with fiducial markers. We will follow its notation as closely as possible for consistency, summarizing the main concepts and presenting some new points relevant for CET samples.

Two main differences of CET samples with respect to plastic embedded sections are: first, gold beads are randomly placed in the $z$-axis instead of being restricted to one or two planes, which avoids ill-posed situations during the projection model fitting. Second, CET samples do not suffer from sample warping by radiation as much as plastic embedded sections, although cryo-sections are an exception to this rule (Fig. 13.6).

Since we are dealing with point-like features, it is easier to formulate the alignment problem using transformations between points instead of transformations between coordinate systems. In order to do that we need to identify the microscope coordinate system $M$ with the sample coordinate system $S$. Moreover, we need to project $M$ into each image projection axis $B_i$ along the optical axis of the microscope by setting $O_i = O_m$, $\boldsymbol{m}_1 = \boldsymbol{u}_i$ and $\boldsymbol{m}_2 = \boldsymbol{v}_i$. Following this notation, the projection model from one point in the sample to a pixel in the $i$th image can be approximated to first order by:

$$p_{ij} = m_i P R(\gamma_i) R(\beta_i) R(\alpha_i) \begin{bmatrix} x_j \\ y_j \\ z_j \end{bmatrix} + \begin{bmatrix} \Delta u_{ii} \\ \Delta v_i \end{bmatrix} \quad (13.1)$$

where $m_i$ is a scaling factor to account for changes in magnification between images, $d_i = (\Delta u_i, \Delta v_i)^{\mathrm{T}}$ represents the image shift with respect to the reference center $O_m$ and $P = \begin{bmatrix} 1 & 0 & 0 \\ 0 & 1 & 0 \end{bmatrix}$ is a projection matrix. $R(\,)$ represents a rotation matrix defined by Euler angles, where $\alpha$ represents the elevation angle that rotates around the first axis, $\beta$ represents the tilt angle that rotates around the second axis, and $\gamma$ represents the azimuth angle that rotates around the optical axis of the microscope.[3] We can find the parameters in this projection model by solving an optimization problem. This procedure is known as bundle adjustment in the literature (Hartley and Zisserman, 2004) and using Eq. (13.1) can be formulated as follows:

$$\underset{\alpha_i \beta_i \gamma_i r_i d_i m_i}{\operatorname{argmin}} \sum_{i=1}^{n_P} \sum_{j=1}^{n_T} w_{ij} \cdot \rho \left( p'_{ij} - p_{ij} \right) \quad (13.2)$$

where ρ is some kind of cost function that penalizes differences between the observations $p'_{ij} = (u_i, v'_{ij})^{\mathrm{T}}$ and the model estimated projected points $\boldsymbol{p}_{ij}$, and $w_{ij}$ is a weight to account for occluded markers. There are two crucial modeling choices in this optimization. Firstly, the type of motions and distortions introduced by the electron microscope must be modeled appropriately. Secondly, the number of markers to be tracked must be sufficient to fit the parameters defined in the first choice. Overfitting is a very common problem and can result in lower quality reconstructions. Lower

[3] In this case, we assume the $y$-axis as the tilt axis.

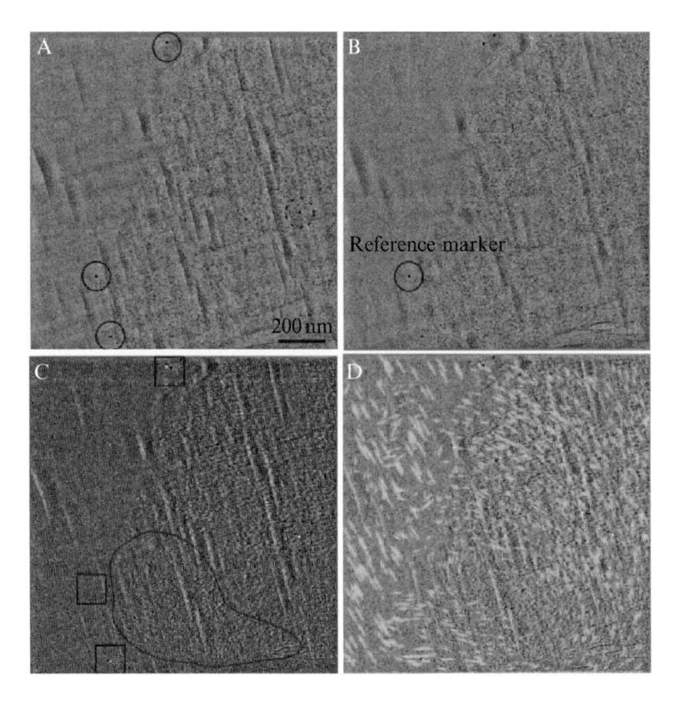


**Figure 13.6** Figure from Castaño-Diez et al. (2007) showing deformations of the vitreous section during the tilt series. Two zero degree images recorded at the beginning and at the middle of the tilt series (after a dose of 20 e/Å$^2$) are compared on the basis of fiducials and cross-correlation. (A) First projection image. The crevasses in the vitreous ice are clearly visible. Three gold fiducials marked with circles are outside the cell. A fourth gold fiducial (marked with the dashed circle) appears in the middle of the cell, although it is located on the carbon support film. (B) Second zero degree projection of the tilt series recorded using identical electron microscope conditions (focus, magnification, etc.). After 20 e/Å$^2$, the crevasses appear much smoother. One of the three markers located close to the cell is used to align the two images (reference marker) in order to calculate the difference image. (C) The difference image between the two zero degree projections. The reference marker is no longer visible; however, the other markers are clearly visible (indicated through the black boxes). From the characteristic black and white pattern it becomes obvious that the fiducials moved anisotropically, since the fiducial marker on the top of the image moved to the left and the other fiducial on the lower part of the image moved to the right. Apparently the section moved also with respect to the carbon film, therefore the fiducial on the carbon film also shows some movement (indicated through the dashed box). The purple region was used for a final reconstruction. (D) Vector field calculated with local cross-correlation visualizing the movement of small regions of the image. (For interpretation of the references to color in this figure legend, the reader is referred to the Web version of this chapter.)

residual error in Eq. (13.2) does not always mean better reconstructions, since it is always possible to reduce the mean residual error by adding new parameters in the model.

Different packages offer different choices for the cost function. When $\rho\left(\begin{bmatrix} a \\ b \end{bmatrix}\right) = a^2 + b^2$ we have a well-known nonlinear least-squares problem. Frank (2006) and Castaño-Diez *et al.* (2006) describe how to solve Eq. (13.2) in this case using alternate linear least-square fits and quasi-Newton methods, respectively. However, it is also well known in optimization literature (Boyd and Vandenberghe, 2004) that least-squares is not robust to outliers. In other words, if the tracking step made any mistake, the fitting of the projection model using least-squares is going to be largely biased by these outliers. Thus, the above methods generally require user intervention to verify the model fitting and manually correct possible outliers.

Given the low SNR in CET samples, it is highly unlikely that the tracking step makes no mistakes. Brandt and Ziese (2006) and Amat *et al.* (2008) focus on fully automating the alignment process, and they leverage on methods developed in robust statistics to use a $\rho$ that is robust to outliers and can be solved numerically almost as efficiently as least-squares. An example of these robust metrics are M-estimators, such as the Huber penalty (Boyd and Vandenberghe, 2004).

As mentioned in Section 3.1, markers can be occluded in different images. If the *j*th marker is occluded in the *i*th image we can set $w_{ij} = 0$ in Eq. (13.2). Partial trajectories of any length help in the projection model fitting process as long as they overlap to guarantee a coherent solution across the tilt series. Table 13.1 shows a simulation on synthetic data to test these ideas. We generated $n_T$ markers for three different cases and projected them across $n_P = 122$ images following a perfect single-axis tilt rotation. We added Gaussian noise to each projected location with zero mean and $\sigma = 5/3$. Results show that it is better to track as many gold markers as possible to improve the projection model fitting. However, failing to track overlapping partial trajectories, can translate in a large error.

The most common acquisition geometry is single-axis tilt, where $\gamma$ is considered as the constant in-plane rotation of the tilt axis and the user

**Table 13.1** Three synthetically generated alignment cases to show two things: first, partial trajectories help fitting the model as long as they overlap to produce a coherent solution across the tilt series

| $n_P$ | $n_T$ | % Occluded marker locations | Mean reprojected error | S.D. reprojected error |
|---|---|---|---|---|
| 122 | 59 | 30 | 0.406 | 0.005 |
| 122 | 29 | 50 | 2.161 | 0.800 |
| 122 | 29 | 0 | 0.646 | 0.070 |

Second, a minimum number of markers is necessary to reduce alignment error due to uncertainty in center bead localization. Each case was run 10 times to obtain statistics.

changes $\beta_i$ incrementally to generate the tilt series. $\alpha$ is only relevant if we suspect that the tilt axis is not perpendicular to the optical axis. However, a number of other acquisition geometries have been proposed in the literature. For example, Lanzavecchia *et al.* (2005) present a conical tomography approach, where each fiducial marker follows an ellipsoidal trajectory instead of straight lines. This motion can be modeled by Eq. (13.1) and solved using the same techniques as explained above.

Software package TxBR (Lawrence *et al.*, 2006, 2009) presents the most general projection model for alignment and reconstruction up to this date. TxBR solves for any general set of orientations in Eq. (13.2) and adds nonlinear terms up to third-order to the bundle adjustment. This model was developed to cope with severe optical aberrations and sample warping in plastic embedded samples with large fields of view. In general, CET samples do not suffer from such severe distortions. However, it can be used in dual-axis to improve reconstruction quality (L.R. Comolli and F. Amat, unpublished data) and in cryo-sections, which present more severe distortions. Moreover, as CCD cameras become larger and people experiment with different acquisition geometries besides single-axis tilt, approaches like the one in TxBR can be very useful. The trade-off of such an approach is that 100 markers need to be tracked per image to avoid overfitting the model, which makes automatic tracking methods critical to process the data efficiently.

## 4. Marker-Free Alignment

Sometimes we cannot align the sample following the methods outlined in the previous section. For example, it might not be possible to add gold beads to the CET sample; or we might not be able to identify enough gold beads to fit all the model parameters and we need to use other features to improve alignment. Furthermore, in some samples the assumption that the gold beads and the sample move as a rigid body might not hold. Ice melting and sample degradation can lead to a decoupling of the dynamics between the fiducials and the observed sample. In these situations, an alignment that tracks specimen features of biological interest is preferable over one based on artificial markers because the goal is to reconstruct the specimen structure itself as faithfully as possible.

All the principles about fitting projection models and robustness from the previous section apply here. The only difference is that we do not have high contrast point-like features to track, which makes the problem harder. Frank (2006, Chap. 6) presents a great review of methods for marker-free alignment. However, most of them were developed for plastic embedded sections. Quoting from Frank (2006, pp. 212)-: "The future challenge will be to find a reliable and accurate method, especially for such cryo-samples that are beyond the capabilities of the alignment methods currently available." In this chapter, we present some recent developments that try to meet those challenges for cryo-samples. We also preserve the taxonomy from Frank (2006, Chap. 6) to classify different marker-free alignment approaches between feature-based and 3D model-based.

## 4.1. Feature-based alignment

Approaches in Castaño-Diez *et al.* (2007, 2010) and Sorzano *et al.* (2009) try to keep as much as possible from the spirit of standard alignment with gold beads. In the absence of real fiducials, these algorithms try to locate series of features in the tilt series that can be assimilated to projections of a unique 3D object. This series of features constitutes a *trail* of *virtual markers* and the set of $(x_{ij}, y_{ij})_i = 1, \ldots, n_T$ coordinates of the *j*th trail can be used as the $u'_{i,j}$ and $v'_{i,j}$ in Eq. (13.2).

This assimilation has to be done carefully, as there is a fundamental difference between a trail of virtual markers and a trail of observed positions of the same gold bead through a series of micrographs: positions of gold markers are a clear target in each CET image, while positions of virtual markers computed by a given algorithm should be understood as putative positions, whose accuracy needs to be carefully analyzed before using them. If this analysis states that some of the located trails have indeed the properties of a trail of gold beads, they can be plugged into the pipeline for projection model estimation described in Section 3.3. Therefore, in order for this methodology to converge with the regular fiducial alignment two steps need to be undertaken: first, the trails need to be collected, and second the trails have to be analyzed. In the following sections, we examine how the Allignator software package (Castaño-Diez *et al.*, 2010) handles these two tasks in practice.

### 4.1.1. Allignator

The basic idea to collect trails of virtual markers is to locate local areas in different micrographs that appear to correspond to different views of the same object. This idea is performed by selecting a region (or patch) inside a micrograph and then searching for a *match* or similar patch in the micrograph corresponding to the next tilt angle. Iterating the process for the whole tilt series and for different starting patches yields the set of trails that will be analyzed in the second step. As simple as it sounds, the understanding of some technical details explained below will increase the chances that the constructed trails are useful.

As in the case of fiducial markers, we use NCC to search for matches in different micrographs. However, as mentioned in Section 3.1, establishing a direct relationship between the magnitude of the NCC and quality of the match is unreliable. In order to improve reliability and reduce the computational burden in subsequent steps, Alligator complements the NCC scoring with a hysteresis check defined as follows: let $p_1$ be a patch in projection image 1 and $p_2$ be the center of the patch in projection image 2 that among all possible patches yields the best NCC with respect to $p_1$. One would expect that reversing the procedure, that is, considering $p_2$ and looking for a patch in image 1 by maximizing the NCC, one would recover the original $p_1$. If the location of the new patch in image 1 is shifted by more than one pixel the algorithm discards the match.

Two practical questions required *ad hoc* solutions based on experience: how to select *anchor points* and the patch size to compute NCC values. First, an anchor point is the initial point to start collecting a trail. Brandt *et al.* (2001a) use a standard corner detector to select anchor points. In principle, such locations could be plugged into Alligator. In practice, it is better to randomly select thousands of anchor points across the reference projection. Even if the number of points needed to generate a good alignment is very dependent on the dataset, a good strategy is to start with 1000 anchor points, and enlarge this number if the subsequent step fails to provide a good set of trails. The patch size is critical because patches that are too small will not contain enough information and those too large will combine information stemming from different objects, making the task of following a single feature along the tilt line more difficult. In practice, patches with a side length ranging up to 200 pixels are found to provide the best results.

Once we have collected thousands of possible trails, we have to analyze them since the vast majority tend to be wrong. In this context, the so called tilt line plot is a very helpful tool to quickly compare pairs of trails to assess their quality. In a tilt line plot, the coordinates of two trails are subtracted and plotted in the same graph. In the absence of tilt-dependent magnification or rotation[4] (terms $m_i$ and $\alpha_i$ in Eq. (13.1)), these should behave as a set of discrete points distributed along a line perpendicular to the tilt axis. Alligator systematizes this idea by scanning all the trail pairs and keeping the ones that show enough quality. The quality of each pair of trails is defined according to two principles: first, the regression error of the experimental tilt line to a projection model must stay below a threshold value defined by the user. Secondly, the pair of trails is checked to be numerically stable to rule out the situations in which the least-squares problem is nearly ill-posed. This situation is extremely common in practice, and failing to handle it will lead to spurious results. The mean distance among the points in a tilt line can be understood as an *stability index*, and it can be used to automate the selection of stable pairs of trails by eliminating the pairs whose stability remains below a given threshold set by the user.

[4] The in-plane rotations and magnification values are removed from the model used to analyze the quality of trails in order to avoid the overfitting that would arise from applying the full model to just a pair of trails.

The algorithm uses the high quality pairs as seeds to try to construct larger sets of trails that regress to the same alignment model and improve the fitted projection model (different models found in a tilt series can reveal regions of the sample that move in different ways but each in a coherent fashion). Once such set of trails has been identified, it is possible for the user to enrich the regression model using the full set of parameters described in Eq. (13.1). Notice that the idea of comparing the relative trajectory of all possible pairs of trails to infer good alignment while avoiding the exponential complexity of the combinatorial problem is similar in spirit to how RAPTOR handles correspondence (Section 3.2.2) for fiducial marker alignment.

### 4.1.2. Alignment from extended-features in TxBR

Exploring the information of a set of virtual markers as explained above and treating them as if they were real fiducials implicitly makes two assumptions: first, the projections of some features in the sample can be tracked from micrograph to micrograph by searching for local image similarities. Second, the center of each located patch corresponds to the projection of the same 3D point inside the feature.

These assumptions will not be met in general since a three-dimensional object appears slightly different from one micrograph to the next. Actually, assigning the center of each patch to the projection of the same 3D position is guaranteed to be completely exact only in the case of noise-free projections from spherical objects such as gold fiducial markers. However, for thin samples with small tilt angle increments the assumptions are valid and the methodology explained for Allignator delivers high accuracy marker-free alignments in practice.

In the case where the assumptions do not hold, Phan *et al.* (2009) present a new approach based on aligning projected outlines of thin, high electron density surfaces present in the sample. These surfaces are the result of the staining processes which deposit material preferentially on membranes. Because the inner structure of cells is rich in membranes these surfaces are prevalent in stained objects. Such surfaces generally project to curves identifiable in the images. Unfortunately, the situation is somewhat different in unstained cryo-samples, since surfaces appear as boundaries between regions of different electron density. Although the mathematical model is essentially the same, locating the boundary is somewhat more difficult, especially if automated methods are required. Boundary detection in CET samples is still ongoing research.

Mathematically, we can model this situation as follows: the contours in the images are the projection of points on the surface which are curves tangent to the electron trajectories. These curves in the object are termed as generating contours. The tangency and projection conditions are well-defined, and we can make a direct formulation of first and second order errors in estimation of these curves and the reprojected curves from information which can be recovered from the images.

This formulation raises the possibility of alignment of series of electron microscope (EM) images via reconstruction of surfaces in the object from the observed contours in the images. If electron trajectories were straight lines the alignment problem would reduce to a problem in projective duality, which is the basis of one of the approaches in the method of occluding contours (Kang *et al.*, 2001). As discussed above, for most CET images, this assumption can be taken to be true, and alignment of EM images could be taken to be a direct application of this theory. For large field images presenting nonlinear distortions, a further optimization step is necessary using the initial estimates provided by projective duality. The basic idea is to employ the error terms obtained from reprojection errors and tangency conditions. The parameters of the error model gives TxBR joint estimates of surface patches and curvilinear projection model. This optimization can be modeled as a variational generalization of the nonlinear bundle adjustment employed by TxBR for fiducial alignment.

## 4.2. 3D model-based approach

The 3D model-based approach deviates from the pipeline shown in Fig. 13.1 and from the idea of tracking features through the tilt series. It is similar to the single particle methodology, where the alignment is carried out in an iterative procedure which uses reprojections of a 3D reconstruction (projection matching). The advantage of this approach is that all specimen features within the region of interest implicitly contribute to the alignment, whereas the methods presented above rely on a limited number of fiducial points that are tracked through the tilt series. Also, when the image contrast is low, the detection of specimen features may be difficult so that an accurate alignment cannot be achieved. The main assumption in the model-based approach is that the specimen is treated as a rigid body and thus distortions of any kind are not being compensated. In this section, we describe the details of the model-based alignment as implemented in the *Protomo* software package (Winkler, 2007). *Protomo* was originally developed for thin paracrystalline specimens (Taylor *et al.*, 1997) and subsequently modified to accommodate unordered specimens (Winkler and Taylor, 2006). In practice, it has proved to be best suited for relatively thin specimens (Liu *et al.*, 2006) but was also successfully applied to thicker cryo-samples (Zhu *et al.*, 2006).

Since the alignment approach in *Protomo* is based on cross-correlation of images and not on tracking point-like features, it is more suitable to define the parameters in terms of transformations of coordinate systems rather than coordinate transformations of fiducial points. Some additional definitions are listed here to describe the tilt geometry as it is defined in the

*Protomo* package. The specimen tilt from the initial, untilted state in which $M$ and $S$ are identical, is described by the transformation of $M$ to $S$:

$$s_j = \sum_{k=1}^{3} r_{i,jk} m_k, \quad j = 1, \ldots, 3$$

The coefficients $r_{i,jk}$ can be written as matrix $R_i$ which is the product of the rotation matrices shown below:

$$R_i = R_0 R_{\mathrm{d}}^{\mathrm{T}} T_i R_{\mathrm{d}}$$

Each of these rotation matrices is associated with parameters of the tilt experiment: $R_{\mathrm{d}}$ and its transpose $R_{\mathrm{d}}^{\mathrm{T}}$ define the direction of the tilt axis. We assume here that $R_{\mathrm{d}}$ rotates the coordinate axis $\boldsymbol{m}_1$ to the tilt axis. $R_{\mathrm{d}}$ can be written as a product of two rotations; the two angles are the tilt azimuth $\psi$ and an elevation angle $\varphi$. The direction of the tilt axis can assume any angle relative to the microscope coordinate system. It is preferable, however, to orient it roughly along one of the coordinate axes of the images to maximize the reconstructed area throughout the tilt series. $T_i$ defines the rotation about the tilt axis and is directly related to the goniometer readings. The associated tilt angle is $\vartheta_i$ for the $i$th projection.[5] An additional rotation $R_0$ is required to account for unknown or inaccurate geometric parameters, that is, the nominal tilt angles may not correspond to the actual tilt if the specimen does not lie flat on the grid.

For simplicity, only one tilt axis has been considered above. It is trivial, however, to extend the description to multiple axes simply by introducing multiple rotation parameters $R_0$ and $R_{\mathrm{d}}$ for each group of projection images corresponding to each tilt axis. Additionally, defining the tilt axis direction as a parameter and not keeping it fixed allows all computations for alignment and backprojection to be carried out with a single interpolation of the raw or preprocessed images, thus avoiding a degradation of resolution by multiple successive interpolations.

The geometric parameters that need to be determined by the alignment are the angles described above, as well as a common origin relative to the specimen structure. The common origin is the point $O_s$ expressed in the coordinates of the coordinate system $B_i$ for each particular projection image. Additional parameters are introduced by the 2D transformation of $B_i$ to the projection of $M$ along the optical axis $\boldsymbol{m}_3$, which relates the recorded images to the microscope coordinate system:

$$C_i = f_i U_i S_i V_i^{\mathrm{T}}$$

[5] The variable names are different from Eq. (13.1) because they do not represent the same quantities.

This transformation accounts for an in-plane rotation of the projected image, linear distortions such as shear and an isotropic scale change $f_i$. $U_i$ and $V_i$ are orthogonal matrices, and $S_i$ is a diagonal matrix which can easily be computed by a singular value decomposition. $C_i$ is initialized as the unit transformation and the total transformation $A_i = R_iC_i$ is reevaluated computationally in two steps: first, fixing $R_i$ and estimating the correction terms $C_i$ (alignment), and subsequently updating the geometric parameters $R_i$ based on the correction terms (geometry refinement).

The alignment is based on cross-correlating each individual image of the tilt series with a separate reference image. Tilting a specimen results in a foreshortening of its projection, which must be compensated before the correlation is performed (Guckenberger, 1982). This compensation is achieved by applying the transformations $A_i$ to the raw images. The interpolation operation will also resample equivalent specimen areas to the same size, which is a requirement if the cross-correlation function is computed via Fourier transforms of the images. The implemented procedure includes an optimization algorithm which, besides determining translational shifts, maximizes the correlation between images and the reference by varying the linear transformations $A_i$ which simultaneously provides updated in-plane rotations, shear and an isotropic scale factor. These reference images are generated from a reprojection of a preliminary map which was computed based on already aligned images in an alignment cycle. An alignment cycle starts with the image of the untilted specimen and successively merges in the projections at higher tilt, until the whole tilt series is aligned.

The result of the alignment is a new set of $A_i$ from which new $R_i$ are evaluated by minimizing the deviation of $S_i$ from a unit matrix. Normally, several alignment cycles with intervening recalculation of the geometric parameters are required to align a tilt series accurately. The progress of the alignment can be monitored by inspecting the diagonal elements of the matrices $S_i$. The iterative process is terminated when the deviations become small, approach unity, and do not change significantly between cycles.

## 5. Reconstruction Using the Projection Model

The choice of projection model not only determines the number of markers to track but also the type of backprojection algorithm we should use. If we just consider a global linear projection model for the whole single-axis tilt series (Eq. (13.1)) the reconstruction is straight forward: we just need to correct for magnification changes, in-plane rotations and shifts in each 2D image separately. The aligned tilt series will behave like a perfect rotation around the tilt axis (Fig. 13.7B) and we can use efficient backprojection algorithms (Mastronarde, 1997).

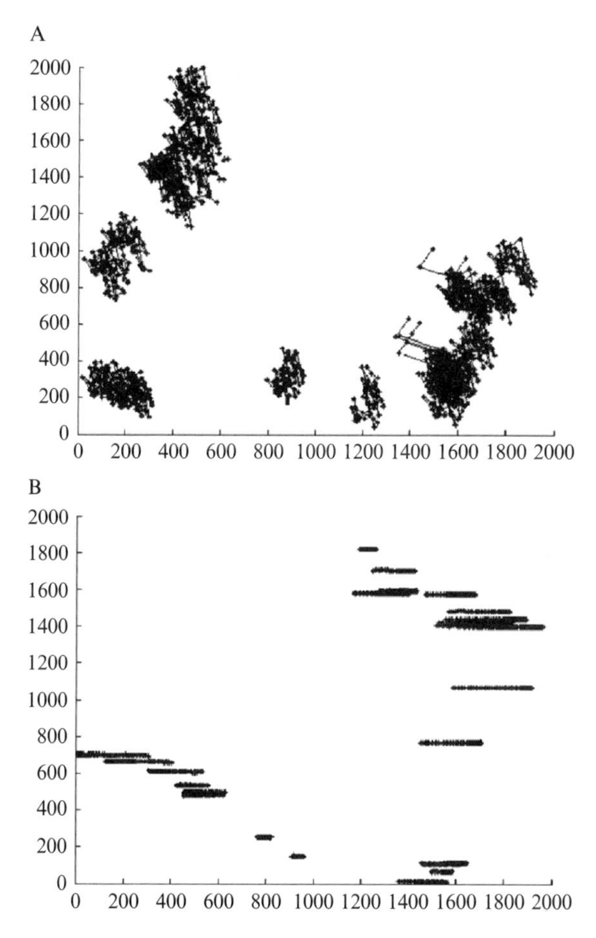


**Figure 13.7** (A) Superimposed tracked trajectories from raw data of a tilt series in a 2K × 2K camera. Connected dots represent locations of the same marker in different images. (B) Same trajectories after alignment. All trajectories follow a smooth straight line. Figure from Amat *et al.* (2008).

However, some distortions present in cryo-section preparations and multiple-axis acquisition schemes, cannot be corrected with linear projection models. In this case, the backprojection needs to compute a more general inverse Radon transform. The freely available packages present two approaches to handle distortions. First, IMOD uses local linear projection models: it divides the field of view into smaller regions and assumes that each of these regions do not present distortions. Therefore, it can reconstruct each region using standard techniques. The main point is to blend reconstructions from adjacent regions seamlessly. Briefly, the program fits a global projection model and then obtains a local projection model for each region as an increment of the global solution. This allows for smooth variation of the backprojection parameters as well as local reconstructions even if the number of gold beads per region is not very large. The reader is referred to Frank (2006, Chap. 5) for a more detailed description.

Second, package TxBR (Lawrence *et al.*, 2006, 2009) presents a method for alignment and reconstruction using nonlinear projection models. The alignment portion of the code provides a joint model of electron trajectories and the orientation of the sample for each tilt image, up to the usual gauge ambiguity given by simultaneous warping of the sample and adjustment of the trajectories. For the purposes of reconstruction the information given by the alignment step is sufficient to handle any geometric distortion present in any CET sample, or generally, any electron tomography sample. TxBR incorporates various methods to impose geometric constraints so that the effects of warping can be pushed over to the electron trajectory model, and the best rotation model for the sample is obtained by methods available in the code.

However, since TxBR backprojects along curved electron trajectories rather than along straight lines, we need to incorporate this information into the filter and the backprojection. The only practical difficulty in handling geometric distortion in the backprojection is fast evaluation of the forward projections, and TxBR incorporates a fast recursion to handle it efficiently. The remaining difficulty is with the filter because of the nature of the 3D distortions. Only distortions in the plane perpendicular to the optical axis ($z$-axis) can be handled by standard warping and filtering techniques applied to images. This is because warping of the image must correspond to the same warping in each $z$-plane. The approach of warping and filtering can be extended to give a more accurate filter. In particular, a different warping must be applied to the image data for each plane of the reconstruction, and then the filter will be more accurate for that $z$-plane. This entails considerable computational cost, and generally the results do not merit this additional time and trouble, therefore 2D image warping is generally sufficient to give good results. Alternatively, rebinning methods based on a method proposed by Defrise and Noo (2002) and Patch (2002) can, in principle, give accurate results, but these, are also costly in computation, and increase reconstruction time beyond acceptable limits.

## 6. Summary

In this chapter, we have presented a summary of the basic steps to successfully align and reconstruct CET samples. It is important to remember that the reconstruction quality can only be as good as the alignment accuracy permits. Therefore, it is crucial to carefully perform all the steps with high accuracy to result in high-resolution tomograms.

The basic four steps in the process are:

1. Tracking of common features across multiple images.
2. Estimation of projection models to refine 3D location of markers, motion of the sample, and imaging parameters during data acquisition.
3. Reconstruction of 3D density map from aligned projections.

Currently there are several freely available packages that automate all the steps for most types of CET samples. A main classification when deciding which package to use is fiducial versus marker-free samples. Another key decision is to choose which projection model better represents the electron trajectories and sample perturbations that occur during the acquisition process (for instance, smoothing of crevasses in cryo-sections), since this is directly related to the number of features that need to be tracked in the sample. The ratio between the number of parameters in the projection models and the number of tracked markers should be at least 5 to 1 to avoid overfitting.

All these advances, in conjunction with others described in this book, such as automated data collection, have made it possible for high-throughput CET systems to generate hundreds of tomograms in the span of a month. Most software packages also allow for user intervention during every step of the process in the case the user wants to have greater control of advanced parameters.

## ACKNOWLEDGMENTS

FA, FM and MH would like to thank Dr. Luis R. Comolli for all the images used in this manuscript to exemplify different challenges of cryo-electron tomography. The authors would also like to thank Dr. David Mastronarde for contributions on the manuscript related to the IMOD software.

## REFERENCES

Amat, F., Moussavi, F., Comolli, L. R., Elidan, G., Downing, K. H., and Horowitz, M. (2008). Markov random field based automatic image alignment for electron tomography. *J. Struct. Biol.* **161,** 260–275.

Boyd, S., and Vandenberghe, L. (2004). Convex Optimization. Cambridge University Press.

Brandt, S. S., and Ziese, U. (2006). Automatic TEM image alignment by trifocal geometry. *J. Microsc.* **222,** 1–14.

Brandt, S., Heikkonen, J., and Engelhardt, P. (2001a). Automatic alignment of transmission electron microscope tilt series without fiducial markers. *J. Struct. Biol.* **136,** 201–213.

Brandt, S., Heikkonen, J., and Engelhardt, P. (2001b). Multiphase method for automatic alignment of transmission electron microscope images using markers. *J. Struct. Biol.* **133,** 10–22.

Castaño-Diez, D., Seybert, A., and Frangakis, A. S. (2006). Tilt-series and electron microscope alignment for the correction of the non-perpendicularity of beam and tilt-axis. *J. Struct. Biol.* **154,** 195–205.

Castaño-Diez, D., Al-Amoudi, A., Glynn, A. M., Seybert, A., and Frangakis, A. S. (2007). Fiducial-less alignment of cryo-sections. *J. Struct. Biol.* **159,** 413–423.

Castaño-Diez, D., Scheffer, M., Al-Amoudi, A., and Frangakis, A. S. (2010). Alignator: A gpu powered software package for robust fiducial-less alignment of cryo tilt-series. *J. Struct. Biol.* **170,** 117–126.

Defrise, M., and Noo, F. K. H. (2002). Improved 2d rebinning of helical cone-beam ct data using johns's equation. *Nucl. Sci. Symp. Conf. Rec.* **3,** 1465–1469.

Elidan, G., McGraw, I., and Koller, D. (2006). Residual belief propagation: Informed scheduling for asynchronous message passing. Proceedings of the Twenty-Second Conference on Uncertainty in AI (UAI). AUAI Press, Boston, MA.

Frank, J. (2006). Electron Tomography: Methods for Three-dimensional Visualization of Structures in the Cell. 2nd edn. Springer.

Guckenberger, R. (1982). Determination of a common origin in the micrographs of tilt series in three-dimensional electron microscopy. *Ultramicroscopy* **9,** 167–173.

Hartley, R. I., and Zisserman, A. (2004). Multiple View Geometry in Computer Vision. 2nd edn. Cambridge University Press.

Herman, G. T. (2009). Fundamentals of Computerized Tomography: Image Reconstruction from Projections. 2nd edition. Springer.

Heymann, J. B., Cardone, G., Winkler, D. C., Steven, A. C., Heymann, J. B., Cardone, G., Winkler, D. C., and Steven, A. C. (2008). Computational resources for cryo-electron tomography in bsoft. *J. Struct. Biol.* **161,** 232–242.

Kak, A. C., and Slaney, M. (2001). Principles of Computerized Tomographic Imaging. Society of Industrial and Applied Mathematics.

Kang, K., Tarel, J. P., Fishman, R., and Coope, D. (2001). A linear dual-space approach to 3d surface reconstruction from occluding contours using algebraic surfaces. International Conference on Computer Vision. pp. 198–204.

Kremer, J. R., Mastronarde, D. N., and McIntosh, J. R. (1996). Computer visualization of three-dimensional image data using imod. *J. Struct. Biol.* **116,** 71–76.

Lanzavecchia, S., Cantele, F., Bellon, P., Zampighi, L., Kreman, M., Wright, E., and Zampighi, G. (2005). Conical tomography of freeze-fracture replicas: a method for the study of integral membrane proteins inserted in phospholipid bilayers. *J. Struct. Biol.* **149,** 87–98.

Lawrence, A., Bouwer, J. C., Perkins, G., and Ellisman, M. H. (2006). Transform-based backprojection for volume reconstruction of large format electron microscope tilt series. *J. Struct. Biol.* **154,** 144–167.

Lawrence, A., Phan, S., and Moussavi, F. (2009). *3d reconstruction from electron microscope images with txbr* ISBI.

Liu, J., Wu, S., Reedy, M. C., Winkler, H., Lucaveche, C., Cheng, Y., Reedy, M. K., and Taylor, K. A. (2006). Electron tomography of swollen rigor fibers of insect flight muscle reveals a short and variably angled s2 domain. *J. Mol. Biol.* **362,** 844–860.

Ma, Y., Soatto, S., Kosecka, J., and Sastry, S. S. (2003). *An Invitation to 3-D Vision: From Images to Geometric Models* Springer.

Mastronarde, D. N. (1997). Dual-axis tomography: An approach with alignment methods that preserve resolution. *J. Struct. Biol.* **120,** 343–352.

Nickell, S., Frster, F., Linaroudis, A., Net, W. D., Beck, F., Hegerl, R., Baumeister, W., and Plitzko, J. M. (2005). Tom software toolbox: acquisition and analysis for electron tomography. *J. Struct. Biol.* **149,** 227–234.

Patch, S. K. (2002). Consistency conditions upon 3d ct data and the wave equation. *Phys. Med. Biol.* **47,** 2637–2650.

Penczek, P., Marko, M., Buttle, K., and Frank, J. (1995). Double-tilt electron tomography. *Ultramicroscopy* **60,** 393–410.

Phan, S., Bouwer, J., Lanman, J., Terada, M., and Lawrence, A. (2009). Non-linear bundle adjustment for electron tomography. World Congress on Computer Science and Information Engineering.

Sorzano, C., Messaoudi, C., Eibauer, M., Bilbao-Castro, J., Hegerl, R., Nickell, S., Marco, S., and Carazo, J. (2009). Marker-free image registration of electron tomography tilt-series. *BMC Bioinform.* **10,** 124.

Taylor, K. A., Tang, J., Cheng, Y., and Winkler, H. (1997). The use of electron tomography for structural analysis of disordered protein arrays. *J. Struct. Biol.* **120,** 372–386.

Torresani, L., Kolmogorov, V., and Rother, C. (2008). Feature correspondence via graph matching: Models and global optimization. Computer Vision ECCV 2008, pp. 596–609.

Trucco, E., and Verri, A. (1998). Introductory Techniques for 3-D Computer Vision. Prentice Hall.

Winkler, H. (2007). 3d reconstruction and processing of volumetric data in cryo-electron tomography. *J. Struct. Biol.* **157,** 126–137.

Winkler, H., and Taylor, K. A. (2006). Accurate marker-free alignment with simultaneous geometry determination and reconstruction of tilt series in electron tomography. *Ultramicroscopy* **106,** 240–254.

Yedidia, J. S., Freeman, W. T., and Weiss, Y. (2005). Constructing free energy approximations and generalized belief propagation algorithms. *IEEE Trans. Inf. Theory* **51,** 2282–2313.

Zhu, P., Liu, J., Bess, J., Chertova, E., Lifson, J. D., Gris, H., Ofek, G. A., Taylor, K. A., and Roux, K. H. (2006). Distribution and three-dimensional structure of aids virus envelope spikes. *Nature* **441,** 847–852.

CHAPTER FOURTEEN

# Correcting for the Ewald Sphere in High-Resolution Single-Particle Reconstructions

Peter A. Leong,* Xuekui Yu,† Z. Hong Zhou,† *and* Grant J. Jensen‡

## Contents



## Abstract

To avoid the challenges of crystallization and the size limitations of NMR, it has long been hoped that single-particle cryo-electron microscopy (cryo-EM) would eventually yield atomically interpretable reconstructions. For the most favorable class of specimens (large icosahedral viruses), one of the key obstacles is curvature of the Ewald sphere, which leads to a breakdown of the Projection Theorem used by conventional three-dimensional (3D) reconstruction programs. Here, we review the basic problem and our implementation of the "paraboloid" reconstruction method, which overcomes the limitation by averaging information from images recorded from different points of view.

* Department of Applied Physics, California Institute of Technology, Pasadena, California, USA
† Department of Microbiology, Immunology and Molecular Genetics, The California NanoSystems Institute, University of California Los Angeles, Los Angeles, California, USA
‡ Division of Biology, Howard Hughes Medical Institute, California Institute of Technology, Pasadena, California, USA

*Methods in Enzymology,* Volume 482
ISSN 0076-6879, DOI: 10.1016/S0076-6879(10)82015-4

# 1. Introduction

X-ray crystallography and nuclear magnetic resonance (NMR) spectroscopy were the first techniques to reveal the atomic structures of biological macromolecules. Electron crystallography then followed, first on "two-dimensional" (2D) crystals (crystals one unit cell thick) (Henderson *et al.*, 1990; Kuhlbrandt *et al.*, 1994) and then on helical (tubular) crystals (Unwin, 2005). To avoid the challenges of crystallization and the size limitations of NMR, it has long been hoped that single-particle cryo-electron microscopy (cryo-EM) would eventually also produce atomically interpretable maps. Steady progress toward this goal has been made, led by reconstructions of large icosahedral viruses, whose 60-fold symmetry, size, and rigid architecture all facilitate precise image alignment (Chapter 7, Vol. 482). Eventually, such efforts will be hampered by the fact that conventional methods assume that EM images are true projections, but in fact they are not: the information delivered by the microscope is actually a mixture of information belonging to a curved surface within the 3D Fourier transform (FT) of the specimen called the Ewald sphere. The mixing occurs when the complex electron wave functions are measured by the CCD or film to produce real images. The severity of the problem increases with specimen thickness, resolution, and electron wavelength.

A method for recovering the full, complex electron wavefunction from focal series was proposed by Schiske (1968). Further discussion then followed through 1990, when the method was reproposed using a different, more intuitive approach (Van Dyck and Op de Beeck, 1990). Saxton, who referred to this class of approaches as the "paraboloid method," later showed it to be equivalent to the original (Saxton, 1994). More recently, the problem was discussed in the context of 3D reconstruction by DeRosier, who outlined four basic strategies to recover all the unique Fourier coefficients by merging focal pairs, images at different tilt angles, or images of ordered (crystalline or helical) objects in reciprocal space (DeRosier, 2000). A different idea for addressing the problem in real space was proposed by Jensen and Kornberg (2000), followed by additional analyses and suggestions by Wan *et al.* (2004). Wolf *et al.* (2006) implemented a version of the paraboloid method in the popular FREALIGN package (Grigorieff, 2007). Simultaneously with the Wolf work, we implemented an iterative version of the same basic method we called *Prec* (for *P*araboloid *rec*onstruction) (Leong, 2009) in three major single particle reconstruction software packages: Bsoft (Heymann, 2001), EMAN (Ludtke *et al.*, 1999), and IMIRS (Liang *et al.*, 2002). The Bsoft and EMAN versions use Cartesian-coordinate systems, while the IMIRS version uses cylindrical coordinates to exploit the advantages of Fourier–Bessel transforms (Klug *et al.*, 1958) (Chapter 5, Vol. 482).

## 2. The Ewald Curvature Problem and Symbols Used

To introduce needed symbols, we will follow DeRosier's derivation of the effects of the Ewald sphere curvature closely (DeRosier, 2000), except that here all Fourier coefficients $F$ are complex and amplitude contrast is included explicitly. Beginning first with the effect of a sample on an incident electron wave and its weak-phase approximation,

$$\frac{A_t(x)}{A_0} = e^{-(\alpha+i\beta)\rho(x)} \approx 1 - (\alpha + i\beta)\rho(x) \tag{14.1}$$

where $A_t(x)$ is the transmitted wave, $A_0$ is the incoming wave, $\alpha$ is the amplitude contrast value, $\beta = \sqrt{1-\alpha^2}$ is the phase contrast value (Erickson and Klug, 1971), $\rho(x)$ is the density of the sample, and i is an imaginary number with magnitude 1; the diffracted wave $F(X)$ takes the form

$$F(X) = \mathrm{FT}[1 - (\alpha + i\beta)\rho(x)] = \delta(X) - (\alpha + i\beta)F_\rho(X) \tag{14.2}$$

where $F_\rho(X)$ is the FT of our sample density.

Considering the sum of a single, symmetric pair of diffracted beams represented by Fourier coefficients $F_L$ and $F_R$ on an Ewald sphere (Fig. 14.1), whose additional path length through the lens with respect to the unscattered beam adds an additional phase shift of $e^{i\chi}$, we have:

$$F(X) = \delta(X) - (\alpha + i\beta)F_L e^{i\chi}\delta(X + X_a) - (\alpha + i\beta)F_R e^{i\chi}\delta(X - X_a) \tag{14.3}$$

where $\chi$ is the wave aberration function at $X_a$ and is defined as

$$\chi(s) = \frac{\pi}{2}C_s\lambda^3 s^4 - \pi\Delta f\lambda s^2 \tag{14.4}$$

in which $\lambda$ is the electron wavelength, $s$ is the spatial frequency, $C_s$ is the spherical aberration coefficient, and $\Delta f$ is the defocus.

The interference of these beams will produce a single complex fringe with a periodicity of $1/X_a$ whose amplitude, $\sigma(x)$, will be

$$\sigma(x) = \mathrm{FT}^{-1}[F(X)] = 1 - (\alpha + i\beta)F_L e^{i\chi}e^{-2\pi i x X_a} - (\alpha + i\beta)F_R e^{i\chi}e^{2\pi i x X_a} \tag{14.5}$$

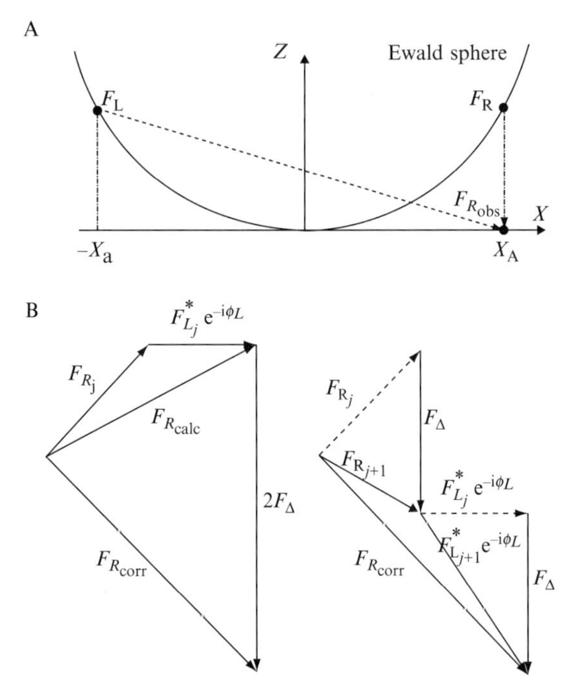


**Figure 14.1** The Ewald sphere and *Prec* algorithm. (A) Fourier coefficients in the transforms of electron microscope images ($F_{R_{obs}}$) are actually combinations of coefficients ($F_L$ and $F_R$) that lie on a spherical surface through the 3D transform of the specimen called the Ewald sphere. (B) *Prec* iteratively recovers the independent values of these coefficients by comparing CTF-corrected observations ($F_{R_{corr}}$) with the calculated sum ($F_{R_{calc}}$) that would have been expected from the right ($F_{R_j}$) and left ($F_{L_j}$) terms of some previous reconstruction, with appropriate phase factors $e^{i\phi_L} = (\alpha + i\beta)^2 e^{i2\chi}$. Half the difference ($F_\Delta$) is then added to $F_{R_j}$ and $F_{L_j}$ to produce the next iteration ($F_{R_{j+1}}$ and $F_{L_{j+1}}$).

The intensity of the wave is recorded as our image

$$|\sigma(x)|^2 \approx 1 - \left[(\alpha + i\beta)F_L e^{i\chi} + (\alpha - i\beta)F_R^* e^{-i\chi}\right]e^{-2\pi i x X_a} \\ -\left[(\alpha + i\beta)F_R e^{i\chi} + (\alpha - i\beta)F_L^* e^{-i\chi}\right]e^{2\pi i x X_a} \tag{14.6}$$

where the $F^2$ terms can be ignored due to the weak-phase approximation.

The FT of our image $F_{obs}(X)$ is then

$$F_{obs}(X) = \delta(X) - \left[(\alpha + i\beta)F_L e^{i\chi} + (\alpha - i\beta)F_R^* e^{-i\chi}\right]\delta(X + X_a) \\ -\left[(\alpha - i\beta)F_L^* e^{-i\chi} + (\alpha + i\beta)F_R e^{i\chi}\right]\delta(X - X_a) \tag{14.7}$$

We see that $F_{R_{obs}}$, the observed Fourier value on the right side at $X = X_a$, is

$$F_{R_{obs}} = -F_L^*(\alpha - i\beta)e^{-i\chi} - F_R(\alpha + i\beta)e^{i\chi} \tag{14.8}$$

Because of the curvature of the Ewald sphere, $F_{\mathrm{L}}$ and $F_{\mathrm{R}}$ are not a Friedel pair (i.e., not complex conjugates), but rather independent Fourier coefficients, mixed by the process of image formation. Thus, conventional methods, which treat $F_{R_{\mathrm{obs}}}$ as if it were the sum of a Friedel pair $F_{\mathrm{L}}$ and $F_{\mathrm{R}}$, do progressively worse as $F_{\mathrm{L}}$ and $F_{\mathrm{R}}$ diverge at higher resolutions.

## 3. The Paraboloid Method in the Context of 3D Reconstruction

The original Fourier coefficients can be recovered by averaging information from multiple images, which each contain different combinations of the unique coefficients. First, images are corrected for the contrast transfer function (CTF). This is performed by multiplying each term $F_{\mathrm{obs}}$ by $-(\alpha - \mathrm{i}\beta)\mathrm{e}^{-\mathrm{i}\chi}$. Unlike conventional CTF corrections, where values around CTF zeros are discarded, here there is no such requirement, since this "complex" CTF-correction (cCTF) is a multiplication by a factor of magnitude 1 rather than a division by a number potentially close to zero. Thus, $F_{R_{\mathrm{corr}}}$, the cCTF-corrected coefficient on the right side, is

$$F_{R_{\mathrm{corr}}} = -F_{R_{\mathrm{obs}}}(\alpha - \mathrm{i}\beta)\mathrm{e}^{-\mathrm{i}\chi} = F_{\mathrm{R}} + F_{\mathrm{L}}^{*}(\alpha - \mathrm{i}\beta)^{2}\mathrm{e}^{-\mathrm{i}2\chi} \tag{14.9}$$

Because each $F_{R_{\mathrm{corr}}}$ is the sum of the correct $F_{\mathrm{R}}$ and a phase-shifted, complex-conjugated $F_{\mathrm{L}}$, at this point it becomes clear how by averaging $F_{R_{\mathrm{corr}}}$ from a number of different images, each measuring the same $F_{\mathrm{R}}$ but different $F_{\mathrm{L}}$'s, the $F_{\mathrm{R}}$'s will add coherently but the sum of $F_{\mathrm{L}}$'s will diminish in comparison. At low resolution, however, where $F_{\mathrm{L}}^{\star} \approx F_{\mathrm{R}}$,

$$F_{R_{\mathrm{obs}}} \approx -F_{\mathrm{R}}(\alpha - \mathrm{i}\beta)\mathrm{e}^{-\mathrm{i}\chi} - F_{\mathrm{R}}(\alpha + \mathrm{i}\beta)\mathrm{e}^{\mathrm{i}\chi} = -2F_{\mathrm{R}}(\alpha\cos\chi - \beta\sin\chi) \tag{14.10}$$

The cCTF-correction then leads to wrong values

$$F_{R_{\mathrm{corr}}} = F_{\mathrm{R}} + F_{\mathrm{R}}(\alpha - \mathrm{i}\beta)^{2}\mathrm{e}^{-\mathrm{i}2\chi} \tag{14.11}$$

since $\chi$ does not vary quickly, causing the second terms to also add coherently and introduce a significant error. Thus at low resolution, it is better to use the simpler, real CTF correction (rCTF), where $F_{\mathrm{obs}}$ is divided by the factor $-2(\alpha\cos\chi - \beta\sin\chi)$. A practical transition point can be found as the spatial frequency at which the cCTF-corrected and the rCTF-corrected reconstructions match best.

After CTF-correcting the raw images, the paraboloid method places the $F_{\mathrm{corr}}$ values in their correct position in Fourier space on the Ewald sphere:

$$F_{R_{\mathrm{PM}}} = \frac{1}{N}\sum_{k}^{N} F_{R^k_{\mathrm{corr}}} = \frac{1}{N}\sum_{k}^{N} F_{R_k} + \frac{1}{N}\sum_{k}^{N} F^*_{L_k}(\alpha - \mathrm{i}\beta)^2 \mathrm{e}^{-\mathrm{i}2\chi_k} \quad (14.12)$$

$$F_{L_{\mathrm{PM}}} = \frac{1}{N}\sum_{k}^{N} F_{L^k_{\mathrm{corr}}} = \frac{1}{N}\sum_{k}^{N} F_{L_k} + \frac{1}{N}\sum_{k}^{N} F^*_{R_k}(\alpha - \mathrm{i}\beta)^2 \mathrm{e}^{-\mathrm{i}2\chi_k} \quad (14.13)$$

where $N$ is the total number of images (indexed by $k$) which contribute to each point.

## 4. The Prec Algorithm

In essence, the paraboloid method therefore "splits" the observed values $F_{\mathrm{obs}}$ into estimates of $F_{\mathrm{R}}$ and $F_{\mathrm{L}}$ by averaging information from a set of images. Once initial estimates are obtained, they can be refined through iteration, since knowledge of any particular coefficient will affect how all the sums it is involved in should be split. In *Prec*'s iterative refinement loop, each $F_{\mathrm{obs}}$ of each image is compared to the expected ("calculated") value $F_{R_{\mathrm{calc}}}$ that is obtained by combining Ewald sphere-related Fourier coefficients from a previous reconstruction:

$$F_{R_{\mathrm{calc}}} = F_{R_j} + F^*_{L_j}(\alpha - \mathrm{i}\beta)^2 \mathrm{e}^{-\mathrm{i}2\chi} \quad (14.14)$$

where the index $j$ represents the $j$th iteration of the reconstruction. The difference between the CTF-corrected observed value for image $k$ and this calculated value is stored as the "error" $2F_{\Delta_k}$:

$$F_{R^k_{\mathrm{corr}}} - F_{R^k_{\mathrm{calc}}} = 2F_{\Delta_k} \quad (14.15)$$

Half of these errors are then added as a refinement to the Fourier component on the right:

$$F_{R_{j+1}} = F_{R_j} + \frac{1}{N}\sum_{k}^{N} F_{\Delta_k} \quad (14.16)$$

The correction can also be immediately added to the left side:

$$F^*_{L_{j+1}}(\alpha - \mathrm{i}\beta)^2 \mathrm{e}^{-\mathrm{i}2\chi} = F^*_{L_j}(\alpha - \mathrm{i}\beta)^2 \mathrm{e}^{-\mathrm{i}2\chi} + F_{\Delta} \quad (14.17)$$

which after rotation, complex conjugation and summation of corrections simplifies to:

$$F_{L_{j+1}} = F_{L_j} + \frac{1}{N}\sum_{k}^{N} F^*_{\Delta_k}(\alpha - \mathrm{i}\beta)^2 \mathrm{e}^{-\mathrm{i}2\chi_k} \tag{14.18}$$

To begin the process, in the first iteration the "reconstruction" to be refined can simply be a set of zeroes. Then the calculated value, $F_{R_{\mathrm{calc}}}$, is also zero and thus the correction applied to the left and right Fourier components ($F_{R_0}$ and $F_{L_0}$) are just the values called for by the paraboloid method, scaled by a simple factor of $\frac{1}{2}$:

$$F_{R^k_{\mathrm{corr}}} = 2F_{\Delta_k} \tag{14.19}$$

$$F_{R_0} = \frac{1}{N}\sum_{k}^{N} F_{\Delta_k} = \frac{1}{N}\sum_{k}^{N} \frac{1}{2} F_{R^k_{\mathrm{corr}}} = \frac{1}{2} F_{R_{\mathrm{PM}}} \tag{14.20}$$

$$\begin{aligned} F_{L_0} &= \frac{1}{N}\sum_{k}^{N} F^*_{\Delta_k}(\alpha - \mathrm{i}\beta)^2 \mathrm{e}^{-\mathrm{i}2\chi_k} = \frac{1}{N}\sum_{k}^{N} \frac{1}{2} F^*_{R^k_{\mathrm{corr}}}(\alpha - \mathrm{i}\beta)^2 \mathrm{e}^{-\mathrm{i}2\chi_k} \\ &= \frac{1}{N}\sum_{k}^{N} \frac{1}{2} F_{L^k_{\mathrm{corr}}} = \frac{1}{2} F_{L_{PM}} \end{aligned} \tag{14.21}$$

Subsequent iterations refine that estimate. Take, for example, any Fourier coefficient $F_{R_0}$ and the contributions to it:

$$F_{R_0} = \frac{1}{N}\sum_{k}^{N} F_{R_k} + \frac{1}{N}\sum_{k}^{N} F^*_{L_k}(\alpha - \mathrm{i}\beta)^2 \mathrm{e}^{-\mathrm{i}2\chi_k} \tag{14.22}$$

where $N$ is the number of images that measured $F_{\mathrm{R}}$.

This can be recast as

$$F_{R_0} \approx \bar{F}_{\mathrm{R}} + \varepsilon \tag{14.23}$$

where $\bar{F}_{\mathrm{R}}$ is the average $F_{R_k}$ and $\varepsilon$ is the residual error which consists of the average of the $F_{\mathrm{L}_k}(\alpha - \mathrm{i}\beta)^2 \mathrm{e}^{-\mathrm{i}2\chi_k}$ terms, which is a random walk with step size of approximately $|F_{L_k}|$. Thus, after the first refinement cycle, the residual error falls off as $\sim \frac{1}{\sqrt{N}}$.

## 5. Implementation of the Prec Algorithm

Three versions of *Prec* have been implemented, one each in the software packages Bsoft, IMIRS, and EMAN, which each have all the functionality required to produce high-resolution reconstructions from raw cryo-EM images. While the mathematical theory is as described above, key differences exist in how the interpolations are handled in the different coordinate systems. Bsoft and EMAN use a Cartesian-coordinate system. Starting with raw cryo-EM images, the Bsoft and EMAN implementations of *Prec* begin by calculating the images' 2D FTs, multiplying them by the cCTF, and then calculating the "$z$-" coordinate (height up the Ewald sphere) for each Fourier coefficient. Taking into account the projection direction, the coefficients from the image are then added to the nearest corresponding lattice points of the "reconstruction" 3D FT with appropriate phase factors. In the Bsoft version, the standard interpolation procedure with weight $w = 1 - d$ (where $d$ is distance in pixels from the measurement to the 3D lattice point) is used. In the EMAN version, any of its various built-in interpolation procedures can be used. After all the data are added to the "reconstruction" 3D FT, each amplitude is divided by the total weight of all the measurements that contributed, and a density map is produced through an inverse 3D FT. Refinement cycles, as implemented in Bsoft, loop through each coefficient of each corrected image transform. The expected value is calculated by summing the coefficients at the nearest corresponding lattice points of the 3D FT of the current reconstruction with appropriate phase factors and complex conjugation. Half the difference between this expected value and the (CTF-corrected) observed value is added to each contributing coefficient.

A different version of *Prec* was implemented within IMIRS. IMIRS uses a cylindrical coordinate system for the reconstruction process, where the 3D reconstruction and its FT are expressed as expansions of cylinder functions, as proposed by Klug *et al.* (1958). We follow the notation used by Crowther *et al.* (1970). The 2D FTs of the raw images are calculated and multiplied by the cCTF as before. The 3D FT of the object is represented in cylindrical coordinates, $Z$, $R$, and $\Phi$. The Ewald sphere of measurements recorded in each image will in general intersect each ring of coordinates in two places. For each intersection of an image Ewald sphere and a ring of the 3D FT, a Fourier coefficient for that location is estimated from the pixels of the FT of the image through bilinear interpolation. Once all the estimates on a particular ring have been calculated, all of them are used to determine the cylindrical expansion terms, $G_n(R,Z)$ through a least squares fit which differs from the conventional IMIRS reconstruction in that the magnitude of the cCTF term is 1 and therefore is not a factor in the weighting of terms. A Fourier–Bessel transform is used next to obtain the $g_n(r,Z)$ terms, which are then used to generate the density map.

Because in this case, the $F_L$ that pairs with each $F_R$ of a randomly spaced intersection of an image Ewald sphere and a Fourier ring does not generally fall upon any ring, a 3D nearest neighbor interpolation was required to estimate its value. Our tests suggested that the losses due to this less accurate nearest neighbor interpolation outweighed the gains obtained by iteration, so that iteration of the cylindrical-coordinate-based version of *Prec* is not recommended. Similarly for the Cartesian-coordinate-based versions of Prec, iterations beyond the first refinement cycle are also not recommended as successive refinements yield minimal gains (see Eq. (14.23)).

## 6. Tests on Simulated Data

All three implementations of *Prec* have been tested on both simulated and experimental data. As an example, a large number of images of the moderate-sized (~300-Å diameter) Foot and Mouth Virus (FMV) (Fry *et al.*, 1993) were simulated with different methodologies, voltages, and signal-to-noise ratios. A complete pdb was generated using the VIPERdb (Shepherd *et al.*, 2006) and then its density was sampled to produce a reference volume using a modified version of *bgex* of the Bsoft package. "Ewald projection" images were simulated by simply summing Fourier coefficients on Ewald spheres using Eq. (14.8) and a complete 1D Whittaker–Shannon interpolation (Shannon, 1949; Whittaker, 1915) in the *Z* direction, followed by an inverse 2D FT. Six data sets of 5000 images each, with acceleration voltages of 15, 25, 50, 100, 200, and 300 kV, respectively, were calculated. FMV reconstructions were then calculated from each data set using the conventional reconstruction programs in Bsoft, IMIRS, and EMAN, which do not correct for curvature of the Ewald sphere. The resolution of each reconstruction was measured by its correlation with the original reference density map in Fourier shells (FSC) and confirmed visually (Fig. 14.2, Bsoft results only). The large number of images ensured that Fourier space was well sampled. The expected increase in resolution as a function of voltage demonstrated the Ewald sphere curvature problem. Analogous reconstructions of the 15 kV data set were then performed with Bsoft, IMIRS, and EMAN implementations of *Prec.* All three programs completely overcame the effects of Ewald sphere curvature (Fig. 14.2, again Bsoft results only). Similarly successful tests have been performed with more accurate sets of simulated images generated using the multislice algorithm (Cowley and Moodie, 1957), with different signal-to-noise ratios, and a larger virus model (the 754-Å diameter Reovirus; Reinisch *et al.*, 2000).

## 7. Application to Experimental Reconstructions

To what extent current experimental reconstructions of icosahedral viruses suffer from uncorrected Ewald sphere curvature remains unclear. While previous analyses (DeRosier, 2000; Jensen and Kornberg, 2000)

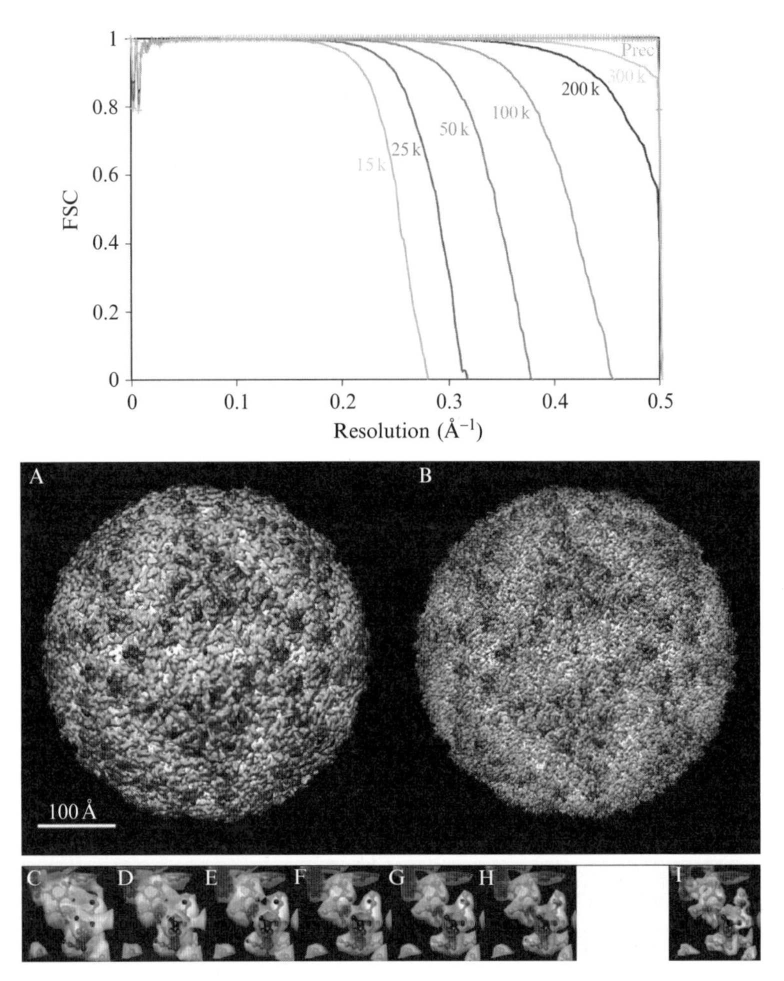


**Figure 14.2** *Prec* overcomes the curvature problem in Ewald projections. (Top) FSC curves for conventional Bsoft reconstructions of the Foot and Mouth Virus from 5000 "Ewald projection" images simulated with the voltages shown, plus a reconstruction from the 15 kV images calculated by the *Prec* program, which completely corrects for the curvature problem. (A and B) Isosurface renderings of the conventional and *Prec* 15 kV reconstructions, respectively. (C–H) Transparent isosurfaces of a single α-helix from the 15, 25, 50, 100, 200, and 300 kV reconstructions, respectively, surrounding the atomic model used to simulate the images. (I) The same helix from the *Prec* 15 kV reconstruction. FSC curves were calculated with *bresolve* (Heymann, 2001) and isosurfaces were rendered with *Chimera* (Pettersen *et al.*, 2004).

suggest that Ewald curvature should already be one of the principal resolution limitations in the most recent, highest resolution (<4 Å) reconstructions, actual reconstructions from simulated images of Reovirus showed that the Ewald curvature did not become severely limiting until ~2.5 Å resolution, and an improvement of an experimental reconstruction through curvature correction has yet to be reported (Leong, 2009). Nevertheless as the size of reconstructed viruses, the number and quality of images that are included in reconstructions, and the precision to which those images can be mutually aligned continue to increase, Ewald curvature correction will eventually become critical.

The Bsoft and IMIRS versions of Prec can be downloaded from www.jensenlab.caltech.edu. The EMAN implementation can be obtained from jiang.bio.purdue.edu.

## ACKNOWLEDGMENTS

We thank Andy Rawlinson, Bernard Heymann, Bill Tivol, Dylan Morris, Yuyao Liang, Wong Hoi Hui, Xiaokang Zhang, Weimin Wu, Wen Jiang, and Nikolaus Grigorieff for helpful discussions about Ewald sphere curvature and the manuscript as well as the Bsoft, IMIRS, and EMAN packages and for providing experimental data. The development and testing of *Prec* was supported in part by NIH grants R01 AI067548 and P50 GM082545 to G. J. J. and R01 GM071940, CA094809, and AI069015 to Z. H. Z.; DOE grant DE-FG02-04ER63785 to G. J. J.; a Searle Scholar Award to G. J. J.; the Beckman Institute at Caltech; and gifts to Caltech from the Parsons Foundation and Agouron Institute.

## REFERENCES

Cowley, J. M., and Moodie, A. F. (1957). The scattering of electrons by atoms and crystals. 1. A new theoretical approach. *Acta Crystallogr.* **10,** 609–619.

Crowther, R. A., Derosier, D. J., and Klug, A. (1970). Reconstruction of 3-dimensional structure from projections and its application to electron microscopy. *Proc. R. Soc. Lond. Ser. A* **317,** 319–340.

DeRosier, D. J. (2000). Correction of high-resolution data for curvature of the Ewald sphere. *Ultramicroscopy* **81,** 83–98.

Erickson, H. P., and Klug, A. (1971). Measurement and compensation of defocusing and aberrations by fourier processing of electron micrographs. *Philos. Trans. R. Soc. Lond. B Biol. Sci.* **261,** 105–118.

Fry, E., Acharya, R., and Stuart, D. (1993). Methods used in the structure determination of foot-and-mouth-disease virus. *Acta Crystallogr. A* **49,** 45–55.

Grigorieff, N. (2007). FREALIGN: High-resolution refinement of single particle structures. *J. Struct. Biol.* **157,** 117–125.

Henderson, R., Baldwin, J. M., Ceska, T. A., Zemlin, F., Beckmann, E., and Downing, K. H. (1990). Model for the structure of bacteriorhodopsin based on high-resolution electron cryomicroscopy. *J. Mol. Biol.* **213,** 899–929.

Heymann, J. B. (2001). Bsoft: image and molecular processing in electron microscopy. *J. Struct. Biol.* **133,** 156–169.

Jensen, G. J., and Kornberg, R. D. (2000). Defocus-gradient corrected back-projection. *Ultramicroscopy* **84,** 57–64.

Klug, A., Crick, F. H. C., and Wyckoff, H. W. (1958). Diffraction by helical structures. *Acta Crystallogr.* **11,** 199–213.

Kuhlbrandt, W., Wang, D. N., and Fujiyoshi, Y. (1994). Atomic model of plant light-harvesting complex by electron crystallography. *Nature* **367,** 614–621.

Leong, P. A. (2009). Computational challenges in high-resolution cryo-electron microscopy. Ph.D. dissertation, California Institute of Technology, Pasadena, California, USA.

Liang, Y. Y., Ke, E. Y., and Zhou, Z. H. (2002). IMIRS: A high-resolution 3D reconstruction package integrated with a relational image database. *J. Struct. Biol.* **137,** 292–304.

Ludtke, S. J., Baldwin, P. R., and Chiu, W. (1999). EMAN: Semiautomated software for high-resolution single-particle reconstructions. *J. Struct. Biol.* **128,** 82–97.

Pettersen, E. F., Goddard, T. D., Huang, C. C., Couch, G. S., Greenblatt, D. M., Meng, E. C., and Ferrin, T. E. (2004). UCSF Chimera - a visualization system for exploratory research and analysis. *J. Comput. Chem.* **25,** 1605–1612.

Reinisch, K. M., Nibert, M., and Harrison, S. C. (2000). Structure of the reovirus core at 3.6 angstrom resolution. *Nature* **404,** 960–967.

Saxton, W. O. (1994). What is the focus variation method—Is it new—Is it direct. *Ultramicroscopy* **55,** 171–181.

Schiske, P. (1968). Zur Frage der Bildrekonstruktion durch Fokusreihen. *In* Proc. 4th Eur. Conf. on Electron Microscopy (Rome). Paper presented at.

Shannon, C. E. (1949). Communication in the presence of noise. *Proc. Inst. Radio Eng.* **37,** 10–21.

Shepherd, C. M., Borelli, I. A., Lander, G., Natarajan, P., Siddavanahalli, V., Bajaj, C., Johnson, J. E., Brooks, C. L., and Reddy, V. S. (2006). VIPERdb: A relational database for structural virology. *Nucleic Acids Res.* **34,** D386–D389.

Unwin, N. (2005). Refined structure of the nicotinic acetylcholine receptor at 4 angstrom resolution. *J. Mol. Biol.* **346,** 967–989.

Van Dyck, D., and Op de Beeck, M. (1990). New direct methods for phase and structure retrieval by HREM. *In* Proceedings of the 12th International Congress on Electron Microscopy (Seattle). Paper presented at:.

Wan, Y., Chiu, W., and Zhou, Z. H. (2004). Full contrast transfer function correction in 3D cryo-EM reconstruction. *In* IEEE Proceedings of ICCCAS 2004 (Chengdu, Sichuan, China). Paper presented at.

Whittaker, E. T. (1915). On the functions which are represented by the expansion of interpolation theory. *Proc. R. Soc. Edinb.* **35,** 181–194.

Wolf, M., DeRosier, D. J., and Grigorieff, N. (2006). Ewald sphere correction for single-particle electron microscopy. *Ultramicroscopy* **106,** 376–382.

CHAPTER FIFTEEN

# Software Tools for Molecular Microscopy: An Open-Text Wikibook

Neil R. Voss,* Clinton S. Potter,* Ross Smith,[†] *and* Bridget Carragher*

## Abstract

We provide a brief description of a Wikibook open-content textbook that was set up with a goal of providing a comprehensive and continually updated list of all of the software packages of interest to the cryo-EM community. While the content of the Wikibook will change over time, here we provide a snapshot of the current state of software tools available, and encourage the members of this community to view the pages, add content, correct errors, and make any other contributions that might be useful.

As can be seen from the numerous chapters in this volume describing sophisticated image processing methods, complex computational analyses are at the heart of nearly every cryo-EM project, and much of the development of the field has been the creation and testing of software packages. Many of the packages have related and overlapping functions and are constantly being updated and enhanced. It can be daunting, therefore, for a newcomer to discover which package or packages are most appropriate for their goal. Towards that end several special issues of the *Journal of Structural Biology* (Aebi et al., 1996; Carragher and Penczek, 2003; Carragher et al., 2007; Smith and Carragher, 2008) have been specifically devoted to descriptions of these applications and several web sites have been set up by community-oriented individuals to provide descriptions and links to the software packages. However, given that software is often in a continuous state of change and new packages appear all the time, maintaining such lists and keeping them up to date places a burden on the individual as well as the institution providing the computer resources. An obvious solution to this

* The National Resource for Automated Molecular Microscopy, Department of Cell Biology, The Scripps Research Institute, La Jolla, California, USA
[†] Center for Health Informatics and Bioinformatics and, Department of Cell Biology, New York University School of Medicine, New York, New York, USA

*Methods in Enzymology*, Volume 482
ISSN 0076-6879, DOI: 10.1016/S0076-6879(10)82016-6

problem is to take advantage of the new generation of publicly supported web sites, generally powered by wiki[1] software. Wikis allow for the easy creation and editing of web pages, and can be set up to allow anyone to contribute and change the entries. A Wiki therefore removes the burden of supporting and updating web pages from a single individual but instead encourages the entire community to contribute to the effort.

Accordingly, in 2007, we created a Wikipedia[2] article to accompany a special issue of the *JSB*. This article provided a consolidated list all of the software packages described in the special issue, as well as several others that were in the public domain. The authors of the packages were contacted and asked to edit and provide additional content to the pages and the community was encouraged to add new software and content on an ongoing basis.

As the utilization and content of the Wikipedia article became established, the curators of Wikipedia transferred it to the Wikibooks[3] site, considered to be a more natural home for this kind of resource. Wikibooks was set up in 2003 to provide resources to host and build open-content textbooks. The Wikimedia Foundation, a nonprofit charitable organization, supports Wikibooks and the content on Wikibooks is covered by the Creative Commons Attribution-ShareAlike 3.0 license.[4] Under this license all contributions remain copyrighted to the authors, but limited freedom is granted to redistribute the content and make derivative works, provided credit is given to the authors and it is released under an identical license. These large public and supported wiki sites provide the infrastructure and support to provide a more permanent home for public resources. This means that the existence of the resource is not subject to any individual's location or the institutional policies of an employer.

The software tools in the Wikibook[5] are loosely and somewhat arbitrarily divided into five separate chapters as follows:

- *General packages*: Packages that offer a comprehensive set of tools to permit the analysis of data in several classes of structural problems.
- *Specific packages*: Packages that offer a comprehensive set of tools to permit the analysis of data in a single class of structural problem. For example, packages specifically focused on objects with helical, icosahedral, crystalline symmetry, etc.
- *Application tools*: Packages that offer a tool or a set of tools to permit the analysis of data in one or more class of structural problems. These have generally been developed to manage one specific step in the structural analysis, for example, CTF correction, particle picking, B-factor correction, etc.

[1] http://en.wikipedia.org/wiki/Wiki
[2] http://en.wikipedia.org
[3] http://en.wikibooks.org/wiki/Main_Page
[4] http://en.wikipedia.org/wiki/Creative_Commons_licenses
[5] http://en.wikibooks.org/wiki/Software_Tools_For_Molecular_Microscopy

**Table 15.1** WikiBook: Software tools for molecular microscopy status as of May 2010

| | Package | Location | Version | Contact | OS | Cost/license | Referencess |
|---|---|---|---|---|---|---|---|
| **General** | Appion | appion.org | 2.0.0 | appion@scripps.edu | Linux<br>Mac<br>Win | Free/OS/<br>Apache 2.0 | Lander et al. (2009) |
| | Bsoft | bsoft.ws | 1.6.0 | Bernard_Heymann@nih.gov | Linux<br>Mac<br>VMS | Free/OS | Heymann and Belnap (2007) |
| | Cyclops | www.bfsc.leidenuniv.nl/software/Cyclops/ | 0.8rc1 | plaisier@chem.leidenuniv.nl | Win | Free/OS<br>GPL | Plaisier et al. (2007) |
| | EMAN | blake.bcm.edu/eman/ | 1.9 | sludtke@bcm.edu | Linux<br>Mac<br>Win | Free/OS<br>GPL/BSD | Ludtke et al. (1999) |
| | EMAN2 | blake.bcm.edu/eman/eman2/ | 2.0RC2 | sludtke@bcm.edu | Linux<br>Mac<br>Win | Free/OS<br>GPL/BSD | Tang et al. (2007) |
| | Eos | www.yasunaga-lab.bio.kyutech.ac.jp/Eos/index.html | 1 | yasunaga@bio.kyutech.ac.jp | Linux<br>Mac<br>Win | Free | Yasunaga and Wakabayashi (1996) |
| | IMAGIC | ImageScience.de/imagic/ | 4D | imagic@ImageScience.de | Linux<br>Mac<br>Win | Commercial | van Heel et al. (1996) |
| | iplt | iplt.org | – | Andreas_schenk@hms.harvard.edu | Linux<br>Mac<br>Win | Free/OS<br>GPL | Philippsen et al. (2007) |
| | MDPP | mdpp.med.nyu.edu/ | 08.200 | smithp01@med.nyu.edu | Linux<br>Mac<br>VMS | Free | Smith and Gottesman (1996) |
| | MRC | www2.mrc-lmb.cam.ac.uk/image2000.html | – | jms@mrc-lmb.cam.ac.uk | Linux<br>Mac<br>Dec | Free/OS<br>Academic | Crowther et al. (1996) |
| | SPARX | sparx-em.org/sparxwiki | 2.0 | pawel.a.penczek@uth.tmc.edu | Linux<br>Mac<br>Win | Free/OS<br>BSD | Hohn et al. (2007) |

(*continued*)

**Table 15.1** *(continued)*

| | Package | Location | Version | Contact | OS | Cost/license | Referencess |
|---|---|---|---|---|---|---|---|
| | Spider | wadsworth.org/spider_doc/spider/docs/documents.html | 18.10 | spider@wadsworth.org | Linux Mac | Free/OS GPL | Frank et al. (1996) |
| | Suprim | ami.scripps.edu/software/suprim/ | 5.5 | amisoftware@scripps.edu | Linux | Free/OS | Schroeter and Bretaudiere (1996) |
| | XMIPP | xmipp.cnb.csic.es/ | 2.4 | xmipp@cnb.csic.es | Linux | Free/OS GPL | Sorzano et al. (2004) |
| **2D crystals** | 2dx | 2dx.org | 3.1.0 | henning.stahlberg@unibas.ch | Linux Mac | Free/OS GPL | Gipson et al. (2007) |
| **Icosahedra** | AUTO 3DEM | cryoem.ucsd.edu/programs.shtm | 2.02 | sinkovit@sdsc.edu | Linux | Free/OS BSD | Yan et al. (2007) |
| **Helices** | BBHP | coan.burnham.org/other-projects/ | 20.0 | coan@burnham.org | Linux | Free | Owen et al. (1996) |
| | IHRSR | – | – | egelman@virginia.edu | Linux | Free | Egelman (2007) |
| | Phoelix | ami.scripps.edu/software/phoelix/ | 1.4 | amisoftware@scripps.edu | Linux | Free/OS Apache 2.0 | Carragher et al. (1996) |
| | RubyHelix | structure.m.u-tokyo.ac.jp/English/software/Ruby-Helix-Page/ruby-helix.html | 1.0 | mkikkawa@m.u-tokyo.ac.jp | Linux Mac | Free/OS GPL | Metlagel et al. (2007) |
| | StokesLab HelicalSW | saturn.med.nyu.edu//research/sb/stokeslab/image-analysis.html | – | stokes@saturn.med.nyu.edu | Linux | Free/OS | – |
| **Single Particle** | Frealign | emlab.rose2.brandeis.edu/software | 8.08 | niko@brandeis.edu | Linux Mac | Free/OS GPL | Grigorieff (2007) |
| | PFT3DR | people.chem.byu.edu/belnap/pft3dr | 2.0 | David_Belnap@byu.edu | Linux VMS | Free OS | Fuller et al. (1996) |
| **Tomography** | EM3D | em3d.tamu.edu | 2.0 | grantser@bio.tamu.edu | Mac Win | Free Academic | Ress et al. (1999) |
| | IMOD | bio3d.colorado.edu/imod | 4.1 | mast@colorado.edu | Linux Mac Win | Free GPL | Kremer et al. (1996) |

| | | | | | | | |
|---|---|---|---|---|---|---|---|
| | | [illegible] | | [illegible] | Linux | Free | Winkler and Taylor (2006) |
| | Raptor | www-vlsi.stanford.edu/TEM/software.htm | 2.1 | famat@stanford.edu | Linux | Free | Amat et al. (2008) |
| | SerialEM | bio3d.colorado.edu/SerialEM/ | 2.8 | mast@colorado.edu | Win | Free Academic | Mastronarde (2005) |
| | TOM | www.biochem.mpg.de/en/rd/baumeister/tom_e/ | – | tom@biochem.mpg.de | Linux Win | Free | Nickell et al. (2005) |
| | TomoJ | u759.curie.u-psud.fr/softwaresu759.html | 2.1 | cedric.messaoudi@curie.fr | Linux Mac Win | Free/OS CeCILL | Messaoudii et al. (2007) |
| | TXBR | confluence.crbs.ucsd.edu/display/ncmir/TxBR | 3.0 | sph@ncmir.ucsd.edu | Linux Mac Win | Free | Lawrence et al. (2006) |
| | UCSF Tomo | msg.ucsf.edu/em/EMNEW2/tomography_page.html | v7.7.4E4 | agard@msg.ucsf.edu | Win | Free Academic | Zheng et al. (2007) |
| | Xplore3D | www.fei.com/LifeSciences/ | – | Robert.snyder@fei.com | Win | Commercial | Schoenmakers et al. (2005) |
| **Data Acquisition** | Leginon | leginon.org | 2.0 | amisoftware@scripps.edu | Linux Mac Win | Free/OS Apache 2.0 | Suloway et al. (2005) |
| | SerialEM | bio3d.colorado.edu/SerialEM/ | 2.8 | mast@colorado.edu | Win | Free Academic | Mastronarde (2005) |
| | TOM | biochem.mpg.de/en/rd/baumeister/tom_e/ | – | tom@biochem.mpg.de | Linux Win | Free | Nickell et al. (2005) |
| | UCSF Tomo | msg.ucsf.edu/em/EMNEW2/tomography_page.html | v7.7.4E4 | agard@msg.ucsf.edu | Win | Free Academic | Zheng et al. (2007) |
| **Particle Selection** | DoG Picker | ami.scripps.edu/software/dogpicker/ | 0.2 | amisoftware@scripps.edu | Linux | Free/OS Apache 2.0 | Voss et al. (2009) |
| | FindEM | code.google.com/p/appion/downloads/list | 2.0 | appion@scripps.edu | Linux | Free/OS | Roseman (2004) |
| | Signature | emlab.rose2.brandeis.edu/software | – | niko@brandeis.edu | Linux Mac Win | Free GPL | Chen and Grigorieff (2007) |
| | SwarmPS | imb.uq.edu.au/index.html?page=44665 | 0.9.2 | d.woolford@imb.uq.edu.au | Linux | Free Academic | Woolford et al. (2007) |

(*continued*)

**Table 15.1** *(continued)*

| | Package | Location | Version | Contact | OS | Cost/license | Referencess |
|---|---|---|---|---|---|---|---|
| | TiltPicker | code.google.com/p/appion/ | 1.0 | vossman77@yahoo.com | Linux Mac Win | Free/OS Apache 2.0 | Voss et al. (2009) |
| **CTF** | ACE | ami.scripps.edu/software/ace/ | 2.3.1 | amisoftware@scripps.edu | MATLAB | Free Apache 2.0 | Mallick et al. (2005) |
| | ACE2 | ami.scripps.edu/software/ace2/ | 1.0 | amisoftware@scripps.edu | Linux Mac | Free/OS Apache 2.0 | – |
| | CTFFind/ CTFTilt | emlab.rose2.brandeis.edu/ | 3.3/1.4 | niko@brandeis.edu | Linux Mac | Free GPL | Mindell and Grigorieff (2003) |
| | EMCTF | ual.es/~jjfdez/SW/emctf.html | 1.1 | jjfdez@ual.es | Linux | Free Academic | Fernandez et al. (2006) |
| | TOMO CTF | ual.es/~jjfdez/SW/tomoctf.html | 1.0 | jjfdez@ual.es | Linux | Free Academic | Fernandez et al. (2006) |
| **Resolution** | FSC | ImageScience.de/fsc/ | – | michael@ImageScience.de | Linux Mac Win | Free | van Heel and Schatz (2005) |
| | RMeasure | emlab.rose2.brandeis.edu/ software | 1.05 | niko@brandeis.edu | Linux Mac | Free/OS GPL | Sousa and Grigorieff (2007) |
| **Denoising** | iMed | coan.burnham.org/ | | coan@burnham.org | Linux | Free Academic | van der Heide et al. (2007) |
| | TOMO AND | ual.es/~jjfdez/SW/tomoand.html | 1.0 | jjfdez@ual.es | Linux Mac | Free Academic | Fernandez and Li (2003) |
| | TOMO BFLOW | ual.es/~jjfdez/SW/tomobflow. html | 1.0 | jjfdez@ual.es | Linux Mac | Free Academic | Fernandez (2009) |

| | | | | | | | |
|---|---|---|---|---|---|---|---|
| **B-factor correction** | bfactor | emlab.rose2.brandeis.edu/software | 1.03 | niko@brandeis.edu | Linux Mac | Free GPL | – |
| | EM BFACTOR | ual.es/~jjfdez/SW/embfactor.html | 1.0 | jjfdez@ual.es | Linux Mac | Free Academic | Fernandez et al. (2008) |
| **Segmen-tation** | CoDiv | coan.burnham.org/ | – | coan@burnham.org | Linux | Free Academic | Volkmann (2002) |
| **Initial Models** | ROTAN | sickkids.ca/research/rubinstein/software/index.html | – | john.rubinstein@utoronto.ca | Linux | Free Academic | Baker and Rubinstein (2008) |
| **Visualization/ Modeling** | Amira | amira.com/ | 5.2.2 | amirasupport@visageimaging.com | Linux Mac Win | Commercial | Stalling et al. (2005) |
| | CoAn/ CoFi | coan.burnham.org/ | – | coan@burnham.org | Linux | Free Academic | Volkmann and Hanein (1999) |
| | Chimera | www.cgl.ucsf.edu/chimera | 1.4.1 | chimera-users@cgl.ucsf.edu | Linux Mac Win | Free Academic | Goddard et al. (2007) |
| | EMData bank | http://www.emdatabank.org/ | – | help@emdatabank.org | Linux Mac Win | Free | Tagari et al. (2002) |
| | EMFit | bilbo.bio.purdue.edu/~viruswww/Rossmann_home/softwares/emfit.php | 5.0 | mr@indiana.bio.purdue.edu | Linux | Free OS | Rossmann et al. (2001) |
| | EM Package | www.biophys.uni-frankfurt.de/frangakis/Amiratools.htm | 1.0.0 | frangakis.group@googlemail.com | Linux Mac Win | Free Academic | Pruggnaller et al. (2008) |
| | Gorgon | gorgon.wustl.edu/ | 2.0.0 | sasakthi.abeysinghe@wustl.edu | Linux Mac Win | Free | Baker et al. (2007) |
| | NORMA | www.igs.cnrs-mrs.fr/elnemo/NORMA/ | 1.0 | karsten.suhre@helmholtz-muenchen.de | Linux | Free | Suhre et al. (2006) |

(*continued*)

**Table 15.1** *(continued)*

| | Package | Location | Version | Contact | OS | Cost/license | Referencess |
|---|---|---|---|---|---|---|---|
| | Open Structure | www.openstructure.org/ | 1.0.0 | Torsten.schwede@unibas.ch | Linux Mac Win | Free LGPL | – |
| | pyMol | pymol.org | – | – | Linux Mac Win | Subscription/ OS | – |
| | Rivem | bilbo.bio.purdue.edu/~ viruswww/Rossmann_home/ softwares/river_programs/ rivem.php | 3.2 | xc@purdue.edu | Linux | Free | Xiao and Rossmann (2007) |
| | Situs | situs.biomachina.org/ | 2.5 | situs@biomachina.org | Linux Mac Win | Free GPL | Wriggers et al. (1999) |
| | Urox | sites.google.com/site/xsiebert/ urox | 2.0.2 | xsiebert@gmail.com | Linux | Free | Siebert and Navaza (2009) |
| **Utilities** | BBHP/ Suprim | coan.burnham.org/other-projects | – | coan@burnham.org | Linux | Free Academic | – |
| | crop | emlab.rose2.brandeis.edu/ software | 1.0 | niko@brandeis.edu | Linux Mac | Free GPL | – |
| | diffmap | emlab.rose2.brandeis.edu/ software | 1.12 | niko@brandeis.edu | Linux Mac | Free GPL | – |
| | em2em | imagescience.de/em2em/index. htm | – | em2em@ImageScience.de | Linux Mac Win | Free | – |
| | Matlab format tools | www.mathworks.com/ matlabcentral/fileexchange/ 27021 | – | fred.sigworth@yale.edu | Linux Mac Win | Free/OS BSD | – |
| | MoniTEM | www.imba.oeaw.ac.at/monitem | 1.0.1 | guenter.resch@imba.oeaw.ac.at | Win | Free | Brunner and Resch (2009) |

*Notes*: OS, open source; Academic, free for academic use; GPL, GNU, General Public License; BSD, Berkeley Software Distribution license; CeCILL, CEA CNRS INRIA Logiciel Libre.

- *Visualization and modeling tools*: Packages that facilitate the interpretation, analysis, or presentation of the results.
- *Utilities*: General utilities like file format conversion, difference mapping, etc.

The information for each software application is formatted into a common template. The template includes a description of the software package, a link to the web site that hosts the software, a contact e-mail address, any relevant publications, and some basic information about the current version, including the supported operating systems and the cost and licensing agreements under which the software is released. Table 15.1 provides some of this information for the ~70 software packages listed in the Wikibook as of May 2010.

While the content of the Wikibook will change over time, this article provides a snapshot of the current state of software tools available, and provides a starting point for new practitioners to the field. We also hope to encourage everyone in the community to view the pages, add content, correct errors, and make any other contributions that might be useful.

## ACKNOWLEDGMENTS

We acknowledge support from the National Institutes of Health (NIH) through the National Center for Research Resources (NCRR) from grants P41-RR17573 (N. V., B. C., and C. P.) and UL1-RR029893 (R. S.). We also gratefully acknowledge those who have contributed to the Wikibook pages, too numerous to mention by name but many of whose contact addresses are listed in Table 15.1.

## REFERENCES

Aebi, U., Carragher, B., and Smith, P. R. (1996). Editorial. *J. Struct. Biol.* **116,** 1.

Amat, F., Moussavi, F., Comolli, L. R., Elidan, G., Downing, K. H., and Horowitz, M. (2008). Markov random field based automatic image alignment for electron tomography. *J. Struct. Biol.* **161,** 260–275.

Baker, L. A., and Rubinstein, J. L. (2008). Angle determination for side views in single particle electron microscopy. *J. Struct. Biol.* **162,** 260–270.

Baker, M. L., Ju, T., and Chiu, W. (2007). Identification of secondary structure elements in intermediate-resolution density maps. *Structure* **15,** 7–19.

Brunner, M. J., and Resch, G. P. (2009). Automated monitoring to reduce electron microscope downtime. *Ultramicroscopy* **109,** 1389–1392.

Carragher, B., and Penczek, P. A. (2003). Analytical methods and software tools for macromolecular microscopy. *J. Struct. Biol.* **144,** 1–3.

Carragher, B., Whittaker, M., and Milligan, R. A. (1996). Helical processing using PHOELIX. *J. Struct. Biol.* **116,** 107–112.

Carragher, B., Potter, C. S., and Sigworth, F. J. (2007). Software tools for macromolecular microscopy. *J. Struct. Biol.* **157,** 1–2.

Chen, J. Z., and Grigorieff, N. (2007). SIGNATURE: A single-particle selection system for molecular electron microscopy. *J. Struct. Biol.* **157,** 168–173.

Crowther, R. A., Henderson, R., and Smith, J. M. (1996). MRC image processing programs. *J. Struct. Biol.* **116,** 9–16.

Egelman, E. H. (2007). The iterative helical real space reconstruction method: Surmounting the problems posed by real polymers. *J. Struct. Biol.* **157,** 83–94.

Fernandez, J. J. (2009). TOMOBFLOW: Feature-preserving noise filtering for electron tomography. *BMC Bioinform.* **10,** 178.

Fernandez, J. J., and Li, S. (2003). An improved algorithm for anisotropic nonlinear diffusion for denoising cryo-tomograms. *J. Struct. Biol.* **144,** 152–161.

Fernandez, J. J., Li, S., and Crowther, R. A. (2006). CTF determination and correction in electron cryotomography. *Ultramicroscopy* **106,** 587–596.

Fernandez, J. J., Luque, D., Caston, J. R., and Carrascosa, J. L. (2008). Sharpening high resolution information in single particle electron cryomicroscopy. *J. Struct. Biol.* **164,** 170–175.

Frank, J., Radermacher, M., Penczek, P., Zhu, J., Li, Y., Ladjadj, M., and Leith, A. (1996). SPIDER and WEB: Processing and visualization of images in 3D electron microscopy and related fields. *J. Struct. Biol.* **116,** 190–199.

Fuller, S. D., Butcher, S. J., Cheng, R. H., and Baker, T. S. (1996). Three-dimensional reconstruction of icosahedral particles—the uncommon line. *J. Struct. Biol.* **116,** 48–55.

Gipson, B., Zeng, X., Zhang, Z. Y., and Stahlberg, H. (2007). 2dx—user-friendly image processing for 2D crystals. *J. Struct. Biol.* **157,** 64–72.

Goddard, T. D., Huang, C. C., and Ferrin, T. E. (2007). Visualizing density maps with UCSF Chimera. *J. Struct. Biol.* **157,** 281–287.

Grigorieff, N. (2007). FREALIGN: High-resolution refinement of single particle structures. *J. Struct. Biol.* **157,** 117–125.

Heymann, J. B., and Belnap, D. M. (2007). Bsoft: Image processing and molecular modeling for electron microscopy. *J. Struct. Biol.* **157,** 3–18.

Hohn, M., Tang, G., Goodyear, G., Baldwin, P. R., Huang, Z., Penczek, P. A., Yang, C., Glaeser, R. M., Adams, P. D., and Ludtke, S. J. (2007). SPARX, a new environment for Cryo-EM image processing. *J. Struct. Biol.* **157,** 47–55.

Kremer, J. R., Mastronarde, D. N., and McIntosh, J. R. (1996). Computer visualization of three-dimensional image data using IMOD. *J. Struct. Biol.* **116,** 71–76.

Lander, G. C., Stagg, S. M., Voss, N. R., Cheng, A., Fellmann, D., Pulokas, J., Yoshioka, C., Irving, C., Mulder, A., Lau, P. W., *et al.* (2009). Appion: An integrated, database-driven pipeline to facilitate EM image processing. *J. Struct. Biol.* **166,** 95–102.

Lawrence, A., Bouwer, J. C., Perkins, G., and Ellisman, M. H. (2006). Transform-based backprojection for volume reconstruction of large format electron microscope tilt series. *J. Struct. Biol.* **154,** 144–167.

Ludtke, S. J., Baldwin, P. R., and Chiu, W. (1999). EMAN: Semiautomated software for high-resolution single-particle reconstructions. *J. Struct. Biol.* **128,** 82–97.

Mallick, S. P., Carragher, B., Potter, C. S., and Kriegman, D. J. (2005). ACE: Automated CTF estimation. *Ultramicroscopy* **104,** 8–29.

Mastronarde, D. N. (2005). Automated electron microscope tomography using robust prediction of specimen movements. *J. Struct. Biol.* **152,** 36–51.

Messaoudii, C., Boudier, T., Sanchez Sorzano, C. O., and Marco, S. (2007). TomoJ: Tomography software for three-dimensional reconstruction in transmission electron microscopy. *BMC Bioinform.* **8,** 288.

Metlagel, Z., Kikkawa, Y. S., and Kikkawa, M. (2007). Ruby-Helix: An implementation of helical image processing based on object-oriented scripting language. *J. Struct. Biol.* **157,** 95–105.

Mindell, J. A., and Grigorieff, N. (2003). Accurate determination of local defocus and specimen tilt in electron microscopy. *J. Struct. Biol.* **142,** 334–347.
Nickell, S., Forster, F., Linaroudis, A., Net, W. D., Beck, F., Hegerl, R., Baumeister, W., and Plitzko, J. M. (2005). TOM software toolbox: Acquisition and analysis for electron tomography. *J. Struct. Biol.* **149,** 227–234.
Owen, C. H., Morgan, D. G., and DeRosier, D. J. (1996). Image analysis of helical objects: The Brandeis helical package. *J. Struct. Biol.* **116,** 167–175.
Philippsen, A., Schenk, A. D., Signorell, G. A., Mariani, V., Berneche, S., and Engel, A. (2007). Collaborative EM image processing with the IPLT image processing library and toolbox. *J. Struct. Biol.* **157,** 28–37.
Plaisier, J. R., Jiang, L., and Abrahams, J. P. (2007). Cyclops: New modular software suite for cryo-EM. *J. Struct. Biol.* **157,** 19–27.
Pruggnaller, S., Mayr, M., and Frangakis, A. S. (2008). A visualization and segmentation toolbox for electron microscopy. *J. Struct. Biol.* **164,** 161–165.
Ress, D., Harlow, M. L., Schwarz, M., Marshall, R. M., and McMahan, U. J. (1999). Automatic acquisition of fiducial markers and alignment of images in tilt series for electron tomography. *J. Electron. Microsc. (Tokyo)* **48,** 277–287.
Roseman, A. M. (2004). FindEM—a fast, efficient program for automatic selection of particles from electron micrographs. *J. Struct. Biol.* **145,** 91–99.
Rossmann, M. G., Bernal, R., and Pletnev, S. V. (2001). Combining electron microscopic with X-ray crystallographic structures. *J. Struct. Biol.* **136,** 190–200.
Schoenmakers, R. H. M., Perquin, R. A., Fliervoet, T. F., Voorhout, W., and Schirmacher, H. (2005). New software for high resolution, high throughput electron tomography. *Microsc. Anal.* **19,** 5–6.
Schroeter, J. P., and Bretaudiere, J. P. (1996). SUPRIM: Easily modified image processing software. *J. Struct. Biol.* **116,** 131–137.
Siebert, X., and Navaza, J. (2009). UROX 2.0: An interactive tool for fitting atomic models into electron-microscopy reconstructions. *Acta Crystallogr. D Biol. Crystallogr.* **65,** 651–658.
Smith, R., and Carragher, B. (2008). Software tools for molecular microscopy. *J. Struct. Biol.* **163,** 224–228.
Smith, P. R., and Gottesman, S. M. (1996). The micrograph data processing program. *J. Struct. Biol.* **116,** 35–40.
Sorzano, C. O., Marabini, R., Velazquez-Muriel, J., Bilbao-Castro, J. R., Scheres, S. H., Carazo, J. M., and Pascual-Montano, A. (2004). XMIPP: A new generation of an open-source image processing package for electron microscopy. *J. Struct. Biol.* **148,** 194–204.
Sousa, D., and Grigorieff, N. (2007). Ab initio resolution measurement for single particle structures. *J. Struct. Biol.* **157,** 201–210.
Stalling, Detlev, Westerhoff, Malte, and Hege, Hans-Christian (2005). Chapter 38, Amira: A highly interactive system for visual data analysis. *In* "The Visualization Handbook," (Charles D. Hansen and Christopher R. Johnson, eds.), pp. 749–767. Elsevier.
Suhre, K., Navaza, J., and Sanejouand, Y. H. (2006). NORMA: A tool for flexible fitting of high-resolution protein structures into low-resolution electron-microscopy-derived density maps. *Acta Crystallogr. D Biol. Crystallogr.* **62,** 1098–1100.
Suloway, C., Pulokas, J., Fellmann, D., Cheng, A., Guerra, F., Quispe, J., Stagg, S., Potter, C. S., and Carragher, B. (2005). Automated molecular microscopy: The new Leginon system. *J. Struct. Biol.* **151,** 41–60.
Tagari, M., Newman, R., Chagoyen, M., Carazo, J. M., and Henrick, K. (2002). New electron microscopy database and deposition system. *Trends Biochem. Sci.* **27,** 589.
Tang, G., Peng, L., Baldwin, P. R., Mann, D. S., Jiang, W., Rees, I., and Ludtke, S. J. (2007). EMAN2: An extensible image processing suite for electron microscopy. *J. Struct. Biol.* **157,** 38–46.

van der Heide, P., Xu, X. P., Marsh, B. J., Hanein, D., and Volkmann, N. (2007). Efficient automatic noise reduction of electron tomographic reconstructions based on iterative median filtering. *J. Struct. Biol.* **158,** 196–204.

van Heel, M., and Schatz, M. (2005). Fourier shell correlation threshold criteria. *J. Struct. Biol.* **151,** 250–262.

van Heel, M., Harauz, G., Orlova, E. V., Schmidt, R., and Schatz, M. (1996). A new generation of the IMAGIC image processing system. *J. Struct. Biol.* **116,** 17–24.

Volkmann, N. (2002). A novel three-dimensional variant of the watershed transform for segmentation of electron density maps. *J. Struct. Biol.* **138,** 123–129.

Volkmann, N., and Hanein, D. (1999). Quantitative fitting of atomic models into observed densities derived by electron microscopy. *J. Struct. Biol.* **125,** 176–184.

Voss, N. R., Yoshioka, C. K., Radermacher, M., Potter, C. S., and Carragher, B. (2009). DoG Picker and TiltPicker: Software tools to facilitate particle selection in single particle electron microscopy. *J. Struct. Biol.* **166,** 205–213.

Winkler, H., and Taylor, K. A. (2006). Accurate marker-free alignment with simultaneous geometry determination and reconstruction of tilt series in electron tomography. *Ultramicroscopy* **106,** 240–254.

Woolford, D., Ericksson, G., Rothnagel, R., Muller, D., Landsberg, M. J., Pantelic, R. S., McDowall, A., Pailthorpe, B., Young, P. R., Hankamer, B., *et al.* (2007). SwarmPS: Rapid, semi-automated single particle selection software. *J. Struct. Biol.* **157,** 174–188.

Wriggers, W., Milligan, R. A., and McCammon, J. A. (1999). Situs: A package for docking crystal structures into low-resolution maps from electron microscopy. *J. Struct. Biol.* **125,** 185–195.

Xiao, C., and Rossmann, M. G. (2007). Interpretation of electron density with stereographic roadmap projections. *J. Struct. Biol.* **158,** 182–187.

Yan, X., Sinkovits, R. S., and Baker, T. S. (2007). AUTO3DEM–an automated and high throughput program for image reconstruction of icosahedral particles. *J. Struct. Biol.* **157,** 73–82.

Yasunaga, T., and Wakabayashi, T. (1996). Extensible and object-oriented system Eos supplies a new environment for image analysis of electron micrographs of macromolecules. *J. Struct. Biol.* **116,** 155–160.

Zheng, S. Q., Keszthelyi, B., Branlund, E., Lyle, J. M., Braunfeld, M. B., Sedat, J. W., and Agard, D. A. (2007). UCSF tomography: An integrated software suite for real-time electron microscopic tomographic data collection, alignment, and reconstruction. *J. Struct. Biol.* **157,** 138–147.

# Author Index

## S

T



U



V



W

## X



## Y



## Z

# Subject Index

**T**

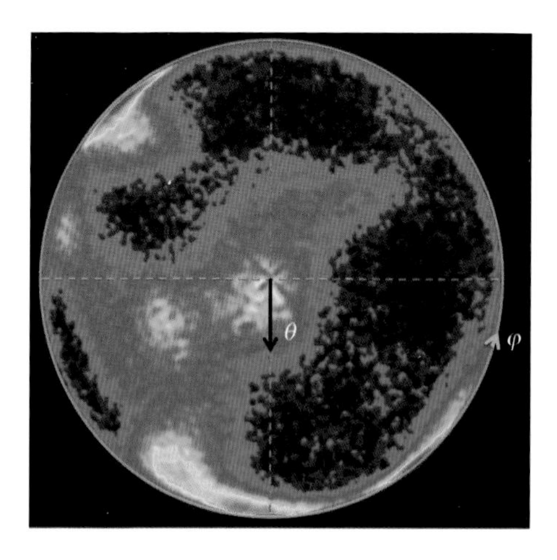


**Pawel A. Penczek, Figure 1.2** Distribution of projection directions $\boldsymbol{\tau}(\varphi,\theta)$ mapped on a half-sphere. The value of each point on the surface of the unitary sphere is equal to the number of projections whose directions given as vectors $\boldsymbol{\tau}(\varphi,\theta)$ (Eq. (1.2)) fall in its vicinity. All directions are mapped on a half-sphere $0 \leq \theta \leq 90$, $0 \leq \varphi < 360$, with directions $90 < \theta \leq 180$ mapped to mirror-equivalent positions $\theta' = 180 - \theta$, $\varphi' = 180 + \varphi$. The 322,688 angles are taken from the 3D reconstruction of a *Thermus thermophilus* ribosome complexed with EF-Tu at 6.4 Å resolution (Schuette *et al.*, 2009) and the occupancies are color-coded with red corresponding to maximum (130) and dark blue to minimum (0). There are no gaps in angular coverage at the angular resolution of the plot.

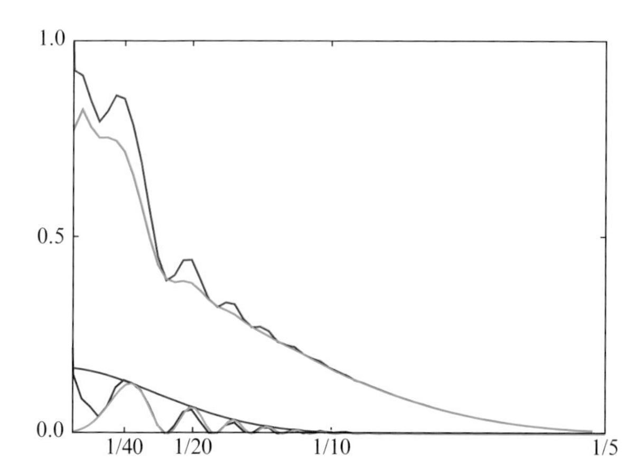


**Pawel A. Penczek, Figure 2.2** Estimation of image formation model characteristics. The PW (red) was computed using particles windowed from a single micrograph (70S ribosomes on carbon support, accelerating voltage 300 kV). The background noise PW (green) was computed from areas surrounding windowed particles. The defocus (2.95 $\mu$m) was estimated using background-subtracted PW (blue) and the envelope function (magenta) modeled by a Gaussian function. The fitted CTF shows excellent agreement with the experimental PW (light blue). Analysis was done using the e2ctf.py utility of EMAN2. The vertical axis is in arbitrary units, and the horizontal axis corresponds to modulus of spatial frequency [1/Å].

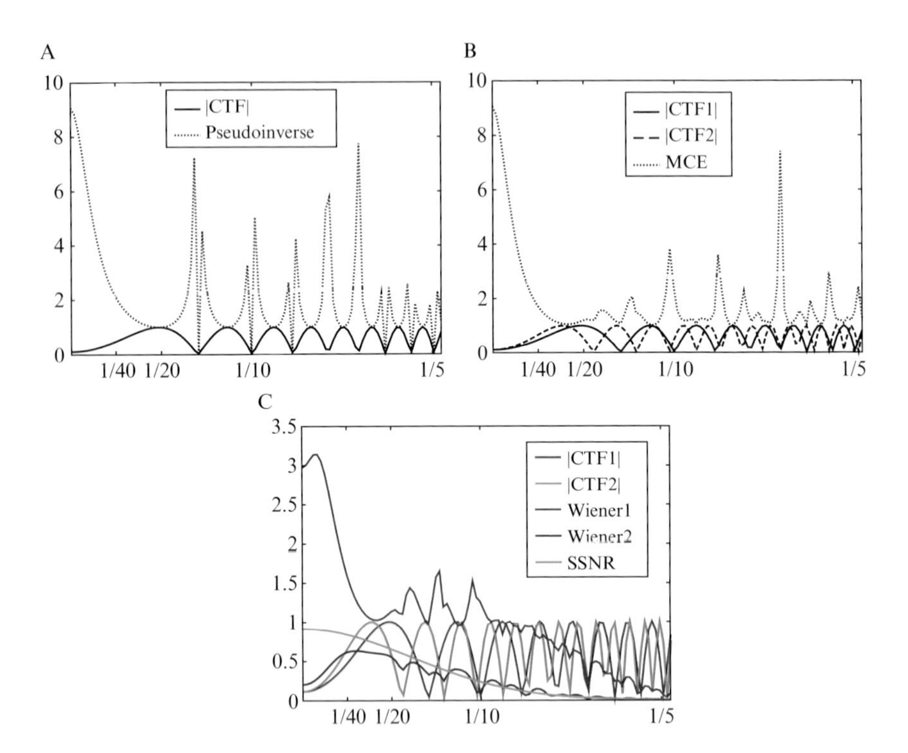


**Pawel A. Penczek, Figure 2.4** Linear cryo-EM image restoration filters. The horizontal axis corresponds to modulus of spatial frequency [1/Å]. (A) CTF (accelerating voltage 300 kV, amplitude contrast 10%, defocus 1.0 $\mu$m) and the pseudoinverse filter (Eq. (2.22)) with $\varepsilon = 0.02$. (B) Two CTFs, first as in (A), the second with defocus 1.6 $\mu$m, and the mean-square error filter (Eq. (2.29)) with the variance of data omitted. (C) Wiener filters (Eq. (2.45)) plotted assuming the SSNR is the same for each CTF and is given by a Gaussian function. For the first filter (magenta), the maximum SSNR is set to 20 while for the second filter (blue), it is set to 0.9. Note that identical filters can be obtained using the CMCE filter (Eq. (2.34)) with regularization function $U$ set to the inverse of the SSNR of the data. For clarity, we plotted moduli of the CTF function.

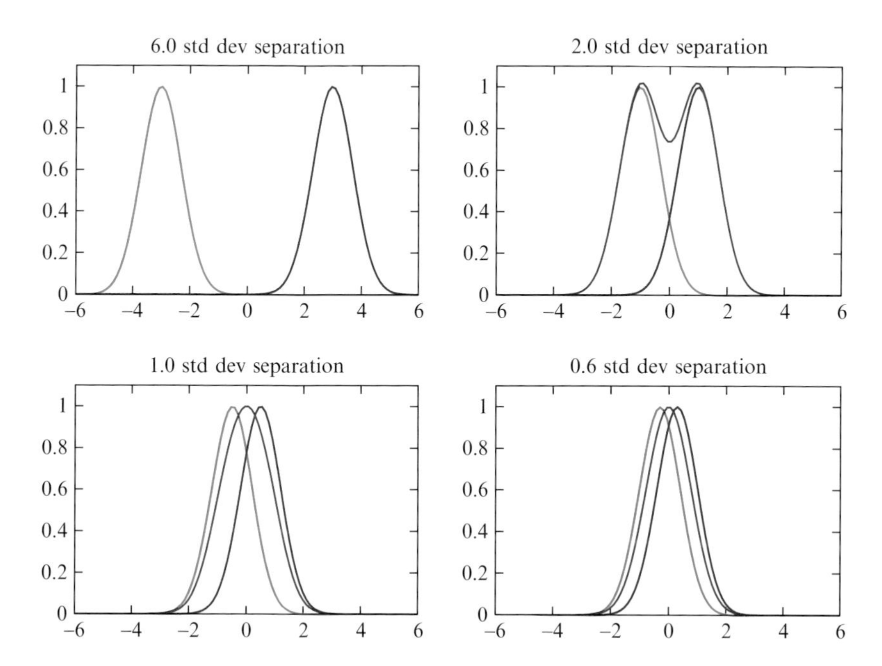


**Pawel A. Penczek, Figure 3.1** Optical resolution is defined as the smallest distance between two points on a specimen that can be distinguished as two separate entities. Assuming the blur introduced by the microscope to be Gaussian with a known standard deviation, the resolution is defined as a distance between points that equals at least one standard deviation. For distances smaller or equal one standard deviation, the observed pattern, that is, sum of two Gaussian functions (green and blue) has an appearance of a pseudo-Gaussian with one maximum (magenta).

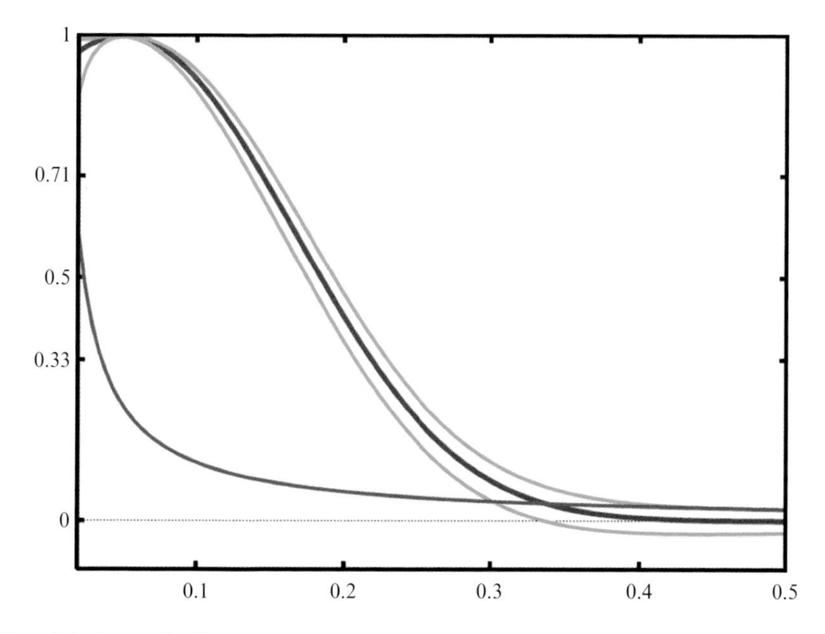


**Pawel A. Penczek, Figure 3.2** Simulated FSC curve (red) with confidence intervals plotted at $\pm 3\sigma$ (blue) (Eq. (3.19)) and $3\sigma$ criterion curve (magenta) (van Heel, 1987).

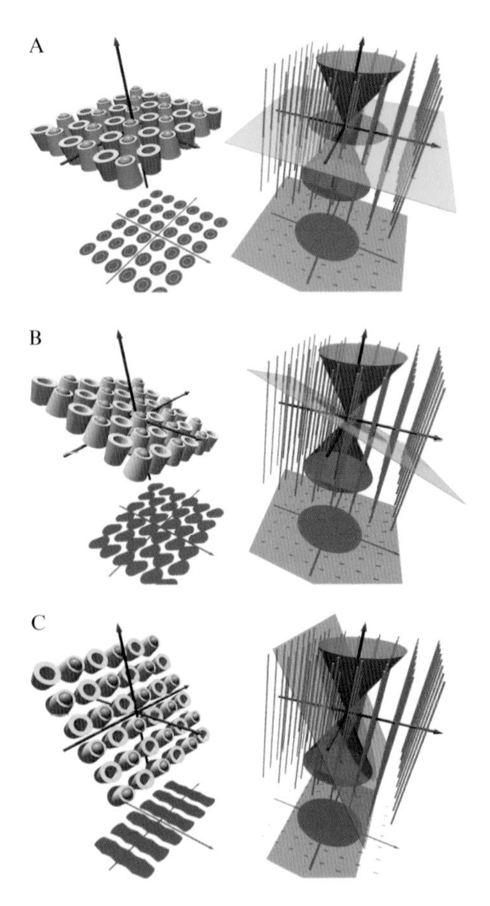


**Andreas D. Schenk *et al.*, Figure 4.1** Cartoon depicting the real-space (left) and Fourier space (right) representations of a 2D crystal. Three sample tilt angles are shown: (A) untilted, (B) 30° tilt, (C) 60° tilt. The imaging process in the microscope produces a projection image of the 2D crystal, as symbolized by the shadow under the crystal on the left side. This image is then corrected for lattice distortions and Fourier transformed. The resulting Fourier transformation corresponds to the central section that is indicated on the right as a plane. The values for amplitude and phase on the diffraction peaks in the Fourier transformation are measured and stored along the lattice lines at the position, where the lattice lines intersect with the plane (right). The crystal tilt angle defines the tilt of that plane, thereby defining the vertical $z^{\star}$ height of the measurement on the individual lattice lines. Each recorded image (left) thereby contributes only one (two due to Friedel Symmetry) measurement to a certain lattice line (right), while these measurements can under certain conditions be used for several lattice lines in case of crystal symmetry. Since the 2D crystals cannot be imaged at tilt angles higher than $\pm 70°$, the lattice line values in the indicated missing cone region in Fourier space cannot be experimentally determined, resulting in the so-called "missing cone" problem.

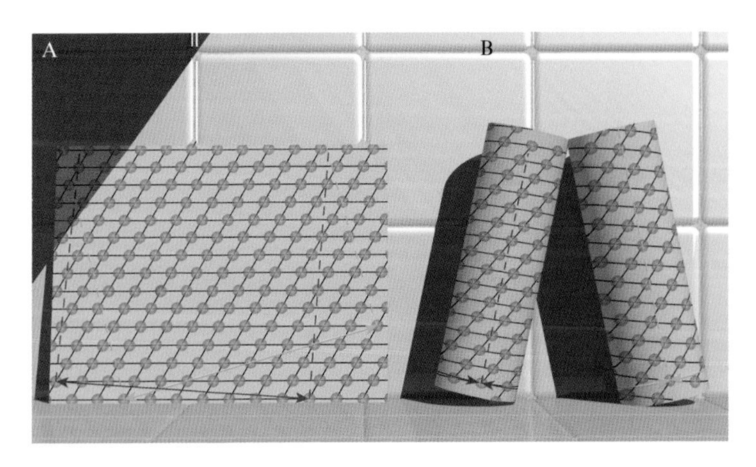


**Ruben Diaz *et al.*, Figure 5.2** Relationship between a planar 2D lattice and a helical assembly. (A) The 2D lattice is characterized by a regular array of points. An infinite variety of lines can be drawn through these points and each set of lines can be assigned a Miller index ($h$,$k$). For example, the black lines shown here could be assigned to the (1,0) and (0,1) directions. Two circumferential vectors are shown in green and red and these can be used to generate two unique helical structures shown in panel B. The dashed red and green lines are parallel to the $z$ axis in the resulting helical structures. (B) Helical lattices result from superimposing lattice points on either end of the circumferential vectors shown in panel A. Each set of lines through the 2D lattice are transformed into a family of helices. The start number ($n$) of each helix corresponds to the number of lines that cross the circumferential vector. The red circumferential vector produces helices with $n = 1$ and $n = 10$. The green circumferential vector produces helices with $n = -4$ and $n = 8$. For a left-handed helix, $n < 0$.

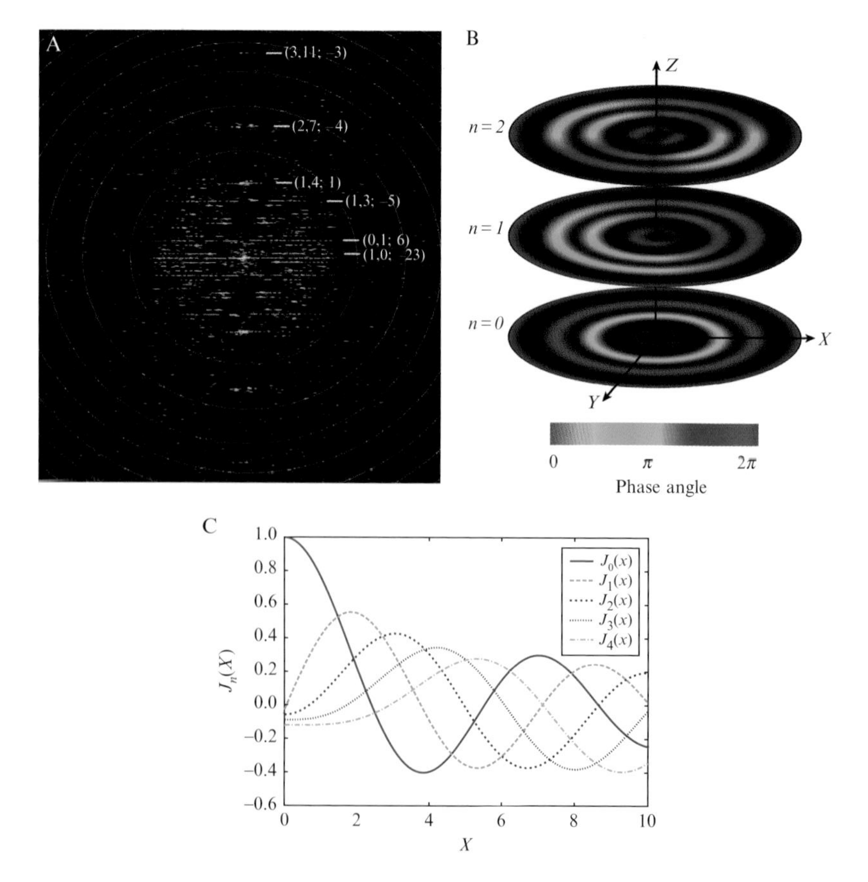


**Ruben Diaz *et al.*, Figure 5.3** Fourier transform of a helical assembly. (A) The 2D Fourier transform from a Ca-ATPase helical tube (e.g., Fig. 5.5A) is characterized by discrete layer lines that run horizontally across the transform. Each layer line corresponds to a helical family (cf., Fig. 5.2) and can be assigned a Miller index ($h$,$k$). The layer line running through the origin is called the equator and has a Miller index of (0,0). The vertical axis is called the meridian and the transform has mirror symmetry across the meridian. The start number of each helix ($n$) is shown next to each Miller index ($h$,$k$; $n$), and this start number determines the order of the Bessel function appearing on that layer line. The red circles indicate the zeros of the contrast transfer function and the highest layer line (3,11) corresponds to 10 Å resolution. (B) 3D distribution of three layer lines from a hypothetical helical assembly with Bessel orders of 0, 1, and 2, as indicated. The $Z$ axis corresponds to the meridian, the $X$ axis corresponds to the equator, and the $Y$ axis is the imaging direction. Thus, the $X$–$Z$ plane would be obtained by Fourier transformation of a projection image (e.g., panel A). The amplitude of the 3D Fourier transform is cylindrically symmetric about the meridian, but the phase (depicted by the color table at the bottom of (B)) oscillates azimuthally, depending on the Bessel order. Thus, the phase along the $n = 0$ layer line (equator) remains constant; the phase along the $n = 1$ layer line sweeps through one period, and the phase along the $n = 2$ layer line sweeps through two periods. (C) Amplitudes of Bessel functions with orders $n = 0$–4. Note that as $n$ increases, the position of the first maximum moves away from the origin.

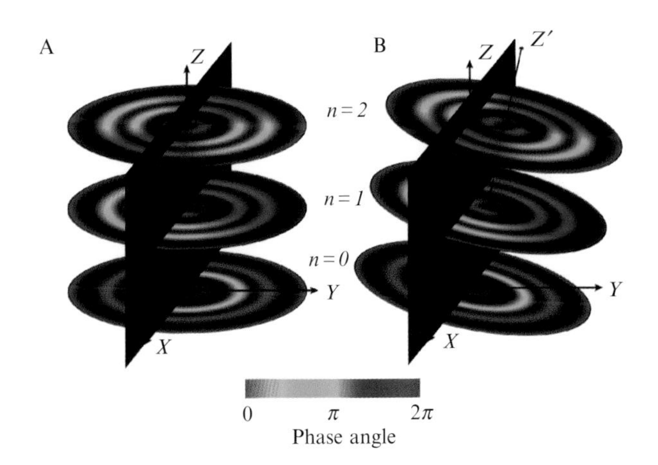


**Ruben Diaz *et al.*, Figure 5.7** Out-of-plane tilt. These diagrams illustrate the relationship between the 3D Fourier transform and the central section that results from the projection along the viewing direction (*Y*). Due to this projection, layer lines are sampled where they intersect the *X*–*Z* plane (black). (A) Untilted helical assembly where the helical axis is coincident with the *Z* axis of the transform and layer lines are sampled at azimuthal angles ($\psi$) equal to 0° and 180°. (B) Helical assembly that is tilted away from the viewing direction, causing sampling of layer lines at $\psi \neq 0°$ and 180°. $Z'$ corresponds to the helical axis and the angle between $Z'$ and $Z$ corresponds to the out-of-plane tilt, $\Omega$. This tilt produces systematic phase shifts that are dependent on the order of the Bessel function along each layer line. Phase are represented by the color table shown at the bottom.

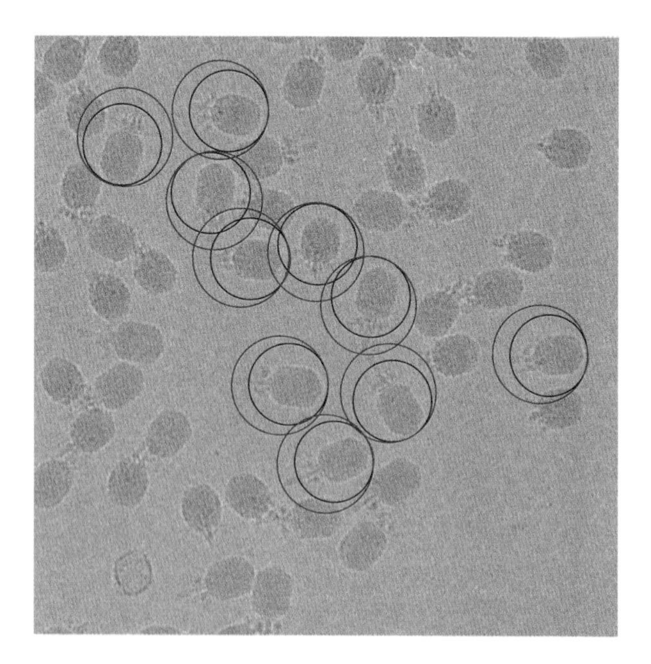

**Jinghua Tang *et al.*, Figure 7.2** Progressive boxing of phage particles. The small blue circles indicate the initial boxing of the particles, with box size chosen to capture head and proximal portion of tail. As the reconstruction progresses, the box size is gradually expanded to include more of the tail (large red circles). Note that the red and blue circles are not concentric and that the top of the phage head remains a constant distance from the edge of the circle.

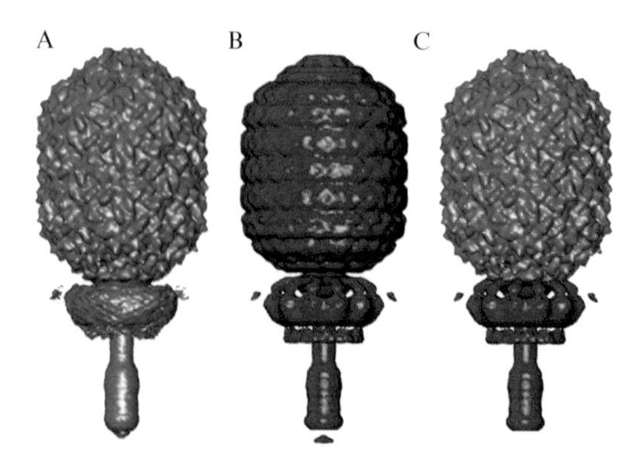


**Jinghua Tang *et al.*, Figure 7.4** Construction of asymmetric, hybrid model. (A) Shaded surface representation of complete phage reconstruction with fivefold symmetry enforced during processing. (B) Same as (A) for phage with 12-fold symmetry. Since the head and tail do not share the same symmetry, these reconstructions smear out the tail and head densities in panels (A) and (B), respectively. (C) Hybrid model obtained by combining head from fivefold reconstruction and tail from 12-fold reconstruction. Colors in the hybrid map highlight contributions from the two symmetrized maps. At this point the symmetry mismatch between the head and tail was unknown and no effort was made to impose a particular rotational alignment between the two segments.

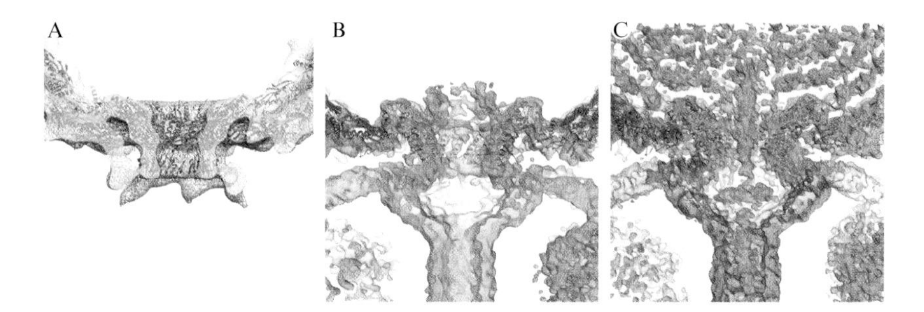


**Jinghua Tang *et al.*, Figure 7.5** Rigid body fit of gp10 connector crystal structure (magenta ribbon model) and gp8 capsid subunit homology model (red), into $\phi$29 density maps (gray). The top portion of the connector fits well into the prohead (A), ghost (B), and virion (C) reconstructions, whereas the lower portion only fits well into the prohead.

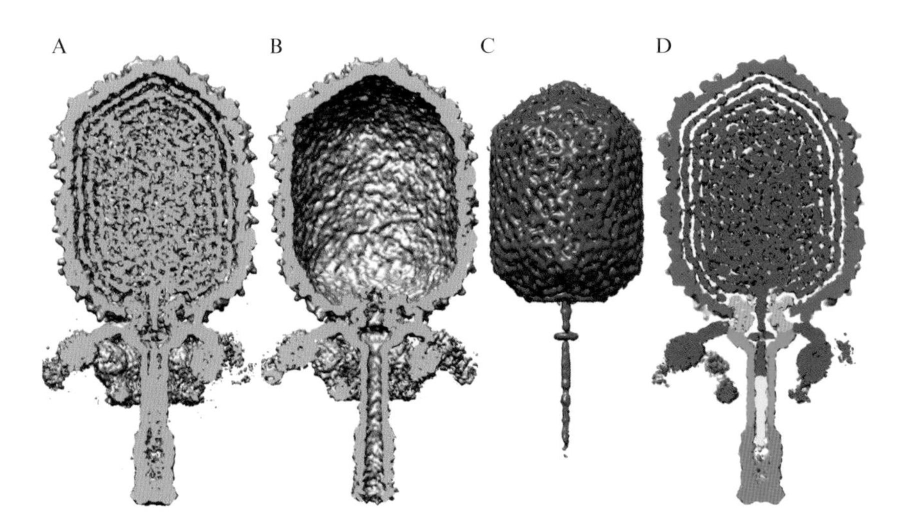


**Jinghua Tang *et al.*, Figure 7.6** Segmentation of viral components in reconstructed phage density map. (A) Shaded surface representation of virion reconstruction with front half of map removed to reveal internal structure. (B) Same as (A) for ghost reconstruction. (C) Difference map obtained by subtracting ghost from virion contains density (red) attributed to dsDNA genome plus the terminal protein, gp3. (D) Segmented map of $\phi$29 virion with components distinguished by color: capsid (gp8), light blue; connector (gp10), yellow; lower collar and tube (gp11), green; knob (gp9), cyan; appendages (gp12*), magenta; terminal protein (gp3) covalently attached to right end of DNA, white; and DNA, red.

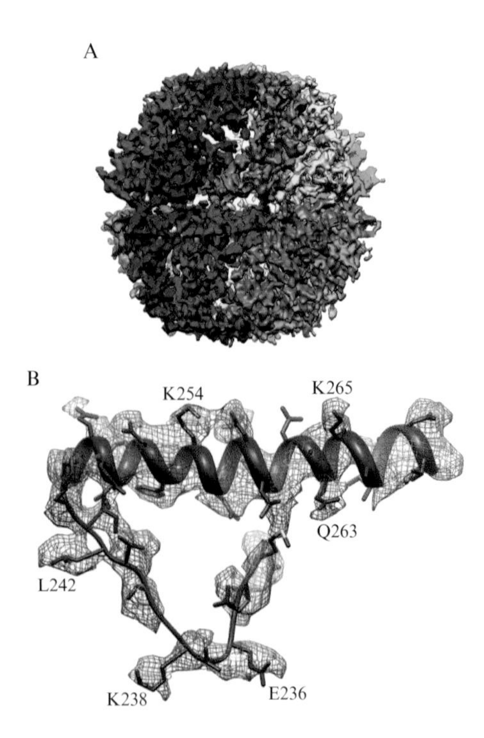


**Yao Cong and Steven J. Ludtke, Figure 8.3** (A) Side view of the 4.7 Å resolution asymmetric cryo-EM map of TRiC in the closed conformation, with the eight distinct subunits in each of the two rings highlighted in different colors. This map reveals the location of an unenforced twofold symmetry between its two rings. By enforcing this twofold symmetry, the map was further extended to 4.0 Å resolution (Cong *et al.*, 2010). (B) A portion of the 4.0 Å resolution reconstruction illustrating the visibility of side chain densities. Side chain densities were sufficient to unambiguously distinguish among the eight highly homologous subunits forming TRiC.

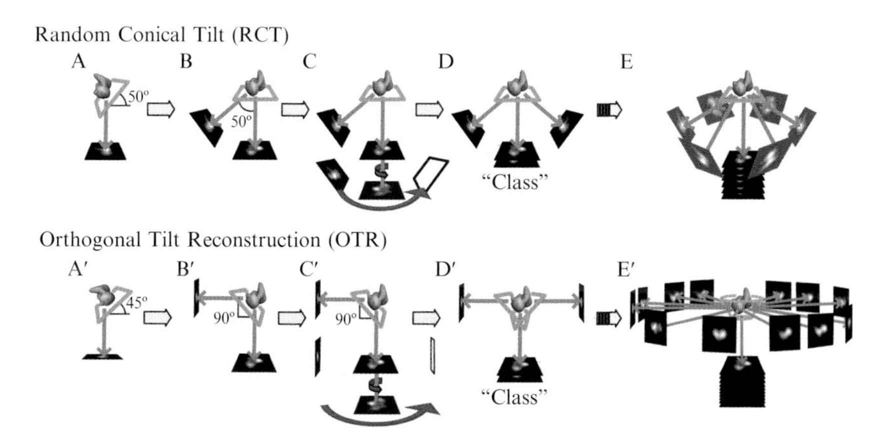


**Andres Leschziner, Figure 9.1** Geometry of the Random Conical Tilt (RCT) and Orthogonal Tilt Reconstruction (OTR) methods. The basic steps in the RCT (top) and OTR (bottom) methods are illustrated in this figure. Throughout the figure, the green object represents a molecule; the orange rhomboid the support; blue arrows the direction of projection (imaging); and red arrows the in-plane rotations applied during alignment and classification. (A and A′) An image is collected with the sampled tilted, either to 50–60° (RCT) or 45° (OTR). (B and B′) A second image is recorded from the same area after the sample has been either returned to 0° (RCT) or tilted to −45° (i.e., 45° in the opposite direction) (OTR). The images collected in A and B (or A′ and B′) constitute a "tilt pair." (C and C′) The 0° (RCT) or −45° (OTR) image from a second pair of images can be aligned to the first one if the images represent the same (but rotated) view of the molecule. The in-plane rotation applied to the image being aligned determines the new spatial location of its tilt mate (represented by the empty black frames). (D and D′) The two images are now members of the same "class" (i.e., they represent the same view and are aligned to each other). As indicated in the previous step, the alignment results in their tilt mates "fanning" out in a cone (RCT) or equator (OTR) around the molecule that gave rise to them. This step also illustrates one of the main differences between RCT and OTR: the two molecules giving rise to the two images in the "class" have the same orientation in RCT (although different in-plane rotations on the support) (D) while their orientations are entirely different in the case of OTR (D′). (E and E′) Once there are enough images in a class to fully sample the desired structure, it can be reconstructed. The orange truncated cone shown in E′ emphasizes the fact that every image in this arrangement has originated from a molecule adopting a different orientation on the support.

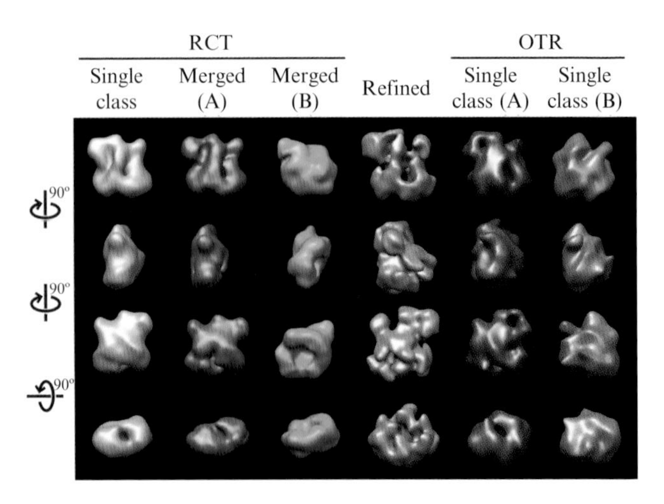


**Andres Leschziner, Figure 9.7** Comparison between RCT and OTR initial reconstructions: surface rendering. This figure shows different views of the two initial RCT reconstructions reported by Wang *et al.* (2007) (dark (A) and medium (B) green) as well as one of the single-class reconstructions (light green) merged into volume A (H. Wang, personal communication); two of our OTR initial models (red (A) and orange (B)) and the final, refined exosome structure from Wang *et al.* (2007) (gray). The "Merged" RCT models are the result of combining 6 (Merged A) and 4 (Merged B) single-class volumes and contain a total of 701 and 633 particles, respectively. The OTR models are both single-class volumes and were generated from 222 (Single Class A) and 240 (Single Class B) particles. All four initial volumes were filtered to 20 Å. The final exosome structure (gray) (EMDB entry EMD-1438) was obtained by refining the Merged A volume against approximately 4000 0° images of the complex in negative stain and is shown at 19 Å as published by Wang *et al.* (2007). The arrows on the left indicate the relationships among the four views shown for each of the volumes. The volumes were displayed using Chimera (Pettersen *et al.*, 2004).

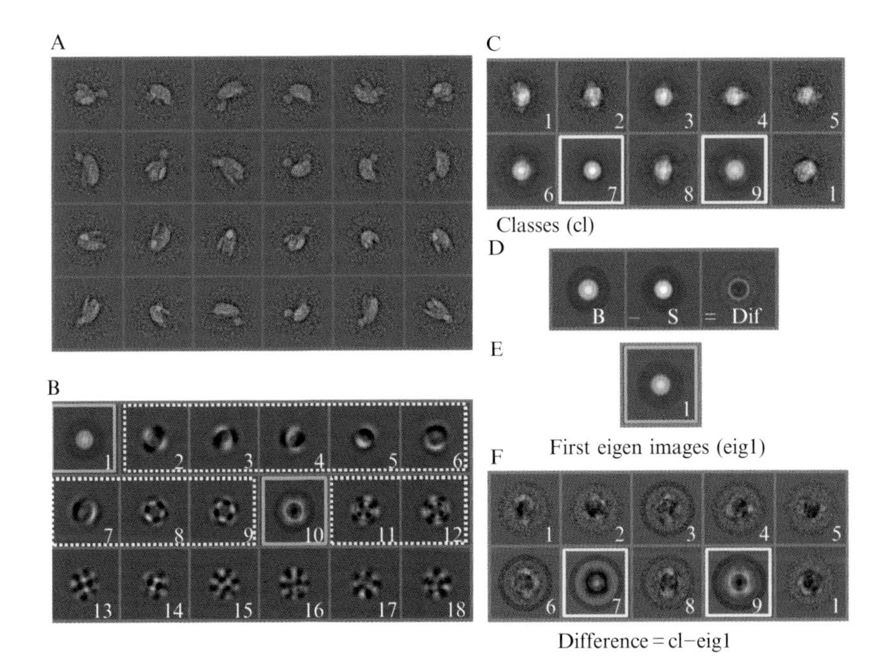


G

| | Big | Small | Superimposed structures | Differences |
|---|---|---|---|---|
| Model | | | | |
| Iteration 1 | | | | |
| Iteration 3 | | | | |

**Elena V. Orlova and Helen R. Saibil, Figure 12.2** Separation of a mixed population of model particles with size variation. (A) Representative images of particles in different orientations. (B) Eigenimages of the mixed data set. Eigenimage 1 is the sum of all images and eigenimage 10 reports on size heterogeneity in the data set. (C) Result of classification using eigenimages 1 and 10 at full weight and eigenimages outlined with the white dotted line downweighted by a factor of 0.001. Eigenvectors above 12 were not used. (D) Difference between rotationally averaged total sums of the original big and small subsets. (E) The first eigenimage from (B) was subtracted from all classes shown in (C) to generate the differences shown in (F). Images outlined in white in (C) and (F) indicate a class formed by small particles (7), which has a negative difference with the first eigenimage, and class 9, comprised of large particles, which has a positive difference with the first eigenimage. (G) Surface representations of the original models (upper row). Middle row, 3D reconstructions obtained after the first round of separation by MSA using the ring eigenimages. The big reconstruction was obtained from the images in class 9 and the small reconstruction from class 7. Bottom row, 3D reconstructions after competitive alignment and angular refinement.

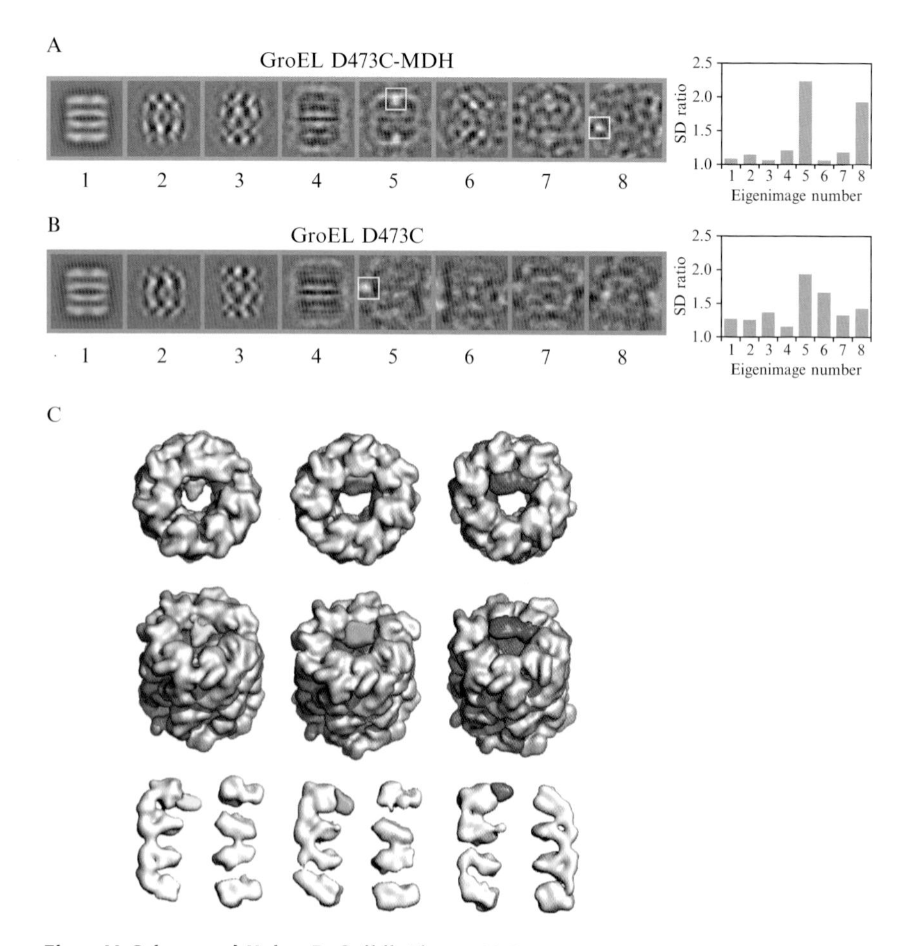


**Elena V. Orlova and Helen R. Saibil, Figure 12.5** Real data with variations in ligand occupancy: chaperonin–substrate complexes. Eigenimages of (A) the initial 8000 image data set of GroEL–MDH complexes and of (B) 6800 images of empty GroEL. Standard deviation ratios calculated as in Fig. 12.4D are plotted in the corresponding graphs on the right. (C) Surface representation of three different asymmetric cryo-EM maps of GroEL complexed with nonnative MDH obtained from sorted images (Elad *et al.*, 2007). Shown are end views of the maps in (top row), tilted views (middle row) followed by central sections (bottom row). The GroEL subunits are generated from the atomic structures of subunit domains fitted into the EM densities. Substrate densities were extracted from the experimental density maps and are highlighted in different colors. Panels (A) and (B) are reproduced from Elad *et al.* (2007) and (C) is reproduced from Saibil (2008).

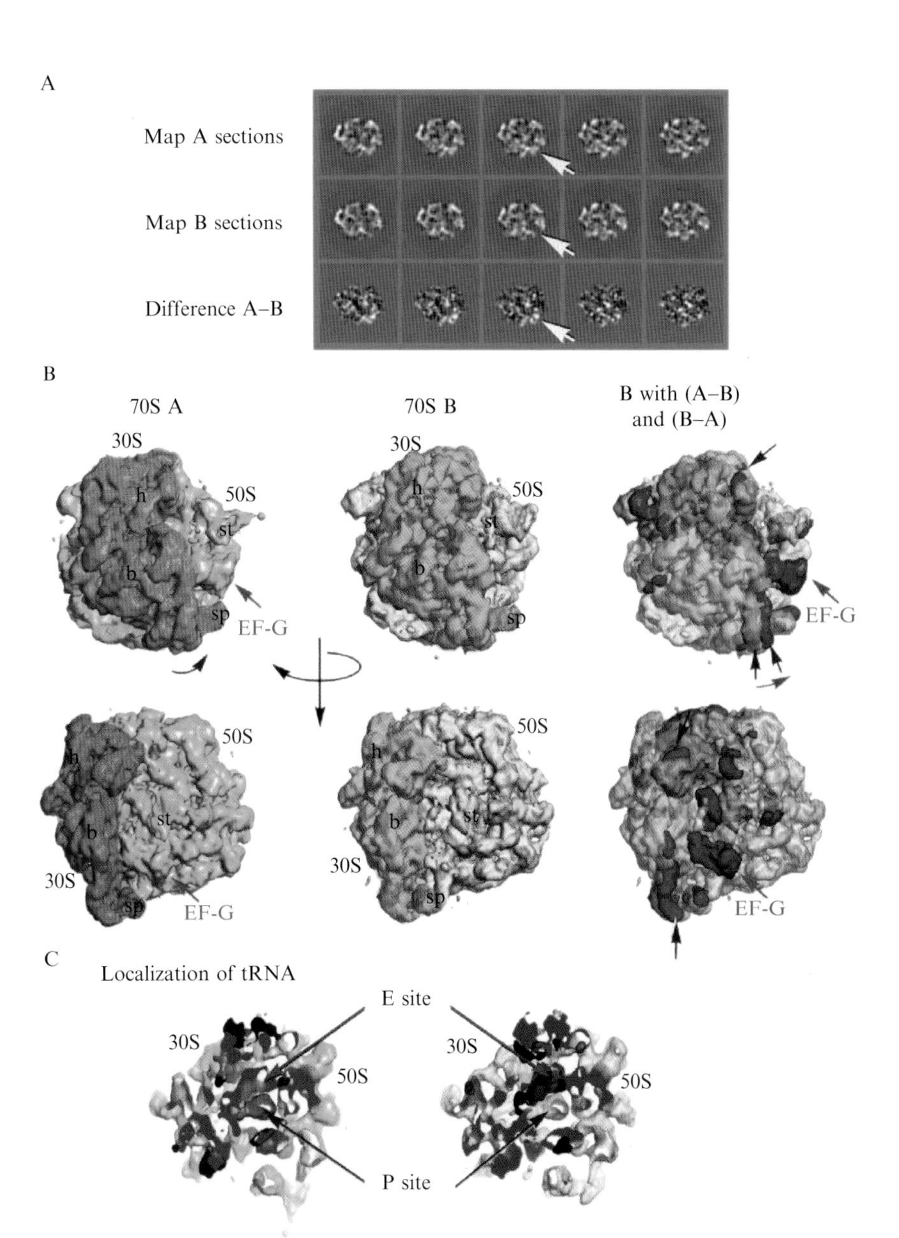


**Elena V. Orlova and Helen R. Saibil, Figure 12.7** Reconstructions of ribosome–EF-G complexes. (A) Sections of the 3D maps and their difference. The 70S ribosome map A reveals extra density in the cleft between large and small subunits, whereas map B does not show such density. White arrows point to the major differences in the sections. (B) In map A, the small subunit is shown in blue and the large subunit in light blue. In structure B, the subunits are orange (small) and gold (large). The difference map between structures A and B (in red) is superimposed on the B map (right column). Red arrows point to the EF-G location. Black arrows point to differences related to movement of the small subunit. (C) Cutaway views of the 70S ribosome with and without EF-G. Red arrows indicate the E-site tRNA that is only occupied in the presence of elongation factor. Blue arrows indicate the P-site tRNA that is occupied in the absence of elongation factor. Labels: h, head; b, body; sp, spur; st, stalk base.